Short Wavelength Laboratory Sources
Principles and Practices

Short Wavelength Laboratory Sources
Principles and Practices

Edited by

Davide Bleiner
University of Bern, Bern, Switzerland
Email: davide.bleiner@iap.unibe.ch

John Costello
Dublin City University, Dublin, Ireland

Francois de Dortan
Academy of Sciences of the Czech Republic, Prague, Czech Republic

Gerry O'Sullivan
University College Dublin, Dublin, Ireland

Ladislav Pina
Czech Technical University, Prague, Czech Republic

and

Alan Michette
Kings College London, London, UK

THE QUEEN'S AWARDS
FOR ENTERPRISE:
INTERNATIONAL TRADE
2013

This work is a publication of the COST Action Group MP0601.

Print ISBN: 978-1-84973-456-1
PDF eISBN: 978-1-84973-501-8

Published by The Royal Society of Chemistry,
Thomas Graham House, Science Park, Milton Road,
Cambridge CB4 0WF, UK

Registered Charity Number 207890

Visit our website at www.rsc.org/books

Printed in the United Kingdom by CPI Group (UK) Ltd, Croydon, CR0 4YY

ESF provides the COST Office through an EC contract

COST is supported by the EU RTD Framework programme

EUROPEAN COOPERATION
IN SCIENCE AND TECHNOLOGY

COST – the acronym for European Cooperation in Science and Technology- is the oldest and widest European intergovernmental network for cooperation in research. Established by the Ministerial Conference in November 1971, COST is presently used by the scientific communities of 36 European countries to cooperate in common research projects supported by national funds.

The funds provided by COST - less than 1% of the total value of the projects - support the COST cooperation networks (COST Actions) through which, with EUR 30 million per year, more than 30 000 European scientists are involved in research having a total value which exceeds EUR 2 billion per year. This is the financial worth of the European added value which COST achieves.

A "bottom up approach" (the initiative of launching a COST Action comes from the European scientists themselves), "à la carte participation" (only countries interested in the Action participate), "equality of access" (participation is open also to the scientific communities of countries not belonging to the European Union) and "flexible structure" (easy implementation and light management of the research initiatives) are the main characteristics of COST.

As precursor of advanced multidisciplinary research COST has a very important role for the realisation of the European Research Area (ERA) anticipating and complementing the activities of the Framework Programmes, constituting a "bridge" towards the scientific communities of emerging countries, increasing the mobility of researchers across Europe and fostering the establishment of "Networks of Excellence" in many key scientific domains such as: Biomedicine and Molecular Biosciences; Food and Agriculture; Forests, their Products and Services; Materials, Physical and Nanosciences; Chemistry and Molecular Sciences and Technologies; Earth System Science and Environmental Management; Information and Communication Technologies; Transport and Urban Development; Individuals, Societies, Cultures and Health. It covers basic and more applied research and also addresses issues of pre-normative nature or of societal importance. For more information the web address is: http://www.cost.eu

This publication is supported by COST.

PREFACE

Alan knew that: *Verba volant, scripta manent*. Spoken words fly away, written ones remain. The strength of a networking instrument, it is a bit its weakness: rapid verbal exchanges may indeed fly away. Frequently meetings with community peers, for exchanging results, problems, solutions, and research priorities established our European identity in science and technology. Especially for young researchers, or research groups without large budgets for regularly attending all the big overseas conferences, COST Actions offer a unique possibility. Support to "fly around" words (and thoughts) all over the continent, through non-archival, non-formal, non-hierarchic exchanges is great idea. The risk is however there, that when the curtain draws only a sequence of stimulating presentations and coffee breaks was made, no major legacy or touchable deliverable being left. Alan knew that.

So what? Indeed answering the *"so-what"* question is the necessary conclusion of any intellectual process. Indeed, answering the *"so-what"* question means sharp-focusing all achievements black-on-white. Alan knew that, too.

I admit it, the outfit of the present monograph has changed as Alan's commitment is irreplaceable. The original idea, as proposed by the Action's father, Prof. Alan Michette of King's College *(the place where the father of electromagnetic theory, J.C. Maxwell, was active long before Alan)*, was to release a textbook. After he passed away in May 2013, the big endeavor that relied fully on his shoulders faced a stall. Thus, I have more modestly offered my service to nail down available contributions for *words to remain*. Alan would expect that. However less ambitious my service has been.

Finally, we have edited an excellent collection of thematic contributions, providing state-of-art updates as well as a complete vision on the field of short-wavelength laboratory sources. The ongoing follow-up Action MP1203, led by Dr. Philippe Zeitoun, can now build-up on solid bases. We did not let any spoken work fly away. *Alan knew that.*

Davide Bleiner
Bern

This book is dedicated to the memory of Prof. Alan Michette, who sadly passed away in 2013 before this book was finished.

THE COST ACTION MP0601 – Short Wavelength Laboratory Sources

It was in September 2006 in Pisa, the city of the leaning tower, that the proposal for the COST Action MP0601 took its final shape and became a reference framework for the community of short wavelength radiation science and technology. It was in that occasion that Alan Michette was unanimously asked to lead what was going to become a truly comprehensive and uniquely effective initiative in a rapidly evolving scientific field.

Pre-Action meeting in Pisa, 4-5 September 2006

Short wavelength radiation has been used in medicine and materials studies since immediately after the 1895 discovery of x-rays. The development of synchrotron sources over the last 25 years, or so, has led to a boom in applications in other areas. More recently, the advent of X-ray free electron lasers is opening new horizons in the exploration of matter.

However, besides the large capital investments required for these large installations (more than 100M€ to build, perhaps about 1M€ per beam-line per year to run), synchrotron beam-lines are often over-subscribed by factors of three or more.

Despite the widely-acknowledged advantages of synchrotron radiation, many opportunities offered by X-rays are hindered by the scale of the installation and by the impossibility to conceive a large scale use of synchrotron radiation.

Indeed there is much work, of great economic, societal or health significance, that can never be done using synchrotron sources. An example is the possibility offered by synchrotron radiation in medical diagnostic imaging which, however, will never become routine at the present level of size and cost. Security inspections at airports and ports – as these inevitably become more extensive, and sophisticated, throughput must be increased significantly to minimize delays. In-field studies of valuable items that cannot be transported to synchrotrons, require the development of compact integrated (source, optics, detector) systems. At the same time, in-field studies of pollutants *in situ*, require similar development of integrated systems but probably at different energies, not to mention routine medical and dental work, where improved sources and associated equipment will lead to reductions in patient dose. Some applications also need continuous, high throughput, like EUV lithography, for future generation micro/nano circuits or radiobiological studies of radiation-induced mutations, just to mention a few.

According to the Web of Science, since 1990 the total number of scientific papers per year has risen by 45%, while there has been an eleven-fold increase in publications concerning X-rays. Over the same period there has been a 30 times increase in papers involving synchrotron sources, but even so fewer than 7% of x-ray applications use synchrotrons.

zAll the above point to the need to develop alternative, cost effective, more accessible sources which can offer at least some of the properties of synchrotrons. This was only the underlying background motivation for starting the Action that was then conceived with a range of objectives to cover all aspects of short wavelength radiation science and technology.

6th Joint Working Groups Meeting in Kraków, 27-28 May 2010

Indeed, in Alan's vision, shared by all participants, the Action's underlying aim was to maintain and enhance Europe's position in the increasingly important research area of short wavelength radiation, in order to widen access to applications that either can only currently be undertaken at large scale facilities such as synchrotrons or cannot be done at such facilities. Applications that will be enhanced include medicine (such as studies related to radiation-induced cancers), materials science, security, environmental science, lithography and chemical analysis. An example of the latter is in determining the chemical constitution of paints for restorative work and other aspects related to cultural heritage and historic artefacts. The wavelength range is natural for nanoscience and nanotechnology and thus there will be significant impact for the development of nanoscale structures and investigations. Similarly, the wavelength range is suitable for nanoscale imaging of specimens in near-natural environments. It has also been demonstrated that short wavelength spectroscopy provides improved detection limits for impurity concentration measurements in metals and alloys.

1st Joint Working Groups Meeting in Pisa, 22-23 November 2007 .

In order to realize the main purpose, the Action was structured to address several secondary objectives that were recognized as crucial in the development if the field. Of paramount importance was the need to increase output fluxes by at least an order of magnitude over existing values, both spectrally integrated and at specific, application dependent, wavelengths. Here a reference application that would benefit from increased flux was identified as short wavelength lithography, throughput being an essential user requirement. Parallel to this, source size was also identified as a fundamental parameter to control. Indeed, applications such as high-resolution (scanning) microscopy and radiobiological probing of cells and sub-cellular structures require small source sizes, typically below 1µm, without compromising flux. This means achieving sources with high brightness and, where possible, collimated emission. In the case of electron impact and similar sources where the primary beams are continuous in time, the above requirements also call for alleviation of target heat loading. Instead, in the case of plasma sources, tipically pulsed, flux increase will require higher repetition and, inevitably, new ways of replenishing the material from which the plasma is formed, the so-called targetry requirements. Wide ranges of target materials and geometries – solid, layered, gaseous, liquid – are possible in plasma sources and, as discussed here, the Action addressed much of this complexity to provide reliable and dependable solutions taking into account parameters such as flux, spectral power density, pulse duration and debris.

4th Joint Working Groups Meeting in Salamanca, 14-15 May 2009 (right)

Another main objective of the Action concerned the sources of ultrashort X-ray pulses. Indeed, a dramatic development in ultrashort-pulse laser-driven X-ray sources was occurring when drafting the programme of the Action and such sources were becoming increasingly important and likely to open new dimensions in research and applications.

Although some developments in short pulse sources will remain facility based, laboratory scale sources are invaluable as test beds for optics (especially non-dispersive), detectors, methodologies (e.g., pump-probe experiments), instrumentation, etc., before much more expensive tests at large-scale sources, and the Action took care of this scenario.

The development of improved compact portable sources was and remains an increasingly strategic objective for development of sources for a variety of applications. Carbon nanotube cathodes, with integrated optics and detectors, may offer efficient in-field measurements for environmental and heritage conservation applications. However, miniaturization of sources based upon other concepts, including laser-driven sources, may soon became a reality and the Action explored underlying principles of these sources to spot bottlenecks and possible solutions.

An aspect of great relevance is the use of focusing optics to collect, manipulate and focus radiation. This is a wide field of research itself and the Action did not neglect the importance of focusing techniques while still retaining the main emphasis of sources.

Finally, significant attention was dedicated to the ongoing effort in generating dependable modeling resources that can drive experimental programmes and establish robust methods of comparing source behaviour with respect to variable parameters.

All this and much more was in the DNA of the Action and was pursued in a friendly environment, by a collective effort involving enthusiastic young researchers, motivated group leaders and distinguished senior scientists, all guided and inspired by our friend Alan.

Alan Michette, Coordinator of the Action

Leonida A. Gizzi
(vice-chair of the Action)

Contents

Source Development

Integrated Systems

Applications

Modelling and Simulation

ATOMIC AND PLASMA PHYSICS SOFTWARE AND DATABASES FOR THE SIMULATION OF SHORT WAVELENGTH SOURCES

F. de Dortan, [1,2] D. Kilbane, [3] J. Vyskočil, [4] P. Zeitoun, [5] A. Gonzalez, [2] A. de la Varga, [2] O. Guilbaud, [6] D. Portillo, [2] M. Cotelo, [2] A. Barbas[2] and P. Velarde[2]

[1]Institute of Physics, Academy of Sciences of the Czech Republic, Prague, Czech Republic
[2]Institute of Nuclear Fusion, UPM, Madrid, Spain
[3]University College Dublin, Belfield, Dublin 4, Ireland
[4]Czech Technical University, Prague, Czech Republic
[5]Laboratoire d'Optique Appliquée, ENSTA ParisTech, Chemin de la huniere, 91671 Palaiseau, France
[6]Laboratoire de Physique des Gaz et des Plasmas, Universite Paris XI, 91761 Orsay, France

1 INTRODUCTION

A large variety of software has been developed for numerical simulations of plasma radiation emission and transport and the design of optics. Many of these codes are not specifically for short wavelengths but they can be helpful in the design of X-ray and extreme ultraviolet (XUV) sources. The astrophysics community has written codes to study the spectra and evolution of stars and gas clouds while civilian and military applicants of atomic processes were aware very early of the need for atomic, plasma and hydrodynamic software to simulate nuclear fission and fusion and their effects. Of more relevance to low cost EUV sources, the large synchrotron community has also generated efficient tools for the simulation of radiation from electrons and the design of short wavelength optics.

The present intention is not to present all the available software, since resources such as Computer Physics Communications[1] and Plasma Gate[2] give much more comprehensive, if never complete, reviews. Instead, some of the more popular programs will be introduced, as they are proven and tested assistance is available from experienced users. The software may be downloaded from the web sites listed in the references; registration is sometimes required.

2 DATABASES

The National Institute of Standards and Technology (NIST) Atomic Spectra Database is a comprehensive compilation of the experimental wavelength, strengths and level energies of spectral lines with links to the original papers.[3] Access to the ionisation energies is available on the same website. The Atomic Molecular Data Service (AMDAS) of the International Atomic Energy Agency (IAEA) is another source of databases and online software.[4]

The OPAL opacity tables[5,6,7] give access to the monochromatic Rosseland opacities of 22 elements of interest for solar astrophysics - hydrogen, helium, carbon, nitrogen, oxygen, fluorine, neon, sodium, magnesium, aluminium, silicon, phosphorous, sulphur, chlorine, argon, potassium, calcium, titanium, chromium, manganese, iron and nickel. The opacities are for solar mixtures but may be easily extended to other mixtures and pure elements. Local Thermodynamic Equilibrium (LTE) is assumed. Some equations of state are also available for the most representative elements of the Sun, namely hydrogen, helium, carbon, oxygen and neon. Some Fortran subroutines are available for interpolation. The tables are computed and maintained at the Lawrence Livermore National Laboratory.[8]

The Opacity and Iron[9,10] Projects are international collaborations to estimate stellar envelope opacities and compute Rosseland mean opacities of elements relevant for astrophysics, essentially the same list as OPAL except for helium, fluorine, phosphorous, potassium and titanium. Data tables can be generated online choosing custom mixture of elements. A global archive can also be downloaded including some routines to compute the opacities online; radiative accelerations are also included. Fine structure atomic data - energy levels, radiative transition probabilities, electron impact excitation cross sections and rates and photo-ionisation cross sections - are also available online.[11,12]

Sesame[13] is a library of tables for the thermodynamic, electric and radiative properties of materials with Fortran subroutines for the use of the libraries, three of which are available:

- Equation Of State (EOS) for over 150 materials (simple elements, compounds, metals, minerals, polymers, mixtures...). This library contains pressure, energy, Helmholtz free energy, thermal electronic and ionic contributions, and sometimes vaporisation, melt and shear tables for temperatures in the range 0-105 eV and densities of 10^{-6} - 10^{4} g/cm³;

- An opacity library where mean Planck and Rosseland opacities as well as ionisation and electron conductive opacities are provided for temperatures above 1eV and for elements with Z=1-30;

- A conductivity library giving mean ionisation, electrical and thermal conductivities, thermoelectric coefficients and electron conductive opacities for elements with Z=1-96.

Many methods are used to obtain the EOS, the aim being to have thermodynamically self-consistent equations that are generated using the most accurate physics to provide the best possible agreement with available experimental data. Fortran subroutines are provided to read the data and compute the thermodynamic variables and their first derivatives at the desired points. Libraries are created and maintained at the Los Alamos National Laboratories where they can be obtained after registration.[14]

3 ONLINE SOFTWARE

The COWAN[15] suite of code on the Los Alamos T4 network[16] gives direct access to the computation of detailed or averaged energy levels including configuration interactions. It is based on the Hartree Fock method,[17,18] and line strength, collisional excitation in distorted waves (DW) or the first order many body theory (FOMBPT) approximations are also available. Collisional ionisation can be computed using the scaled hydrogenic,[19] binary encounter[20] (billiard-like collision taking into account target momentum) or distorted wave[21] approximations. Photo-ionisation cross sections as well as auto-ionisation rates are also calculated.

A web interface of the FLYCHK code[22], is available at the National Institute of Standards and Technologies (NIST). It generates atomic level populations and charge state distribution as well as overall radiative losses for low- to mid-Z elements under non-LTE conditions. The full version of this code, which is discussed more fully in section 3.3, is available to download.[23,24]

The Centre for X-Ray Optics (CXRO) at the Lawrence Berkeley National Laboratory provides online access to the computation of transmission factors of neutral materials (either gaseous or solids, and pure or compounds).[25] It is also possible to load the scattering factor tables in order to create custom databases for local calculations. It is also possible to compute online the reflectivities of materials or multilayers, whatever the composition and density. Transmission gratings efficiency can also be calculated.[26]

4 SOFTWARE TO BE DOWNLOADED OR REQUESTED

4.1 Software for Atomic physics

Multi Configuration Hartree Fock (MCHF) codes were initially developed by Charlotte Froese Fischer.[27] Many versions are available on the Computer Physics Communications (CPC) Program Library[28] and on the internet to compute structure, wave functions and energy levels as well as radiative transition rates. Many subsequent programs are also available to determine the collisional rates between the computed levels within a wide range of approximations, for example ATSP2K on CPC or at NIST.[29]

Single or Multi Configuration Dirac Fock (SCDF[30], MCDF[31,29]) codes rely on the fully relativistic Dirac equation. Many versions exist, all permitting structural computation with level energies and radiative transition rates. Some also give access to collisional transition and photo-ionisation cross sections and auto-ionisation rates within the same package; otherwise external programs have to be used subsequently. Many of these present multiple optimisation procedures to increase the precision of the energies and wave-functions or rates useful for extremely detailed spectroscopy. This may not be useful for sources emitting many closely spaced lines but may help describe better X-ray laser amplification of single lines. The Multi Configuration Dirac Fock and General Matrix Element (MCDFGME)[32] package is more complete, including Born collisional excitation, photo-ionisation and auto-ionisation cross sections and rates.

Autostructure[33,34] is a Fortran program for computing atomic and ion energy levels, radiative and auto-ionisation rates and photo-ionisation cross sections. It is based on Superstructure[35] and performs calculations in orbital angular momentum / spin (Russell Saunders or LS-) or intermediate (jj-) coupling[36] using non-relativistic or semi-relativistic wave-functions. Radial functions use a model potential, either Thomas-Fermi (TF) or Slater-Type-Orbital (STO). The data can be post processed to generate di-electronic recombination rate coefficients for doubly excited state populations and satellite line emission modelling. This code has been used successfully to model di-electronic recombination of tin ions relevant to EUV lithography.[37]

The COWAN[15,38,39] suite of Fortran 77 codes is a package of computer programs using the MCHF method to compute level energies and structures, radiative transition wavelengths and probabilities, electron impact excitation, photo-ionisation cross sections and auto-ionisation rates. It provides full access to all the intermediate information and variables such as radial wave-functions, centre of gravity configuration energies, radial Coulomb and spin-orbit integrals. It is also possible to perform least squares fits to experimental energy levels for spectroscopic use.

The Relativistic Atomic Transition and Ionization Properties (RATIP)[40] suite of programs calculates relativistic atomic transition, ionization and recombination properties. It is particularly suitable for open-shell atoms and ions and is capable of calculating energy levels, transition probabilities, Auger parameters, photo-ionisation cross sections and angular parameters, radiative and di-electronic recombination rates and many other atomic properties. It was developed as a scalar FORTRAN 90/95 code and all computations are performed within the MCDF framework as implemented in GRASP92 and GRASP2K packages[41,42]. Successful applications of RATIP include atomic photo-ionisation and electron spectroscopy, study of highly charged ions,[43] spectroscopy of heavy and super-heavy elements,[44] the generation of atomic data for astrophysics and plasma physics[45] and the search for time reversal violating interactions in atomic systems.[46]

The Hebrew University – Lawrence Livermore Atomic Code (HULLAC)[47,48] is a complete suite of Fortran atomic codes including structure and transitions. It is a consistent atomic model using the same wave functions for computing all atomic processes relevant to plasma spectroscopy. These wave functions are obtained by solving the Dirac equation in a parametric potential. A factorisation /interpolation method allows faster computation of collisional cross sections. The initial version of the code has been recently rewritten to increase the size of matrices and allow not only detailed levels but also configuration averaging both relativistic and non-relativistic. It now has easy user input, a collisional radiative solver, the ability to use the mixed transition arrays and to model the effects of external radiation fields; spectral software is also included. HULLAC is being used by many groups around the world and may be obtained from its authors.

The Flexible Atomic Code (FAC)[49,50] is a complete suite of atomic codes including structure and transitions, mostly similar to HULLAC. It solves the relativistic Dirac equation, using a single central parametric potential to compute the orbitals. Radiative decay, collisional excitation and ionisation, auto-ionisation and photo-ionisation cross sections and rates are computed within the distorted wave approximation. Coulomb Born approximation with exchange (between incoming and target electrons) and binary encounter dipole[20] approximations can also be used to compute the collisional ionisation more quickly. Results can be either detailed or averaged in relativistic configurations. Detailed levels may also be split into magnetic sublevels and the transitions computed between these. FAC is written in FORTRAN 77 for the physics and C for the architecture. Two interfaces are available, one online, the other a PYHTON scripting interface which allows easier extensive use of the program. It is very stable and straightforward to install. A collisional radiative solver and spectrum computation are available but not documented and difficult to use and so most users develop their own solver and radiative emission and transmission software.

4.2 Equation of State and Opacities

ABINIT[51,52] is a widely used package allowing to compute the total energies, charge densities and electronic structures of a combination of a nucleus and electrons using density functional theory (DFT). ABINIT also include options to optimise the geometry according to the DFT forces and stresses, to perform molecular dynamics simulations using these forces, or to generate dynamical properties (vibrations and phonons), dielectric properties, mechanical properties and thermodynamic properties. Some possibilities of ABINIT go beyond DFT, for example the many-body perturbation theory, which uses the single particle Green´s function G and the screened Coulomb interaction W (GW approximation), and Time-Dependent DFT. The codes have been written in Fortran 90 by more than 200 developers who upgrade and maintain it under the GNU/GPL license.

IONMIX,[53] a Fortran 77 code available on the CPC website[28] for determining the EOS and radiative properties of LTE and non-LTE plasmas composed of many different elements, is designed for plasmas with high electron temperatures, $T_e \geq 1eV$, and low to moderate ion densities $N_i(cm^3) \le 10^{20}(T_e/Z)^3$. Steady-state ionisation and excitation populations are determined by either the Saha-Boltzmann equation or by collisional radiative resolution using hydrogenic ion approximation. Bound-bound, bound-free, free-free and electron scattering processes are taken into account to compute multi-group Planck and Rosseland mean opacities. Auto-ionisation and reverse process are not taken into account.

BADGER[54] is an EOS library written in Fortran 90 and available on the CPC website.[28] It models plasmas more correctly in the low temperature, high density regime, especially inertial confinement plasmas. It computes single or double (different electronic and ionic) temperatures plasmas using the Thomas Fermi model.[56] Ion EOS data, for Z=1-86, is obtained either from the ideal gas model or from a quotidian EOS (semi-empirical fitting of available experimental data for higher densities and lower temperatures) with scaled binding energies.[55] Electron EOS data may be computed from the ideal gas model or from an adaptation of the screened hydrogenic model with angular momentum splitting.[57] The ionisation and EOS calculations can be carried out assuming LTE or non-LTE mode using a variant of the Busquet equivalent temperature method.[58] The code is written as a software library to be directly linked to external codes without having to generate tabular data. It also includes a wrapper code that allows the generation of data tables if in-line implementation is impractical.

4.3 Computation of Synthetic Spectra and Spectroscopic Analysis

Cretin[59,61] is a multi-dimensional non-LTE radiation transfer code. It combines atomic kinetics and radiation transport in a self-consistent way and is suitable for modelling laboratory plasmas with a wide range of applications, including non-uniform materials, complex geometries (planar, cylindrical and spherical in 1D, Cartesian and cylindrical in 2D, Cartesian in 3D) and non-negligible optical depths. Atomic data is generated externally for Cretin using either average atom or unresolved transition arrays (UTA) models for applications that focus on gross energetics, or fully detailed models such as the flexible atomic code (FAC)[49] when spectroscopic comparisons are required. Cretin has been successfully applied in a variety of applications including inertial confinement fusion (ICF), magnetic fusion, X-ray lasers and laser produced plasmas. Recently it was successfully applied to match experimental data for laser-driven systems and hohlraums.[60]

FLYCHK,[22,23] as briefly described above, is a straightforward and rapid tool which provides ionization and population distributions of homogeneous plasmas with accuracy sufficient for most initial estimates; in many cases is applicable for more sophisticated analysis. FLYCHK solves rate equations for level population distributions by considering collisional and radiative atomic processes. The code is designed to be straightforward to use and yet is general enough to apply for most laboratory plasmas. It can also be applied to low-to-high Z ions in either steady-state or time-dependent situations. Plasmas with arbitrary electron energy distributions and single or multiple electron temperatures can be studied as well as radiation-driven plasmas. To achieve this versatility and accuracy in a code that provides rapid response schematic atomic structures are used, such as scaled hydrogenic cross-sections and look-up tables. The code also employs the *jj* configuration averaged atomic states and oscillator strengths calculated using the Dirac–Hartree–Slater model[62] for spectrum synthesis. Numerous experimental and analytic comparisons performed have shown that FLYCHK provides meaningful estimates of ionization distributions, well within a charge state for most laboratory applications.

PRISM-SPECT[63,64] is a commercial spectral analysis code, designed to simulate the atomic and radiative properties of laboratory and astrophysical plasmas. For user specified thermodynamic conditions, it computes emission, absorption and ionisation of LTE and non-LTE plasmas. Population and spectra can also depend on an external radiation field and non-Maxwellian electron distributions. The resolution can be either stationary or time-dependent. A strong emphasis has been put on a graphic package to help in setting up the problem, visualizing the results, the progress of computations and spectroscopic analysis. All the elementary atomic processes and their reverses are taken into account. Doppler, natural, auto-ionisation, collisional electronic- and ionic- Stark broadening are employed to compute the line profile. Experimental energy levels and oscillator strength are used when available. An atomic database for elements Z=1-18 is included; other elements may be obtained on request.[64]

4.4 Radiative Hydrodynamics

MEDUSA[65] is a one dimensional Lagrangian radiation hydrodynamics code, developed for simulations of laser inertial fusion. Electrons and ions are treated as having different temperatures but behaving as perfect gases. Radiative losses are through Bremsstrahlung only. Thermonuclear reactions are included in the computation and different ion species may be specified at the same time. The geometry can be planar, cylindrical or spherical. The code is written in standard Fortran (64) and is available from CPC.[28]

MULTI[66,67] is a one dimensional Lagrangian radiation hydrodynamics code also available from CPC. It was developed for inertial confinement fusion and related laser experiments. The hydrodynamic equations are combined with a multigroup method for radiation transport. It uses external tables for EOS data, Planck and Rosseland opacities. The original geometry was planar, but a large community of users has improved it and versions with femtosecond laser interactions, cylindrical and spherical geometries have been implemented. It is written in Fortran 77 and R94 (a language developed for easier interfacing and translated to C).MULTI 2D[67,68] is a two dimensional version designed for indirect inertial fusion applications and diagnosis of experiments. The geometry is cylindrical and hydrodynamics is Lagrangian. EOS and opacities are interpolated from external tables. It is written in C and is available from CPC.[28]

Z*BME (Z* Blackbox Modelling Engine)[69] is a commercial system tool based on the adaptation of the two dimensional radiation-magneto-hydrodynamic code Z*[69] to facilitate numerical modelling of EUV plasma sources by non-specialists. It can simulate both discharge and laser produced plasma sources and generate two dimensional maps of thermodynamic conditions of the plasma as well as ionisation maps, radiative losses, conversion efficiency maps and time-dependent limited bandwidth EUV emission power. An imbedded particle in cell (PIC) code provides time dependent fast ion production maps. It includes its own tables for thermodynamic equilibrium properties (EOS, ionisation) and radiative properties. Non-LTE ionisation and radiation transport can also be computed. At the end of simulations, spectra and heat fluxes can also be output. It is integrated into a specific computation environment to provide a turn-key simulation instrument without specialist knowledge in numerical simulation; thus Z*BME is used in many laboratories. Numerous simulations have been conducted to simulate EUV sources for nanolithography, describing, for example, debris production from laser interactions with mass limited droplets, conversion efficiency of xenon and tin based plasma sources, thermal load of discharge source electrodes and non-stationary and non-LTE effects on plasma ionisation. The code is available from EPPRA sas (Villebon sur Yvette, France)

ARWEN[70] is a two dimensional hydro-radiative code with adaptive mesh refinement (AMR) including laser interactions. It is an Eulerian code with flux limited electron thermal conduction and multiple groups of photon frequencies (multigroup) radiation transport. It uses external data files for EOS and multigroup opacities. Laser energy deposition can be treated with a simple model or assuming refraction effects with a 2D ray-tracing subroutine. It includes two temperatures (electron and ion), time-dependent ionisation and gain computation. Planar and cylindrical geometries are available. The AMR allows the resolution to increase only in regions where the gradients increase over a defined limit and leads to uniform numerical errors while saving computation time. A graphical interface is included to display 2D time-dependent results as videos and a standard interface in hierarchical data format (HDF) is also available for easy post-processing by other codes. ARWEN is written in Fortran and C++ and is available from Pedro Velarde (DENIM/UPM, Madrid).

4.5 X-ray Lasers Hydrodynamics and Output

EHYBRID[71] is a 1.5 dimensional (one dimensional plus autosimilar expansion in transverse direction) hydroradiative Fortran code coupled with level population dynamics, able to model X-ray lasers dynamics. It provides extensive information about the plasma dynamics and the gain of the modelled X-ray laser. Non-normal incidence of the pumping laser pulse can be taken into account as well as very short laser pumping pulses. It uses external EOS files or a perfect gas approximation if EOS is not available. It also uses external atomic data files to compute the transient ionisation and the evolution of lasing populations. EHYBRID is available from Geoffrey Pert (York University, UK)

4.6 High Harmonic Generation

QPROP[72,73] is a time-dependent Schrödinger (or Kohn-Sham) solver, designed for the study of atoms or other spherical systems in intense laser fields. It was developed to study laser-atom interactions in the non-perturbative regime where non-linear phenomena such as above-threshold ionisation, high-harmonic generation and dynamic stabilisation occur. Two methods of resolution are available : full resolution with density functional theory[74] (DFT, also called Kohn-Sham equations) where the number of perturbed electrons is chosen by the user or faster but limited resolution with a "frozen core" of electrons simulated by an external potential a single-active (perturbed) electron. In DFT, wave functions are expanded in spherical harmonics. The time-dependent Schrodinger equations can only be solved for a linear laser field polarisation. In the single-active non-relativistic approximation, the time-dependent Schrodinger equation is solved in three dimensions and the laser polarisation can be either linear or elliptical. The emitted photon spectrum is obtained by Fourier-transformation of the dipole acceleration. QPROP is written in C++ and maintained at Rostock University.

4.7 Plasma Physics (PIC codes)

LPIC++[75,76] is a one dimensional electromagnetic and relativistic particle in cell (PIC) code for simulating highly intense and ultra-short laser-plasma interactions, including ionization dynamics and collisions. It is written in C++ and is designed to work on Linux-Unix type parallel architectures and has been extended to include radiation friction force effects using Landau-Lifschitz approximation.[77] A 2D version is also available.

EPOCH[78,79] is a multidimensional (1D, 2D, 3D) PIC code for general plasma simulations. It is an electromagnetic relativistic code which features arbitrary laser pulse and plasma slab shapes, higher-order particle shapes, and a moving window algorithm for simulating large domains. It includes Monte Carlo algorithm for simulation of quantum electrodynamic (QED) effects at ultrahigh intensities such as emission of gamma-rays and electron-positron pair creation. It has a rich set of output options, with provisions for visualization in Matlab, VisIt and IDL. The code is written in Fortran 90 and parallelised using the MPI library. Input is specified with text files.

4.8 Propagation of X-ray Lasers and the Effect of Coherent X-rays on Materials

The free electron laser atomic, molecular, and optical physics program package FELLA[80,81] is a suite of codes to study the interaction of optical laser radiation and X-rays with atoms and molecules, to explore the ultrafast and the ultra-small. The motivation was to study atoms and molecules using X-ray free electron lasers (XFEL) in combination with intense optical lasers. It can be used in more general cases such as interaction induced by intense synchrotron radiation. The atomic physics program treats the electron structure of atoms in Hartree-Fock-Slater approximation and their interaction with photons of one or two energies. The molecular physics programs treat the X-ray absorption by laser-aligned molecule and the optical physics code computes the propagation of laser and X-rays through gaseous media. FELLA is written in Fortran 95 and maintained at Argonne National Laboratory.

4.9 Optics Design

IMD[82,83] is an IDL application which can calculate specular and non-specular (diffuse) X-ray to infrared performances of an arbitrary multilayer structure, i.e., a structure consisting of any number of layers, any thickness and any materials. Reflectance, transmittance, absorbance, electric field intensities, phase shifts as well as the amplitude and phase components of the reflectance (psi and delta functions) for ellipsometry are computed with an algorithm which includes interfacial roughness and diffusion. A stochastic model of film growth and erosion can be used to account for the evolution of interfacial roughness through the film stack. Specular and non-specular optical functions can be calculated as functions of incidence angle, wavelength, polarisation, spectral and angular resolution, as well as parameters that describe the multilayer structure. It includes a database of optical constants for over 150 materials but also allows material constants to be input. IMD uses a simplified graphical interface to help define the properties of the multilayers and also to visualise and analyse in 1D or 2D the properties of the designed optic.

ZEMAX[84] is a commercial optical design program. It is widely used to design and analyse optical systems performing sequential ray tracing through optical elements, non sequential ray tracing for analysis of straight light and physical optics beam propagation. It can model the propagation of rays through elements such as lenses, mirrors and diffractive optical elements. It models the effect of optical coatings and generates spot diagrams and ray-fan plots. The optics propagation feature can be used for problems where diffraction is important, including the propagation of laser beams, holography and coupling of light into single-mode optical fibres. It can be useful in optimizing performance and reducing aberrations. Although the program is designed for visible light users can specify the physical characteristics for other wavelength.

RAY[85,86] is BESSY synchrotron's ray tracing program to calculate synchrotron radiation beamlines. It is a FORTRAN design tool also for general optics applications in IR, UV, soft and hard X-rays which may work on Windows as well as Linux platforms. It can compute transmission, reflection, dispersion or diffraction as well as rocking curves, photon flux, resolving power and polarisation of/on nearly all kind of optics geometries such as zone plates, slits, foils, mirrors, spherical or planar gratings, crystals, multilayers and graded multilayers... The code starts with a source volume with defined emission characteristics and computes its modification by optical elements, taking into account the phase space to obtain wavefront and coherence characteristics. Various orders of reflections on gratings being treated. It is designed user-friendly and easy to learn.

4.10 Image Analysis

XTRACT[87] is a suite of Windows based programs specifically designed for the reconstruction and structure recovery of samples studied by X-ray phase contrast imaging. It contains tools for pre-processing of recorded data as well as phase retrieval and cone beam computed tomography modules, software for X-ray image simulation, analysis and processing. More than 20 different algorithms are available for phase and / or amplitude extraction. It can treat most standard formats of grey-scale images and can be obtained from the Australian National Science Agency CSIRO.[88,89]

References

[1] http://www.journals.elsevier.com/computer-physics-communications/

[2] http://plasma-gate.weizmann.ac.il/directories/free-software/

[3] http://www.nist.gov/pml/data/asd.cfm

[4] http://www-amdis.iaea.org/

[5] C. A. Iglesias and F. J. Rogers, Astrophys. J. 464, 943 (1996)

[6] F. J. Rogers, F. J. Swenson, and C. A. Iglesias, Astrophys. J. 456, 902 (1996))

[7] Chen, Xue-Fei; Tout, Christopher A. Chinese J. of Astron. and Astrophys., V.7, n.2, pp. 245-250 (2007).

[8] http://opalopacity.llnl.gov/opal.html

[9] Seaton M.J., 1987, J. Phys. B 20, 6363

[10] Badnel et al., MNRAS, 360, 458 (2005))

[11] http://op-opacity.obspm.fr:8080/opacity/

[12] http://opacities.osc.edu/rmos.shtml

[13] B. I. Bennett, J. D. Johnson, G. I. Kerley, and G. T. Rood, "Recent Developments in the Sesame Equation-of-State Library," Los Alamos Scientific Laboratory report LA-7130 (1978).

[14] http://t1web.lanl.gov/doc/SESAMEdbasetxt.html.

[15] The Theory of Atomic Structure and Spectra (Los Alamos Series in Basic and Applied Sciences) Robert D. Cowan

[16] http://aphysics2.lanl.gov/tempweb/lanl/

[17] V. Fock, Z. Physik 61, 126 (1930}

[18] D. R. Hartree and W Hartree, Proc. Roy. Soc. A156, 9 (1935)

[19] R. E. H. Clarck et al, Ap. J., 381 (1991) 597

[20] L. Vriens, Phys. Rev. 141 (1966) 88 and Y.-K. Kim and M. E. Rudd, Phys. Rev. A 50 (1994) 3954

[21] I. I. Sobelman, V. A. Vainstein and E. A. Yukov, "Excitation of Atoms and Broadening of Spectral Lines", Nauka, Moscow (1979)

[22] H. K. Chung, High Energy Density Physics, 1 (2005) 3

[23] http://www-amdis.iaea.org/FLYCHK/

[24] http://nlte.nist.gov/FLY/

[25] B.L. Henke, E.M. Gullikson, and J.C. Davis. X-ray interactions: photoabsorption, scattering, transmission, and reflection at E=50-30000 eV, Z=1-92, At. Data Nucl. Data Tables Vol. 54 (no.2), 181-342 (July 1993).

[26] http://henke.lbl.gov/optical_constants/

[27] Computational Atomic Structure: An MCHF Approach, C. Froese Fischer, T. Brage, P. Jönsson (Institute of Physics, Bristol, 1997)

[28] http://cpc.cs.qub.ac.uk/

[29] http://nlte.nist.gov/MCHF/

[30] A. L. Ankudinov et al., Comput. Phys. Comm. 98 (1996) 359

[31] F. A. Parpia, C. Froese Fischer and I. P. Grant, Comput. Phys. Comm. 175 (2006) 745

[32] https://dirac.spectro.jussieu.fr/mcdf/mcdf_code/mcdfgme_accueil.html

[33] N. R. Badnell, J. Phys. B 19, 3827 (1986))

[34] http://amdpp.phys.strath.ac.uk/tamoc/

[35] W. Eissner, M. Jones and H. Nussbaumer, CPC 8, 270 (1974))

[36] W. C. Martin and W. L. Liese, "Atomic Spectroscopy", originally in chapter 10 of "Atomic, Molecular and Optical Physics Handbook, G. W. F. Drake Ed., AIP Press, Woodburry, NY (1996)

[37] N. R. Badnell, A. Foster, D. C. Griffin, D. Kilbane, M. O'Mullane and H. P. Summers, J. Phys. B: At. Mol. Opt. Phys. 44, 135201 (2011)

[38] ftp://plasma-gate.weizmann.ac.il/pub/software/dos/cowan/

[39] http://www.tcd.ie/Physics/people/Cormac.McGuinness/Cowan/

[40] S. Fritzsche Computer Physics Communications, 183, 1525 (2012)

[41] F. A. Parpia, C. F. Fischer, I. P. Grant, Comput. Phys. Commun, 94, 249 (1996)

[42] P. Jönsson, X. He, C. Froese Fischer, I. P. Grant, Comput. Phys. Commun. 177, 597 (2007)

[43] S. Fritzsche et al, J. Phys. B 38 (2005) S707; S. Fritzsche et al, Phys. Rev. A 78 (2008) 032703; S. Fritzsche et al, Phys. Rev. Lett., 103 (2009) 113001

[44] M: Sewtz et al., Phys. Rev. Lett., 90 (2993) 163002

[45] C. Z. Dong et al, MNRAS, 369 (2006) 1735

[46] J. Bieron et al, J: Phys. B, 37 (2004) L305; J. Bieron et al., Phys. Rev. A, 80 (2009) 012530

[47] Busquet M., Klapisch M., Bar-Shalom A., J. Quant. Spectr. Rad. Transfer 71, 225 (2001)

[48] Busquet M., A.Bar-Shalom, Klapisch M., Oreg J., J.Phys. IV France 133, 973 (2006)

[49] M. F. Gu, Can. J. Phys., 86 (2008) 675

[50] http://sprg.ssl.berkeley.edu/~mfgu/fac/

[51] X. Gonze et al, Comput. Phys. Comm. 180 (2009) 2582

[52] http://www.abinit.org/

[53] J. MacFarlane, Comput. Phys. Comm. 56 (1989) 259

[54] T.A. Helmes and G. A. Moses, Comput. Phys. Comm. 183 (2012) 2629

[55] C. Bhattacharya et al., J. Appl. Phys., 102 (2007) 064915

[56] R. M. More, Adv. At. Mol. Phys., 21 (1985) 305

[57] G. Faussurier, C. Blancard and P. Renaudin, HEDP, 4 (2008) 114

[58] M. Busquet et al., HEDP, 5 (2009) 270

[59] H. A. Scott, Journal of Quantitative Spectroscopy and Radiative Transfer, 71, 689 (2001)

[60] H. A. Scott and S. B. Hansen, High Energy Density Physics, 6, 39 (2010)

[61] http://www.llnl.gov/def_sci/cretin

[62] Scofield J H 1972 1974 At. Data Nucl Data Tables 14 121

[63] J. J. MacFarlane et al, High Energy Density Phys. 3 (2007) 181

[64] http://www.prism-cs.com/index.htm

[65] J. P. Christiansen et al, Comput. Phys. Comm. 7 (1974) 271

[66] R. Ramis, R. Schmalz and J. Meyer-Ter-Vehn, Comput. Phys. Comm. 49 (1988) 475

[67] http://138.4.113.100/multi/index.html

[68] R. Ramis, J. Meyer-Ter-Vehn and J. Ramirez, Comput. Phys. Comm. 180 (2009) 977

[69] S. V. Zakharov, V. G. Novikov and P. Choi, "Z* code for DPP and LPP source modelling" P.223 in "EUV sources for Lithography", Ed. V. Bakshi, SPIE Press (2005); S. V. Zakharov, P. Choi, A. Y. Krukovskiy, V. G. Novikov, V. S. Zakharov, EUV Source Workshop Vancouver BC, Cannada, May 25, 2006

[70] F. Ogando and P. Velarde, J. Quant. Spectrosc. Radiat. Transf. 71, 541 (2001).

[71] G. J. Pert, J. Fluid Mech., 131 (1983) 401

[72] D. Bauer and P. Koval, Comput. Phys. Comm., 174 (2006) 396

[73] http://www.physik.uni-rostock.de/physik-forschung/arbeitsgruppen/qtmps/qprop/

[74] W. Kohn, L. J. Sham, Phys. Rev. 140 (1965) A1133

[75] R. Lichters et al, Phys. Plasmas 3(9), 3425 (1996), A. Kemp, R. Pfund and J. Meyer-Ter-Vehn, Phys. Plasmas 11 (12), 5648 (2004)

[76] http://lichters.net/download.html

[77] L.D. Landau, E.M. Lifshitz Quantum Mechanics Sec. Ed. Pergamon, Oxford (1965), p. 491

[78] C. S. Brady and T. D. Arber Plasma Physics and Controlled Fusion, 53, 015001 (2011)

[79] http://ccpforge.cse.rl.ac.uk/gf/project/epoch/

[80] C. Buth and R. Santra, Phys. Rev. A, 75 (2007) 033412, Christian Buth and Robin Santra, "FELLA (Free Electron Laser Atomic, Molecular, and Optical Physics Program Package), Version 1.3.0, Argonne National Laboratory, Argonne, Illinois, USA, with contributions by Mark Baertschy, Kevin Christ, Chris H. Greene, Hans-Dieter Meyer, and Thomas Sommerfeld (2007)

[81] https://wiki-ext.aps.anl.gov/amo/index.php/Main_Page

[82] D. L. Windt, Computers in Physics, 12, 360-370 (1998))

[83] http://www.rxollc.com/idl/

[84] http://www.radiantzemax.com/

[85] F. Schäfers, "The BESSY Raytrace Program RAY" in "Modern developments in X-ray and Neutron Optics", Ed. by A. Erko, M. Idir, T. Krist and A. Michette, Springer series in Optical Sciences, vol. 137 (2007)

[86] https://www.helmholtz-berlin.de/forschung/grossgeraete/nanometeroptik/methods/software_en.html

[87] H. O. Moser et al, in 'Developments in X-Ray Tomography VI', Proceedings of SPIE Vol. 7078, 707814, (2008)

[88] http://www.ts-imaging.net/WebHelp/X-TRACT/X-TRACT.htm

[89] http://www.ts-imaging.net/Services/

MODELLING OF PLASMA-BASED SEEDED SOFT X-RAY LASERS

E. Oliva, T. T. Thuy Le and P. Zeitoun

Laboratoire d'Optique Appliquée, ENSTA ParisTech, Ecole Polytechnique ParisTech, CNRS, Chemin de la Hunière, 91761, Palaiseau, France

1 INTRODUCTION

The development of bright sources of coherent soft X-rays is of major current interest. Free-electron lasers (FELs), seeded plasma-based soft X-ray lasers and high harmonic generation (HHG) sources are being applied in diverse fields such as biology[1] and physics.[2] However, only FELs provide the energy (~10 μJ) and the ultra-high intensity (>10^{16} W / cm^2) needed for breakthrough experiments such as single-shot imaging of samples with characteristic evolution times in the femtosecond range.[3] On the other hand, plasma-based soft X-ray lasers have demonstrated the highest energy per pulse (10 mJ),[4] but the spatial coherence is weak and the pulse durations are hundreds of picoseconds; these two drawbacks prevent such sources from achieving the ultra-high intensity needed for the most demanding applications.

The technique of seeding high order harmonics in a plasma amplifier overcomes these problems, as the amplified high harmonics conserve the good wavefront and spatial coherence of a short pulse while extracting the energy stored in the amplifier.[5–11] When seeding in gas amplifiers the resulting beam is fully polarized and has, as expected, a high degree of coherence[6] and a diffraction-limited wavefront. However, the pulse duration is still too long for applications (around 5 ps) and the energy is still not sufficient (about 1 μJ). The next logical step is to seed denser plasmas, created from solid targets. As the density is higher, the amplifier can store more energy and the output pulse duration should decrease due to collisional line broadening.[12] Nevertheless, the energies obtained in experiments are less than 100 nJ and duration is still of the order of picoseconds.[11]

In order to explain these results and design an amplifier to deliver several millijoules in hundreds of femtoseconds, a thorough knowledge of the creation and evolution of the plasma amplifier and the propagation of the seed through the plasma is needed. In this chapter, the codes used to understand the processes involved on the amplification of coherent X-ray radiation in plasmas will be described along with the results obtained, opening the way towards ultra-intense (~10 mJ, ~200 fs) fully coherent soft X-ray lasers.

2 SOFTWARE

Modelling the full setup of a multi-stage, plasma-based seeded soft X-ray laser is a huge enterprise as it involves several physical processes, including laser absorption and propagation in plasmas, plasma hydrodynamics, atomic physics and electromagnetism. These have different spatial and temporal characteristic scales, from some femtoseconds for the amplification of high harmonics in the plasma to several nanoseconds during the creation and evolution of the plasma amplifier – about six orders of magnitude. Solving the complete multidimensional problem would require complicated software (in terms of computational science and physics) running in large parallel computers. As there are many different parameters which play a crucial role in the design of these experiments, it is completely impractical to use such an approach. Thus, some simplification is needed to reduce the complexity and computational cost of the tools.

The first simplification consists of dividing the problem into different physical regimes, which can then be solved separately instead of in a coupled way. This means that each problem is treated with optimal techniques and approximations which reduce the computational cost and enhance the accuracy of the results. In addition to this, as there are fewer free parameters, it is easy to optimize several parts of the system separately, such as the laser parameters to create the amplifier and the high harmonic energy and duration. In the case treated here, the creation and evolution of the amplifier (plasma hydrodynamics) is decoupled from the other processes. Some parameters of interest which depend on the atomic physics can be calculated by post-processing the hydrodynamic data and effectively coupling with the high harmonic electromagnetic field propagation through the amplifier in order to model the amplification of the seeded harmonics and the evolution of atomic populations in time via the Maxwell-Bloch equations.[13]

3 THE ARWEN CODE

The creation of a plasma from a solid target and its evolution in time is modelled using the two dimensional hydrodynamic code ARWEN, which also takes into account electron thermal conduction and radiation transport.[14] It has been developed and is maintained by the Instituto de Fusión Nuclear of the Universidad Politécnica de Madrid, Spain. It has been applied in several different fields, such as inertial confinement fusion,[15] laboratory astrophysics[16] and plasma-based seeded soft X-ray lasers.[17,18,19] ARWEN solves the radiation hydrodynamics equations (1)–(8) with thermal conduction

$$\frac{\partial \rho}{\partial t} + \nabla \rho \, \vec{u} = 0 \tag{1}$$

$$\frac{\partial \rho \vec{u}}{\partial t} + \nabla \rho \, \vec{u} \, \vec{u} = -\nabla(\overrightarrow{P_m} + \overrightarrow{P_r}) \tag{2}$$

$$\frac{\partial \rho E_m}{\partial t} + \nabla[\rho E_m \, \vec{u} + (\overrightarrow{P_m} + \overrightarrow{P_r})\vec{u}] = S_E + \nabla \overrightarrow{q_c} + \nabla \overrightarrow{q_r} \tag{3}$$

$$\Omega \nabla \vec{I} + \vec{\kappa} \vec{I} = \varepsilon \tag{4}$$

$$\nabla \overrightarrow{q_c} = -\nabla(\overrightarrow{k_e} \nabla T) \tag{5}$$

$$\nabla \vec{q_r} = \int \left(\kappa \vec{I} - \varepsilon \right) d\upsilon \qquad (6)$$

$$E_r = \frac{1}{c} \int I \, d\Omega \, d\upsilon \qquad (7)$$

$$P_r = \frac{1}{3} E_r \qquad (8)$$

In these equations, ρ is the density, $\mathbf{u}$ is the velocity vector, $\mathbf{P}_m$ and $\mathbf{P}_r$ are the matter and radiation pressure tensors, E_m is the energy of the matter (kinetic and internal energy of the plasma), S_E is the laser energy source i.e. the laser energy absorbed by the plasma, $\mathbf{q}_c$ and $\mathbf{q}_r$ are the heat fluxes due to conduction and radiation, c is the speed of light, $\mathbf{I}$ is the radiation intensity, Ω is the solid angle in which radiation propagates, κ and ε are the opacity and the emissivity of the plasma, $\mathbf{k}_e$ is the thermal conductivity, v is the frequency of the radiation and E_r is the energy of the radiation field.

The numerical solutions of equations (1)–(8) require a range of different approaches, including integral equations and hyperbolic and elliptical partial derivative equations (PDEs). Thus it is necessary to treat each subset of the equations with a solver adapted to that particular problem.

The compressible Navier-Stokes equations (1–3), including radiation pressure and radiation heat transfer, are solved with an Eulerian scheme using a high-order Godunov method with an approximate Riemann solver.[20] The flux-limited thermal electron conduction equation (5) and the radiation transport equations (4,6) are treated with multi-grid[21] and Sn (discrete ordinates method, where the solid angle is discretized into several regions) multi-group methods[22,23] respectively. Matrix-free solvers for thermal conduction are available in the new version of the code.

As usual when working with compressible Navier-Stokes equations, a closing relation, i.e., an equation of state (EOS), is needed to solve the system. The ARWEN code does not compute the EOS each timestep for each mesh cell as it would be too expensive computationally. Instead, the EOS data is created once and written into a table. The ARWEN code reads the appropriate value from the table and, if necessary, interpolates it, which is faster than computing the EOS itself.

The EOS data are obtained from the quotidian equation of state (QEOS),[24] and the multipliers are adjusted to fit experimental data as Hugoniot curves. Opacities are computed with the JIMENA[25] and BiGBART codes, also developed at the Instituto de Fusión Nuclear.

Compressible fluid-like media with high energy density (such as laser produced plasmas) can develop structures including shockwaves and hydrodynamic instabilities which, to be resolved properly and reduce the numerical errors, need a very fine mesh in the modelling. Nevertheless, these structures appear only in a small part of the computational domain, most of the flow being regular. The majority of the flow is therefore well resolved using a coarse grid, with some small regions needing a finer grid. As using the finest grid in the whole domain will result in a waste of computational time, and so special meshing techniques must be used. The ARWEN code and all its packages are based on the adaptive mesh refinement (AMR) technique.[26,27] This consists of putting patches of finer grids only in the regions where they are needed, producing a uniform numerical error and saving computational time. The structure of levels and grids is created and controlled by the C++ library BoxLib.

4 ATOMIC PHYSICS

As explained above, the atomic physics related to the creation of gain inside the amplifier are decoupled from the hydrodynamic calculations (a newer version of the ARWEN code will couple temporal ionization dynamics and thus the gain calculation with the standard hydrodynamic model). Hence parameters of interest such as gain and saturation fluence must be calculated by post-processing the hydrodynamic data given by ARWEN. To simplify the calculations, a simple three-level atomic model was used, as shown in figure 1.

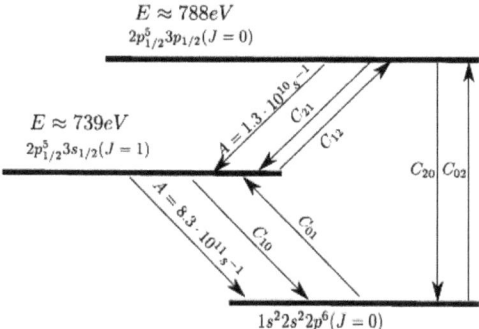

Figure 1 *Simplified Grotrian scheme and transitions of neon-like iron used in the post-processing calculations, using published data.*[28,29]

The lasing transition is $2p_{1/2}^5 3p_{1/2} \, J = 0 \rightarrow 2p_{1/2}^5 3s_{1/2} \, J = 1$ in neon-like iron (Fe^{16+}) at $\lambda = 25.5$ nm. The fundamental level (0) is assumed to be much more populated than the lower (1) and upper (2) levels and so its population can be assumed to be constant. To assure the creation of a population inversion between levels 1 and 2, the radiative transition between the level 2 and the fundamental level must be forbidden. In addition to this, level 1 must decay quickly to the fundamental level. With these conditions, electron collisional excitation can create the population inversion.[30,31,32]

In this model, the populations of levels 1 and 2 are calculated by solving the stationary rate equations

$$\frac{dN_i}{dt} = \sum_j C_{ij} n_e N_j + \sum_j A_{ij} N_i, \qquad (9)$$

$$\frac{dN_i}{dt} = 0, \qquad (10)$$

where N_i is the population of level i, n_e is the electron density, A_{ij} is the Einstein spontaneous emission coefficient of the transition and the C_{ij} are the collisional (de)excitation rates between levels i and j. The electron collisional rates are computed using Van Regemorter's formula,[33] equation (11) using the electron density and temperature given by the ARWEN code and assuming a Maxwellian distribution of electron speeds. The detailed balance principle allows the inverse process rate to be computed, equation (12),

$$C_{ij} \approx 1.6 \times 10^{-5} \frac{f_{ij} \langle g \rangle}{\Delta E_{ij} \sqrt{kT_e}} e^{-\Delta E_{ij}/k_B T_e}, \tag{11}$$

$$C_{ji} \approx \frac{\gamma_i}{\gamma_j} C_{ij} e^{\Delta E_{ij}/k_B T_e}, \tag{12}$$

where f_{ij} is the oscillator strength, $\langle g \rangle$ is the Gaunt factor 0.2 in this case), k_B is the Boltzmann constant, T_e is the electron temperature, ΔE_{ij} is the energy difference between the levels and γ^i is the degeneracy of level i.

Once the populations of the levels involved in the transition have been determined, the small signal gain, equation (13), and the saturation fluence, equation (14), both at the line centre, can be calculated

$$g_0^{ss}(v = v_0) = \left(N_2 - \frac{\gamma_2}{\gamma_1} N_1 \right) \sigma_{stim}(v = v_0), \tag{13}$$

$$F_{sat}(v = v_0) = \frac{h v_0}{\sigma_{stim}(v = v_0)}, \tag{14}$$

where v_0 is the frequency at the line centre, σ_{stim} is the stimulated emission cross-section and h is the Planck constant. In order to easily compute the cross-section, it is necessary to make a final assumption about the line profile throughout the plasma amplifier. In this case, only Doppler broadening is taken into account and the resulting formulae are

$$\sigma_{stim}(v) = \Phi(v) \frac{\lambda^2}{8\pi} A_{21}, \tag{15}$$

$$\Phi_D(0) = \lambda \left(\frac{m_i}{2\pi k_B T_i} \right)^{1/2}, \tag{16}$$

where $\Phi(v)$ is the Doppler spectral profile of the line, $\Phi_D(0)$ is the value of the Doppler spectral profile at the line centre, m_i is the ion mass and T_i is the ion temperature. Using this analysis, it is possible to estimate the amplification of the intensity and energy as a function of length in the plasma by solving the ordinary differential equation

$$\frac{dE}{dx} = g_0 I = \frac{g_0^{ss}}{1 + E/E_{sat}} E \tag{17}$$

where E is the energy, g_0 is the gain and E_{sat} is the saturation energy (the product of F_{sat} and the area). The solution of this equation for different values of small signal gain and saturation energy is shown in figure 2, demonstrating the characteristic shape of typical amplification curves. In the first few hundred micrometres there is no apparent amplification. In this part of the amplifier, the broad harmonic linewidth decreases until it fits the amplifier bandwidth, at which point exponential amplification begins – the steep parts of the curves. In this regime, the beam is strongly amplified until it reaches the saturation energy, after which the amplified beam is intense enough to depopulate the upper level of the transition appreciably, destroying population inversion and thus reducing amplification. This saturation regime starts, for the situation reported in figure 2, when the beam has passed through 1-2 mm of plasma, when the gain-length product has a characteristic value of around 18.

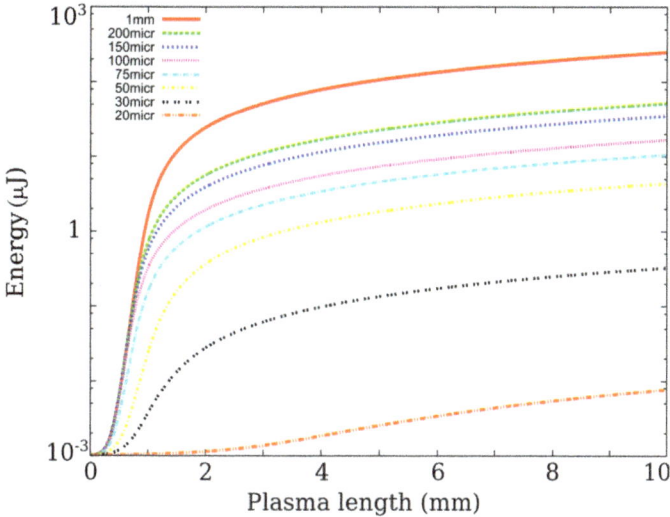

Figure 2 *Energy amplification as a function of plasma length for different plasma widths..*

4.1 Propagation Through Inhomogeneous Media

Laser created plasmas are strongly inhomogeneous media with large electron density gradients which induce refractive index gradients, causing the seeded beam to deviate from a straight trajectory. This results in the seeded beam to exit the gain zone, inhibiting amplification. The understanding of this effect is crucial in order to effectively amplify the beam.

Since seeded soft X-ray lasers are directional (namely the propagation direction through the plasma) it is possible to apply ray-tracing techniques to model the refraction (and amplification) of the seeded beam. As an example, figure 3 shows the results of post-processing of the ARWEN data given by three-dimenional ray-tracing code SHADOX (developed by the Laboratoire d'Optique Appliquée, France, and the Instituto Superior Tecnico, Portugal).

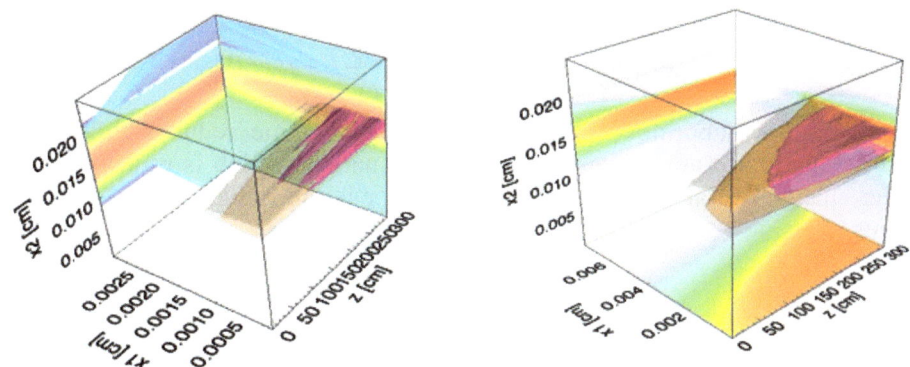

Figure 3 *Amplification and spatial shape of the seeded beam through (left) an inhomogeneous target and (right) a more homogeneous target. The simulations were done using the 3D ray-tracing code SHADOX.*

Figure 3 shows the evolution of the beam through the plasma for different electron density gradients. The gradient is steeper in the figure on the left, resulting in filamentation and degradation of the beam. The beam in the figure on the right is much more homogeneous, as the electron density varies slowly and thus refraction effects are less important.

4.2 Maxwell-Bloch simulation of the amplification of radiation.

As explained in the preceding sections, post-processing the hydrodynamic data given by ARWEN gives valuable information on how to seed correctly in order to minimize refraction effects and about the energy and spatial shape of the output beam. However, it is not possible to obtain information about the temporal profile of the output beam, although estimating the line broadening can give some limiting values about the duration of the pulse[12] but not a detailed picture of the temporal energy distribution. In addition the dynamic populations and dynamic effects induced in the beam (such as Rabi oscillations) cannot be studied with such post-processing since the equations are solved for the steady state.

In order to take these temporal effects into account, it is necessary to couple the atomic physics results with the electromagnetic field and solve the equations both spatially and temporally. Because of the different timescales involved in the processes, from tens of femtoseconds (high harmonic pulse duration) to picoseconds (plasma response), Maxwell-Bloch type equations must be used.

The Maxwell-Bloch equations can be deduced from the Maxwell wave equation for electric fields propagating through plasmas,

$$\Delta \vec{E} - \frac{1}{c^2}\frac{\partial^2 \vec{E}}{\partial t^2} - \frac{\omega_p^2}{c^2}\vec{E} = \mu_0 \frac{\partial^2 \vec{P}}{\partial t^2} \tag{18}$$

where ω_p is the plasma frequency, μ_0 is the vacuum permeability and **P** is the polarization density of the medium. Assuming a monochromatic field of frequency ω_0, and in the slowly varying envelope approximation (neglecting second derivatives of the electric field and first derivatives of the polarization), equation (18) can be rewritten, for the direction of propagation, z, as

$$\frac{2i\omega_0}{c^2}\left(\frac{\partial E}{\partial t}+c\frac{\partial E}{\partial z}\right) = -\frac{\omega_0^2}{c^2}\left(-\frac{\omega_p^2}{\omega_0^2}E+\mu_0 c^2 P\right). \tag{19}$$

Changing from the laboratory frame to the photon frame reduces this partial differential equation to an ordinary differential equation, which is much easier to solve. The change of variable is $\tau = t - z/c$ where τ is the reduced time. Defining $\xi = c\tau$ gives

$$\frac{\partial E}{\partial \xi} = \frac{i\omega_0}{2c}\left[\mu_0 c^2 P - \left(\frac{\omega_p}{\omega_0}\right)^2 E\right]. \tag{20}$$

The amplification (first term in square brackets) and damping (second term) of the electric field through the plasma is described by equation (20). As usual, a constitutive relation (a relation between two physical quantities, specific of a material, that approximates its response to external fields) which relates polarization and material properties is needed. In this case, polarization is given by

$P = \text{Tr}(\rho d)$ where ρ is the density operator, d the atomic electric dipole and Tr denotes the trace of the matrix. The Zeeman coherences can be neglected and the resulting equation for polarization is[34]

$$\frac{\partial P}{\partial \tau} = \Gamma - \gamma P - \frac{i z_{21}^2}{\hbar} E \left(N_2 - N_1 \right) \tag{21}$$

where a source term Γ has been added to model the spontaneous emission. This term has a vanishing correlation time and it is normalized to obtain (via the Wiener-Khinchine theorem) the correct Lorentzian line width.[35] The depolarization term γ is given approximately by the electron-ion collision frequency. Z_{21} is the dipole matrix element.

Finally, the populations N_1, N_2 appearing in equation (21) are computed using standard rate equations, explained in section 4, but taking into account the temporal dependence and adding a term coupling the populations and the electric field

$$\frac{\partial N_2}{\partial t} = \sum_k C_{k2} N_k + \frac{\text{Im}\left(E^* P\right)}{2\hbar} \tag{22}$$

$$\frac{\partial N_1}{\partial t} = \sum_k C_{k1} N_k - \frac{\text{Im}\left(E^* P\right)}{2\hbar}. \tag{23}$$

The Einstein A coefficients are included in the C coefficients. This set of equations (20)-(23) are the so-called Maxwell-Bloch equations. When solved properly (using high order Runge-Kutta methods for example), the amplification of the seeded pulse and the evolution of the temporal and spatial shapes of the pulse are obtained, as shown in figure 4.

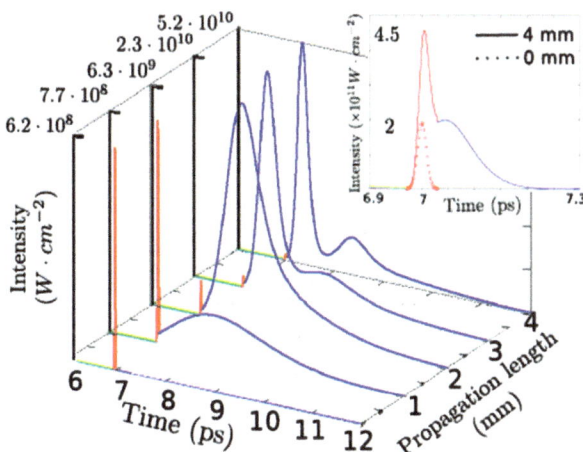

Figure 4 *Temporal shape of the seeded high harmonic through the plasma amplifier. The typical structure of harmonic and wake (representing Rabi oscillations and coherent decay) is observed.*[36–39]

5 VISUALIZATION

The data given by the programs described above consists of arrays of numbers. Although these numerical data are needed to make quantitative physical predictions, in order to describe the temporal evolution of the system and gain a deep knowledge of the problem some visualization tool is necessary. Images and/or videos are valuable tools since spatial structures, such as jets or gradients, and their evolution in time can be easily recognized.

Although most programs have their own tools to check the simulations and to study the physics, sometimes it is better to use more powerful visualization. This adds one more step to the analysis as the output data must be written in a format compatible with the visualization tool. This can be done by modifying the program itself (changing the output subroutines) or by post-processing the data. The solution adopted depends on each particular case — how difficult it is to modify the source code, how many simulations are already done, and so on.

As an example, figure 5 shows a 3D image of the electron density and gain of a plasma simulated using ARWEN; the image was made using VisIt.[40] As explained, the 2D data given by ARWEN was post-processed to obtain the gain. Then, both sets of data were rewritten in the format required by VisIt, making it possible to create 3D images, combine different sets of data (in this case gain and electron density) and find the best conditions (angle, illumination, transparency, etc.) to best understand the physics involved.

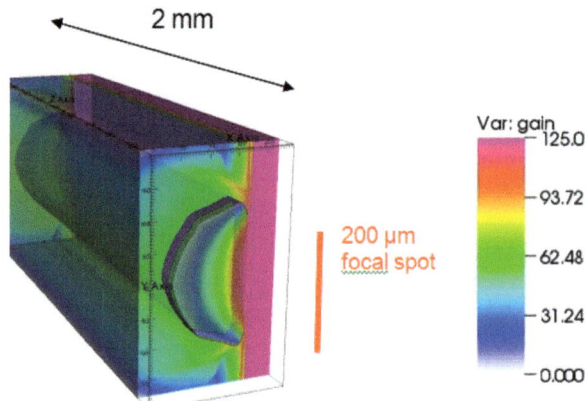

Figure 5 *3D image of electron density and gain of a plasma simulated with ARWEN. The gain zone is shown embedded into the plasma. Some structures such as lateral jets can be seen. The figure was created using VisIt.*[40]

6 CONCLUSIONS

In this chapter we have reviewed the modelization of seeded soft X-ray lasers using plasma amplifiers from solid. Since these sources present different temporal and spatial scales from nanoseconds (plasma characteristic time), picoseconds (atomic system) and femtoseconds (HOH electric field envelope), specific multiscale tools are needed. The creation and evolution of the plasma amplifier is studied with a 2D hydrodynamic code, optimized for this kind of problem. Atomic parameters, as gain and saturation fluence, are obtained postprocessing hydrodynamic data. Then, amplification is studied in different ways : a 3D ray-tracing code is used to study the role of refraction and focusing in the amplification process, whereas a Maxwell-Bloch code is used to model the temporal evolution of the seeded radiation and its coupling with the atomic system. Powerful visualization tools are also used to properly display the 3D properties of the system. The results obtained with these computational tools provide an invaluable aid to comprehend experimental results and design and optimize new schemes, by identifying the key parameters of the experiments.

References

1 R.E. Burge, M.T. Browne, P. Charalambous, G.E. Slark and P.J. Smith, *Opt. Lett.*, 1993, **18**, 661.

2 J. Itatani, J. Levesque, D. Zeidler, H. Niikura, H. Pépin, J.C. Kieffer, P.B. Corkum and D.M. Villeneuve, *Nature*, 2004, **432**, 867.

3 H.N. Chapman, *Nature Mater.*, 2009, **8**, 299.

4 B. Rus, T. Mocek, A.R. Präg, M. Kozlová, G. Jamelot, A. Carillon, D. Ros, D. Joyeux, and D. Phalippou, *Phys. Rev.*, 2002, **A66**, 063806.

5 T. Ditmire, M.H.R. Hutchinson, M.H. Key, C.L.S. Lewis, A. MacPhee, I. Mercer, D. Neely, M.D. Perry, R.A. Smith, J.S. Wark and M. Zepf, *Phys. Rev.*, 1995, **A51**, R4337.

6 P. Zeitoun, G. Faivre, S. Sebban, T. Mocek, A. Hallou, M. Fajardo, D. Aubert, P. Balcou, F. Burgy, D. Douillet, S. Kazamias, G. de Lachèze-Muriel, T. Lefrou, S. le Pape, P. Mercère, H. Merdji, A.S. Morlens, J.P. Rousseau and C. Valentin, *Nature*, 2004, **431**, 426.

7 Y. Wang, E. Granados, M.A. Larotonda, M. Berrill, B.M. Luther, D. Patel, C.S. Menoni and J.J. Rocca, *Phys. Rev. Lett.*, 2006, **97**, 123901.

8 N. Hasegawa, T. Kawachi, A. Sasaki, M. Kishimoto, K. Sukegawa, M. Tanaka, R.Z. Tai, Y. Ochi, M. Nishikino, K. Nagashima and Y. Kato, *Phys. Rev.*, 2007, **A76**, 043805.

9 Y. Wang, E. Granados, F. Pedaci, D. Alessi, B. M. Luther, M. Berrill and J.J. Rocca, *Nature Photonics*, 2008, **2**, 94.

10 F. Pedaci, Y. Wang, M. Berrill, B. Luther, E. Granados and J.J. Rocca, *Opt. Lett.*, 2008, **33**, 491.

11 Y. Wang, M. Berrill, F. Pedaci, M.M. Shakya, S. Gilbertson, Z. Chang, E. Granados, B.M. Luther, M.A. Larotonda and J.J. Rocca, *Phys. Rev.*, 2009, **A79**, 023810.

12 D.S. Whittaker, M. Fajardo, P. Zeitoun, J. Gautier, E. Oliva, S. Sebban and P. Velarde, *Phys. Rev.*, 2010, **A81**, 043836.

13 Y. Castin and K. Mølmer, *Phys. Rev.*, 1995, **A51**, R3426.

14 F. Ogando and P. Velarde, *J. Quant. Spectrosc. Radiat. Transf.*, 2001, **71**, 541.

15 P. Velarde, F. Ogando, S. Eliezer, J. Martínez-Val, J. Perlado and M. Murakami, *Laser Part. Beams*, 2005, **23**, 43.

16 P. Velarde, D. García-Senz, E. Bravo, F. Ogando A.R. Relaño, C. García and E. Oliva, *Phys. Plasmas*, 2006, **13**, 092901.

17 K. Cassou, P. Zeitoun, P. Velarde, F. Roy, F. Ogando, M. Fajardo, G. Faivre and D. Ros, *Phys. Rev.*, 2006, **A74**, 045802.

18 E. Oliva, P. Zeitoun, S. Sebban, M. Fajardo, P. Velarde, K. Cassou and D. Ros, *Opt. Lett.*, 2009, **34**, 2640.

19 E. Oliva, P. Zeitoun, P. Velarde, M. Fajardo, K. Cassou, D. Ros, S. Sebban, D. Portillo and S. le Pape, *Phys. Rev. E*, 2010, **82**, 056408.

20 P. Colella and H.M. Glaz, *J. Comput. Phys.*, 1985, **59**, 264.

21 S.F. McCormick and R.E. Ewing, *Multigrid Methods*, 1987, (Philadelphia: SIAM).

22 S. Chandrasekhar, *Radiative Transfer*, 1960, (Dover Publications: New York).

23 K.D. Lathrop, *Nucl. Sci. Eng.*, 24, 381 1966.

24 R.M. More, K.H. Warren, D.A. Young and G.B. Zimmerman, *Phys. Fluids*, 1988, **31**, 3059.

25 E. Minguez and R. Falquina, *Laser Part. Beams*, 1992, **10**, 651.

26 M.J. Berger and P. Colella, *J. Comput. Phys.*, 1989, **82**, 64.

27 C.A. Rendleman, V.E. Beckner, M. Lijewski, W. Crutchfield and J.B. Bell, *Comput. Visualization Sci.*, 2000, B, 147.

28 T. Shirai, Y. Funatake, K. Mori, J. Sugar, W. L. Wiese, and Y. Nakai, J. Phys. Chem. Ref. Data 19, 127 1990.

29 J.A. Koch, B.J. MacGowan, L.B. Da Silva, D.L. Matthews, J.H. Underwood, P.J. Batson, R.W. Lee, R.A. London and S. Mrowka, *Phys. Rev.*, 1994, **A50**, 1877.

30 B.L. Whitten, A.U. Hazi, M.H. Chen and P.L. Hagelstein, *Phys. Rev.*, 1986, **A33**, 2171.

31 W.H. Goldstein, B.L. Whitten, A.U. Hazi and M.H. Chen, *Phys. Rev.*, 1987, **A36**, 3607.

32 P.V. Nickles, V.N. Shlyaptsev, M. Kalachnikov, M. Schnurer, I. Will and W. Sandner, *Phys. Rev. Lett.*, 1997, **78**, 2748.

33 H.V. Regemorter, *Astrophys. J.*, 1962, **136**, 906.

34 A. Sureau and P.B. Holden, *Phys. Rev.*, 1995, **A52**, 3110.

35 O. Larroche, D. Ros, A. Klisnick, A. Sureau, C. Möller and H. Guennou, *Phys. Rev.*, 2000, **A62**, 043815.

36 I.R. Al Miev, O. Larroche, D. Benredjem, J. Dubau, S. Kazamias, C. Möller and A. Klisnick, *Phys. Rev. Lett.*, 2007, **99**, 123902.

37 C.M. Kim, K.A. Janulewicz, H.T. Kim and J. Lee, *Phys. Rev.*, 2009, **A80**, 053811.

38 C.M. Kim, J. Lee and K.A. Janulewicz, *Phys. Rev. Lett.*, 2010, **104**, 053901.

39 E. Oliva, P. Zeitoun, M. Fajardo, G. Lambert, D. Ros, S. Sebban and P. Velarde, *Phys. Rev. A*, 84,013811 (2011).

40 https://wci.llnl.gov/codes/visit/

FIELD COHERENCE OF EUV SOURCES

Olivier Guilbaud

Laboratoire de Physique des Gaz et Plasma, Unité Mixte CNRS Université Paris-Sud, France

1 INTRODUCTION

As sources of electromagnetic radiation provide shorter wavelengths, optical instrumentation and techniques must keep pace. Such evolution take the form of extrapolation or adaptation of concepts well known at longer wavelengths; the ability to measure the wavefront properties of sources has the same importance in the extreme ultraviolet (EUV) range as in the visible and, for this reason, Hartmann and Shack-Hartmann sensors have been adapted to very short wavelengths.[1] For example, they have been used to characterize high order harmonic beams and synchrotron beamlines, and are being tested for use at advanced X-ray sources such as the Linac Coherent Light Source (LCLS) X-ray free electron laser (XFEL). However, the characteristics of short wavelength radiation also require the development of new techniques in order to measure novel properties of the source. An example is the RABBITT (reconstruction of attosecond beating by interference of two photon transitions) technique[2] which was developed to measure the relative spectral phase of high order harmonics in the context of attosecond pulse train generation and uses IR-XUV photo-ionisation as a coherent nonlinear process from which the attosecond structure can be extracted. Adapting the general theory of coherence to shorter wavelengths is straightforward, and methods for measuring the coherence properties of the source are usually direct adaptations of classical techniques, namely wavefront and amplitude division interferometers. However, this does not mean there are no challenges to short wavelength coherence measurement and utilisation.

These challenges include
- Technical challenges. Interferometry requires optics with excellent flatness with respect to the wavelength, which becomes more demanding to achieve as the wavelength decreases. In addition, grazing or multilayer optics must be used to provide high reflectivities. A specific difficulty arises when considering amplitude division interferometers, since EUV beamsplitters will be needed.
- *Modelling challenges*. The diversity of short wavelength sources is such that they encompass all the classical approximations usually made in theoretical optics. Some sources can be considered as fully coherent (e.g., high order harmonics),

whereas EUV lines emitted by laser produced or discharge plasmas can be considered as spatially incoherent at the source and with a random temporally stationary field. It is more significant that many sources lie between these two limiting cases. The XFEL and transient soft X-ray lasers rely on (self) amplification of spontaneous emission, or (S)ASE, leading to partially coherent features and a breakdown of the stationary hypothesis.

Despite these difficulties, there are many potential applications of the coherent properties of short wavelength sources, including the following. First, because of the short wavelength, interferometric methods will provide high sensitivity on the evolution of surfaces.[3] EUV radiation can penetrate very dense plasmas and any shift in the interferometric pattern will provide information on the local refractive index and thus on the plasma electron density. Next, holography[4,5] and coherent diffraction at short wavelengths are powerful tools for producing time resolved images of nano-samples or macromolecules that are difficult to crystallize. Also, interference patterns produced by powerful sources can be used for permanent nano-patterning of surfaces or to generate a spatially modulated transient excitation of a sample (transient grating pump-probe methods).

The basics of coherence theory are well established[6] and so only a summary is given in the following section. Illustration related mainly to the development of high order harmonics and plasma based soft X-ray lasers will be given. The physics of such sources is detailed in other chapters of this book. The last parts of the present chapter provide a general framework of coherence properties applicable to all sources through the notion of spectral coherence.

2 SPATIAL AND TEMPORAL COHERENCE

2.1 Coherence Functions

Consider the complex representation of an optical field, $U(r,t)$, in the scalar approximation. From a theoretical point of view, the concept of coherence is related to correlations between different parts of the field. If the intensity, $|U(r,t)|^2$, is relatively smooth then phase correlations are the central aspect. The coherence function Γ is defined by:

$$\Gamma\left(r_1,r_2,t_1,t_2\right)=\left\langle U_1^*(r_1,t_1)U_2(r_2,t_2)\right\rangle, \tag{1}$$

where the bracket $\langle\rangle$ represents an ensemble average og all the field realizations. In the context of classical optical sources with pulse durations extremely long compared to the characteristic field evolution time (in this context, a random evolution), the field is assumed to be stationary and ergodic. The ensemble average is then equivalent to an average over an integration time long compared to this characteristic The stationarity of the field results in the time dependence of Γ depending only on the time difference $\tau = t_2 - t_1$,

$$\Gamma(r_1,r_2,t_1,t_2)=\Gamma(r_1,r_2,\tau). \tag{2}$$

Various useful functions related to experimentally measurable quantities can now be derived from Γ; for example, the average local intensity I is given by Γ when $\tau = 0$. In interferometry the fields U_1 and U_2 in the definition of Γ are related to an incoming field and interfere on a detector such as a CCD camera, producing fringes with visibility proportional to the modulus of Γ.

The spatial coherence function, defined by :

$$\Gamma(r_1, r_2) = \Gamma(r_1, r_2, \tau = 0), \tag{3}$$

can be used, for example, to predict the fringe visibility produced in a Young's slits experiment when the slits are located at r_1 and r_2. If r_1 is fixed ($\equiv 0$), the modulus of the spatial coherence function globally decreases as $|r_2|$ increases. The width of the function $|\Gamma(0, r_2)|$ is called the spatial coherence length, Λ_T .Various definitions of the width exists, including width at half maximum, width at 1/e or RMS width.

The temporal coherence function, defined by putting $r_1 = r_2 = r$ is:

$$\Gamma(r, \tau) = \Gamma(r, r, \tau), \tag{4}$$

It allows, for example, the prediction of the evolution of fringe visibility in a Michelson experiment (as needed for measuring temporal coherence) when the path difference $\delta = c\tau$ is changed. The width of the function $|\Gamma(r, \tau)|$ is called the temporal coherence time, τ_c, which again is defined in a variety of ways. Since fringe visibility also depends on the intensity balance between the two interfering beams, a normalised coherence function, the degree of coherence, is introduced,

$$\gamma(r_1, r_2, \tau) = \frac{\Gamma(r_1, r_2, \tau)}{\left[I(r_1) I(r_2, \tau) \right]^{1/2}}. \tag{5}$$

The spectral density $S(r, \omega)$, where ω is the angular frequency, is defined as the average squared modulus of the Fourier transform of the field U. The stationarity of the field then leads to the Wiener-Khintchine theorem[7]

$$\Gamma(r, \tau) = \int S(r, \omega) \exp(-i\omega\tau) d\omega \tag{6}$$

From a practical point of view, measuring the temporal coherence function in terms of τ provides access to the average spectral profile of the source through a Fourier transform.

2.2 EUV and Soft X-ray Spatial Coherence Measurement

The spatial coherence function of a source can be reconstructed by illuminating a set of Young's slits with different separations. The evolution of the fringe visibility at the centre of the fringe field, as a function of the slit separation, directly reflects the evolution of the spatial coherence function across the beam. An example is shown in figure 1, where this procedure has been applied to EUV beams (λ=32,8nm) emitted by a high order harmonic generation source, an unseeded soft X-ray laser, and this soft X-ray laser seeded with the harmonic beam[8]. This method is simple and efficient but interpretation of the results must be done with care. The spatial coherence is probed at two points at both side of the optical axis, and it has to be assumed that the temporal coherence length is large compared to the path difference involved in the interference process.

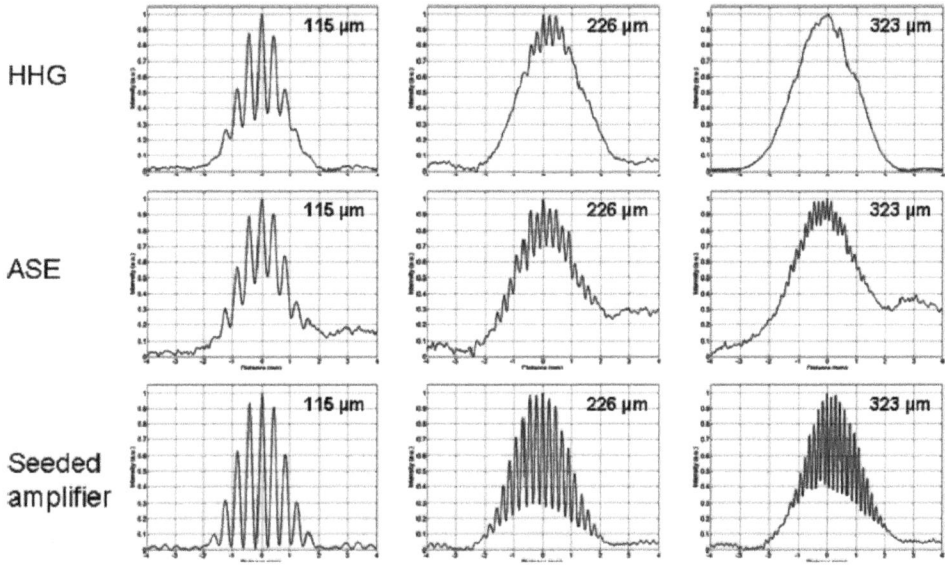

Figure 1 *Spatial coherence measurement with variable separation Young's slits (125,226 and 323 μm). Top: high order harmonics beam. Middle: plasma-based soft X-ray laser. Bottom: plasma-based soft X-ray laser seeded with harmonics.*[8]

2.3   EUV and Soft X-ray Temporal Coherence Measurement

A range of Michelson-like interferometric systems have been developed for the EUV and soft X-ray range. The instrument presented in figure 2 is a wavefront division interferometer where a Fresnel bimirror splits the incoming beam into two parts. The small angle between the mirrors leads to an overlap of the two secondary beams, and a detector placed in this overlapping region will record fringes. The system is designed to work at a small grazing angle α to obtain high reflectivity ($\alpha = 6°$). The insert in figure 2 gives more details on the mirror shape; the dihedron geometry allows the introduction of a path difference between the beams by moving vertically one of the mirrors by an amount z. This operation does not change the spatial superposition of the beams, and thus any observed fringe visibility decrease will be associated to a temporal coherence loss and not to an uncontrolled spatial coherence loss. For this wavefront division interferometer, to obtain high fringe visibility the spatial coherence length has to be higher than the width of the interference field. If the source is spatially incoherent, the interferometer has to be placed far from it.

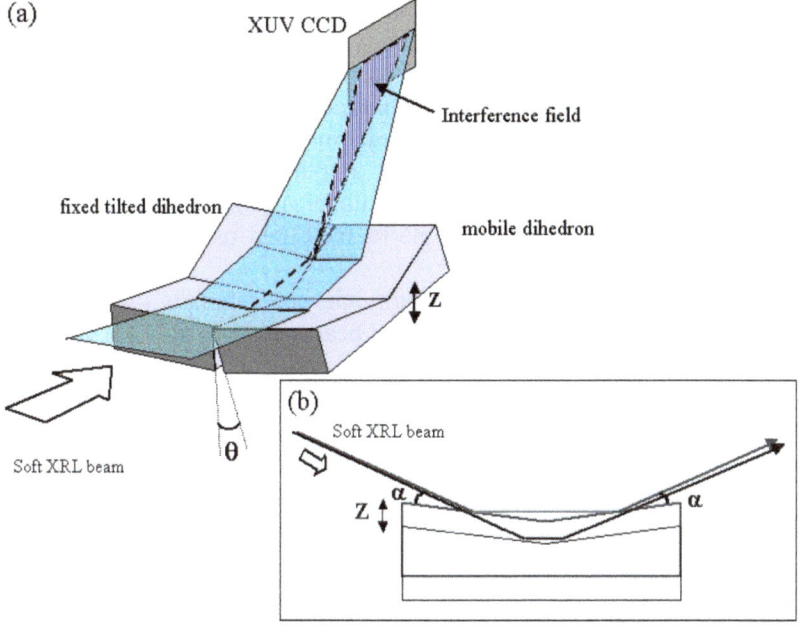

Figure 2 *Variable path difference interferometer based on the wavefront division principle. The insert shows the geometry of the fixed and mobile dihedra that enable a change in path difference without changing the overlap conditions of the interfering beams.*

3 COHERENCE PROPERTIES OF HIGH ORDER HARMONICS

3.1 Spatial, Temporal and Phase Coherence

The high order harmonics of an intense infrared laser beam inherit most of its coherence properties; the spatial coherence length covers the beam dimension of a single harmonic. Wavefront division interferometers are then particularly suitable and make interferometry experiments with this type of source more straightforward. A simple procedure can be used to measure the spatial coherence length of the harmonic beam after spectral selection. The contrast of the interference fringes produced in a classical Young's double slit experiment is measured as a function of slit separation as shown in figure 1.

The coherence time of each harmonic is of the order of the pulse duration, a few tens of femtoseconds. The pulse is close to the Fourier limit, i.e., the lower limit for the pulse duration for a given spectral content of the pulse. However the temporal coherence is not constant along the beam profile. In the generation medium, the source term is the harmonic single atom response or dipole. It can be expressed as a superposition of two quantum contributions called "first" and "second" electron trajectories[9]. Each of this contribution to the total dipole has a phase φ with a linear dependence with the IR intensity I : $\varphi = \alpha I$ where α is depending on the electron trajectory. Spatial and temporal variation of I will then lead to an increased spatial divergence and to a spectral broadening of the harmonic. First trajectory for which α is small will be less divergent and spectrally narrow compared to second trajectory contribution for which α is many times greater.

This intensity dependence of the dipole phase and time dependent ionisation of the medium can affect the pulse length of a single harmonic leading to a small difference from the Fourier limit. The important point is that the harmonic field E can still be described in a deterministic form, using the spectral amplitude $A(\omega)$ and phase $\varphi(\omega)$, $E(\omega) = A(\omega)\exp(i\varphi(\omega))$, and this description does not change significantly from one pulse to the next. The spectral phase can in principle be measured using sophisticated frequency resolved optical gating (FROG) techniques.[10]

An interesting feature of high order harmonic generation is the mutual coherence properties of sources generated by two separated beams coming from the same laser chain. This property led to the development of internal frequency conversion interferometer, which has the potential of significantly simplify EUV interferometric devices.[11] In addition, two consecutive harmonic orders have a stable phase relationship leading in the temporal domain to an attosecond pulse train structure in a similar way to modelocking in infrared lasers. Once this relative phase difference is known, manipulation techniques like EUV chirped mirrors can be developed to minimize the duration of each pulse of this train.

3.2　Applications of High Order Harmonic Coherence

Applications of the coherence properties of high order harmonics have been extensively explored. These include EUV transient grating experiments which are under consideration to explore surface and solid wave dynamics at very high wave vectors.[12] Two harmonic beams crossing at a large angle θ on a surface will generate a fringe pattern of period $\lambda_{HHG}/2\sin(\theta/2)$. If powerful enough, this intensity modulation will excite a wave with an imposed wavevector. The decay of this wave can be followed through the diffraction of a third beam (also coming from an harmonic source) on this transient grating.

Demonstration of coherent or lens-less imaging of nano-system using high order harmonics, has also been demonstrated recently.[13] Numerical reconstruction techniques are required to retreave the object. They not only allow the identification of small details in the object due to use of short wavelength but also allows 3D reconstruction.

It should be also noted that high order harmonics without spectral selection will of course cover a large range of wavelengths and will thus be associated with a very short temporal coherence time. It has been suggested that this property might be useful for the detection of buried defects in EUV lithography masks.[14] if technics similar to the optical coherence tomography (OCT) were developed.

4　COHERENCE OF PLASMA BASED EUV AND SOFT X-RAY LASERS

4.1　General Properties

Plasma based soft X-ray lasers operate in the same spectral range as high order harmonic sources. In these devices, a population inversion is obtained between two levels of highly charged ions in a hot dense plasma. The plasma generation and pumping require a high current and a fast discharge in a capillary or high energy and/or high intensity lasers.[15]

Several technical solutions have been developed to meet these demands, but they all use collisional excitation for pumping and rely on ASE. The short gain lifetime resulting from the over-ionization of lasing ions, cooling by plasma electrons and plasma expansion precludes the use of an optical resonator (cavity) to give a highly coherent beam. However, the gain is sufficiently high to reach saturation in one pass of the lasing medium and so powerful EUV beams can be obtained, with spatial coherence and beam divergence determined by the propagation conditions through the gain medium. The orders of magnitude of the divergence, $\Delta\theta$, and the spatial coherence length, Λ_T, at the output of the source are given by

$$\Delta\theta \sim a/L, \quad \Lambda_T \sim \lambda L/a, \tag{7}$$

where a and L are the width and length of the gain region; the quantity $N = a^2/\lambda L$ is known as the Fresnel number. The lasing lines are highly monochromatic due to the moderate ion temperatures and densities and to the gain narrowing process[16]. The relative spectral width, $\Delta\lambda/\lambda$, is $\sim 10^{-5}$–10^{-4}, leading to long coherence times τ_c, but it is only measurement of the temporal coherence function that can give access to the line shape and width of such a highly monochromatic source.

Depending on the pumping parameters, short wavelength lasers can exhibit very different coherence properties because of the large ranges of possible Fresnel numbers, the ratio of the pulse duration and coherence time can vary between ~1 and ~100.

4.2   Capillary Discharge Soft X-ray Lasers

In capillary discharge soft X-ray lasers ($\lambda = 46.9$ nm) the gain medium is ideally very long, ≈ 30 cm, with a Fresnel number close to unity, low divergence and a spatial coherence length determined by the size of the beam. Spatial coherence measurements using Young's slits have confirmed this ideal,[17] but also indicate that an additional effect must be introduced to explain a higher than predicted spatial coherence length. The annular shape of the beam intensity profile and some wavefront sensor analysis[17] give strong indication that the refraction of the beam in the plasma due to electron density gradients enhances the beam spatial filtering and mode selection. Such sources are now used for soft X-ray microscopy, micro-ablation and nano-patterning because of their high power and high spatial coherence.[18] The temporal coherence is of great importance for these patterning technics[19] because they require long path difference and high monochromaticity. This crucial quantity has been measured using the variable path difference wavefront division interferometer[20] presented in section 2. The experimental setup is presented in figure 3. The fringe visibility has been measured as a function of the path difference in the interferometer. The results are presented in figure 4. Coherence lengths of up to 700 μm for long capillaries have been measured and the fringe visibility evolution is clearly Gaussian, meaning that the laser line spectral profile is also Gaussian. The spectral width deduced from the visibility through the Wiener-Khintchine theorem matches reasonably well with radiative transfer model predictions taking into account Doppler broadening and gain narrowing.

From a coherence point of view, the capillary discharge plasma can be modelled as a stationary source. Because the pumping evolution is slow compared to the population dynamics, the gain and pulse durations are very long (nanoseconds) compared to the coherence time (picoseconds). The stationary source hypothesis, required to apply the Wiener-Khintchine theorem to the ASE random field, is thus reasonable.

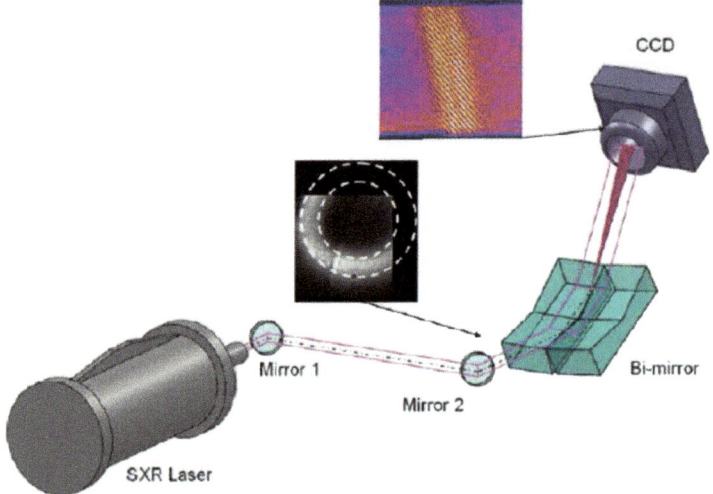

Figure 3   *Measurement of the temporal coherence of a capillary discharge soft X-ray laser. The output is aligned with the axis of a bi-dihedron interferometer using two multilayer mirrors that also remove the out of band plasma radiation. The wave front division interferometer produces two beams that interfere on the CCD. The central image shows the far field doughnut shape of the laserbeam; the size of the detector did not allow the entire beam to be recorded.*

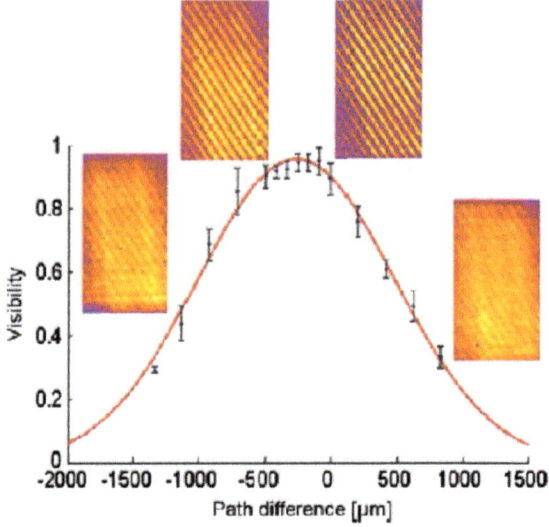

Figure 4   *Fringe visibility as a function of the path difference introduced between the two arms of the interferometer. The curve is a Gaussian fit of the data.*

4.3 Laser Based Soft X-ray Lasers

Using the same physical principles, it is possible to generate EUV laser with almost the opposite coherence properties, namely a high Fresnel number ($L < 1$ mm) and a pulse duration T not significantly longer than the coherence time τ_c. Transient Collisional Excitation[21] and optical field ionization (OFI)[22] soft X-ray lasers are typical examples of this situation. Note also that an inbetween situation exists with the quasi steady state X-ray laser: the Fresnel number is still high but the pulse duration is much longer than τ_c.

The spatial coherence of transient soft X-ray lasers has been measured;[23] using comparison of experimental and theoretical diffraction patterns generated with different assumptions about the spatial coherence of the source. The value obtained, a few micrometers, was two orders of magnitude smaller than the source size.

The temporal coherence of this type of source has also been characterised with wavefront[24] and amplitude[25] division interferometers. The coherence time (1–3 ps) is of the order of the pulse duration (3–6 ps) measured with a streak camera. The situation $T/\tau_c \sim 3$ has important consequences. First, the radiation cannot be assumed to be a stationary random signal and the Wiener-Khintchine theorem should thus be used with caution. Simulations of this kind of soft X-ray laser ASE field with a Bloch Maxwell code[26] show the presence of important intensity spikes in the temporal profile with timescales only a few times smaller than T (figure 5). This random spike structure also exists in the spatial domain with typical dimension related to the spatial coherence length. Such patterns are clearly observed in experimental far fields of transient (figure 6) and OFI soft X-ray lasers, both of which have T/τ_c ratios of several times unity[27].

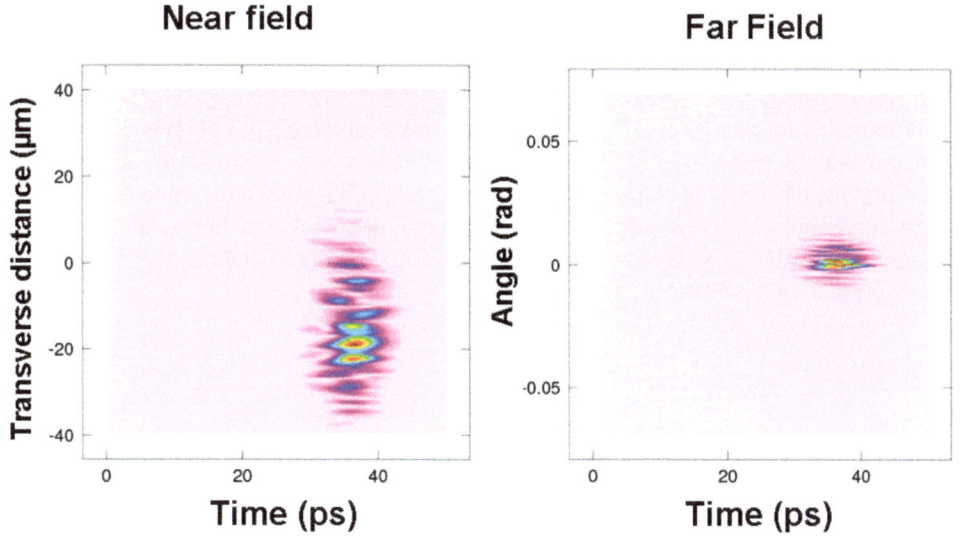

Figure 5 *Bloch-Maxwell simulation of the transient soft X-ray laser field as functions of time and gain medium transverse distance.*

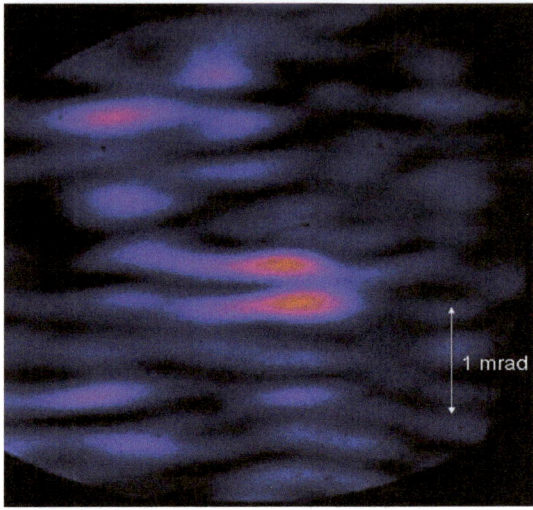

Figure 6 *Single shot far-field image of a transient soft X-ray laser (nickel like molybdenum, λ = 18.9 nm).*

In conclusion, the non-stationary properties of the field associated with partial spatial coherence has significant implications. It should be noted that, in the temporal domain, un-seeded XFELs also show similar spiking structures due to the initial shot noise of the self amplified spontaneous emission (SASE) process; reviews of the SASE XFEL coherence and statistical properties have been published.[28,29]

4.4   Seeded Soft X-ray Lasers

It is now possible to obtain soft X-ray lasers with very high levels of coherence by seeding the plasma with a high order harmonic beam.[30,31] Using the instruments and methods describe previously, it has been possible to demonstrate full spatial coherence of the beam[8] and fully temporally coherent pulses[32] with temporal coherence and line width very close to the un-seeded values, as shown in figure 7.[33] It should be noted that an ASE background is always present even with seeded operation.

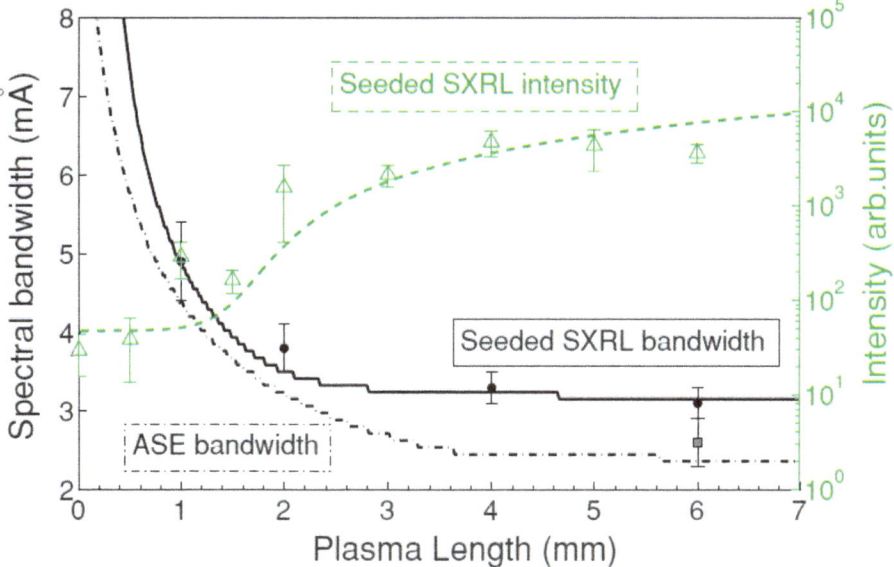

Figure 7 *Spectral line width of seeded and un-seeded OFI soft X-ray lasers (λ = 32.8 nm) as a function of the amplifying plasma length.[33]*

4.5 Some Applications of Soft X-ray Laser Coherence

The high pulse intensities of plasma-based soft X-ray lasers enables a large variety of plasma probing experiments where interferometry is used as a promising technique to find out the plasma electron density. Demonstration experiments have been successfully achieved using both wavefront and amplitude division interferometers. High efficiency interferometers using diffraction gratings instead of beam splitters have been developed for this purpose.[34,35,36]

5 SPECTRAL COHERENCE OF SOFT X-RAY SOURCES

5.1 Position of the problem

In the previous sections of this chapter sources with very different spatial coherence properties were discussed, and so the phenomenological modelling of the emitted fields has to be entirely different:

- High harmonic radiation can be modelled as fully coherent beams and can be approximated by quasi-Gaussian profiles corrected by astigmatism if required;
- Highly spatially incoherent sources such as quasi-steady state lasers are best described using the classical Van Cittert-Zernicke model;
- For all intermediate cases, i.e., partially spatially coherent sources, more sophisticated models are required. A useful one is the Gaussian-Schell model.[37]

    The same concepts can be used for the temporal properties. At one extreme, the field is continuous, stationary and essentially random in time but correlation properties can be found. At the other extreme, the field is finite in time and has a well defined, deterministic structure. In the intermediate case, the field is not only temporally partially coherent, but also not stationary. To cover all these situations in the same theoretical framework the concept of spectral coherence is useful. This can be illustrated using a simple but very general idea, the partially coherent Gaussian pulse model.[38,39]

5.2   The Spectral Correlation

The following introduction is based on the formalism introduced by Lajunen *et al.*,[39] in which $U(r,t)$ represents a random non-stationary scalar wave field (for example the field of one X-ray laser pulse). The Fourier transform of the field, $\tilde{U}(r,\omega)$, can be a single shot spectrum of a soft X-ray laser. The mutual coherence function for a non-stationary field is defined by:

$$\Gamma\left(r_1,r_2,t_1,t_2\right)=\left\langle U_1^*(r_1,t_1)U_2(r_2,t_2)\right\rangle,\tag{8}$$

where the angle brackets denote ensemble averaging over all possible field configurations (for example the average of many X-ray laser shots in interferometric experiments). With this formalism, the pulse temporal envelope is $I(r,t) = \Gamma(r,r,t,t)$.

Let us now define the cross spectral correlation function by:

$$W\left(r_1,r_2,\omega_1,\omega_2\right)=\left\langle U_1^*(r_1,\omega_1)U_2(r_2,\omega_2)\right\rangle.\tag{9}$$

The spectral envelop is then $S(r,\omega) = W(r, r,\omega,\omega)$; this corresponds experimentally to the average spectrum of a soft X-ray laser, obtained through the accumulation of many pulses on a spectrometer. The generalized Wiener-Khintchine theorem[7] establishes a link between Γ and W:

$$\Gamma\left(r_1,r_2,t_1,t_2\right)=\iint W(r_1,r_2,\omega_1,\omega_2)\exp(-i(\omega_1 t_1-\omega_2 t_2))d\omega_1 d\omega_2.\tag{10}$$

As with classical stationary fields it is possible to define normalized values of Γ and W. Let us introduce γ and μ the mutual degree and spectral degree of coherence functions:

$$\gamma\left(r_1,r_2,t_1,t_2\right)=\frac{\Gamma\left(r_1,r_2,t_1,t_2\right)}{\left[I(r_1,t_1)I(r_2,t_2)\right]^{1/2}}\tag{11}$$

$$\mu\left(r_1,r_2,\omega_1,\omega_2\right)=\frac{W\left(r_1,r_2,\omega_1,\omega_2\right)}{\left[S(r_1,\omega_1)S(r_2,\omega_2)\right]^{1/2}}\tag{12}$$

5.3 Partially Coherent Gaussian Pulse Model

We will use this formalism to derive a simple but very general source model. Consider a source with an average spectrum of central pulsation ω_0 and spectral width Ω. The spectral envelop (or shape) of the spectrum can be modelled with a gaussian shape:

$$S(z,\omega) = W_0 \exp\left(-\frac{(\omega-\omega_0)^2}{\Omega^2}\right) \tag{13}$$

An infinite number of field structure having this spectrum exist. The Partially Coherent Gaussian Pulse model (PCPGM)[38] makes some assumptions on the value of W which are sufficiently simple to enable practical calculations but still general enough to cover a large variety of situations. In this 1D model, the choice of the W or μ function is such that the spectrum as a shape equal to (13). This leads to the following structure for the spectral degree of coherence:

$$\mu(z_1,z_2,\omega_1,\omega_2) = W_0 \exp\left(-\frac{(\omega_1-\omega_2)^2}{2\Omega_C^2}\right)\exp\left(i\frac{\omega_2 z_2 - \omega_1 z_1}{c}\right) \tag{14}$$

where Ω_c is a metrics of the spectral interval bellow which the different spectral component are correlated. For a same average spectrum, we can consider different fields with different values of Ω_c, ranging from zero to infinity. To understand the consequence of this quantity let us derive the field intensity envelope from this model. It can be established that the corresponding intensity envelop is:

$$I(z,t) = I_0 \exp\left(-\frac{(t-z/c)^2}{T^2}\right) \tag{15}$$

where $T = \left(\Omega^{-2} + \Omega_c^{-2}\right)^{1/2}$. A null value of Ω_c corresponds to an infinite lasting stationary field, an infinite value to a fully coherent, Fourier limited pulse. A non-zero finite values will finally corresponds to a non-stationary fields.

To appreciate the effect on any temporal coherence measurement, one can calculate the modulus of the normalized degree of coherence γ :

$$|\gamma(z_1,z_2,t_1,t_2)| = \exp\left[-\frac{[(t_1-t_2)-(z_1-z_2)/c]^2}{2T_C^2}\right] \tag{16}$$

with $T_c = 2T\Omega_c/\Omega$ being a measure of the temporal coherence of the pulse.

The coherence properties of a pulse with spectral width Ω and duration T can now be modelled with the PGCPM, and can be related to coherence quantities Ω_c and T_c:

$$\Omega_C = \left(T^2 - 1/\Omega^2\right)^{-1/2}, \quad T_C = 2\left(\Omega^2 - 1/T^2\right)^{-1/2} \qquad (17)$$

5.4   Application to a Real Experiment: Determination of the Line Width of a soft x-ray laser from its Coherence Time

As an example, let us consider the problem of the estimation of the spectral width Ω of a non-stationary source from a set of interferograms recorded at different delays τ between the two arms of a Michelson interferometer. In this arrangement, as described in section 3, the CCD camera has an integration time much longer than the (unknown) pulse duration T. From the interferograms, the average visibilities $V(\tau)$ are extracted. For a stationary field ($T \to \infty$), the Wiener Khintchine theorem is valid and a Fourier transform of $V(\tau)$ would give the spectral profile $S(\omega)$.

For other values of T, Figure 8 shows the evolution of $V(\tau)$ for pulses having the same spectral width but different durations. The coherence time T_c is maximum for the stationary case and minimum for the Fourier-limited case. The difference between these two values is about of 40%, and thus the determination of the line width from the measured coherence time of an incoming pulse with unknown duration will lead to an error of this magnitude. However, if the pulse duration is known, a correction can be added to obtain a more precise estimation using relations (17). Note that even an order of magnitude estimate of T is sufficient to provide a reasonably precise value of Ω, that is below the experimental uncertainties in measuring the visibility or shot to shot fluctuations.

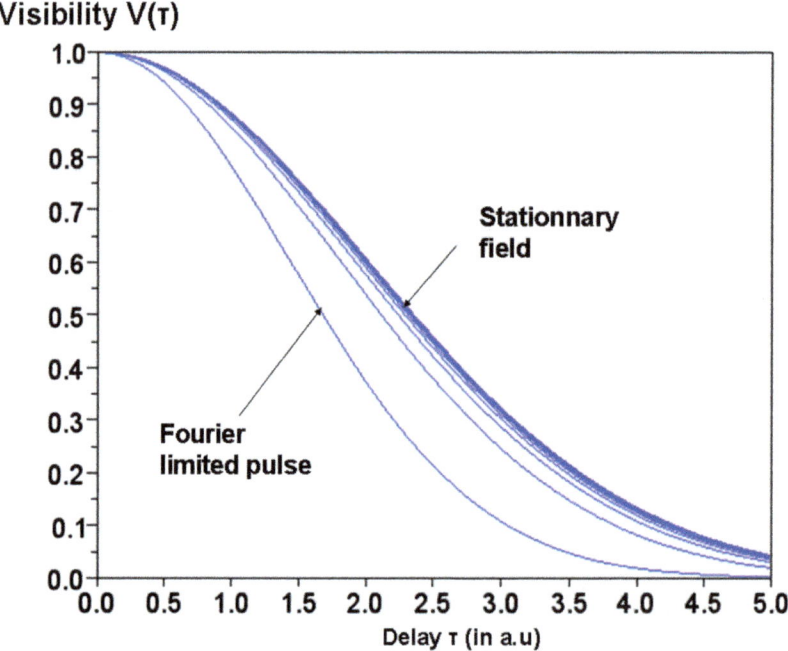

Figure 8   *Simulation of the fringe visibility as a function of the delay τ between interfering pulses for the same spectral width with different pulse durations T.*

6 SUMMARY AND CONCLUSION

EUV coherence measurement and control is a challenging task which has dramatically progressed in recent years, mainly motivated by the development of sources with novel properties and their applications Non-stationary fields are an important aspect of this domain, for which numerous theoretical tools are now available. SASE XFELs have not been treated extensively in this chapter, but these devises share many common theoretical and instrumental aspects with other sources. Recently, the principle of single shot measurements of temporal and spatial coherence has been investigated using LCLS X-ray free electron laser[40,41]. This technique is based on the analysis of the large angle diffraction pattern from a scattering sample, which involves path differences corresponding to greater than the expected coherence time. This idea will also be of use for other sources, in order to analyse single shot field structures or to mitigate shot to shot fluctuations. In the visible range, static Fourier-transform Michelson interferometry using echelon mirrors is already operational, and an interesting development of this technique has been proposed by Chilla, Marconi and Rocca.[42] In this device, gratings are not only used as beam splitters but they also contribute to the generation of pulses with tilted energy fronts. This enables a complete record of the temporal coherence function in a single image and potentially in a single shot. Finally, many concepts presented here are also relevant for determining field polarisation structures, but this is beyond the scope of this chapter.

References

1   J. Gautier et al., Optimization of the wavefront of high order harmonics, Eur. Phys. J. D **48**, 459 (2008).

2   P.M. Paul et al., Observation of a train of attosecond pulses from high harmonic generation, Science **292**, 1689 (2001).

3   N. Hasegawa et al., Proc. SPIE 8140, X-ray lasers and coherent X-ray sources: Development and Applications IX, 81400G (2011).

4   R.L. Sandberg et al., Tabletop soft-x-ray Fourier transform holography with 50 nm resolution, Optics Lett., **34** 1618 (2008).

5   R.A. Bartels et al., Generation of spatially coherent light at extreme ultraviolet wavelengths, Science **297**, 376 (2002).

6   M. Born and E. Wolf, Principles of Optics, 7th edition, Cambridge University Press (1999).

7   D.A. McQuarrie, Statistical Mechanics, Harper and Row, New York, 1976.

8   J-P Goddet et al., Demonstration of a spatial filtering amplifier for high-order harmonics, Optics Lett. **32**,1498 (2007).

9   M. Bellini et al., Temporal Coherence of Ultrashort High-Order Harmonic Pulses, Phys Rev. Lett. **81** 287 (1998).

10   Y. Mairesse and F. Quere, Frequency-resolved optical gating for complete reconstruction of attosecond bursts, Phys. Rev. A **71**, 011401(R) (2005).

11   S. Dobosz et al, Internal frequency conversion extreme ultraviolet interferometer using mutual coherence properties of two high-order-harmonic sources, Rev. Scientif. Instrum. **80**, 113102 (2009).

12   R. I. Tobey et al, Transient grating measurements of surface acoustic waves in thin metal films with extreme ultraviolet radiation, Appl. Phys. Lett. **89**, 091108 (2006).

13   A-S. Morlens, J. Gautier et al., Submicrometer digital in-line holographic microscopy at 32 nm with high-order harmonics, Optics Letters, **31** 3095 (2006).

14   S. DeRossi et al., Probing multilayer stack reflectors by low coherence interferometry in extreme ultraviolet, Applied optics **47**, 2109 (2008).

15   H. Daido, Review of soft x-ray laser researches and developments, Rep. Prog. Phys. **65**, 1513 (1998).

16   J.A. Koch et al., Experimental and theoretical investigation of neonlike selenium x-ray laser spectral linewidths and their variation with amplification, Phys Rev A **50**, 1877 (1994).

17   S. Le Pape et al., Electromagnetic-Field Distribution Measurements in the Soft X-Ray Range: Full Characterization of a Soft X-Ray Laser Beam, Phys. Rev. Lett. **88**, 183901 (2002).

18   C. Menoni et al. in Proceedings of the international conference on soft x-ray lasers, Paris, S.Sebban editor, EDP science (2012).

19   L. Urbanski et al. , Defect-tolerant extreme ultraviolet nanoscale printing, Optics Lett. **37**, 3633 (2012).

20   L. Urbanski et al., Spectral linewidth of a Ne-like Ar capillary discharge soft-x-ray laser and its dependence on amplification beyond gain saturation, Phys. Rev. A **85**, 033837 (2012).

21   Y.V. Afanas'ev and V.N. Shlyaptsev, Formation of a population inversion of transitions in Ne-like ions in steady-state and transient plasmas, Sov. J. Quantum Electron **19**, p1606 (1989).

22   B.E. Lemoff et al. , Femtosecond pulse driven, electron excited XUV lasers in eight-times-ionized noble gases, Opt. Lett. **19**, p 569 (1994).

23   H. Tang et al., Jpn journal of Phys. (2002).

24   O. Guilbaud et al., Longitudinal coherence and spectral profile of a nickel-like silver transient soft X-ray laser, Eur. Phys. J. D **40**, 125 (2006).

25   R. F. Smith et al., Opt. Lett. **28**, 2261 (2003).

26   I. AlMiev et al, Phys. Rev. Lett. **99**, 123902 (2007).

27   O. Guilbaud et al., Origin of microstructures in picosecond soft x-ray lasers, Europhys. Lett **74**, p823 (2006).

28   G. Geloni et al., Coherence properties of the european XFEL, NJP **12** 035021 (2010).

29   I.A. Vartanyants and A. Singer, Coherence properties of hard x-ray synchrotron sources ans x-ray free electron lasers, NJP **12**, 035004 (2010).

30   Ph. Zeitoun et al., Nature **431**, 466 (2004).

31   See E. Oliva in this book.

32   O. Guilbaud et al., Fourier-limited seeded soft x-ray laser pulse, Optics Lett. **35**, 1326 (2011).

33   F. Tissandier et al., Observation of spectral gain narrowing in a high-order harmonic seeded soft x-ray amplifier, Phys. Rev. A **81**, 063833 (2010).

34   H. Tang et al., Diagnostics of laser-induced plasma with soft X-ray (13.9 nm) bi-mirror interference microscopy Applied. Phys B 78, p975 (2004).

35   L.B. Da Silva et al.,Phys. Rev. Lett. **74** p 74 (1995).

36   J. Grava et al.,Dynamics of a dense laboratory plasma jet investigated using soft x-ray laser interferometry, Phys. Rev. E **78**, 016403 (2008).

37   R. Tommasini et al., Coherence properties of an amplified spontaneous emission laser: experiments on a 10 Hz vacuum-ultraviolet H-laser, Opt. Comm. **180**, 277 (2000).

38   P. Paakkonen et al., Partially coherent gaussien pulses, Opt. Comm. **204**, 53 (2002).

39   H. Lajunen et al., JOSA A **21**, 2117 (2004).

40 I.A. Vartanyants et al., Cohrence Properties of Individuals Femtosecond Pulses of an X-Ray Free Electron Laser, Phys Rev Lett. **107**, 144801 (2011).

41 C. Gutt et al, Single shot spatial and temporal coherence properties of the SLAC Linac Coherent Light Source in the hard x-ray regime, Phys. Rev. Lett 108, 024801 (2012).

42 J. Chilla et al, Soft-x-ray interferometer for single-shot laser linewidth measurements, Optics Lett. **21**, 955 (1997).

REACHABLE EXTREME ULTRAVIOLET WAVELENGTHS ACCORDING TO ELEMENTS / ATOMIC DATA

D. Kilbane,[1] F. de Dortan,[2] G. O'Sullivan[1] and V. Zakharov[3]

[1]University College Dublin, Belfield, Dublin 4, Ireland
[2]Department of Ultraintense Lasers, Division of High Power Systems/PALS Centre, Institute of Physics of the Czech Academy of Science, Na Slovance 2. 182 21 Prague 8 Czech Republic
[3]EPPRA sas, Villebon-sur-Yvette, 91140, France

1 INTRODUCTION

In order to realize extreme ultraviolet lithography (EUVL), tin and xenon laser produced plasmas (LPPs) and discharge plasmas have been identified as the most powerful radiation sources at a wavelength of 13.5 nm,[1,2,3,4] including Sn^{8+}–Sn^{13+} emission due to 4d–4f and 4p–4d transitions and Xe^{10+} emission due to 4d–5p transitions.[1,4,5,6,7,8,9] The future for EUVL, however, is operation at shorter wavelengths and this has been the focus of many recent research efforts, e.g. using gadolinium and terbium at 6.75 nm.[10,11,12,13,14] At shorter wavelengths, in the soft X-ray (SXR) region such as the "water window" (2.3–4.4 nm), live biological sampling is desirable.[15,16] Sources currently in operation in this region are strong quasi-monochromatic emitters from nitrogen ions at $\lambda = 2.879$ nm and $\lambda = 2.478$ nm, arising from $1s^2$–1s2p in N^{5+} and 1s–2p in N^{6+}, respectively, and broadband emission in the wavelength range 2–4 nm from argon gas targets.[15] Broadband emission sources are used in contact microscopy while the quasi-monochromatic sources are appropriate for biological imaging. Development of compact, high repetition rate, table top SXR sources using gas puff targets, which have the advantage of being debris free, offer a laboratory *in situ* alternative to free-electron lasers and synchrotrons and can be used in a range of experiments, e.g. microscopy, spectroscopy and metrology. Extreme ultraviolet (EUV) and SXR sources have been identified, at several wavelengths,[17,18] with the possibility of being used (i) in next generation EUVL in the event of suitable highly reflective mirrors becoming available, and (ii) for compact table top SXR microscopy. It is proposed that these radiation sources could be generated by laser produced plasmas in which solid targets are irradiated with intense laser pulses.

In 1981, O'Sullivan and Carroll performed the first systematic recording of the 4d–4f emission in the elements caesium through to lutetium using the laser-produced plasma technique.[19] These spectra consisted of relatively narrow regions of resonance-like emission which became more complex and moved to shorter wavelengths with increasing nuclear charge Z.[19,20] In order to explain such complicated spectra, Mandelbaum and co-workers[21] applied the unresolved transition array (UTA) approach developed by Bauche-Arnoult, Bauche and Klapisch[22,23,24,25] and concluded that interactions between the $4p^6 4d^{N-1} 4f$ and $4p^5 4d^{N+1}$ configurations are responsible for narrowing the transition arrays and their superposition in adjacent ion stages. Extending this work, detailed theoretical calculations in elements with $Z = 57$ (lanthanum) to 89 (actinium) have been carried out and the results, which will be discussed in the remainder of this chapter, indicate possible future EUV and SXR sources.[17,18]

2 EUV AND SXR EMISSION SPECTRA OF IONS WITH Z = 57–89

4d–4f and 4p–4d transitions in ion stages with open 4d sub-shells give rise to the strongest lines occurring in the EUV. Detailed calculations were performed with the flexible atomic code (FAC)[26] which uses a fully relativistic approach based on the Dirac equation, thus allowing its application to ions with large values of nuclear charge. Configuration interactions (CI) were included using the following basis set: $4p^6 4d^N$, $4p^6 4d^{N-1} nl$ and $4p^5 4d^{N+1}$, where $n \le 8$, $l \le 3$ and $1 \le N \le 10$, resulting in 330 theoretical EUV/SXR emission spectra which have previously been presented in full.[17,18] Two examples, namely gadolinium (Z = 64) and terbium (Z = 65) are presented in figure 1; there is significant interest currently in these elements for next generation lithography sources. CI in tin ions lead to shifts to shorter wavelengths in the 4d–4f and 4p–4d transitions as well as a narrowing of the spectral width,[27] whilst in xenon ions it lead to a large increase in the number of transitions.[8] The narrowing of the peaks in gadolinium and terbium ions implies that strong emission can be achieved over a wide range of plasma conditions, a result that is true for all elements considered.

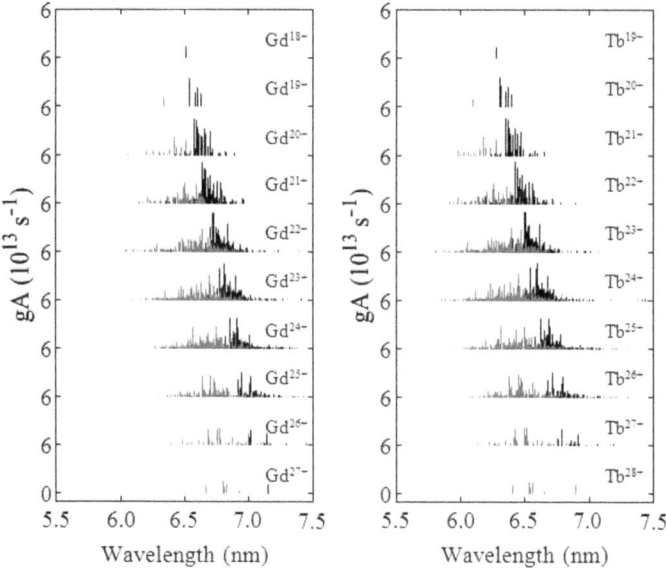

Figure 1 Gd^{18+}–Gd^{27+} *and* Tb^{19+}–Tb^{28+} *spectra computed using the FAC code. Black denotes 4d–4f transitions and gray denotes 4p–4d transitions. The gA scale has the same range for each ion stage (0 - 6×10^{13} s^{-1}).*

As was observed for the lanthanides,[17,18,19,20,21] both 4d–4f and 4p–4d emission moves to shorter wavelength with increasing Z. For brevity, figure 2, in which A is the Einstein coefficient for spontaneous emission and g is the statistical weight of the upper level, displays only the maximum peak emissions (in terms of gA values) as functions of wavelength: (a) 4d–4f and 4p–4d in the lanthanide series, (b) 4d–4f in elements with Z = 72–89 and (c) 4p–4d in elements with Z = 72–89. The dependence of peak transition energies on atomic number Z is presented in figure 3. This shows that with increasing Z the 4d–4f and 4p–4d emission peaks clearly separate. However, the fact that the maximum gA value remains comparable for each transition type over the range of lanthanides and $Z = 72$–89 implies that these elements are potential radiation sources in the wavelength range $\lambda = 2.5$–6 nm.

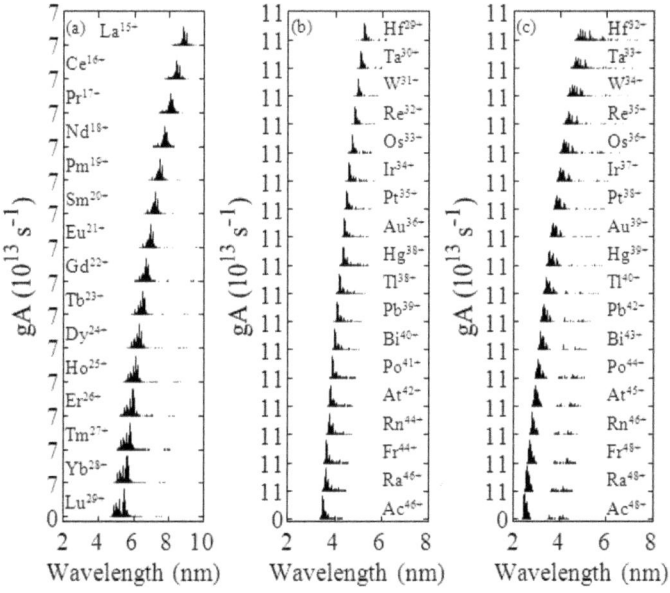

Figure 2 *Maximum peak emission from unresolved transition arrays: (a) 4d–4f and 4p–4d in the lanthanide series, (b) 4d–4f in elements with Z = 72–89, and (c) 4p–4d in elements with Z = 72–89. The gA scale has the same range for each ion stage ((a) $0 – 7\times10^{13}$ s^{-1}, (b) $0 - 11\times10^{13}$ s^{-1} and (c) $0 - 11\times10^{13}$ s^{-1}).*

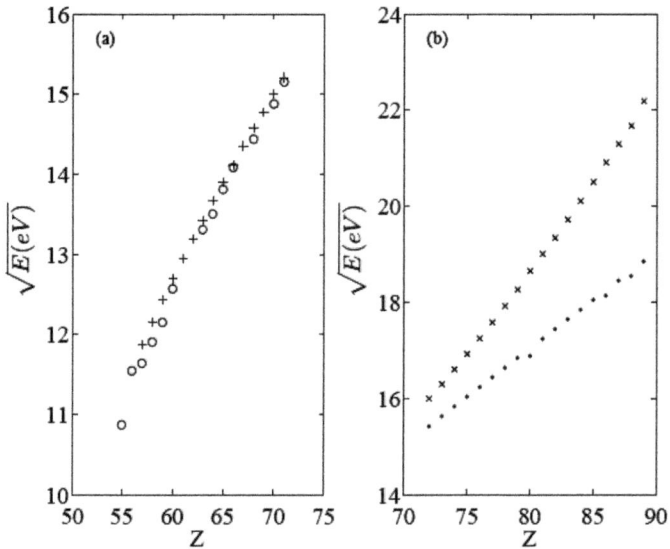

Figure 3 *Dependence of unresolved transition array energies on atomic number Z: (a) 4d–4f (crosses[18] and open circles[19]) and (b) 4d–4f (dots) and 4p–4d (crosses).*

3 UNRESOLVED TRANSITION ARRAY STATISTICS OF IONS WITH $Z = 57$–89

In order to quantify the emission from the 4d–4f and 4p–4d transitions in $Z = 57$–89 ions, the unresolved transition array (UTA) approach developed by Bauche-Arnoult, Bauche and Klapisch[22,23,24,25] was adopted. The average and variance of the transition energies ($\bar{E}$ and σ^2 respectively) can be expressed as the gA-weighted sums

$$\bar{E} = \frac{\sum_{j,i<j} g_j A_{ji} E_{ij}}{\sum_{j,i<j} g_j A_{ji}} \tag{1}$$

and

$$\sigma^2 = \frac{\sum_{j,i<j} g_j A_{ji} (\bar{E} - E_{ij})^2}{\sum_{j,i<j} g_j A_{ji}} \tag{2}$$

where A_{ji} is the Einstein coefficient for spontaneous emission from level j to level i, and g_j is the statistical weight of the upper level. The mean wavelength $\bar{\lambda}_{gA}$ and the spectral width $\Delta\lambda_{gA}$ of the transition array can be defined as follows

$$\bar{\lambda}_{gA} = 10^7 / \bar{E} \tag{3}$$

$$\Delta\lambda_{gA} = \sqrt{8\ln 2} \times 10^7 \, \sigma / \bar{E}^2, \tag{4}$$

where $\bar{E}$ and σ are expressed in cm^{-1} and $\bar{\lambda}_{gA}$ and $\Delta\lambda_{gA}$ in nanometres.

Encouraged by the fusion community, Radtke and co-workers[28] first identified the $4p^6 4d$–$4p^5 4d^2$ and $4p^6 4d$–$4p^6 4f$ transitions in tungsten using an electron-beam ion trap and theoretical calculations employing the multi-configurational relativistic HULLAC code[29]. The resulting spectra were successfully analyzed using the UTA framework which demonstrated that these spectra can be adequately described using this approach. This statistical analysis was applied to each of the 4d–4f and 4p–4d UTAs separately, and the ranges in mean wavelength $\bar{\lambda}_{gA}$ and spectral width $\Delta\lambda_{gA}$ values are given in table 1. From these values, and the results presented in figure 2, it is possible to identify strong emitters for almost all wavelengths in the approximate range $\lambda = 2.5$–6 nm. Indeed, in some cases it is possible that a target of mixed composition may give broad-band emission across this wavelength range encompassing the "water window" and soft X-ray regions. These emitters complement the currently used sources presented in table 2.

4 SUMMARY

Extreme ultraviolet and soft X-ray sources have been identified, using the FAC relativistic code, as emission arising from 4d–4f and 4p–4d transitions in palladium-like to rubidium-like ions of lanthanum through actinium. The spectra of these strong emitters are seen to separate and move to shorter wavelength as the nuclear charge increases. The emission was comparable from each source and was characterized using the unresolved transition array model. In future applications such as microscopy, spectroscopy and lithography, these radiation sources may be generated by employing the laser-produced plasma technique.

Table 1 *Calculated mean wavelength* $\bar{\lambda}_{gA}$ *and spectral width* $\Delta\lambda_{gA}$ *ranges of the 4d–4f and 4p–4d UTAs of lanthanum through actinium ions. Wavelengths throughout the table are given in nanometres.*

Element	4d–4f		4p–4d	
	$\bar{\lambda}_{gA}$ range	$\Delta\lambda_{gA}$ range	$\bar{\lambda}_{gA}$ range	$\Delta\lambda_{gA}$ range
Lanthanum	8.959–9.359	0–0.497	8.639–9.153	0.209–0.890
Cerium	8.413–9.006	0–0.615	8.365–8.780	0.306–0.951
Praseodymium	8.005–8.671	0–0.660	7.891–8.425	0.205–0.972
Neodymium	7.635–8.356	0–0.692	7.520–8.095	0.369–1.010
Promethium	7.335–8.063	0–0.728	7.187–7.781	0.328–1.029
Samarium	7.026–7.785	0–0.753	6.888–7.483	0.396–1.050
Europium	6.754–7.524	0–0.785	6.619–7.205	0.409–1.077
Gadolinium	6.506–7.276	0–0.893	6.369–6.935	0.436–1.086
Terbium	6.277–8.035	0–0.909	6.148–6.714	0.437–1.043
Dysprosium	6.067–7.772	0–1.252	5.929–6.471	0.435–1.063
Holmium	5.870–7.526	0–1.247	5.722–6.235	0.480–1.075
Erbium	5.687–7.142	0–1.141	5.526–6.000	0.472–1.076
Thulium	5.515–6.895	0–1.080	5.338–5.780	0.433–1.084
Ytterbium	5.353–6.666	0–0.932	5.160–5.570	0.414–1.097
Lutetium	5.201–6.454	0–0.883	4.988–5.367	0.461–1.106
Hafnium	5.056–6.260	0–0.855	4.824–5.172	0.517–1.113
Tantalum	4.919–6.085	0–0.824	4.665–4.987	0.524–1.121
Tungsten	4.789–5.935	0–0.800	4.456–4.803	0.485–1.112
Rhenium	4.674–5.817	0.134–0.751	4.301–4.629	0.302–1.106
Osmium	4.556–5.635	0.134–0.735	4.151–4.386	0.352–0.921
Iridium	4.444–5.444	0.177–0.709	4.004–4.285	0.334–1.094
Platinum	4.336–5.277	0.203–0.685	3.862–4.129	0.266–1.102
Gold	4.233–5.129	0.205–0.664	3.723–3.978	0.251–1.103
Mercury	4.134–4.993	0.238–0.629	3.588–3.831	0.235–1.098
Thallium	4.039–4.865	0.234–0.601	3.457–3.690	0.217–1.093
Lead	3.947–4.743	0.248–0.584	3.328–3.554	0.133–1.085
Bismuth	3.859–4.624	0.254–0.579	3.191–3.420	0–1.071
Polonium	3.774–4.513	0.259–0.560	3.074–3.292	0–1.058
Astatine	3.692–4.297	0.264–0.566	2.969–3.230	0.111–1.209
Radon	3.612–4.197	0.269–0.549	2.858–3.108	0.104–1.191
Francium	3.535–4.087	0.274–0.530	2.750–2.991	0.098–1.172
Radium	3.461–3.985	0.278–0.480	2.645–2.876	0.092–1.150
Actinium	3.389–3.883	0.283–0.474	2.541–2.766	0–1.128

Table 2 *Table-top radiation sources in the EUV and SXR regions.*

Source	Wavelength (nm)	Transition	Emission type
Xe^{10+}	13.5	4d–5p	Broad EUV
Sn^{8+}–Sn^{13+}	13.5	4d–4f,4p–4d	Broad EUV
Ar^{9+}	3.656–3.888	$2s^2 2p^5$–$2s^2 2p^4 3d$	Broad SXR[15,30]
Ar^{10+}	3.353–3.631	$2s^2 2p^4$–$2s^2 2p^3 3d$	Broad SXR[15,30]
Ar^{11+}	3.065–3.274	$2s^2 2p^3$–$2s^2 2p^2 3d$	Broad SXR[15,30]
N^{5+}	2.8787	$1s^2$–$1s2p$ ($^1S_0 - {}^1P_1^o$)	Quasi-monochromatic SXR[15,31]
N^{5+}	2.9084	$1s^2$–$1s2p$($^1S_0 - {}^3P_2^o$)	Quasi-monochromatic SXR[15,31]
N^{6+}	2.4779	$1s$–$2p$($^2S_{1/2} - {}^2P_{3/2}^o$)	Quasi-monochromatic SXR[15,31]
N^{6+}	2.4785	$1s$–$2p$($^2S_{1/2} - {}^2P_{1/2}^o$)	Quasi-monochromatic SXR[15,31]
C^{4+}	3.37342	$1s^2$–$1s2p$($^2S_{1/2} - {}^2P_{3/2}^o$)	Monochromatic SXR[31,32]
C^{4+}	3.37396	$1s^2$–$1s2p$($^2S_{1/2} - {}^2P_{1/2}^o$)	Monochromatic SXR[31,32]
O^{6+}	2.16020	$1s^2$–$1s2p$($^1S_0 - {}^1P_1^o$)	Monochromatic SXR[33]

ACKNOWLEDGMENT

This work was supported by Science Foundation Ireland under Principal Investigator research grant 07/IN.1/I1771.

References

1 S.S. Churilov, Y.N. Joshi, J. Reader and R.R. Kildiyarova, *Phys. Scr.*, 2004, **70**, 126.
2 V. Bakshi (ed.), *EUV Sources for Lithography*, 2006, SPIE Press, Bellingham.
3 G. O'Sullivan, A. Cummings, C.Z. Dong, P. Dunne, P. Hayden, O. Morris, E. Sokell, F. O'Reilly, M.G. Su and J. White, *J. Phys. Conf. Ser.*, 2009, **163**, 012003.
4 A. Sasaki, A. Sunahara, H. Furukawa, K. Nishihara, S. Fujioka,T. Nishikawa, F. Koike, H. Ohasni and H. Tanuma, *J. Appl. Phys.*, 2010, **107**, 113303.
5 N. Böwering, M. Martins, W.N. Partlo and I.V. Fomenkov, *J. Appl. Phys.*, 2004, **95**, 16.
6 S.S. Churilov and A.N. Ryabtsev, *Phys. Scr.*, 2006, **73**, 614.
7 M. Poirier, T. Blenski, F. de Gaufridy de Dortan and F. Gilleron, *J. Quant. Spectrosc. Radiat. Transfer*, 2006, **99**, 482.
8 F. de Gaufridy de Dortan, *J. Phys. B: At. Mol. Opt. Phys.*, 2007, **40**, 599.
9 J. Zeng, C. Gao and J. Yuan, *Phys. Rev. E*, 2010, **82**, 026409.
10 S.S. Churilov, R.R. Kilidiyarova, A.N. Ryabtsev and S.V. Sadovsky, *Phys. Scr.*, 2009, **80**, 045303.
11 T. Otsuka, D. Kilbane, J. White, T. Higashiguchi, N. Yugami, T. Yatagai, W. Jiang, A. Endo, P. Dunne and G. O'Sullivan, *App. Phys Lett.*, 2010, **97**, 111503.
12 T. Otsuka, D. Kilbane, T. Higashiguchi, N. Yugami, T. Yatagai, W. Jiang, A. Endo, P. Dunne and G. O'Sullivan, *App. Phys Lett.*, 2010, **97**, 231503.
13 D. Kilbane and G. O'Sullivan, *J. Appl. Phys.*, 2010, **108**, 104905.

14 A. Sasaki, K. Nishihara, A. Sunahara, H. Furukawa, T. Nishikawa and F. Koike, *App. Phys Lett.*, 2010, **97**, 231501.

15 P.W. Wachulak, A. Bartnik, H. Fiedorowicz, P. Rudawski, R. Jarocki, J. Kostecki and M. Szczurek, *Nucl. Instrum. Methods Phys. Res. B*, 2010, **268**, 1692.

16 H. Legall, G. Blobel, H. Stiel, W. Sandner, C. Seim, P. Takman, D. H. Martz, M. Selin, U. Vogt, H. M. Hertz, D. Esser, H. Sipma, J. Luttmann, M. Höfer, H. D. Hoffmann, S. Yulin, T. Feigl, S. Rehbein, P. Guttmann, G. Schneider, U. Wiesemann, M. Wirtz and W. Diete, *Optics Express*, 2012, **20**, 18362.

17 D. Kilbane and G. O'Sullivan, *Phys. Rev. A*, 2010, **82**, 062504.

18 D. Kilbane, *J. Phys. B: At. Mol. Opt. Phys.*, 2011, **44**, 165006.

19 G. O'Sullivan and P.K. Carroll, *J. Opt. Soc. Am.*, 1981, **71**, 227.

20 P.K. Carroll and G. O'Sullivan. *Phys. Rev. A*, 1982, **25**, 275.

21 P. Mandelbaum, M. Finkenthal, J.L. Schwob and M. Klapisch, *Phys. Rev. A*, 1987, **35**, 5051.

22 C. Bauche-Arnoult, J. Bauche and M. Klapisch, *Phys. Rev. A*, 1979, **20**, 2424.

23 C. Bauche-Arnoult, J. Bauche and M. Klapisch. *Phys. Rev. A*, 1982, **25**, 2641.

24 J. Bauche, C. Bauche-Arnoult, E. Luc-Koenig, J-F. Wyart and M. Klapisch, *Phys. Rev. A*, 1983, **28**, 829.

25 C. Bauche-Arnoult, J. Bauche and M. Klapisch, *Phys. Rev. A*, 1985, **31**, 2248.

26 M.F. Gu, *Astrophys. J.*, 2003, **582**, 1241.

27 W. Svensden and G. O'Sullivan, *Phys. Rev. A*, 1994, **50**, 3710.

28 R. Radtke, C. Biedermann, J.L. Schwob, P. Mandelbaum and R. Doron, *Phys. Rev. A*, 2001, **64**, 012720.

ABSORPTION OF SHORT PUMPING PULSES FOR GRAZING INCIDENCE PUMPED X-RAY LASERS

D. Ursescu

National Institute for Lasers, Plasma and Radiation Physics, Atomistilor Str. 409, Bucharest-Magurele, Romania[a]

1 AIM OF GRAZING INCIDENCE LASER PUMPING IN X-RAY LASERS

The development of coherent short wavelength sources is a key point for a broad range of applications encompassing biology, medicine, nanotechnologies and fusion. X-ray lasers are such sources with major advantages of peak brilliance, quasi-monochromaticity and compactness.

The experimental demonstration of a transient collisionally excited X-ray laser (TCE XRL) was a milestone in XRL development.[1] The total pump energy needed in the first experiments was reduced by several orders of magnitude, to about 10 J, by using a nanosecond pre-pulse to create a pre-plasma and a picosecond normal incidence pulse to produce the population inversion in the multiply ionized plasma. This result offered average size laser laboratories the possibility to install and improve XRL systems very quickly.

In the optimization race to provide similar soft XRLs with table-top dimensions, a further significant step was made by changing the incidence angle of the main pumping pulse from near normal incidence to grazing incidence pumping (GRIP).[2] This method reduced the required pumping energy to values below 1 J. As a consequence, table-top XRL systems could be demonstrated with several orders of magnitude increased repetition rates (~10 Hz). This approach has been successful in producing XRL wavelengths as short as 10.9 nm[3] with pump energies of less than 2 J. Moreover, it has also been demonstrated[4] that the efficiency and peak brilliance of such a GRIP XRL can be significantly improved; up to 3 μJ energy with 1 J pump energy and 10^{28} photons/s/mm^2/mrad^2/0.1% bandwidth was obtained with a Ni-like molybdenum XRL. On the theoretical side, the effect of the non-normal incidence angle was analyzed in[5,6] in an attempt to scale the pumping scheme to sub-10 nm wavelengths, including the water window, by means of the EHYBRID code. This is a 1.5 dimensional (MHD) -hydrodynamic code coupled with level population dynamics and provides extensive information about the plasma dynamics and the gain for the modelled XRL. The evaluation of energy deposition in the plasma in the code is extrapolated from experimental

---

[a] Part of this work was performed at PHELIX Department, GSI-Darmstadt, Planckstr. 1, Darmstadt-Wixhausen, Germany

data. This is not straightforward as there are many laser and plasma parameters to be taken into account in general and there is little experimental data. In addition, in the special case of an ultra-intense laser pulse interacting with a plasma (GRIP XRL pumping), the nonlinear inverse Bremsstrahlung (NIB) absorption mechanism[12,13] together with the incidence angle on the plasma play key roles and make the extrapolation more difficult.

Understanding the coupling of the NIB mechanism with the incidence angle will aid the optimization of the XRL pump energy usage and also can significantly contribute to the design of shorter wavelength XRLs pumped at non-normal incidence. The model developed here could be applied to other ultra-intense pulse experiments such as plasma mirrors, where the plasma created with pre-pulses behaves like a mirror for the main part of a laser pulse. This requires the maximization of the plasma reflectivity, rather than absorption as for the case of XRL pumping studied here.

The evaluation of absorption of ultra-intense non-normal incidence pulses in plasmas is presented here from the modelling viewpoint. This is to simplify the extrapolation of the absorption data, taking into account both the NIB mechanism and the incidence angle of the laser pulse.

The study performed here in this direction is therefore a complementary study to previous work,[5,]6 with the aim of providing a more detailed description of the propagation and absorption of short pulses in plasmas. Section 2 describes the model used and points out the effect of the NIB mechanism on the short-pulse energy deposition in the pre-plasma. The results show that small GRIP angles can produce an unwanted reduction of the absorption but also indicate means to control it. In section 3 a comparison of the model with some experimental data is provided. The results are compared with an analytical approach[7,8] and with the experimental results in a non-normal incidence pumped Ni-like zirconium XRL.[9,10,11] In addition, a 7.3 nm wavelength samarium GRIP XRL is analysed to show that efficient pumping can be obtained by tuning the GRIP angle, and the NIB absorption via the short-pulse duration and the pre-plasma scale length.

A reference laser wavelength of 1054 nm was used in all the calculations. Ray-tracing is used for the study of the pump laser propagation in plasma. For the absorption model, frequency-type parameters are used, as also the inverse Bremsstrahlung (IB) absorption process can be understood in terms of resonant processes. The plasma electron frequency, associated to charge density oscillations in the plasma, is given by the relation:

$$\omega_{pe} = \sqrt{\frac{n_e e^2}{m\,\epsilon_0}}$$

Where n_e is the electron density of the plasma, e is the electric charge, m is the mass of the electron, and ϵ_0 is the permittivity of free space. It is worth noting that the plasma frequency depends only on the electron density, and does not depend on parameters such as temperature. The IB absorption becomes strong when the laser reaches places where plasma frequency equals the laser frequency. As a consequence, one defines the plasma electron critical density as the density where the plasma frequency equals the laser frequency under consideration.

2 PROPAGATION AND ABSORPTION MODEL

2.1 Non Linear Inverse Bremsstrahlung Model

For TCE XRLs, the main mechanism for short-pulse laser absorption is inverse Bremsstrahlung (IB). The plasma electron density is assumed to vary in space following an exponential decay function with a scale length determined by a nanosecond long pre-pulse and, as a consequence, it is much longer than the laser wavelength; in this way, other absorption mechanisms such as Brunel effect and p-polarized radiation the resonance absorption process are negligible.

From the kinetic point of view, IB is determined by the collision frequency of the electrons in the plasma. At high laser intensities, $>10^{14}$ W/cm^2, the large electric field distorts the thermal distribution of the electrons, thus changing the electron-ion collision frequency v. In these cases, the electron mean-square thermal velocity v_t^2, which appears in the collisional rate through the plasma temperature factor, is increased due to the averaged squared electron oscillation velocity v_{lf}^2 in the laser electric field which now is comparable to v_t^2,

$$v_{\text{eff}}^2 = v_t^2 + v_{lf}^2 \tag{1}$$

where $v_{lf}^2 (eV) = 1.5\ 10^{13} I (W cm^{-2}) \lambda^2 (\mu m)$.

This parameter regime is known as NIB due to the fact that the correction to the usual IB formula is made through a factor which is laser field dependent; absorption is proportional to:

$$v = 2.91\ 10^6 \frac{Z \cdot f \cdot n_e \cdot \log \Lambda}{T_{eV}^{3/2}} \ (s^{-1}) \tag{2}$$

where the electron temperature T_{eV} is in electron-volts, n_e (cm^{-3}) is the electron density, Z is the average ion charge and $f = f\left(v_{lf}^2/v_t^2\right)$ is the correction factor for NIB: $f\left(\frac{v_{lf}^2}{v_t^2}\right) = \frac{3}{\pi} \int_0^\pi d\alpha \int_{-1}^1 dy . y^2 . \cos^2 \alpha . e^{-\frac{v_{lf}^2 y^2 \cos^2 \alpha}{v_t^2}}$ for classical non-relativistic treatment of the electrons and is solved using numerical integration. It can be approximated by $f \approx (1 + \frac{v_{lf}^2}{v_t^2})^{-3/2}$ with $\frac{v_{lf}^2}{v_t^2} = \frac{1.5\ 10^{13}\ I \lambda^2}{T_e(eV)}$ in agreement with the previous formula to within 10% (see equation 3.34 from [12]). The quantum / relativistic treatment is more complicated but the values obtained differ by at most 10% of the classical non-relativistic case.

Figure 1 shows the IB correction factor f as a function of pre-plasma temperature for different laser intensities, assuming a Maxwellian distribution of the electron velocities. The range for the laser intensities is typical of the main pulse in the TCE XRL schemes. The plasma temperature is represented on a logarithmic scale. At 10^{13} W/cm^2 there is little influence from the correction factor on the collision rate, while at 10^{15} W/cm^2 there is a significant decrease in the correction factor at temperatures below 100 eV. This factor can drastically affect the absorption of the short pulse at low plasma temperatures and strong laser fields. At low pre-plasma temperatures the correction factor can decrease the collision rate to a few percent of its initial value.

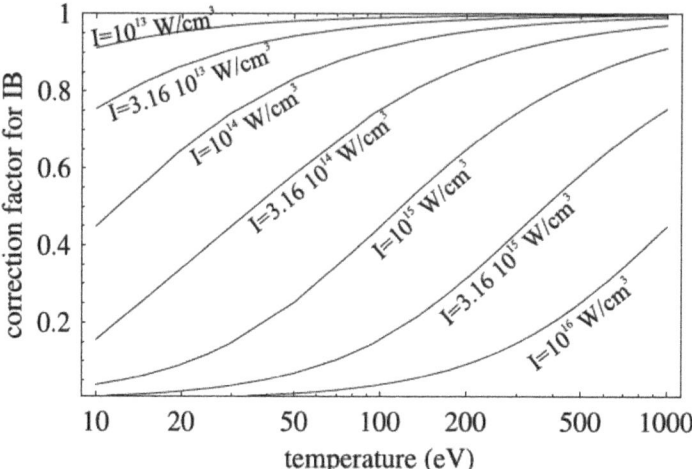

Figure 1 *The inverse Bremsstrahlung correction factor for different pre-plasma temperatures and various laser intensities.*

The electron velocity distribution in the plasma was assumed to be Maxwellian for figure 1. If this assumption is incorrect the correction factor can be over-estimated by up to a factor of two[12] and, as a consequence, the absorption of laser radiation in the plasma is also over-estimated. The additional factor to correct the absorption for a non-Maxwellian distribution is

$$f_{\text{L}} = 1 - \frac{0.553}{1 + \left(0.27 v_{\text{t}}^2 / Z v_{\text{lf}}^2\right)^{3/4}},\tag{3}$$

depending also on the average ionization state of the plasma, Z. Taking into account the dependence of the collision rate on Z and the correction factor – see equation (2) – further analysis is simplified by using a global factor Z_f in the collision rate

$$Z_f = Z.f.f_L\tag{4}$$

This factor has to be included in any evaluation of the propagation of laser radiation in plasmas. Numerical values of Z_f can then be understood as being equal to the average ionization of the plasma when the NIB process is negligible or as an "effective" average ionization of the plasma that is always smaller than the "true" average ionization of the plasma due to NIB effect.

2.2 The Plasma Complex Refractive Index

The propagation of a laser beam in a plasma depends on reflection, refraction and absorption processes as described using the complex refractive index $\tilde{n}$,[13]

$$\tilde{n} = \sqrt{\varepsilon} = n_{\text{r}} + i n_{\text{a}}\tag{5}$$

$$\varepsilon = 1 - \frac{\omega_{\text{pe}}^2}{\omega(\omega + i v)}\tag{6}$$

where ε is the plasma dielectric function, ω_{pe} is the electron plasma frequency, ω is the laser angular frequency, v is the electron-ion collision frequency, n_r is the real part of the refractive index and n_a is the complex part of the refractive index which determines absorption in the plasma.

The normalized frequencies $\bar{v} = v/\omega$ and $\bar{\omega}_{pe} = \omega_{pe}/\omega$ are used to rewrite the real and the complex parts of the refractive index as

$$n_r = \frac{\bar{v}\bar{\omega}_{pe}^2}{\sqrt{2}\left(1+\bar{v}^2\right)}\left(\sqrt{-1+\frac{\bar{\omega}_{pe}^2}{1+\bar{v}^2}+\sqrt{1+\frac{\bar{\omega}_{pe}^4-2\bar{\omega}_{pe}^2}{1+\bar{v}^2}}}\right)^{-1} \tag{7}$$

$$n_a = \frac{1}{\sqrt{2}}\sqrt{-1+\frac{\bar{\omega}_{pe}^2}{1+\bar{v}^2}+\sqrt{1+\frac{\bar{\omega}_{pe}^4-2\bar{\omega}_{pe}^2}{1+\bar{v}^2}}}. \tag{8}$$

Description of the refractive index in terms of normalized frequencies is advantageous due to the reduced number of parameters involved, but it is not transparent in terms of the MHD parameters. In order to come closer to such a description it is necessary to convert the normalized frequencies into electron density and temperature. A possible approach to perform this conversion is detailed here further. First it is assumed an exponential decay profile for the electron density of the plasma that allows one to compute the trajectory of the rays in the plasma using a light propagation model. Then, taking into account the plasma parameters along the trajectory of the ray, one computes the absorption scale length of the plasma, hence the energy deposited by the pump laser along the ray path. In turn, this energy deposition is used to compute the evolution of the plasma temperature, ionization degree and the change in the electron density distribution. Finally, using these updated parameters for the plasma description, new trajectories of the rays in the plasma and energy deposition are computed using (7) and (8).

2.3 Light Propagation Model

The results of the analytical modelling of the laser produced pre-plasma using nanosecond laser pulses have been used to describe the pre-plasma MHD,[7,8] assuming an exponential profile for the plasma electron density. The equation of the propagation of a laser beam in the plasma can be written as [17,18]:

$$\frac{d}{ds}\left(n\frac{d\boldsymbol{r}}{ds}\right) = \nabla n,$$
$$\partial_{x(s)}n_{ref}(x(s)).(\partial_s x(s)^2 - 1) + n_{ref}(x(s)).\partial_s\,\partial_s\,x(s) = 0$$
$$\partial_{x(s)}n_{ref}(x(s)).\partial_s x(s)\partial_s z(s) + n_{ref}(x(s)).\partial_s\,\partial_s\,z(s) = 0$$

where $ds = \sqrt{dx^2+dy^2+dz^2}$ is the path element and $\boldsymbol{r} = \{x,y,z\}$ is the position vector while n is the complex refraction index.

With the z axis along the plasma line focus and the x axis perpendicular to the target, the refractive index depends only on x and the propagation is described by a set of two coupled differential equations:

$$\left(\partial_s x(s)\right)_{s=0} = -\cos\alpha \tag{9}$$
$$\left(\partial_s z(s)\right)_{s=0} = \sin\alpha \tag{10}$$

the incidence angle on the target is α and the initial entrance point in the plasma is chosen to be the coordinate origin:

$$x(0) = 0 \tag{11}$$
$$z(0) = 0 \tag{12}$$

This boundary value problem (BVP) can be solved for an arbitrary initial plasma scale length and temperature and for a given incidence angle of the main pulse onto the pre-plasma, which is the essence of the GRIP approach for XRLs.

For a low collision frequency ($\bar{\nu} \ll \overline{\omega_{pe}}$) there are several analytical solutions of the refraction/absorption problem for both linear and exponential density profiles.[5,6] For the exponential case, the analytical formula has been reported,[14] but only for grazing angle propagation. In the work presented here a range of angles, from grazing to normal incidence, was analysed and so the use of a numerical approach was preferred for solving the BVP, using Mathematica 5.0. The BVP solution describes the path of a ray in the plasma, and so the next step is to compute the intensity of the ray at any point, using the complex part n_a of the refractive index.

2.4 Absorption Model in Grazing Incidence

The local absorption scale length d_a can be defined as

$$d_a = \frac{\lambda_0}{4\pi n_a} \tag{13}$$

where λ_0 is the laser wavelength and the attenuation of the wave intensity is exponential with the 1/e decay determined by d_a. The absorption scale length is a function of the normalized plasma and collision frequencies. It decreases strongly at low collision frequencies, but is several orders of magnitude larger than the laser wavelength, as shown in figure 2 for $Z_f = 8$. It can be seen that the absorption scale length can take values over several orders of magnitude depending on the normalized electron collision rate and that it is smaller towards the critical density of the plasma.

Figure 3 shows the region $d_a = 0$–200 μm in more detail; the typical plasma scale length is in the range 20–50 μm. From figures 2 and 3 it is clear that at high densities, close to critical density, the absorption is very effective over short distances whereas in less dense plasmas the laser pulse can propagate over distances of the order of millimetres without strong reduction in intensity. The key role in this absorption is due, as expected, to the electron collision frequency to which it is very sensitive.

The change in intensity along the path defined by the BVP solution is obtained by integrating

$$I(s_0) = I_0 \exp\left(-\int_0^{s_0} \frac{ds}{d_a[x(s)]}\right) \tag{14}$$

where I_0 is the initial laser intensity and s_0 is a given point along the optical path.

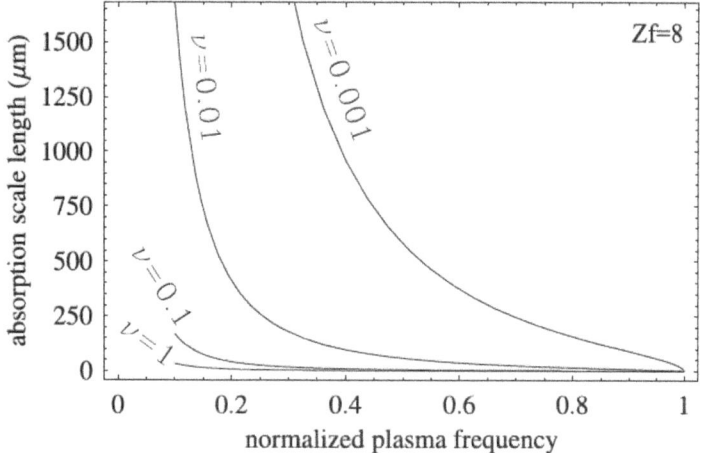

Figure 2 *Absorption scale length as a function of normalized plasma frequency for different normalized collision rates.*

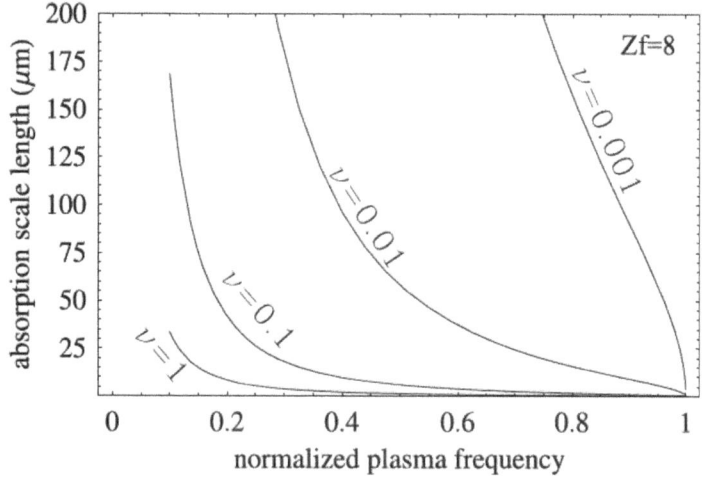

Figure 3 *Detail of the absorption scale length as a function of normalized plasma frequency for different normalized collision rates (only the refractive index varies).*

Any ray in the plasma can be described using $x(s)$, $z(s)$ and $I(s)$ which provide the position and transmitted intensity depending on the input plasma temperature, scale length and density at different incident angles.

2.5 Parametric Analysis of the Absorption

The parametric analysis described above and below was implemented, unless explicitly mentioned, by considering a fixed plasma temperature (30 eV) and density scale length (30 μm), typical values for XRL pre-plasmas, and by varying the incidence angle from 10–80° in steps of 10°. For the normalized collision length an average charge state times inverse Bremsstrahlung correction factor of $Z_f = 1$ was used for convenience below. The laser is turned on at the position with the coordinates {0,0} and the subsequent propagation of the light in the plasma is computed. The zero point on the x axis corresponded to an electron density of 10^{19} cm^{-3} , two orders of magnitude below the critical density, so the laser absorption in the plasma below this density can be neglected.

2.5.1 Effect of Incidence Angle on the Absorption

Figure 4 shows the penetration depth of the laser radiation in the plasma. It varies from 125 µm for 10° incidence angle on the plasma down to 50 µm at 80°, equivalent to 2-4 scale lengths of the plasma penetration region (30 µm). In the other direction the propagation takes place in a region of 60–770 µm depending on the incidence angle.

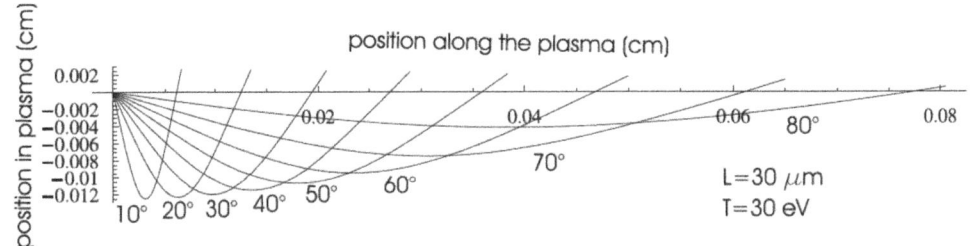

Figure 4 *Path of the incident laser light in the plasma for different laser incidence angles; horizontal and vertical axes use the same scale.*

However, while the propagation in the plasma is significantly longer at large incidence angles, the absorption is reduced as shown in figure 5. Here, the transmission of the laser pulse along the x axis is presented. The ray enters the plasma at the point $(x, \text{transmission}) = (0,1)$ and the total energy which has been deposited in the plasma can be evaluated from the transmission value at the same $x = 0$ coordinate at the other end of the curve. For large angles, the energy deposited in the plasma represents only 60% of the total energy (in the mentioned conditions $Z_f = 1, T_e = 30eV, N_e$ scale length $= 30\mu m$). This shows that, under certain conditions, the coupling of the laser energy into the plasma will be inefficient for large incidence angles.

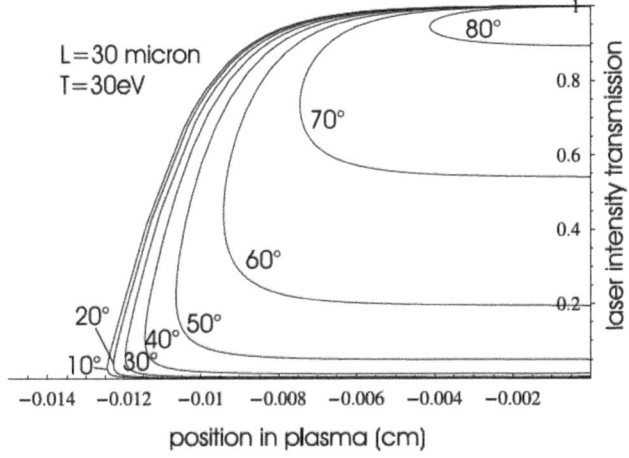

Figure 5 *Transmitted short-pulse intensity as a function of the depth in the plasma for different incidence angles.*

2.5.2 Effect of Plasma Scale Length on Absorption

Figure 6 shows the calculations for a fixed incidence angle of 70° and different plasma scale lengths at a constant electron temperature of 30 eV and $Z_f = 6$. In this case the laser beam reaches the same plasma density region (1.2×10^{20} cm^{-3}) for all scale lengths. As a check of the ray tracing code, it can be noted that the electron plasma density at the turning point is in good agreement with the analytical formula[15] which is applicable to any incidence angle for a linear electron density profile,

$$\cos \alpha = \sqrt{\frac{n_e}{n_{ec}}} \qquad (15)$$

where n_e and n_{ec} are electron density at the turning point and the critical density, respectively, and α is the incidence angle on the target.

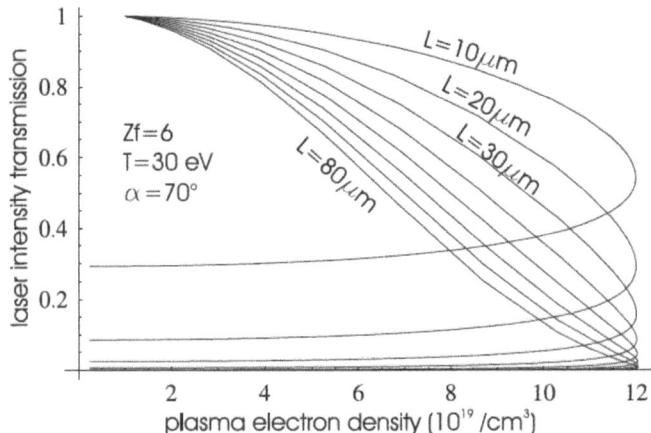

Figure 6 *Transmission in the plasma as function of the plasma density for several plasma scale lengths.*

The pre-plasma scale length depends on the pre-pulse intensity, duration and on the relative delay between the short and long pulses. It should be at least a few tens of micrometres in order to obtain sufficient propagation of the XRL in the gain region. Otherwise the amplified spontaneous emission is deflected away from the narrow optimal gain region and does not reach saturation. Figure 6 demonstrates that the scale length of the plasma can also have a significant effect on the absorption, with up to 30% of the energy of the laser being transmitted through the plasma for steep gradients. This effect is attributed to the fact that the path in the plasma is significantly shorter in this case, as clearly shown in figure 7 where the ray tracing of the main pulse in the plasma was carried out under the same conditions as in figure 6.

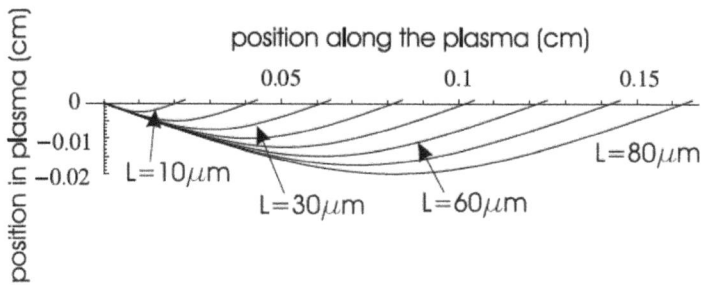

Figure 7 *Ray tracing in the plasma for different plasma scale lengths.*

2.5.3 Effect of NIB Correction Factor on Absorption

Figure 8 shows the effect of the Z_f factor on the transmitted intensity through the plasma as a function of position at a constant incident angle of 70° for a 30 μm scale length and 30 eV electron temperature. While the propagation trajectory of the laser pulse remains the same, there is a large effect on the transmitted intensity, which increases to more than 50% for small Z_f factors indicating an inefficient use of the main pulse energy.

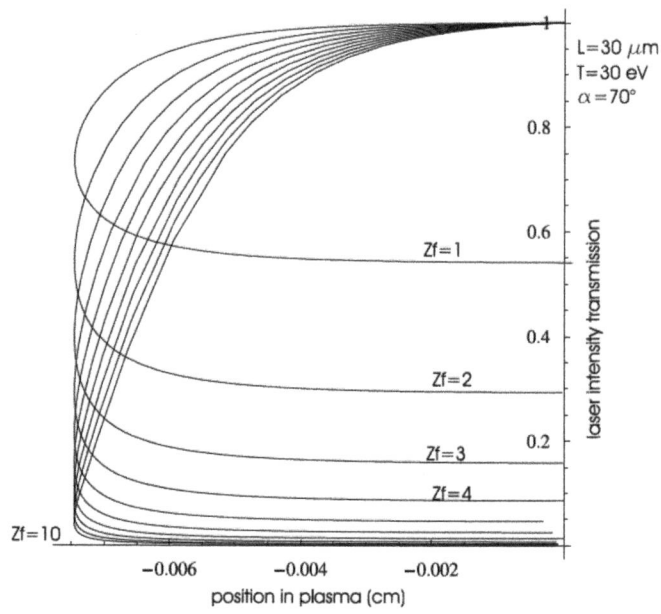

Figure 8 *Transmitted intensity as a function of depth in a plasma for several Z_f values.*

2.5.4 Effect of Temperature on Absorption

A final parametric study is presented in figure 9. The plasma scale length for this case was 30 μm, with $Z_f = 6$. The transmitted intensity is represented as a function of the position in the plasma for the temperature range 30–150 eV. The transmitted energy increases at higher temperatures to up to 70% of the incident energy. This effect can be easily understood according to the electron-ion collision rate formula, equation (2), which shows a $T^{-3/2}$ dependence; as a direct consequence there is a strong reduction of the absorption through the IB mechanism when the plasma temperature is increased. For long wavelength XRLs the pre-plasma temperature is in the range of a few tens of electron-volts. When — for short wavelength XRLs — higher average ionization plasmas are required, the pre-plasma temperature must reach hundreds of electron-volts and this consequence becomes especially important.

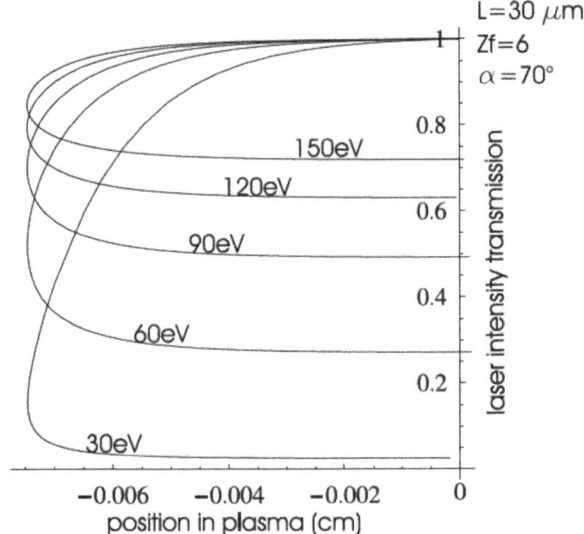

Figure 9 *Transmitted intensity as a function of the depth in a plasma for different plasma temperatures.*

3 COMPARISON OF EXPERIMENT WITH SIMULATION

In this section a numerical approach to evaluate the absorption of the main pulse in the plasma is described. Former analytical computation by Li and Zhang[7,8] of the temperature at the end of the main pulse in TCE XRLs[7,8] do not take into account the reduced absorption of the pre-pulse due to NIB and assume that all the main pulse energy is absorbed. These approximations concerning the plasma opacity are still very useful in the case of normal incidence on the target if all the rays reach critical density regions and are absorbed. The analytical model predicts an almost linear increase of temperature with the main pulse energy and an increase of the scale length of the plasma, roughly as the square root of the deposited energy. This is a consequence of the implicit assumption that all the pulse energy is converted into thermal energy of the electrons. No dependence of the absorption with the main pulse duration can be predicted by this model approximation.

In the work described here, the scaling law for the main pulse final temperature is revised. The correction to the model is introduced through the IB correction factor and through the opacity calculation using the complex refractive index as described in the previous sections. The effect of these corrections is illustrated for experiments with a TCE Ni-like zirconium XRL at 45° and 72° incidence angles and an energy of 3J on target.[10,11]

The first step of the calculations is to derive the temperature dependence of the main pulse transmission through the plasma directly using the scaling laws of the main pulse temperature and the plasma scale length as provided by the analytical model of [7]. The result is plotted in figure 10.

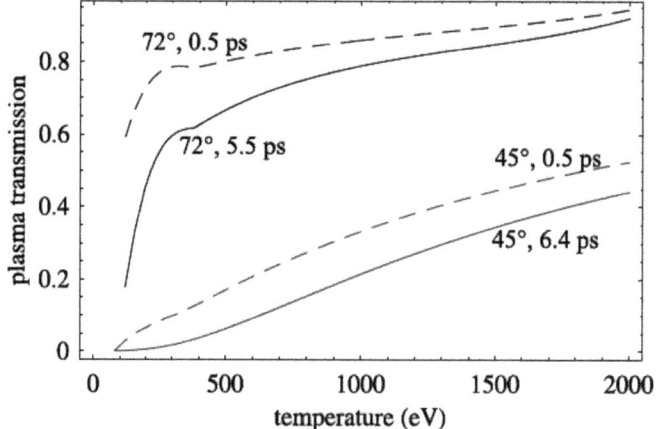

Figure 10 *Calculated transmission of a 0.5 ps pre-pulse and a 6.5 ps main pulse through a plasma at an incidence angle of 45° (blue)[9] together with a 0.5 ps pre-pulse and a 5.5 ps main pulse at 72° (black).[10]*

The IB absorption correction factor depends on the assumed electron temperature. As mentioned previously, this correction factor can be influenced by the non-Maxwellian electron velocity distribution,[12] which is included in the calculation of the results shown in figure 10. The electron-electron thermalisation time is in the order of 10 ps, longer than the main pulse. Hence it can be assumed in the first-order approximation for the IB correction factor that the temperature associated with the electron distribution function is unchanged during the main pulse interaction with the plasma.

The second step of the calculation involves solving

$$\frac{dT(t)}{dt} = \gamma\left[1 - \Xi\big(T(t)\big)\right] \tag{16}$$

where $T(t)$ is the temperature of the plasma at time t, which ranges between 0 and the duration of the main pulse, and Ξ is the transmission of the plasma as a function of temperature as shown in figure 10. Equation (16) implies that the increase in the temperature is proportional to the absorbed energy, $1 - \Xi$. The coefficient γ is obtained from the analytical model[7,8] assuming full absorption of the pulse,

$$\gamma = \frac{T_{main} - T_{free}}{t_{main}} \tag{17}$$

where T_{free} and T_{main} are the plasma temperatures after the free expansion period and after the main pulse and t_{main} is the duration of the main pulse. For an opacity of unity, linear dependence[7,8] is recovered due to the choice of γ.

The solution of equation (16) for $T(t)$ as a numerical function was obtained using an interpolation of the points in figure 10 for $\Xi(T)$. The results for a 5.5 ps main pulse in 72° configuration are presented in figure 11. The plasma temperature during the main pulse reaches 1250 eV at the end of the main pulse, in contrast to 4000 eV as predicted by the linear model with plasma opacity equal to unity,[7,8] also shown in figure 11.

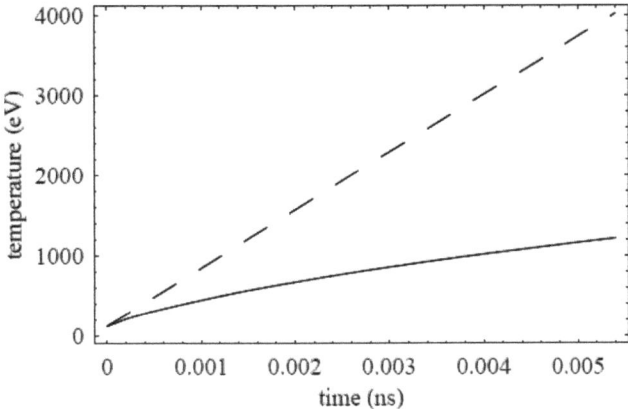

Figure 11 *Temperature evolution of the plasma in time according to the analytical model of [7] (dashed line) and according to the present model which includes the IB correction factor and opacity for a 5.5 ps pulse in 72° configuration (straight line).*

For experimental validation of the calculations, data for a Ni-like zirconium XRL with a 45° short-pulse incidence angle on target[9] are compared with the semi-analytical results, obtained as described above. The analysis concerns only the absorption of the short pulse in the plasma and the kilo-electronvolt emission registered with a cross-slit camera. The temperature was computed at various main pulse durations for comparison with the signal of the cross-slit camera. The longer the pulse, the higher the temperature; for the longest pulse of 6.4 ps the temperature reached with NIB and GRIP absorption is 2250 eV, while according to the linear model the pulse duration has almost no influence on the predicted temperature of 3000 eV.

The experimental signal obtained on a camera with constant quantum efficiency through an aluminium filter (low energy cut off 600 eV) is shown in figure 12 together with the theoretical prediction at 45° incidence angle. The theoretical plasma emission was computed assuming that it is proportional to the fourth power of the temperature blackbody radiation and shows reasonable agreement with the experimental values.

4. ANALYSIS OF A TRANSIENT COLLISIONALLY EXCITED SAMARIUM X-RAY LASER

This section illustrates the plasma parameters and the absorption mechanism effect for a TCE samarium XRL (operating at 7.3 nm) using the analytical-numerical approach described in the previous sections. Samarium was chosen in order to support the development of a sub-10 nm XRL by shedding light on the parameters which influence the absorption mechanism of the short pumping pulse in the pre-plasma.

The experimental parameters were taken to be those of a saturated TCE samarium XRL;[16] a 0.28 ns pre-pulse at a wavelength of 1.054 μm and a focused intensity of 10^{12} W/cm^2 was followed by a 1 ps pulse and intensity 3×10^{15} W/cm^2 at the same wavelength. The incidence angles on the target were 0° for the pre-pulse and 10° for the second pulse. The transmission of the plasma as a function of position along the optical path is shown in figure 13a. The pre-pulse plasma scale length was 20 μm and the temperature at which the ray-tracing analysis was performed was about 1 keV, in keeping with the analytical model. The total transmission of the plasma was about 16% in this case.

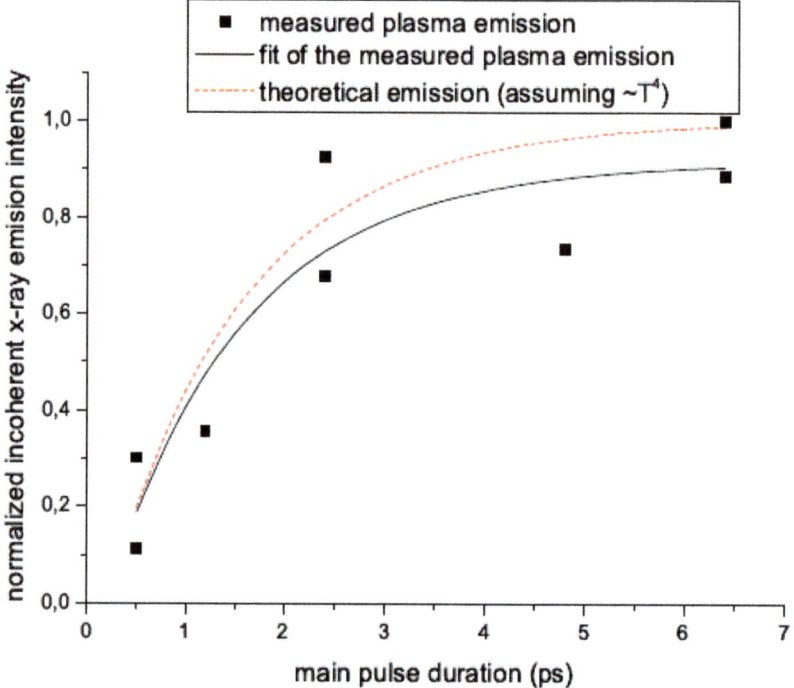

Figure 12 *Comparison of the cross-slit camera measurement with theory for an incidence angle of 45°.*

Changing the incidence angle to 45° increases the transmission of the plasma to 32% (figure 13b) and so the short pulse energy is not used optimally. One way of reducing the transmission of the plasma is to reduce the short pulse intensity and hence the NIB effect. In figure 13c the short pulse duration was increased to 5 ps, reducing the transmission to 24%, and then increasing the pre-pulse duration to 800 ps modifies the plasma scale length considerably, from 20 μm to 50 μm. Hence the optical path in the plasma is significantly increased and the transmission is almost suppressed, as shown in figure 13d where 98% of the short pulse is absorbed.

This analysis shows that blindly applying the GRIP method to a normal-incidence pumped Ni-like samarium XRL reduces the efficiency due to reduction of the short-pulse absorption (as shown in the figure 13b), or even worse for small glancing angles when the pre-plasma becomes transparent to the short pulse. Increasing the short-pulse duration, which reduces the NIB effect, is not enough to cure the problem as shown in figure 13c, although it is efficient in longer wavelengths XRLs, as shown by the presence of the inverse 3/2 power of temperature in the electron-ion collision rate, equation (2). A specific treatment that can be applied for samarium XRLs is to modify the plasma scale length together with the short pulse duration if a 45° angle is used, as shown in figure 13d.

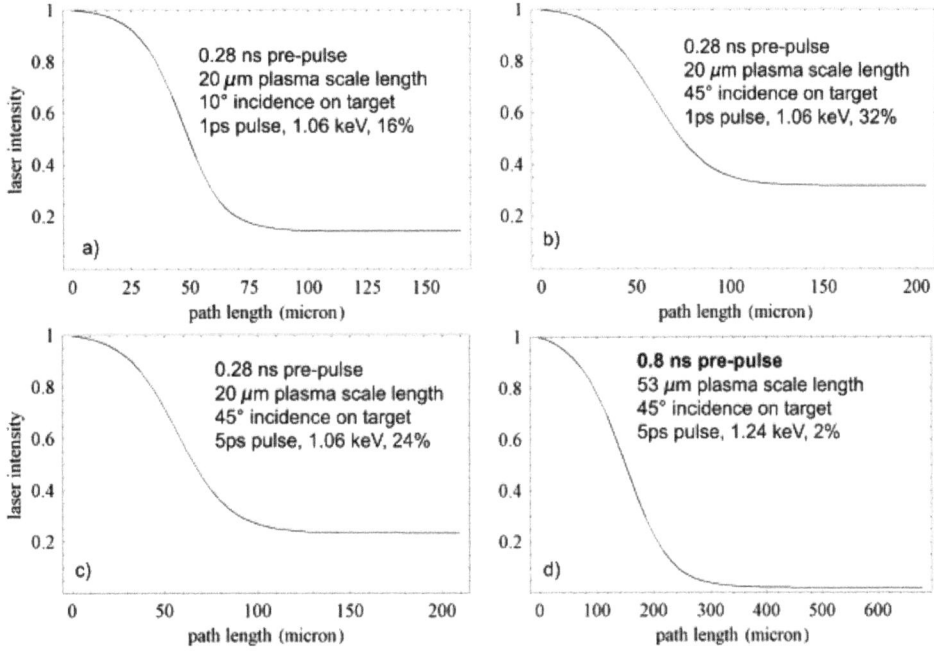

Figure 13 *(a) The short pulse energy transmitted through a plasma as a function of the position along the optical path for a set of parameters corresponding to experiment.[16] (b), (c) and (d) show how modifications of the parameters can lead to different absorption in the plasma.*

5 CONCLUSIONS

The study presented here has proposed a way to introduce the NIB mechanism in the complex refractive index formalism for understanding the propagation of short light pulses in plasmas. It was applied to the specific case of GRIP XRL efficiency analysis. However, it could also be useful for a broader range of applications related to the behaviour of short pulses in plasmas, such as plasma mirrors and the propagation of ultra-short light pulses in plasma channels.

The complex refractive index formalism was applied to two specific XRL arrangements. In the first, GRIP XRLs, the results show that at very small grazing angles little energy is deposited in the plasma due to the reduced opacity. Slightly larger grazing angles, in the range 25–30° instead of 18° for a wavelength of 1.054 μm, and longer, but still short, pulses can circumvent the opacity problem and increase the XRL efficiency since the pumping energy is then efficiently deposited in the pre-plasma. In the second case, sub-10 nm XRL systems, the required pump energy is significantly higher, around 20 J for a Ni-like lanthanum XRL to a few kilojoules for water-window XRLs. Increasing the opacity of the plasma by tuning the plasma parameters can significantly increase the temperature of the plasma formed by a short pulse and reduce the pumping energy requirements, making shorter wavelength XRLs possible at limited pump energies.

ACKNOWLEDGMENTS

The author acknowledges enlightening discussions of the results with Thomas Kühl, his doctoral studies supervisor, Paul Neumayer, Geoff Pert and James Dunn. Also acknowledged are travel and meeting opportunities provided through Lasernet (EU Contract HPRI-CT-2000-40016), LASELAB Europe (EU Contract RII3-CT-2003-506350) and Romanian Research Project PN2-CNCSIS-RP6/2007.

References

1 P.V. Nickles, V.N. Shlyaptsev, M. Kalachnikov, M. Schnurer, I. Will and W. Sandner. *Phys. Rev. Lett.*, 1997, **78**, 2748.
2 R. Keenan, J. Dunn, P.K. Patel, D.F. Price, R. F. Smith, and V.N. Shlyaptsev, *Phys. Rev. Lett.*, 2005, **94**, 103901.
3 J.J. Rocca, Y. Wang, M.A. Larotonda, B.M. Luther, M. Berrill and D. Alessi, *Opt. Lett.*, 2005, **30**, 2581.
4 K. Cassou, S. Kazamias, D. Ros, F. Plé, G. Jamelot, A. Klisnick, O. Lundh, F. Lindau, A. Persson, C.G. Wahlström, S. de Rossi, D. Joyeux, B. Zielbauer, D. Ursescu and T. Kühl, *Opt. Lett.*, 2007, **32**, 139.
5 G.J. Pert, *Phys. Rev.*, 2006, **A73**, 033809.
6 G.J. Pert, *Phys. Rev.*, 2007, **A75**, 023808.
7 Y.J. Li and J. Zhang, *Phys. Rev.*, 2001, **E63**, 036410.
8 Y.J. Li, X. Lu and J. Zhang, *Phys. Rev.*, 2002, **E66**, 046501.
9 P. Neumayer, W. Seelig, K. Cassou, A. Klisnick, D. Ros, D. Ursescu, T. Kuehl, S. Borneis, E. Gaul, W. Geithner, C. Haefner and P. Wiewior, *Appl. Phys.*, 2004, **B78**, 957.
10 D. Ursescu, B. Zielbauer, T. Kühl, P. Neumayer and G. Pert, *Phys. Rev.*, 2007, **E75**, 045401.
11 T Kuehl, D. Ursescu, V. Bagnoud, D. Javorkova, O. Rosmej, K. Cassou, S. Kazamias, A. Klisnick, D. Ros, P. Nickles, B. Zielbauer, J. Dunn, P. Neumayer, G. Pert and PHELIX-team, *Laser and Particle Beams*, 2007, **25**, 93.
12 A.B Langdon, *Phys. Rev. Lett.*, 1980, **44**, 575.
13 S. Eliezer, *The interaction of high-power lasers with plasmas*, 2002, (Bristol: IOP Publishing).
14 J. Kuba, D. Benredjem, C. Moller and L. Drska, *8th International Conference on X-Ray Lasers*, 2002, **641**, 46.
15 G.J Tallents. *J. Phys.*, 2003, **D36**, R259.
16 R.E. King, G.J. Pert, S.P. McCabe, P.A. Simms, A.G. MacPhee, C.L.S. Lewis, R. Keenan, R.M.N. O'Rourke, G.J. Tallents, S.J. Pestehe, F. Strati, D. Neely and R. Allott, *Phys. Rev.*, 2001, **A64**, 053810.
17 M. Born and E. Wolf, Foundations of geometrical optics, in *Principles of Optics*, pages 116-141, Cambridge University Press, 1999
18 J. Kuba, D. Benredjem, C. Moller and L. Drska, Analytical Ray-tracing of a transient X-ray laser: Ni-like Ag laser at 13.9nm, X-RAY LASER 2002: 8th International Conference on X-ray Lasers, 641: 46-51, 2002.

THEORETICAL ANALYSIS AND EXPERIMENTAL APPLICATIONS OF X-RAY WAVEGUIDES

I. Bukreeva,[1,2,3] D. Pelliccia,[4] A. Cedola,[1] A. Sorrentino,[5] F. Scarinci,[6] M. Ilie,[6] M. Fratini,[1] V.E. Asadchikov,[2] V.L. Nosik[2] and S. Lagomarsino[1]

[1]CNR Istituto Processi Chimico-Fisici, UOS Roma, c/o Dip. Fisica, Università Sapienza, P.le A. Moro, 200185 Rome, Italy
[2]Shubnikov Institute of Crystallography, Leninsky prospekt 59, Moscow, 119333 Russia
[3]Lebedev Physics Institute, Leninsky prospekt 53, Moscow, 119991, Russia
[4]School of Physics, Monash University, Victoria 3800, Australia
[5]ALBA Synchrotron Light Source, 08290 Cerdenyola del Vallès, Barcelona, Spain
[6]CNR Istituto Fotonica e Nanotecnologie, V. Cineto Romano, 42, 00156 Rome, Italy

1 INTRODUCTION

Planar and two dimensional X-ray waveguides (WGs) are optical elements of considerable interest both from a fundamental point of view, and for their use in microprobe experiments, i.e., microimaging, microdiffraction, microfluorescence, etc. After the first demonstration of the feasibility to provide a sub-micrometre beam with planar WGs[1,2] and with 2D WG,[3] many different schemes and many applications have been demonstrated. For a short review see Lagomarsino and Cedola (2004).[4]

WGs have properties that make them worthy of further development and studies: (i) they work over a large energy range, from soft to hard X-rays; (ii) unlike capillaries, which are made of glass, WGs can have different cladding materials that can be optimized for the required energy; (iii) the roughness of the reflecting surfaces can be controlled down to fractions of a nanometre; (iv) WGs can provide beam dimensions from micrometres down to a few nanometres; (v) they are relatively easy to fabricate, and quite inexpensive; (vi) they inherently provide coherent beams, because they are based on the propagation of resonance modes; (vii) multiple WGs can be fabricated close to each other, in order to create interference patterns;[5] and (viii) WGs can be split to provide holographic devices.[6,7]

The outgoing beam, diffracted at the exit slit of the WG, can be exploited for high-resolution X-ray microscopy and micro-diffraction imaging[8,9] and in coherent X-ray diffraction imaging (CXDI) experiments.[10,11] For X-ray microscopy, the nano-sized channel is used as a secondary source to magnify an object onto a detector screen, with an attainable resolution limited only by the actual size of the WG channel. On the other hand, with the CXDI technique,[12] resolution increase has been demonstrated, in principle bounded only by the fundamental diffraction limit. The degree of coherence of the incident wave is an extremely important factor to consider in CXDI experiments. In practice, in most of the experiments, the beam that arrives onto the sample has encountered several optical elements on its path from the source, e.g., monochromators, mirrors, slits, etc., that lead to a coherence degradation at the sample position.[13–15] This lack of coherence distortion can severely disturb the wavefront distortion convergence of the phasing process, as discussed in reference 15. In this respect, a waveguide offers one of the best optical elements for producing coherent and divergent beams, both properties being extremely important for CXDI experiments.[7,10,16,17,18]

This chapter presents a theoretical analysis of electromagnetic field propagation in planar waveguides and experimental results obtained with synchrotron and laboratory sources. The basic equations of electromagnetic modes in a front-coupling planar WG are presented in section 2, and section 3 reports a comparison between computer simulation and experimental results in the general case of tapered waveguides. Then an original algorithm to retrieve the amplitude and phase of the electromagnetic field at the exit of the planar WG using an iterative method is presented (section 4). A theoretical analysis and computer simulation for WG with an additional periodic structure on the side reflecting walls of the WG are described in section 5, and experimental results are discussed in chapter 6 proving that spatial resolutions of few hundred nanometres in phase contrast imaging can be obtained with a laboratory source. Conclusions are drawn in section 7.

2. PLANAR WAVEGUIDE. GUIDED-MODE PROPAGATION ANALYSIS

A WG can be described schematically as a very narrow channel (typical size of tens to hundreds of nanometres) in which the incoming radiation is trapped, as in an optical resonator. The x-ray radiation propagates in the channel through resonant modes, providing nanometer size coherent X-ray beams at the exit of the WG.

The incident radiation can be coupled into the guiding layer of X-ray WGs in two different ways: Resonant Beam Coupling (RBC) and front coupling (FC). RBC[1,2] takes place in a three layer WG, with the incident beam at grazing angle transmitted by the very thin upper layer and trapped by the intermediate guiding layer (see figure 1a). With this scheme the incoming beam of several tens of micrometres can be compressed down to nanometre dimensions and significant effective gain ($N = 100$) can be achieved.[19] In the FC scheme (figure 1b)[17] the incoming radiation is directly side-coupled with the WG aperture, and the spatial acceptance is therefore equal to the WG gap. Moreover, generally different modes are excited simultaneously in the WG. A variant of the front coupling scheme can be adopted with a pre-reflection of the incoming beam just in front of the WG entrance at one of the WG resonance angles (see figure 1c). In this way only one particular mode will be selected. In this geometry the spatial acceptance of the WG is twice the WG gap.

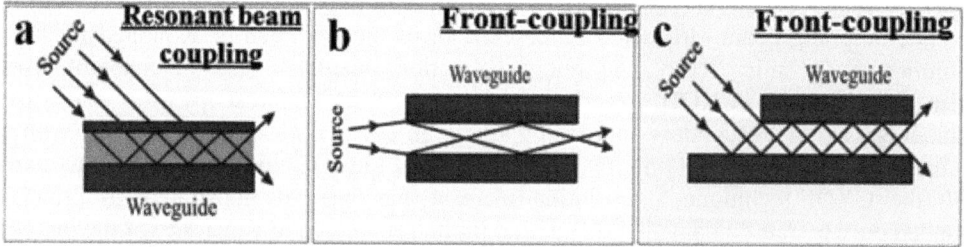

Figure 1 *Step-index multimode WGs in geometry (a) Resonant beam coupling (b) Front coupling (c) Front coupling with pre-reflection.*

Plane monochromatic wave illuminating the entrance aperture of the front coupling waveguide excites a set of eigen-modes inside the vacuum gap

$$U(x,0) = \sum_{m=1}^{m_{max}} a_m \Psi_m (x) \tag{1}$$

where $m_{max} \approx 2d\theta_c/\lambda$ is the maximum number of modes supported by the waveguide; d is the width of the vacuum guiding layer, $\theta_c = (2\delta)^{1/2}$ is the critical angle where δ is the refractive index decrement of the cladding layer, which has a complex refractive index $n = 1 - \delta - i\beta$, where β is the absorption index.

The coefficients a_m of the expansion can be estimated using overlap integral

$$a_m = \frac{\int U(x,0)\psi_m (x)dx}{\sqrt{\int \psi_m^2 (x)dx}}. \tag{2}$$

At the distance z the field profile changes and can be written as

$$U(x,z) = \sum_m a_m \psi_m (x) \exp[i\gamma_m z] \tag{3}$$

where γ_m is the propagation constant of the guided mode.[20]

3. TAPERED WAVEGUIDES

3.1 Basic Equations for Tapered Waveguides

In the general case the width of the WG guiding layer is not constant, but can be narrower or wider towards the exit aperture with respect to the entrance. In other words it can be tapered. Focusing tapered WGs are able to concentrate beams to a very small focal spot – the theoretical full width half maximum (FWHM) is about 10 nm[21] and as a result they can provide elevated output flux enhancement.[22]

The distribution of the electromagnetic field in a tapered WG is given by the solution of the Helmholtz equation

$$\nabla^2 U(r,\varphi) + n^2 k^2 U(r,\varphi) = 0 \tag{4}$$

where $k = 2\pi/\lambda$ is the wave vector, λ is the wavelength of the incoming radiation, and $U(r,\varphi)$ is the scalar wave field satisfying the boundary conditions

$$U(r,0) = U(r,\alpha) = 0 \tag{5}$$

at the interface vacuum gap and claddings, where (r,φ) are polar coordinates and α is the angle of tapering (see figure 2b). Any penetration of the field in the cladding material is neglected; this approximation can be used when the penetration depth, $\rho = 1/k\sqrt{\theta_c^2 - \theta^2}$, of the field in the cladding material is much less than the gap width d or, in other words, when the condition $\rho/d \approx \lambda/d\sqrt{\theta_c^2 - \theta^2} \ll 1$ is satisfied, where θ is the grazing incidence angle of the radiation. For hard X-rays at $\lambda \sim 0.1$ nm, small angles, $\theta < \theta_c$, and a relatively wide vacuum gap ($d \geq 200$ nm) the contribution of the evanescent field to the propagation of X-rays in the WG can be neglected to a first approximation. For large angles, $\theta \approx \theta_c$, and narrower vacuum gap, $d \approx d_c = \lambda/2\theta_c$, where d_c is the critical gap width supporting only one mode,[21] the penetration of radiation into the cladding layers has to be taken into account.

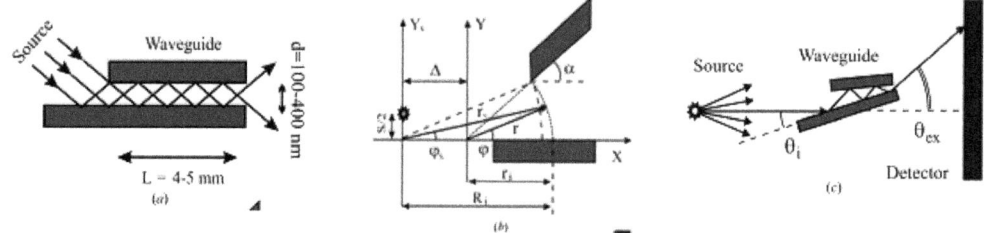

Figure 2 *Schemes of front coupling WGs with pre-reflection; (a) constant gap WG; (b) tapered WG; the polar systems of coordinates (r,φ) and (r₃,φ₃) used for analytical calculations are shown in the figure; (c) schematic drawing indicating incidence, θᵢ, and exit, θₑₓ, angles (reprinted with permission).* [23]

A detailed derivation of the electromagnetic field in the guiding layer has been given previously.[23] Here, the important point is that in the general case several modes are excited simultaneously into the WGs, and that there are some restrictions to the tapering angle and to the incoming field in order to have essentially single mode excitation. In the general case of a tapered WG the resonance conditions are expressed as

$$\theta_m = \frac{m\pi}{kr_i\alpha}, \qquad m = 1, 2, 3, \ldots \tag{6}$$

where r_i is defined in figure 2b.

Taking into account that at the entrance aperture of the WG the gap d is equal to $r_i\alpha$, equation (6) is equivalent to the well-known resonance condition

$$\theta_m = \frac{m\lambda}{2d}. \tag{7}$$

In the general case, the wavefront curvatures of the incident wave and of the eigenmodes at the entrance aperture of the WG are different. Therefore even at resonance conditions, the phase mismatch at the entrance can excite more than one mode.[21] Strong coupling of the incident beam into the particular single mode $\Psi_m(r,\varphi)$ can be provided only for tapering angles α satisfying the condition

$$|\alpha \pm \beta| \leq \frac{\lambda}{2d} \tag{8}$$

where $\beta = d/R_i$ is the WG numerical aperture and R_i is the distance between the source and the entrance aperture of the WG (figure 2b). The plus sign holds for converging WGs, and the minus sign for diverging ones.

If the resonance conditions, equation (6), or the inequality, equation (8), are not satisfied, all modes are contextually excited in the guiding layer, though with different coefficients. The propagation of modes in the guiding layer is accompanied by damping of radiation due to absorption at the walls. Neglecting the penetration of field in the claddings the coefficient of transmission of the m-th mode for a tapered WG can be estimated as the product of the Fresnel coefficients R_{Fr} of successive reflections of the incident plane wave with resonance angle θ_m – see equation (7) –

$$R_m \approx \prod_{n=1}^{N} R_{Fr}\left(\theta_m \pm n\alpha\right) \tag{9}$$

where N is the maximum number of reflections. For a WG without tapering ($\alpha = 0$), the coefficient of transmission is[18]

$$R_m \approx R_{Fr}^N\left(\theta_m\right) \approx R_{Fr}^{L\theta_m/d}\left(\theta_m\right). \tag{10}$$

Taking into account equation (9) the total field profile is given by

$$U\left(r,\varphi\right) = \sum_{m=1}^{m_{max}} c_m R_m \Psi_m(r,\varphi). \tag{11}$$

Finally, the field at the detector can be calculated as the Fourier transform of the field at the exit aperture of the WG, given by equation (11).

3.2 Comparison of Calculated Profiles and Experimental Results

The analytical treatment presented above has been compared with a number of experimental results taken at both laboratory and different synchrotron radiation sources. The X-ray propagation problem has been accounted by modelling a suitable X-ray source. For a synchrotron X-ray facility the beam illuminating the WG entrance aperture can be considered as a plane wave due to the large source-to-WG distance and the small aperture of the WG. A laboratory X-ray source is modelled by a set of elementary spherical radiators with the initial phase randomly distributed within the interval 0–2π. The obtained field is the average over realizations of the random process. The wave from an elementary radiator after reflection in front of the entrance aperture of the WG gives rise to standing waves that, inside the guiding layer, can be written as a linear combination of modes. The distribution of intensity for an extended lab source is then found as a sum of contributions from independent radiators. The spectrum of the radiation from the X-ray tube (K_α and K_β lines for a copper target, $E \approx 8$ keV) including a continuous wide Bremsstrahlung spectrum (3–30 keV) is taken into account.

WGs measured in experiments were fabricated from silicon slabs, using lithographic processes and wafer bonding.

3.2.1 Synchrotron Radiation. Experimental characterization of the X-ray WGs has been carried out on different beam lines at various synchrotron facilities, in order to compare the results obtained under different illumination conditions and for different X-ray energies. The results show the far field diffraction pattern measured as a function of the waveguide rotation angle, denoted by θ_i, and the exit angle θ_{ex}, as measured on the detector. The 2D experimental maps are compared with simulated ones, in order to extract the WG parameters.

The results in figure 3a were obtained at the beam line Fluo@ANKA (Karlsruhe Institute of Technology) at an energy of 11 keV, selected by a double multilayer monochromator with a bandwidth $\Delta E/E \approx 10^{-2}$. No additional optics were used, resulting in nearly plane and pink beam illumination onto the WG entrance. The CCD detector, with effective pixel size 1 μm was located 15 cm from the exit aperture of the WG, corresponding to an exit angle resolution of 6.7 μrad. The measured intensity distribution, when compared to a computer simulation (figure 3b), indicates a diverging WG with an entrance gap of 360 nm and an exit aperture of 1600 nm.

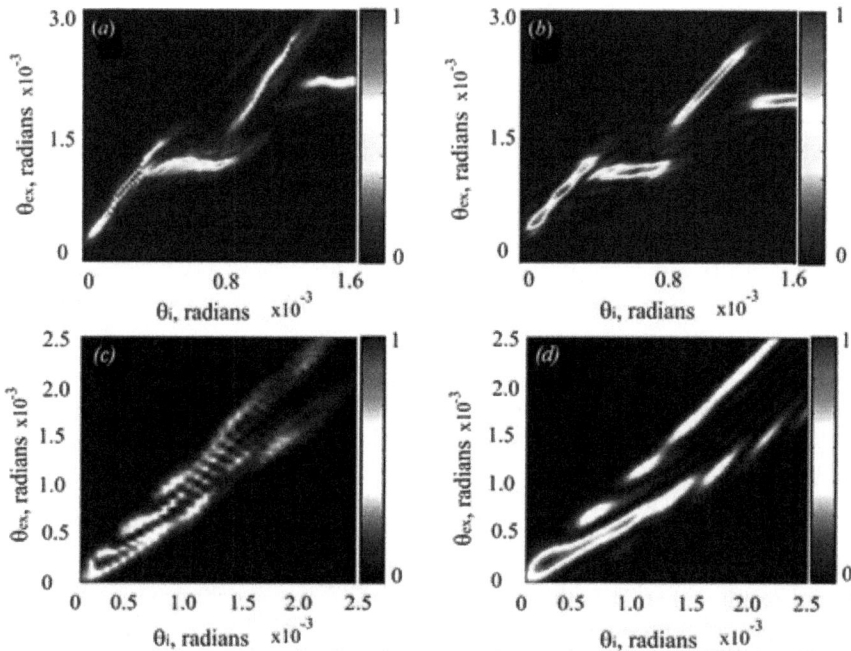

Figure 3 *(a) Experimental results from ANKA. (b) Computer simulation for a diverging tapered WG with a gap of 360 nm at the entrance and 1600 nm at the exit. (c) Experimental results from the ESRF. (d) Computer simulation for a diverging tapered WG with gap of 150 nm at the entrance and 670 nm at the exit (reprinted with permission).[23]*

Completely different illumination conditions were achieved at the ID1 beam line of the European Synchrotron Radiation Facility (ESRF) in Grenoble, using 11 keV photons, provided by a Si(111) double crystal monochromator. Therefore a much narrower bandwidth, $\Delta E/E \approx 10^{-2}$, was employed. In addition a stack of refractive lenses was used upstream of the WG to create a converging wavefront at the WG entrance. A CCD camera with a nominal pixel size of 2 μm was put at a distance of 0.6 m from the WG. The experimental results and the related computer simulation are shown in figures 3c and 3d. Good agreement is found for a diverging WG, with entrance and exit gaps of 150 nm and 670 nm, respectively, and a tapering angle of about 0.2 mrad.

A further experiment was carried out at the microfluorescence beamline at Elettra (Trieste, Italy). In this case the beam was monochromatic (a double crystal Si (111) monochromator was used) and parallel, i.e., no focusing optics were inserted between the monochromator and the WG. A CCD camera with nominal pixel size of 12.3 μm and measured point spread function (PSF) of about 20 μm was placed 330 mm downstream of the WG. The measurements were taken at a photon energy of 7665 eV.

The 2D intensity distribution is shown in figure 4a, and the result of a computer simulation for the silicon WG with a vacuum gap of 430 nm is shown in figure 4b. Clear modal structure is seen in both the data and the simulation, indicating extremely good coherence of the exit beam. Figure 4c shows the intensity distributions along the axis $0-\theta_i$ for the experimental data (dotted line) and the simulation (full line).

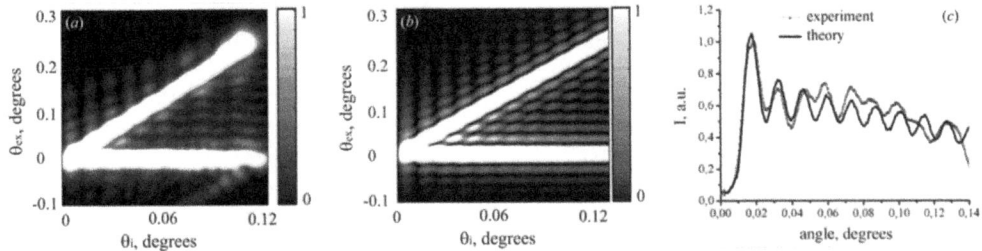

Figure 4 *Diffraction pattern on the detector as a function of incidence, θ_i, and exit, θ_{ex}, angles ($\lambda = 0.161$ nm): (a) experimental result; (b) computer simulation; (c) distribution of intensity along the axis $0-\theta_i$ for experimental data (dashed line) and simulation (bold line). WG length L = 4.6 mm, gap d = 430 nm (reprinted with permission).[23]*

An important consequence of the high degree of coherence is the possibility to employ iterative algorithms to characterize the exit field of a WG by the sole measurement of its far-field diffraction pattern. This topic will be discussed in section 4.

3.2.2 Laboratory X-ray Microfocus Source. Measurements of the tapered WGs were also performed with a laboratory source. A microfocus source with a nominal spot size of 15 μm and a copper target (Cu K_α, $\lambda = 0.154$ nm, bandwidth $\Delta\lambda/\lambda \approx 10^{-3}$) was used. The WGs were put at a distance of 3 cm from the source. A 30 μm wide slit was placed in front of the WG entrance to limit the background. The far-field intensity distribution was measured with a CCD camera with a pixel size of 12.3 μm and measured PSF of 20 μm, placed at a distance of 41 cm downstream of the WG exit. In this condition the WGs are illuminated with a diverging and rather monochromatic X-ray beam. The bandwidth is in fact limited to the natural width of the $K_{\alpha1}$ and $K_{\alpha2}$ lines, while the broad Bremsstrahlung contribution has a very limited influence, for two reasons. First, the multiple reflections inside the channel act as a low-pass filter for the incoming bandwidth and, second, the indirect detection mechanism (via a scintillating screen coupled to the CCD chip) is optimized for the detection of the copper characteristic K lines and its efficiency significantly decreases at high X-ray energy.

The experimental diffraction maps, together with the corresponding computer simulations, are shown in figure 5. Figures 5a (experimental) and 5b (simulated) show the results for a 4.5 mm long diverging tapered WG, with the vacuum gap expanding from 150 nm to 330 nm with an angle of 40 μrad. Figures 5c (experimental) and 5d (simulated) show the intensity distributions for a WG of length 4.9 mm and a gap narrowing from 305 nm to 200 nm. The coherence length of the tabletop microfocus source at the entrance of the WG is $h \approx 300$ nm. Comparing this value to the double aperture of the WG $2d_{in} = 300$ nm shown in figures 5a and 5b, it is possible to conclude that in this case the illumination is totally coherent for the WG. On the other hand, for the WG shown in figures 5c and 5d, the X-ray coherence length of $h \approx 300$ nm is less than the double aperture of the WG, $2d_{in} = 610$ nm, thus broadening of the diffraction pattern takes place due to the finite size of the source.

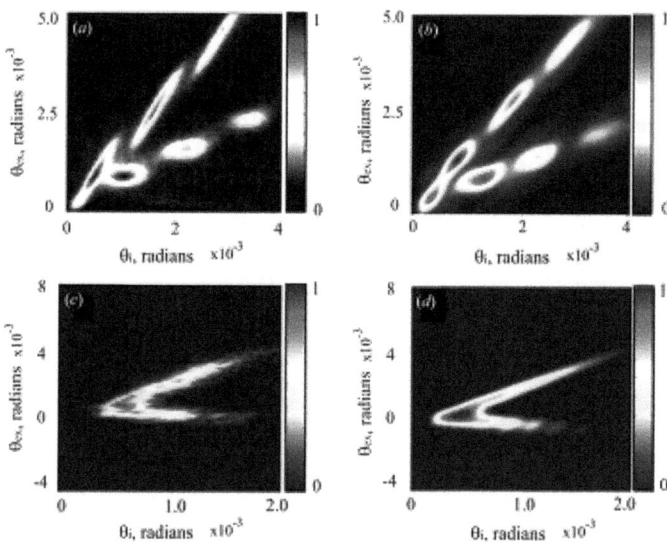

Figure 5 *Far-field diffraction patterns at the detector for the table-top source (0.154 nm) (a,c) experimental results, (b,d) computer simulations. The computer simulations indicate that the WG in a) and b) has a length $L = 4.5$ mm, a tapering angle $\alpha = 40$ μrad, an entrance gap $d_{in} = 150$ nm and an exit gap $d_{ex} = 330$ nm. The WG shown in c) and d) has $L = 4.9$ mm, $\alpha = -22$ μrad, $d_{in} = 305$ nm and $d_{ex} = 200$ nm (reprinted with permission).[23]*

In the computer simulation the X-ray spectrum of the copper target including the K lines and the white Bremsstrahlung was taken into account. The effect of the Bremsstrahlung can be seen in figure 5b, as a weak halo around the bright spots, in good agreement with the experimental result shown in figure 5a. The intensity of the halo is comparable to the level of the background, and does not give any additional information. In figures 5c and 5d the grey intensity level is set in order to exclude the background, giving a larger dynamic range to display the relevant features in the bright spots.

It is worth noting that when the incident beam is partially incoherent the output intensity pattern does not display interference oscillations as expected.

4. RETRIEVAL OF THE X-RAY WAVE FIELD AT THE WAVEGUIDE EXIT

So far, methodologies to simulate the wave propagation inside a WG channel and the coupling between the channel and the outside regions have been discussed. These in turn have been used to simulate the experimental results and characterize the measured devices. A complementary approach is based on the direct retrieval of the wave field at the exit face of the waveguide, using iterative algorithms. Such phase retrieval algorithms,[24] start from the measured far-field diffraction pattern (Fourier space constraints) and known constraints on the extent and nature of the field to be retrieved (real space constraints) to iteratively approach the solution satisfying both sets of constraints.

In the current case the problem of retrieving a complex function from the modulus of its Fourier transform is considered: such problems do not have unique solutions in one dimension.

In fact, indicating by $|\tilde{\psi}(q)|^2$ the measured 1D intensity distribution in the far-field (Fourier space), let $\psi(x)$ be the conjugate real-space wave function. From the properties of the Fourier transform (FT), the intensity distribution is unchanged by the following transformations[25]

$$\psi(x) \rightarrow e^{i\alpha}\psi(x) \tag{12a}$$

$$\psi(x) \rightarrow \psi(x+x_0) \tag{12b}$$

$$\psi(x) \rightarrow \psi^*(-x). \tag{12c}$$

This set of ambiguities, especially the third, generally prevents the confident application of iterative phase retrieval methods to one-dimensional problems. Therefore, in order to employ such algorithms on the one-dimensional diffraction patterns produced by planar WGs, a procedure must be determined to reduce the ambiguities. Recently[26] a method was proposed for *a posteriori* reduction of the ambiguities based on the correlation analysis of the solution of a large number of runs of an iterative phase retrieval algorithm with different random starting phases.

The procedure works as follows. First, a standard Hybrid Input-Output (HIO) algorithm is applied to the measured 1D far-field intensity. Denoting by

$$\psi_k^j = \left|\psi_k^j\right| \exp\left(i\varphi_k^j\right)$$

The key operation of the method is the calculation of the correlation coefficient between any two reconstructions indicated by ψ^j and ψ^l. The approach was to calculate the real correlation coefficient between the corresponding arguments φ^j and φ^l. The choice is motivated by the presence of ambiguity of equation (12c) which suggests that for each run both possibilities φ_k^j and $-\varphi_{N-k}^j$ must be taken into account. The latter is obtained by a cyclic shift applied to the inverse of each argument.

The procedure starts by choosing a reference reconstruction, typically the one that shows the highest average correlation (i.e., the average of the correlation coefficient with any other reconstruction). Subsequently the transformations described in equation (12) are applied to the other reconstructions in order to maximize the correlation. The solutions that display a high value of the correlation, e.g., $\max\left(c_n^j, \bar{c}_m^j\right) \geq 0.95$ are retained. Finally, to reduce noise, the average of the N_1 acceptable reconstructions is calculated:

$$\psi = \frac{1}{N_1}\sum_j \psi^j \tag{13}$$

The procedure has been tested on the measurement shown in figure 4. The algorithm was repeated $N = 500$ times for each angle, starting from different random phases. This provided N estimates $\psi^j(x,\theta)$ of the complex wave field as a function of the position along the exit face of the WG and the incidence angle. The method described above was then used to find the average solution . Then, by invoking the continuity of the wave field the 2D maps shown in figure 6 were obtained by stitching together the independent 1D reconstructions.

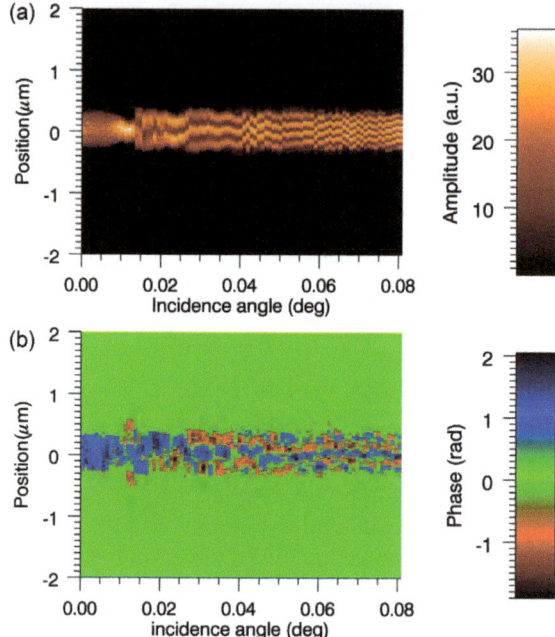

Figure 6 *(a) Reconstructed amplitude and (b) phase of the complex wave field at the exit face of the waveguide, as a function of the incidence angle of the waveguide with respect to the incident beam (reprinted with permission).[26]*

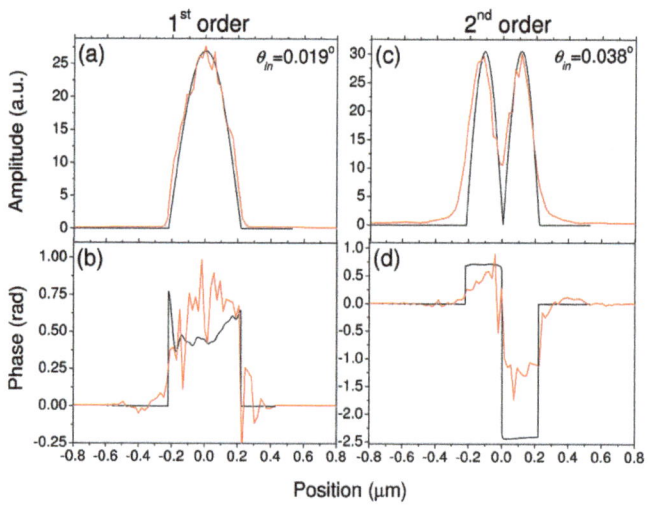

Figure 7 *Comparison between calculated (black line) and reconstructed (red line) wave fields at the exit face of the WG. (a) Amplitude and (b) phase of the first order. (c) Amplitude and (d) phase of the second order. The calculations do not take into account the absorption of the cladding which creates a smooth decay of the field in the cladding (reprinted with permission).[26]*

The reconstructed wave field clearly shows the presence of harmonics characterized by oscillatory behaviour inside the WG channel and an exponential decay outside (figure 7). In general, the propagation of the wave field in the WG is multi-modal. Under such conditions the amplitude and phase profiles of the field at the WG exit depend on the propagation distance. Therefore, in order to calculate the expected field profile the WG length must be known exactly.[23] Nevertheless the main features of the field, such as its extent and phase jumps, depend only on the wavelength and the WG gap d and therefore they can be calculated with much more confidence. In particular, the phase profile of a multimode guided field is characterized by intervals of nearly constant phase divided by phase jumps.[18] This feature is well reproduced by the reconstructed phase showing a nearly binary structure. The value of the gap obtained by this method is in excellent agreement with the value calculated by the previous methods. It is worth noting that the iterative phase retrieval procedure does not give any information about the propagation inside the channel, but only the characteristic of the wave field at the exit. On the other hand it does not require any preliminary assumption about the WG shape and size. In this way the phase retrieval method is fully complementary to the analytical approach. Moreover, the knowledge of the wave field may be extremely useful in applying coherent methods, already proposed in the 2D case,[11] to the simpler 1D situation, more favourable when low power table-top sources can be used.

5 STRUCTURED WAVEGUIDES

It has been shown in the previous sections that the degree of coherence at the WG exit depends on the incoherent mode mixing and the degree of coherence at the WG entrance. On the other hand, by choosing optimized channel geometries, the incoherent contributions can be removed more efficiently than with a simple planar propagating channel. In particular, an additional periodic structure on the side reflecting walls of the WG can efficiently filter out the asymmetric and high-order modes and provide a highly coherent exit beam even when only partially coherent illumination is employed. This is because the incident coherent field profile self repeats in the guiding layer at propagation distances known as fractional Talbot distances. If some additional structure imposes a longitudinal periodicity on the field, not all wave field k-vectors are allowed to propagate. Experimental proof and theoretical explanation of this for X-rays have been given previously.[27–29]

A typical scheme of the FCWG (Front Coupling WG) is shown in figure 1b. X-rays, coupled in the channel from the side, propagate in the vacuum guiding layer between two cladding slabs. Simulations have been carried out for a WG with vacuum gap $d = 100$ nm illuminated by a plane incident wave with wavelength $\lambda = 0.154$ nm. The self-imaging effect[30,31] for a WG with a self-repeating distance of the field profile equal to $z_T = D^2/4\lambda$, where $D = 2d$ is the period of the phase grating, is shown in figure 8b. The computer simulation used was based on the parabolic wave equation (PWE) numerical solution.[32]

The particular case of Talbot effect[33] takes place when a monochromatic plane wave is subjected to a phase shift of π rad when passing through a phase grating with transversal period D (see figure 8a). The incident field profile self repeats at propagation distances $z_T = D^2/4\lambda$, where λ is the wavelength and z is the coordinate along the optical axis. The parameters used in this particular case are: period of the grating $D = 2d = 200$ nm, a wavelength of 0.154 nm, 0.5 duty cycle and zero absorption in the cladding material.

In figures 8a and 8b the simulated intensity distribution in the WG guiding layer and behind the phase grating are compared. Note that the periodicities of self imaging with the phase grating and the WG are the same if the period of the grating is equal to double the width of the WG vacuum gap, i.e. $D = 2d$. A relevant difference in the propagation of guided modes compared to the free-space propagation of the field behind the phase grating is that propagation of the field in free space is lossless, whereas modes propagating in the

WG decrease due to multiple reflections. In addition the WG supports only a finite number m_{max} of modes, 12 for the current example.

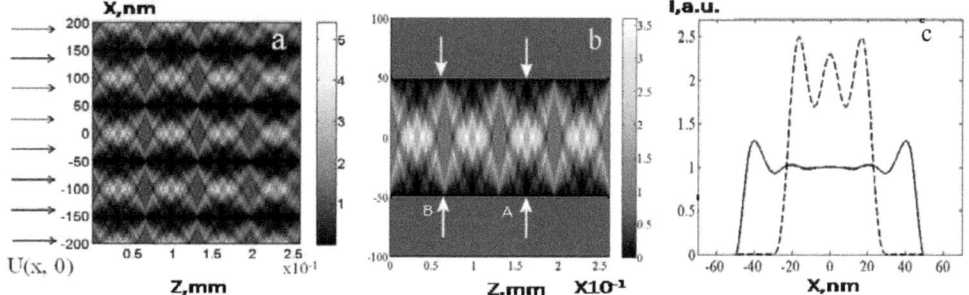

Figure 8 *Simulated x-ray intensity distribution (a) behind a phase grating with a period of D = 2d = 200 nm, λ = 0.154 nm, 0.5 duty cycle and π rad phase shift and (b) in a FCWG with a vacuum gap of d = D/2. (c) Lateral intensity distribution at the odd (dashed line) and even (full line) fractional Talbot distance (reprinted with permission).*[28]

Virtual transversal periodicity in the WG exists due to the reflecting side walls.[34,35,36] Note that at fractional Talbot distances $mL/2$, where m is an odd integer, the image of the field at the WG entrance is demagnified by a factor of two, as in the case of the phase diffraction grating[37,38] [34,35] - see figures 8a and 8b. Therefore a suitable choice for the WG length can halve the size of the exit beam. Figure 8c shows cross sections of the electromagnetic field at odd (dashed line, arrows A in figure 8b) and even (full line, arrows B) fractional Talbot distances.

Figure 8 refers to a WG with smooth reflecting sidewalls; the influence of roughness on X-ray beams transmitted by the WG has been analysed previously.[39,40,41,42] If an optical element (for example a sidewall grating) imposing an additional longitudinal periodicity on the field is introduced, the angular spectrum of the transmitted field is modified. The spectrum is given by the product of two terms, one related to the transversal, the other to the longitudinal properties.[43,44] Considering a combination of both situations, the imposed periodicities in transverse and longitudinal directions can be treated independently.[45] The transverse properties depend on the width of the guiding layer, and the longitudinal ones on the modulation period introduced in the WG. The simple planar WG with constant gap has a discrete set of spatial frequencies defined in equation (1). The longitudinal component of the wave vector in the parabolic approximation is quantized,

$$k_z^m = 2\pi \sqrt{\left(\frac{1}{\lambda}\right)^2 - \left(\frac{m}{p_z}\right)^2} \approx 2\pi \left(\frac{1}{\lambda} - \frac{m^2}{z_T}\right), \quad m = 0, \pm 1, \pm 2, ..., \tag{14}$$

with the Talbot self-image distance $z_T = 2p_z^2/\lambda$. The sidewall grating imposes additional longitudinal periodicity on the field with period p_z, and according to the Montgomery equations, the k-vectors of the wave field obey the condition[43,44]

$$u_z(x) = u_{z+p_z}(x) \Rightarrow u_0(x) = \sum_m A_m \exp\left(ik_x^m x\right). \tag{15}$$

The k_x components have a discrete angular spectrum given by

$$k_x^l = 2\pi\sqrt{\left(\frac{1}{\lambda}\right)^2 - \left(\frac{l}{p_z}\right)^2}, \quad l = \pm1, \pm2, \ldots \tag{16}$$

The coefficients A_m depend on the groove shape. The transmitted field satisfies the resonance conditions both for the WG, equation (14), and for the grating, equation (16). Mathematically these conditions can be written as

$$\left(k_x^m\right)^2 + \left(k_z^l\right)^2 = \left(\frac{2\pi}{\lambda}\right)^2, \tag{17}$$

or a graphical explanation based on the Ewald sphere can be used.[45] Consequently, choosing the parameters appropriately, it is possible to suppress some modes, resulting in selective space filtering of the field passing through the structured WG, both for coherent and incoherent illumination. Because of the small resonance angles for the WG, the period of the side-wall grating can be significantly large with respect to the wavelength. In figure 9a coherent mode mixing is shown for a simple planar WG with vacuum gap $d = 100$ nm and gold cladding. The WG is illuminated with a plane monochromatic wave, parallel to the optical axis, with wavelength 0.1 nm. Figure 9b shows the intensity distribution for a planar WG with the same parameters as in figure 9a but with an additional longitudinal periodicity (structured WG) created with a step-like (Ronchi) grating on the sidewalls of the guiding layer. The grating has period equal to the fractional Talbot distance $d_{\mathrm{eff}}^2/\lambda = 110$ μm and duty cycle 0.3.

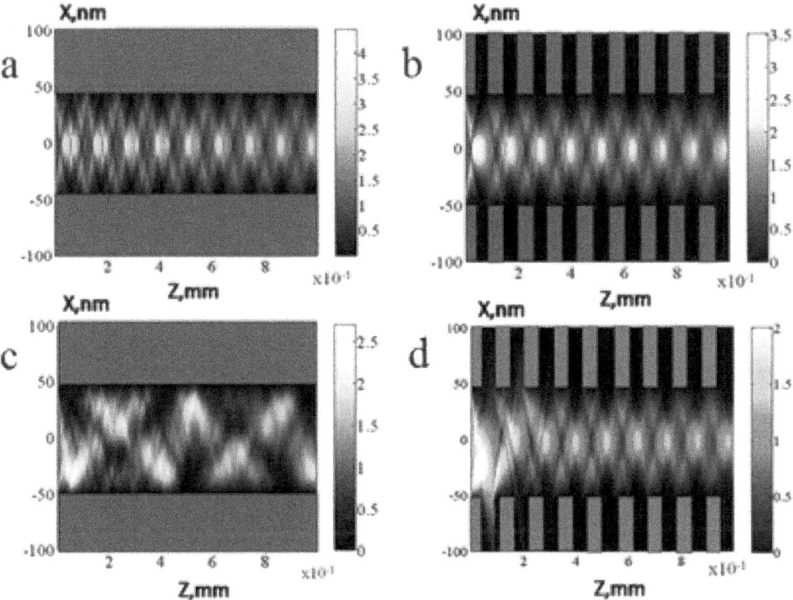

Figure 9 *Simulated intensity distributions in the guiding layer of the WG for a coherent source – (a) plane WG and (b) structured WG – and an incoherent source – (c) plane WG and (d) structured WG (reprinted with permission).*[28]

The calculations show that the structured WG with the given parameters preserves the low-order symmetric modes of the planar WG. The incoherent illumination case is presented in figures 9c and 9d, which show the distribution of intensity in the guiding layer for the normal and structured WG respectively, illuminated by a spatially incoherent beam provided by a random X-ray source of size $S = 15$ μm located at a distance $r = 1$ mm from the WG. The WG was tilted at an angle $\alpha \sim \lambda/2d \approx 5 \times 10^{-4}$ rad (i.e., of the order of the angular width of the WG mode) with respect to the optical axis. In this case a large set of modes was incoherently excited in the guiding layer. In fact, the coherence length $h = \lambda r/S$ of the incoming X-ray beam at the entrance aperture of the WG is ≈6.7 nm, much less than the 100 nm width of the guiding layer. The visibility of the resonances is therefore strongly reduced. In the simulation, the incoherent source was taken as the sum of elementary radiators with random phases. The obtained quasi-incoherent field is the average over many realizations of the random process. Figure 9d shows that, unlike the planar WG which transmits the partially coherent beam over a long distance, the periodic WG works as a filter for the coherent part, allowing only coherent radiation and lo- order symmetric modes to be transmitted.

This reflects in the transverse distribution of intensity at the WG exit, shown in figure 10 for the planar WG (dashed line) and the periodic WG (full line) for the case of asymmetric incoherent illumination. The distribution for the planar WG is wide and strongly asymmetric due to asymmetric modes in the spatial spectrum of the WG, while the distribution for the structured WG has a Gaussian-like symmetric shape due to suppression of asymmetric and high-order symmetric modes in the spectrum.

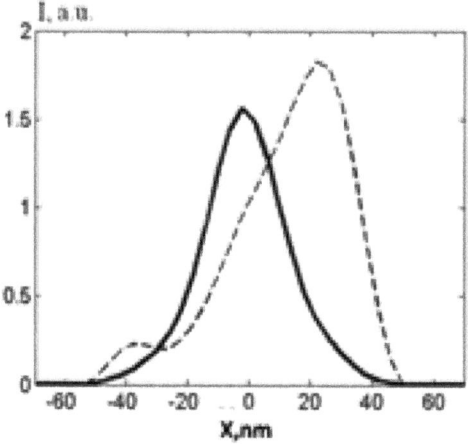

Figure 10 *Transverse intensity distribution at the WG exit, in case of asymmetric incoherent illumination for planar (dashed line) and structured (full line) WGs (reprinted with permission).*[28]

The total intensity at the exit of the periodic WG is reduced with respect to the simple planar WG, because the former filters out the incoherent part of the radiation and suppresses some modes. It should be mentioned that the WG filtering properties are similar to the hollow-WG resonators with external flat mirrors developed for filtering of incoherent radiation.

The size of the beam at the waveguide exit can be reduced by tapering the reflective walls (see figure 11). The filtering properties of the waveguides are provided by a proper choice of the grating period.

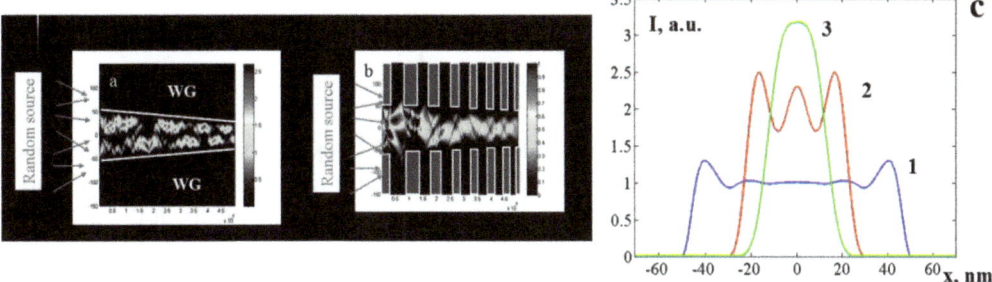

Figure 11 *(a) Propagation of the incoherent wave in a tapered WG and (b) in a tapered WG with a grating. The entrance gap is 100 nm and that of the exit is 60 nm. (c) Curves 1 and 2 are the lateral intensity distributions corresponding to the intensity distribution given in figure 8c, and curve 3 is the lateral intensity distribution at the exit of the tapered structured WG.*

Despite the very narrow width (100 nm) of the guiding channel, the period of the grating is much larger than the wavelength, with a characteristic value of about 100 μm. Therefore the realization of such devices does not present significant technological difficulties using standard tools for micro- and nano-fabrication.

6. PHASE CONTRAST IMAGING WITH PLANAR WAVEGUIDES AND A LABORATORY SOURCE

The previous sections have shown that WGs are effective optical elements able to provide nanometre scale coherent beam even with laboratory sources. To demonstrate the experimental potential of WGs phase contrast imaging has been carried out using a known test pattern to measure the spatial resolution.[46] Figure 12 shows the experimental arrangement using an X-ray microfocus source (NOVA 600 from Oxford Instruments), with a copper anode (wavelength $\lambda = 0.154$ nm) and a nominal source size of $w = 15$ μm, to illuminate the WG at about 1 cm from the source. The sample was at a distance z_1 from the WG exit, and the detector at a distance z_2 from the sample.

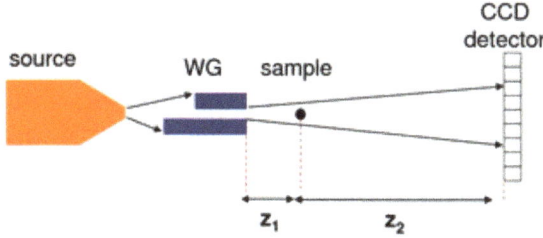

Figure 12 *Experimental setup for projection X-ray microscopy with a WG.*

In the configuration of figure 12 a magnified image of the sample is projected onto the detector. The magnification factor is $M = (z_1 + z_2)/z_1$. In this way the WG acts as a sub-micrometre sized secondary source, improving both spatial resolution and coherence properties of the incident beam.[47] In projection geometry it can be shown[48] that an estimate for the overall resolution is

$$r_{\text{tot}} = \sqrt{\left(s\frac{z_2}{z_1+z_2}\right)^2 + \left(PSF\frac{z_1}{z_1+z_2}\right)^2}, \qquad (18)$$

where s and PSF are, respectively, the FWHMs of the WG exit intensity distribution and the detector point spread function. If $s \ll PSF$, as in the case for WGs, the best resolution is obtained for $z_1 \ll z_2$. On the other hand, large z_2 means low flux density at the detector, and low z_1 implies a limited field of view.

In propagation-based phase-contrast imaging the propagation distance plays a crucial role. Three main imaging regimes are usually considered: near-field, Fresnel and Fraunhofer. Being in one or another depends on the ratio R between the defocusing distance, defined as $z_{\text{def}} = z_1 z_2 / (z_1 + z_2)$, and the Fresnel distance $f = d^2 / 4\lambda$, where d is the typical dimension of the illuminated sample features. The current case corresponds to the near-field regime, where $R \ll 1$ and the maximum contrast is obtained at the sample edge discontinuity.[49] In the experiments reported here different planar asymmetric front-coupling WGs with vacuum core and Silicon claddings were used.[23] The length of the WG devices was 5 mm. The virtual source was then a line and the magnification effect took place in one dimension only.

It should be noted that the system also provided a broad energy filter for the incoming radiation. The WG works in total reflection conditions, thus the high-energy Bremsstrahlung from the source was not reflected but mostly absorbed by the cladding. The transmitted radiation is thus Cu K_α ($(\Delta\lambda/\lambda\sim10^{-3})$) with extremely low high-energy background. Finally, an estimate of the transverse coherence length $l_c \approx \lambda L / w$ of the radiation at the WG entrance was ≈ 100 nm, using a WG with nominal gap size of about 300 nm. Thus this configuration provided a partial coherent illumination at the WG entrance. Nevertheless the degree of coherence at the sample, because of the small dimension of the beam at the WG exit, was enough to show phase contrast effects. The sample stage, moveable along the longitudinal direction, allowed the geometrical magnification to be changed. A CCD detector with nominal pixel size of 12.3 μm and a measured PSF of 18 μm was used.

Table 1 gives some relevant experimental parameters: the WG-sample distance z_1, the corresponding defocusing distance D, approximately equal to z_1, the coherence length l_c at the sample position, the magnification M and the nominal spatial resolution r_{tot} (see equation 18). The total distance z_1+z_2 was fixed at 269 mm.

Table 1 *Relevant experimental parameters for the phase contrast imaging experiments: z1, WG-sample distance; D, defocusing distance; l_c, coherence length at the sample position; M, magnification factor; r_{tot}, spatial resolution.*

z_1(mm)	D(mm)	l_c (μm)	M	r_{tot} (nm)
1.20	1.19	0.92	224	210
1.70	1.69	1.31	158	230
2.70	2.67	2.08	100	270

Figures 13a, 13b and 13c show the measured and calculated X-ray phase contrast profiles at the three z_1 distances of table 1, and figure 13d is a scanning electron microscope image of the test pattern.

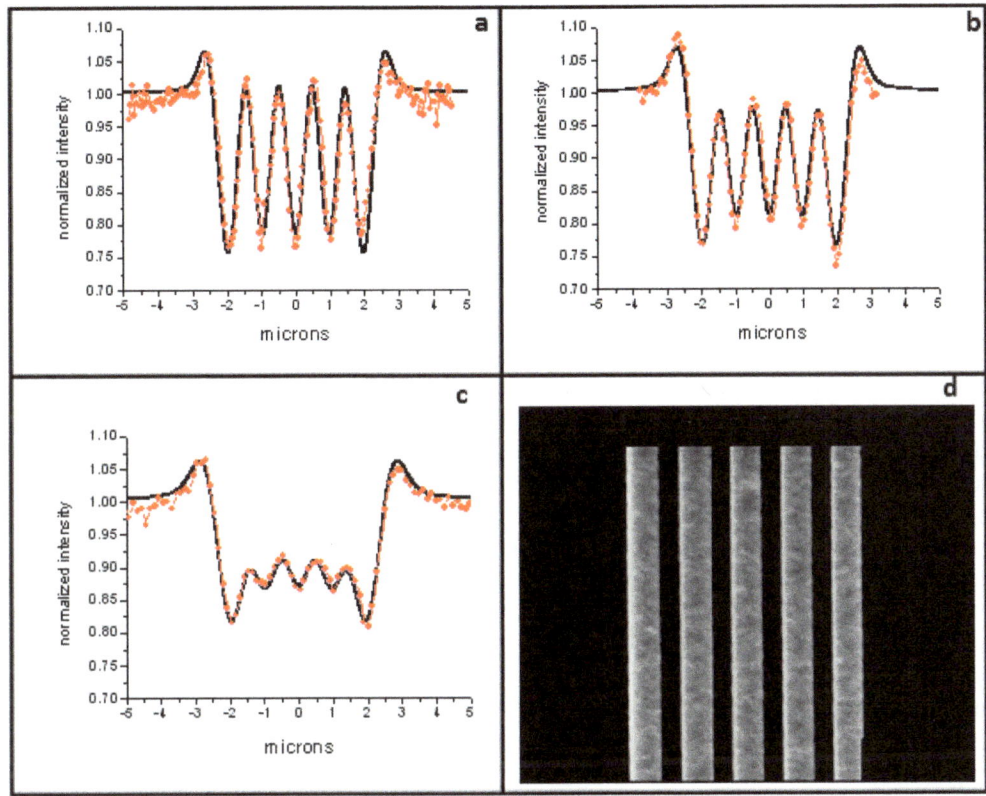

Figure 13 *Measured intensity profiles (red dots) for $z_1 =$ (a) 1.2 mm, (b) 1.7 mm and (c) 2.7 mm compared with the corresponding analytical calculations (black continuous lines). (d) SEM image of the test structure.*

Due to the limited field of view at the sample plane a scan of the sample in the WG beam was necessary. Therefore the corresponding measured intensity profiles are merged sequences of five intensity profiles. Moreover, in order to minimize the statistical error, each intensity profile is the average of five measures. The acquisition time for each measurement was 100 s. For all distances the defocusing distance was significantly smaller than the Fresnel distance (6 mm) for the structure used. Therefore, all images were taken in the near-edge regime. The numerical calculations (black lines in figure 13) are based on the Fresnel-Kirchhoff integral,[19] taking into account the actual resolution of the system estimated using equation (18). The size s of the WG secondary source was approximated by an effective Gaussian distribution. The very good agreement between the measured and the calculated profiles confirms that a spatial resolution of the order of 200 nm is attainable even with laboratory sources.

7 CONCLUSIONS

This chapter has presented a comprehensive description of beam propagation in plane and linearly tapered standard and structured waveguides with different illuminating sources (synchrotron radiation facilities and laboratory sources), and has demonstrated the applicability of X-ray WGs in laboratory environments for phase-contrast imaging with spatial resolutions of few hundred nanometres.

The theoretical and experimental properties of X-ray beams exiting from air gap waveguides have been presented for the case of front coupling in combination with a pre-reflection scheme. Both constant-gap and tapered waveguides have been taken into account. On the basis of this theoretical analysis, a computer code was developed which provides a fast and simple comparison of the experimental results. Simple equations also provide a way to estimate the essential parameters of the WG under examination without computer simulations, i.e. the gap and angle of tapering. During fabrication, some tools can be used to control the WG surface (scanning electron microscopy, atomic force microscopy), but the best way to characterize WG properties is with direct X-ray measurements. It has been shown that a computer simulation on the basis of experimental data can provide the structural parameters of the WG, with good agreement with experimental results taken both with synchrotron radiation, and with a laboratory micro-focus source.[23] The possibility of reconstructing a one-dimensional complex wave field from a single measurement of its diffraction pattern, obtained in the far field, was alos demonstrated. The procedure involved the repeated application of an iterative phase retrieval algorithm, with different random starting phases. The ambiguities inherently present in the one-dimensional problem were reduced by *a posteriori* treatment of the single reconstructions. The method has been verified on the experimental diffraction pattern measured in the far field of an X-ray waveguide. Both amplitude and phase were reconstructed in good agreement with the theoretical prediction and with the independent analytical method.

Furthermore, the use of waveguides was proposed with added longitudinal periodicity on the reflecting side walls of the guiding layer to suppress the asymmetric and the high order modes and to introduce selective band gaps in the angular spectrum.

In practice the structured WG only allows low-order coherent modes to be transmitted, from both incoherent and partially coherent sources. This feature can be particularly useful for imaging with laboratory X-ray sources whenever coherence is critical for image quality. Moreover, mode filtering by WGs can be crucial in CXDI experiments with synchrotron radiation, when optical elements, used for beam focusing, can corrupt the degree of coherence of the field impinging on the sample.

Finally, using a test pattern, it was demonstrated that phase-contrast imaging with laboratory X-ray sources equipped with a WG can provide spatial resolution in one direction of about 200 nm.

References

1 Y.P. Feng, S.K. Sinha, E.E. Fullerton, G. Grubel, D. Abernathy, D.P. Siddons and J.B. Hastings, *Appl. Phys. Lett.*, 1995, **67**, 3647.
2 S. Lagomarsino, W. Jark, S. Di Fonzo, A.Cedola, B. Mueller, P. Engstrom and C. Riekel, *J. Appl. Phys.*, 1996, **79**, 4471.
3 F. Pfeiffer, C.David, M. Burghamme, C. Riekel and T. Salditt, *Science*, 2002, **297**, 230.
4 S. Lagomarsino and A. Cedola, in *Encyclopedia of Nanoscience and Nanotechnology*, (ed. HS Nalwa, American Scientific Publishers), 2004, p681.

5 C. Ollinger, C. Fuhse, A. Jarre and T. Salditt, *Physica B*, 2005, **357**, 53.

6 C. Fuhse, C. Ollinger and T. Salditt, *Phys. Rev. Lett.*, 2006, **97**, 254801.

7 K. Giewekemeyer, H. Neubauer, S. Kalbfleisch, S. P. Krüger and T. Salditt, *New J. Phys.*, **12**, 035008.

8 S. Lagomarsino, A. Cedola, P. Cloetens, S. Di Fonzo, W. Jark, G. Soullié and C. Riekel, *Appl. Phys. Lett.*, 1997, **71**, 2557.

9 S. Di Fonzo, W. Jark, S. Lagomarsino, C. Giannini, L. De Caro, A. Cedola and M. Muller, *Nature*, 2000, **403**, 638.

10 L. De Caro, C. Giannini, A. Cedola, D. Pelliccia, S. Lagomarsino and W. Jark, *Appl. Phys. Lett.*, 2007, **90**, 041105.

11 L. De Caro, C. Giannini, D Pelliccia, C. Mocuta, T.H. Metzger, A. Guagliardi, A. Cedola, I. Bukreeva and S. Lagomarsino, *Phys. Rev. B*, 2008, **77**, 081408.

12 J.W. Miao, T. Ishikawa, B. Johnson, E.H. Anderson, B. Lai and K.O. Hodgson, *Phys. Rev. Lett.*, 2002, **89**, 088303.

13 S.K. Sinha, M. Tolan and A. Gibaud, *Phys. Rev. B*, 1998, **57**, 2740.

14 I.A. Vartanyants and I.K. Robinson, *J. Phys.: Condens. Matter*, 2001, **13**, 10593.

15 G.J. Williams, H.M. Quiney, A.G. Peele and K.A. Nugent, *Phys. Rev. B*, 2007, **75**, 104102.

16 H.M. Quiney, K.A. Nugent and A.G. Peele, *Opt. Lett.*, 2005, **30**, 1638.

17 M.J. Zwanenburg, J.F. Peters, J.H.H. Bongaerts, S.A. de Vries, D.L. Abernathy and J.F. van der Veen, *Phys. Rev. Lett.*, 1999, **82**, 1696.

18 D. Pelliccia, I. Bukreeva, M. Ilie, W. Jark, A. Cedola, F. Scarinci and S. Lagomarsino, *Spectrochim. Acta B*, 2007, **62**, 615.

19 W. Jark, A. Cedola, S. Di Fonzo, M. Fiordelisi, S. Lagomarsino, N.V. Kovalenko and V.A. Chernov, *Appl. Phys. Lett.*, 2001, **78**, 1192.

20 D. Marcuse, *Theory of Dielectric Optical Waveguides*, 1991, (San Diego, Academic Press).

21 C. Bergemann, H. Keymeulen and J.F.van der Veen, P*hys. Rev. Lett.*, 2003, **91**, 204801.

22 S. Panknin, A.K. Hartmann and T. Salditt, *Opt. Commun.*, 2008, **281**, 2779.

23 I.Bukreeva, D. Pelliccia, A. Cedola, F. Scarinci, C. Giannini, L. De Caro and S. Lagomarsino, *J. Synchrotron Rad.*, 2010, **17**, 61.

24 J.R. Fienup, *Appl. Opt.*, 1982, **21**, 2758.

25 R.H.T. Bates, *Optik*, 1982, **61**, 247.

26 D. Pelliccia, D.M. Paganin, A. Sorrentino, I. Bukreeva, A. Cedola and S. Lagomarsino, *Opt. Lett.*, 2012, **37**, 262.

27 T.A. Bobrova and L.I. Ognev, *Technical Physics Letters*, 2000, **26**, 1027.

28 L.I. Ognev, *X-Ray Spectr.*, 2002, **31**, 274.

29 I. Bukreeva, A.Cedola, A.Sorrentino, D.Pelliccia, V.Asadchikov and S.Lagomarsino, *Opt. Lett.*, 2011, **36**, 2602.

30 O. Bryngdahl, *J. Opt. Soc. Am.*, 1973, **63**, 41.

31 R. Ulrich, *Nouv. Rev. Opt.*, 1975, **6**, 253.

32 Y.V. Kopylov, A.V. Popov and A.V. Vinogradov, *Opt. Commun.*, 1995, **118**, 619.

33 H.F. Talbot, *Phil. Mag.*, 1836, **9**, 401.

34 S.V. Kukhlevsky and G. Nyitray, *Opt. Commun.*, 2003, **218**, 213.

35 S.V. Kukhlevsky, *X-Ray Spectr.*, 2003, **32**, 223.

36 S.V. Kukhlevsky and G. Nyitray, *J. Modern Opt.*, 2003, **50**, 2043.

37 C. David, B. Nohammer, H.H. Solak and E. Ziegler, *Appl. Phys. Lett.*, 2002, **81**, 3287.

38 K. Giewekemeyer, H. Neubauer, S. Kalbfleisch, S.P. Krüger and T Salditt, *New J. Phys.*, 2010, **12**, 035008.

39 T.A. Bobrova and L.I. Ognev, *JETP Letters,,* 1999, **69**, 10, 734.

40 L.I. Ognev, *Tech. Phys. Lett.*, 2010, **36**, 133.

41 L.I. Ognev, *Technical Physics*, 2010, **55**, 1628.

42 M. Osterhoff and T. Salditt, *Opt. Commun.*, 2009, **282**, 3250.

43 J. Jahns and A. W. Lohmann, *Appl. Opt.*, 2009, **48**, 3438.

44 S.F. Helfert, B. Huneke and J. Jahns, *J. Eur. Opt. Soc. Rapid Pub.*, 2009, **4**, 09031.

45 A.W. Lohmann, H. Knuppertz and J. Jahns, *J. Opt. Soc. Am. A*, 2005, **22**, 1500.

46 D. Pelliccia, A. Sorrentino, I. Bukreeva, A. Cedola, F. Scarinci, M. Ilie, A.M. Gerardino, M Fratini and S. Lagomarsino, *Opt. Express.*, 2010, **18**, 15998.

47 L. De Caro, C. Giannini, S. Di Fonzo, W. Jark, A. Cedola and S. Lagomarsino, *Opt. Commun.*, 2003, **217**, 31.

48 L. De Caro, A. Cedola, C. Giannini, I. Bukreeva and S. Lagomarsino, *Phys. Med. Biol.*, 2008, **53**, 6619.

49 P. Cloetens, R. Barrett, J. Baruchel, J.-P. Guigay and M. Schlenker, *J. Phys. D Appl. Phys.*, 1996, **29**, 133.

TABLE-TOP SOFT X-RAY Ar^{+8} LASERS EXCITED BY CAPILLARY Z-PINCHES

J. Szasz,[1] M. Kiss,[1] I. Santa,[1] S. Szatmari[2] and S.V. Kukhlevsky[1]

[1]Department of Physics, University of Pecs, Hungary
[2]Department of Experimental Physics, University of Szeged, Hungary

1 INTRODUCTION

Coherent soft X-ray sources with wavelengths in the range $\lambda \approx 0.1$–50 nm are required in many areas of science and technology. Synchrotrons, free-electron lasers, laser-induced plasmas and high-harmonic generators are examples of effective soft X-ray sources, [1,2,3,4] but such sources are usually very expensive and complex. Hence many small laboratories, universities and institutes are interested in more practical, simpler, compact (table-top) and inexpensive sources of soft X-rays. One of the most practical soft X-ray lasers is currently considered to be the Ar^{+8} laser excited by a capillary discharge z-pinch. [5,6,7,8,9,10,11,12,13,14,15,16]

Z-pinch discharges, which previously have been investigated mainly as drivers for thermonuclear fusion, are used in Ar^{+8} lasers for generation and excitation of the hot (temperature $T_e \sim 0.1$ keV) and dense ($N_e > 10^{18}$ cm^{-3}) argon plasma as the active laser medium. In this laser pumping scheme a hot and highly ionized plasma active medium with a diameter of about 500 μm is produced by high-current electric pulses with short rise times (a few tens of nanoseconds) flowing axially through a capillary channel filled with low pressure argon. The plasma temperature and density increase through fast radial compression of the plasma column by the magnetic field created by the current itself. The lasing takes place in the 46.9 nm line of $2p^5 3p$ ($J = 0$)–$2p^5 3s$ ($J = 1$) transition of neon-like argon (Ar^{+8}) through collisional excitation of the ion by hot electrons. Recently considerable effort has been devoted to reducing the laser size from laboratory to table top. In the laboratory-size laser,[5] a capillary z-pinch with a peak current $I \sim 40$ kA is produced by a water capacitor with $C \sim 5$ nF that is charged to a high voltage of up to about 600 kV by a pulsed Marx generator. The output pulse energy of such a laser is up to about 1 mJ. Table-top lasers have been developed by using low-inductance coaxial discharge configurations that decrease the voltage and current necessary for the laser excitation.[6,10] In order to achieve this, a water or ceramic capacitor is charged to a relatively low voltage (100–300 kV) by a Marx-generator or a simple single-stage power unit providing the peak current (10–22 kA) required for saturated laser operation. Laser amplification in the most compact and effective Ar^{+8} laser[17] was obtained in a 21 cm long aluminium oxide ceramic capillary with inside diameter 3.2 mm filled with pre-ionized argon at a pressure of about 0.9 mbar. Laser pulses with energy around 13 μJ were generated at repetition rates of up to 12 Hz. The laser beam profile typically had an annular shape with an angular divergence of about 7 mrad.

The present review concentrates on research and development of a practical, table-top Ar^{+8} laser excited by relatively low current (< 22 kA) and voltage (< 200 kV) capillary z-pinch discharges. In accordance with previous studies, the main motivation in the research and development of table-top Ar^{+8} lasers is the reduction of the overall size from laboratory to table top. Other important motivations include the peculiarities of the physics of capillary Ar^{+8} lasers in the different operation regimes. Laboratory size lasers excited by capillary z-pinches with a peak current of around 40 kA using a high voltage of about 600 kV can produce laser pulses with high energy, up to ~1 mJ, but with low beam quality. The beam profiles are typically annular with divergences of around 7 mrad and low transverse coherence. Most laser applications require a transversally coherent beam with a Gaussian intensity distribution and angular divergence less than about 1 mrad. Laser beams with such parameters can be obtained by using small apertures to define very small ($\sim 1/10^2$) areas of the 1 mJ. The energy of such a coherent, low-divergence beam would be ~10 μJ. A laser beam with such parameters could be obtained directly by using a relatively low voltage (100–200 kV) and peak current (10–22 kA).

In the present study, a transversally coherent beam with a Gaussian intensity distribution, angular divergence less than about 1 mrad and energy ~10 μJ was generated using a long (0.5 m) capillary; Ar^{+8} lasers are usually excited by z-pinch discharges in short (≈ 20 cm) capillaries. The longer plasma column provides a gain-length product GL of about 16, as required for saturated laser operation, at a relatively low gain of $G \approx 0.3$ cm^{-1}. Such a low gain can be produced at relatively low plasma temperature and density using a low voltage and current. For comparison, Ar^{+8} lasers[10,17] excited by capillary z-pinches in short (21 cm) capillaries by using a peak current of 22 kA and a voltage of 100 kV are saturated at high gains, $G \approx 0.7$ cm^{-1}. Such gains require high plasma temperature and density, meaning high voltage and current. The refraction of amplified radiation by radial electron density gradients in the high density column of radius $R_{pl} \approx 0.15$ mm results in the high divergence (≈ 7 mrad) and annular profile of the laser beam. The advantage of the low-density long plasma column is the potential decrease of beam divergence and increased transverse coherence. Indeed, the use of a long ($L \approx 0.5$ m) plasma column of radius $R_{pl} \approx 0.25$ mm with low electron density and radial gradients would provide almost refraction-free laser operation in the super-fluorescent mirror-less mode, where the beam angular divergence ($\varphi \sim R_{pl}/L \approx 0.5$ mrad) is determined by the radius R_{pl} and length L_{pl} of the plasma column rather than by the refraction of amplified rays[11].

It is generally accepted that operation of Ar^{+8} lasers requires the use of an external low-current circuit to provide pre-ionization of the argon gas before the main excitation current of about 20 kA. The crucial role of the pre-ionization current, of ≈ 20 A and duration ≈ 5 μs generated by the external circuit, has been demonstrated previously.[5,6,7,8,9,10,11,12,13,14,15,16,17] The present study demonstrates Ar^{+8} laser operation without using any external low-current pre-ionization circuit. Instead, the pre-ionization of the argon gas was provided by automatic pre-ionization via the so-called gliding discharge on the internal surface of the capillary driven by the main excitation circuit.[18,19,20] Such a technique considerably simplifies the laser device; the use of the relatively low voltage (200 kV) of the main excitation pulse allowed the use of a gas (SF_6 or air at atmospheric pressure) as an electrical insulator in the Marx generator instead of the unpractical biodegradable transformer oil. The reduction of the voltage and current of the main excitation pulse also reduced the ablation of the electrodes[21,22,23] and capillary walls[24,25,26] increasing their lifetimes.

The rest of the chapter is organized as follows. Section 2 describes the basic physical operation principles of Ar^{+8} lasers. The main physical processes in the laser medium (capillary z-pinch plasma) are described. The main plasma parameters, such as the temperature, density, length, diameter and homogeneity required for the saturated operation of the Ar^{+8} lasers are analyzed. The methods and techniques providing the experimental realization of the plasma parameters required for Ar^{+8} laser operation are described in section 3. The experimental results are also presented with a brief discussion of perspectives for future research.

2 THEORETICAL BACKGROUNDS OF Ar^{+8} LASERS

The present section briefly describes the basic physical processes of Ar^{+8} lasers.

2.1 Lasing Mechanism (Atomic Kinetic Code)

Here the lasing mechanism in Ar^{+8} lasers is described by means of an atomic kinetic code.[27] In such lasers, the lasing takes place on the $2p^5 3p(J = 0)$–$2p^5 3s(J = 1)$ transition of neon-like argon at a wavelength of 46.9 nm (figure 1).

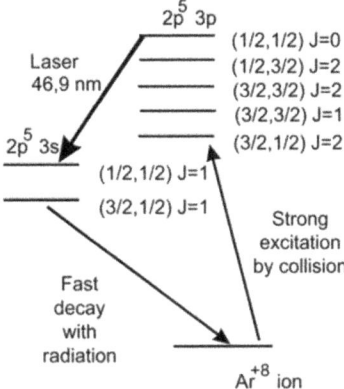

Figure 1 *Lasing in the 46.9 nm line of the $2p^5 3p(J = 0)$–$2p^5 3s(J = 1)$ transition of neon-like argon (Ar^{+8})*

The pure gain G_P is related to the upper and lower laser state population densities N_u and N_l by

$$G_P = \frac{\lambda_{ul}^4}{8\pi c \Delta\lambda_{ul}} A_{ul}\left(N_u - \frac{g_u}{g_l}N_l\right), \tag{1}$$

where λ_{ul} is the wavelength of the transition, c is the light velocity, $\Delta\lambda_{ul}$ is the line half width, which is determined by Doppler and Stark effects, A_{ul} is the probability of the spontaneous transition and g_u and g_l are statistical weights of the energy levels. The pure gain is reduced by the refraction index due to the electron density gradient in the radial direction. In a plasma, the refraction index n is related to the electron density N_e, $n = \left[1 - (N_e / N_{ec})\right]^{1/2}$, where $N_{ec} = \pi m_e c^2 / e^2 \lambda_{ul}^2$ is the critical electron density at which the plasma frequency equals the X-ray frequency. Here, e and m_e are the electron charge and mass. The gain when reduced by the refraction losses in the cylindrical plasma column is given by

$$G = G_{\mathrm{P}} - F\left(r_{\mathrm{pl}} \sqrt{\frac{N_{\mathrm{ec}}}{N_{\mathrm{e0}}}} \right)^{-1}, \qquad (2)$$

where N_{e0} is the maximum electron density in the plasma column, r_{pl} is the plasma column radius and $F = 2.347$ is the factor associated with the geometry of the cylindrical column.

The upper and lower laser state population densities N_{u} and N_{l} can be calculated with high accuracy by solving the system of coupled rate equations – see equation (3) – for the 27 levels ($j = 0, 1, ..., 26$) of the $2p^6$, $2p^5 3s$, $2p^5 3p$ and $2p^5 3d$ energy configuration of Ar^{+8}. The rate equation for the population of an energy level j of neon-like argon is

$$\frac{dN_j}{dt} = -N_j \left[\sum_{i<j} \left(A_{ji} + P_{ji}^{\mathrm{d}} \right) + \sum_{i>j} P_{ji}^{\mathrm{e}} \right] + \sum_{i>j} N_i \left(A_{ji} + P_{ji}^{\mathrm{d}} \right) + \sum_{i<j} N_i P_{ij}^{\mathrm{e}}, \qquad (3)$$

where A, P^{e}, and P^{d} represent the spontaneous radiative decay and collisional excitation and de-excitation rates, respectively. The level populations are significantly affected by radiation trapping in optically thick lines. The probability of escape method can be used to approximate the trapping effect by effective reduction of the original radiative decay rates A by an escape factor $E(\tau)$. The effective decay rate of a relevant transition is given by $E(\tau)A$, where τ is the opacity of the radiative transition between the levels i and j; $\tau_{ij} = k_{ij} r_{\mathrm{pl}}$, where $k_{ij}(\lambda)$ is the absorption coefficient. The population of the levels j described by equation (3) has a transient character. Nevertheless, the quasi steady-state solution of the system of coupled equations is a very good approximation since equilibrium between the energy levels is readily established due to the short lifetimes of the dominant radiative transitions. It should be noted that increasing the number of energy levels in the calculations from 27 to 89 by including the high Rydberg states $2s2p^6 3\ell$ ($\ell = $ s,p,d: 10 levels) and $2p^5 4\ell$ ($\ell = $ s,p,d,f: 52 levels) has a negligible effect on the gain value.

To solve of the system of rate equations the density of Ar^{+8} ions in the ground state N_0^{+8} has to be determined. This density is a few orders of magnitude higher than that of the excited ions ($N_0^{+8} \gg N_{j>0}^{+8}$). The total density of Ar^{+8} ions in all states, $N^{+8} = \sum_j N_j^{+8}$, is approximately equal to N_0^{+8}. The density N^{+8} is determined by the system of coupled rate equations for the balance between ionization and recombination of all the Ar^{+k} ions,

$$\frac{dN^k}{dt} = N_e \left[N^{k-1} I^{k-1} + N^{k+1} \left(R_{rr}^{k+1} + R_{dr}^{k+1} + R_{cr}^{k+1} \right) - N^k I^k - N^k \left(R_{rr}^k + R_{dr}^k + R_{cr}^k \right) \right], \qquad (4)$$

where k ($= 0,1,...,18$) is the ionization stage, I is the ionization rate and R_{rr}, R_{dr} and R_{cr} are the radiative, dielectronic and collisional recombination rates, respectively. The quasi steady state solution ($dN^k/dt = 0$) of the system of coupled equations (4) is reasonable approximation. The ionization and recombination via excited states are usually neglected. Calculation of the density N_0^{+8} under such an approximation can be separated from the solution of the system of equations (3).

The solution of the system of coupled equations (4) gives the densities of the ions in different ionization states. Fractional ion abundances $N^k / \sum N^k$ calculated for different electron kinetic temperatures by using the simplest atomic data (R_{rr}, R_{dr} and R_{cr}) published in the literature[28,29] are shown in figure 2. Note that the Ar^{+8} ions exist in the temperature range of about 50–200 eV with a maximum density at about 100 eV. It should be noted that the use of more comprehensive atomic data[30,31,32] for R_{rr}, R_{dr} and R_{cr} in the model produces a shift to lower temperatures of about 20 eV in the distributions of figure 2, the Ar^{+8} ions existing in the temperature range of about 40–200 eV with a maximum density at around 60–80 eV. These values are consistent with the previous calculations,[33,34] which give the temperature range of about 30–200 eV with the maximum density at ~ 80 eV.

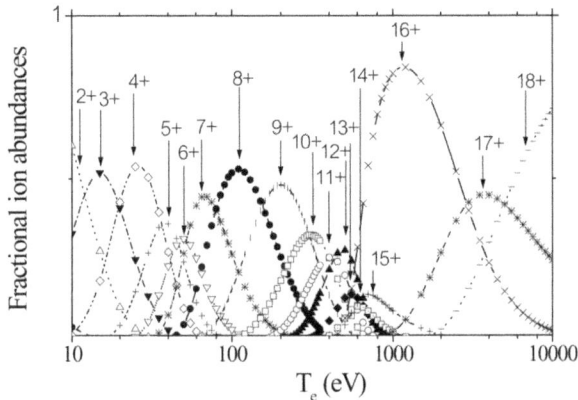

Figure 2 *Fractional ion abundances calculated vs electron kinetic temperatures.[27]*

The gain G in the 46.9 nm line of $2p^53p(J = 0) - 2p^53s(J = 1)$ transition of neon-like argon calculated using equations (1–4) as a function of the electron kinetic temperature T_e and density N_e are shown in figure 3 for two plasma radii. A negative gain corresponds to absorption of the radiation. Figure 3 shows that positive gain exists in the relatively narrow region $T_e \approx 60$–150 eV and $N_e \approx 0.5$–10×10^{18} cm^{-3}. The gain in the plasma column with radius 0.3 mm is considerably lower than in the column of 0.15 mm radius due to radiation trapping in the optically thick lines. Using the more comprehensive atomic data that produce the 20 eV shift to lower temperatures, positive gain exists for $T_e \approx 40$–200 eV and $N_e \approx 0.4$–10×10^{18} cm^{-3}. Ar^{+8} lasers are usually excited by z-pinch discharges in short, ≈ 20 cm, capillaries. A long, ≈ 50 cm, plasma column provides a gain-length product $GL \approx 16$, as required for saturated laser operation, at the relatively low gain of about 0.3 cm^{-1}.

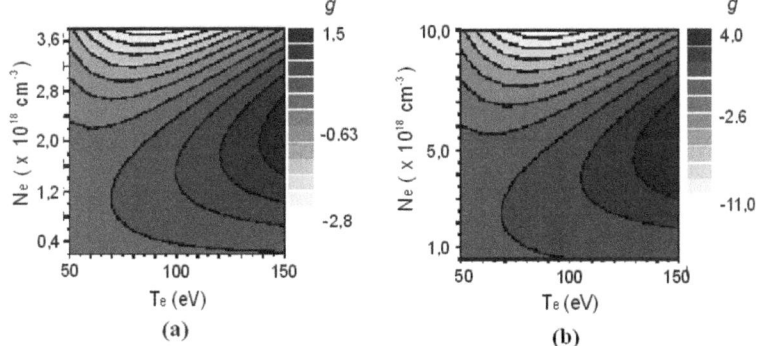

Figure 3 *Gain in the $2p^53p(J = 0)$–$2p^53s(J = 1)$ transition ($\lambda = 46.9$ nm) of Ar^{+8} as a function of the plasma temperature and density for two plasma radii: (a) 0.3 mm and (b) 0.15 mm[27]*

Figure 3 shows that such a low gain can be produced at relatively low plasma temperature and density ($T_e \approx 60$ eV, $N_e \approx 0.5 \times 10^{18}$ cm^{-3}). For comparison, Ar^{+8} lasers[17] excited by a capillary z-pinch in short (≈ 20 cm) capillaries are saturated at higher gains of about 0.7 cm^{-1}. Such a gain requires higher plasma temperature, $T_e \approx 80$–100 eV, and $N_e \approx 10^{18}$ cm^{-3}, as can be seen from figure 3.

The voltage and current required for the production of the plasmas with such temperatures and densities are analyzed in the following section using the magneto-hydrodynamic (MHD) model.

2.2 The Magneto-Hydrodynamic Model

Z-pinch discharges, previously investigated mainly as drivers for thermonuclear fusion, are now used for the generation and excitation of hot ($T \sim 100$ eV) dense ($N_e \sim 10^{18}$ cm^{-3}) plasmas to provide an active medium for Ar^{+8} lasers. The hot and highly ionized plasma active medium, typically with a diameter of around 0.5 mm, is produced by high-current pulses with short rise-times (a few tens of nanoseconds) flowing axially through a capillary channel filled with low-pressure argon. The plasma temperature and density are increased through the fast radial compression of the plasma column by means of the magnetic field created by the current itself. A z-pinch discharge compressed by the Lorentz force $\mathbf{J} \times \mathbf{B}$, where $\mathbf{J}$ is the electrical current density and $\mathbf{B}$ is the magnetic field induced by the current, is shown schematically in figure 4.

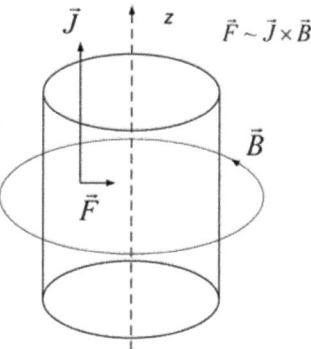

Figure 4 *A z-pinch discharge compressed by the Lorentz force* $\boldsymbol{F} = \boldsymbol{J} \times \boldsymbol{B}$.

The temporal and spatial evolution of the plasma parameters in a capillary z-pinch are usually described with good accuracy by using the standard one-fluid, one-dimensional, two-temperature magneto-hydrodynamic (MHD) model.[35,36,37,38.39,40] The system describing the spatial and temporal distribution of the electron (T_e) and ion (T_i) temperatures and the plasma density (ρ) is provided by the following standard MHD coupled equations,

$$\frac{d\rho}{dt} = -\rho \frac{1}{r} \frac{\partial}{\partial r}(r v_r) \tag{5}$$

$$\rho \frac{dv_r}{dt} = -\frac{\partial P}{\partial r} - \frac{1}{c} j_z B_\phi - \frac{\partial}{\partial r} \Pi_{rr} - \frac{1}{r}\left(\Pi_{rr} - \Pi_{\phi\phi}\right), \tag{6}$$

$$N_e \frac{d\varepsilon_e}{dt} = -\frac{P_e}{r} \frac{\partial}{\partial r}(r v_r) - \frac{1}{r} \frac{\partial}{\partial r}(r q_e) - Q_r + Q_e, \tag{7}$$

$$N_i \frac{\mathrm{d}\varepsilon_i}{\mathrm{d}t} = -\frac{P_i}{r}\frac{\partial}{\partial r}(rv_r) - \frac{1}{r}\frac{\partial}{\partial r}(rq_i) - \Pi_{rr}\frac{1}{r}\frac{\partial}{\partial r}(rv_r) - \frac{v_r}{r}(\Pi_{\phi\phi} - \Pi_{rr}) + Q_i, \tag{8}$$

$$E_z^* = \frac{j_z}{\sigma} - \frac{N}{B_\phi}\frac{\partial T_e}{\partial r}, \tag{9}$$

$$\frac{\mathrm{d}}{\mathrm{d}t}\frac{B_\phi}{r\rho} = \frac{c}{r\rho}\frac{\partial}{\partial r}E_z^*, \tag{10}$$

$$\frac{\mathrm{d}}{\mathrm{d}t} \equiv \frac{\partial}{\partial t} + v_r\frac{\partial}{\partial r}, \tag{11}$$

where $\varepsilon_e = (3/2) T_e$, $\varepsilon_i = (3/2) T_i$, v_r is the plasma radial velocity, E and B_ϕ are the electrical and magnetic fields, $\rho = m_e N_e + m_i N_i \sim m_i N_i$ denotes the plasma density, $P = P_e + P_i = N_e T_e + N_i T_i$ is the plasma total pressure, Π_{ij} denotes the ion viscose tensor (the electron viscosity is neglected), q_e and q_i are respectively the radial heat fluxes of electrons and ions, the value $Q_e = Q_{Joul} - Q_i$ takes into account the Joule heating by the electric current and the heat transfer by the electron-ion elastic collisions, Q_r describes the energy losses by excitation/de-excitation of ions, recombination and Bremsstruhlung processes, j_z is the electric current density, σ is the electric conductivity, and N is the Nernst coefficient. The other symbols have conventional meanings. The system of equations (1)–(11) is solved for the boundary conditions corresponding the non-ablative ceramic capillary. The model considers hydrodynamic flow including shock waves, heat conduction, electron–ion and ion–electron heat exchanges, magnetic field dynamics, magnetic forces, ohmic heating, radiative cooling and ionization. The Nernst and Ettinghausen effects[35,38] are also included into the model. The classical transport coefficients are used. The average ion charge of the plasma is calculated using results of ionization models or simply using the data of figure 2, corresponding to the coronal equilibrium ionization model. The radiation energy loss is computed using the radiation model with widely available atomic data. The MHD equations are solved numerically in a Lagrangian grid. The full electric current $I(t)$ is determined by $I(t) = I_0 \sin(t/T_0)$, where I_0 and T_0 are the current amplitude and the current wave period, respectively. For more details of the model see the studies[2,35,36,37,38,39,40].

The model described above is almost perfect for a detailed description of the physics of capillary z-pinches. Nevertheless, the zero-dimensional, one-fluid, one-temperature MHD model, which is much simpler and transparent for experimentalists, can also be used to understand the plasma evolution and optimally arrange the discharge system for Ar^{+8} laser operation with acceptable accuracy. There are two kinds of zero-dimensional model, namely the standard "snow-plough" model and the so-called simple model. The snow-plough model, which is described in plasma textbooks[41], is less accurate than the simple model[42,43,44], since the latter takes into account radiative energy losses, plasma viscosity and ionization processes. The results of the simple model are usually in acceptable agreement with the standard one-dimensional, one-fluid, two-temperature MHD model presented above.

A brief description of the simple model is as follows. The plasma is assumed to have axial symmetry, and plasma contraction is described by the radial coordinate $R = R(t)$. The model assumes that the plasma is initially uniformly ionized to a kinetic temperature T_{e0}, with a uniformly distributed current density inside the column. The densities and temperatures of the electrons and ions are assumed to be uniform during the collapse. The plasma compression velocity is the determined by the speed $v_r = dR(t)/dt$ of an external sheet of plasma determined by momentum conservation and taking the plasma viscosity. The plasma temperature T is calculated using the energy balance equation, which considers ohmic heating, viscosity, heating by pressure forces and radiation losses. The radiation losses are calculated using the conditions of coronal equilibrium. The radiative energy losses due to Bremsstralung are also taken into account. The electron density is linked to the ion density through the main stage of ionization, determined using the results of the coronal equilibrium model. The electrical current of the excitation pulse is described by the Kirchhoff standard laws of electrical circuits. The voltage in the plasma column is determined from the current using the classical (Spitzer) plasma conductivity and the plasma inductance. More details on the simple z-pinch model can be found elsewhere.[42,43]

The results of the zero-dimensional, one-fluid, one-temperature MHD model can be used to understand the plasma evolution and optimally arrange the discharge system for Ar^{+8} laser operation with acceptable accuracy. Figures 5–7 show the plasma parameters, calculated using the simple model for different initial argon pressures P_0, namely 0.7 mbar, 0.3 mbar and 0.1 mbar. In figures 5–7, R is the plasma radius, T is the plasma temperature and I is the electrical current. The capillary length was 45 cm and the internal diameter was 3.1 mm. The plasma was excited by the discharge of a 6 nF capacitor charged to 200 kV through a circuit of inductance 500 nH. The calculations show that plasmas with temperature about 60 eV and density about 0.5×10^{18} cm^{-3} can be produced, at a time depending on the gas pressure, using relatively low voltage and current of about 200 kV and 20 kA for pressures of around 0.1–0.5 mbar over short times ($\approx$1.5–2 ns).

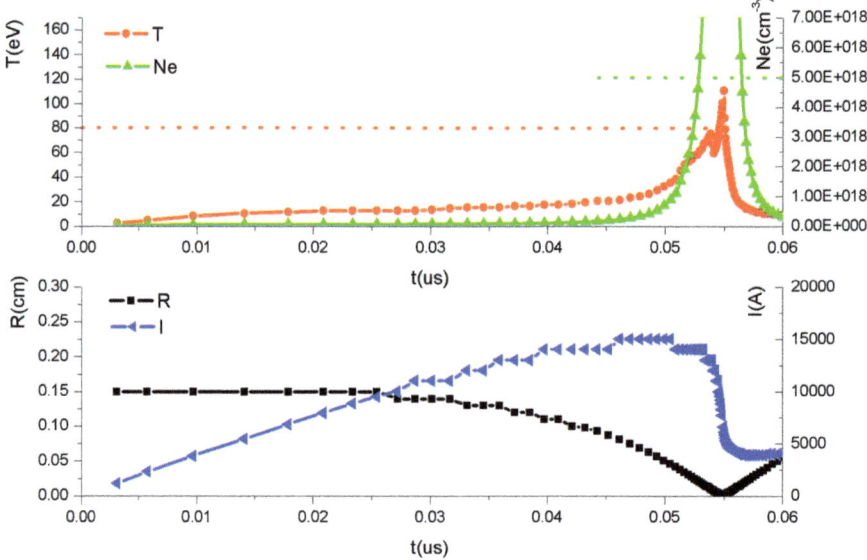

Figure 5 *Plasma parameters calculated using the simple model an initial argon pressures of 0.7 mbar.*

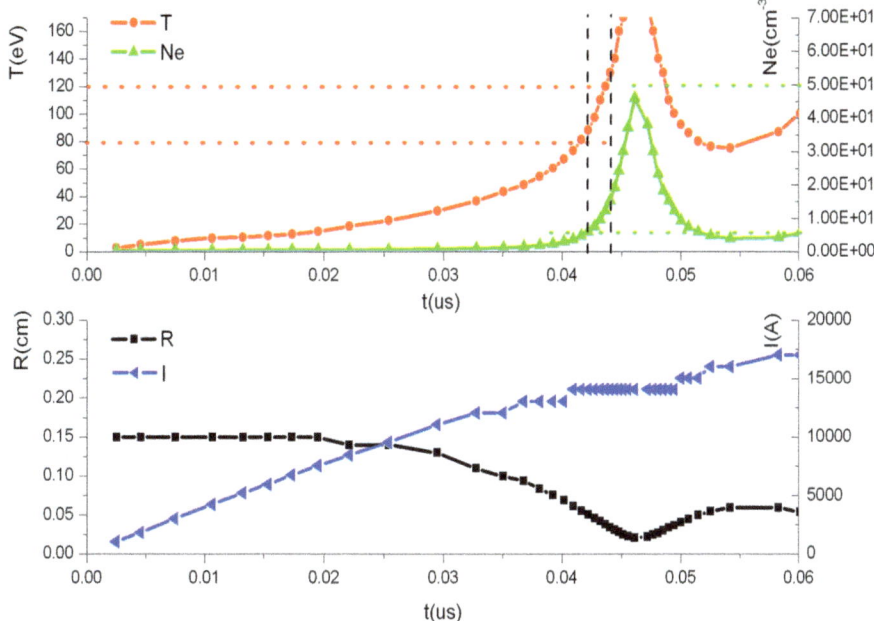

Figure 6 *Plasma parameters calculated using the simple model an initial argon pressures of 0.3 mbar.*

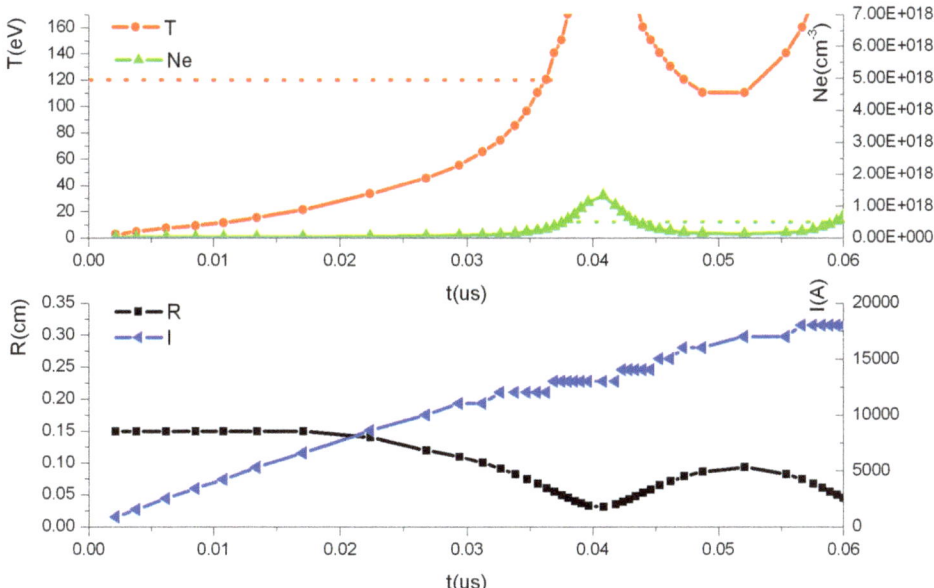

Figure 7 *Plasma parameters calculated using the simple model an initial argon pressures of 0.1 mbar.*

The physics of the simple model and its results are extremely transparent. The results are in acceptable agreement with the experimental data presented in Section 3.3. For comparison figure 8 shows the radial distributions of the parameters N_e, T_e, N_i, T_i, j_z, v_r and

B of the lasing plasma, calculated using the one-dimensional, one-fluid, two-temperature MHD model for similar experimental conditions − $P_0 = 0.3$ mbar; (a) $I = 24$ kA and (b) $I = 19$ kA]. In comparison to the simple model, the 1-D, 1-fluid, 2-temperature MHD model provides additional and more accurate information about the radial distribution of the plasma column parameters. The above-described one-dimensional model confirms the advantage of the low-density, long plasma column associated with the potential decrease of the laser beam divergence and the increase of the beam transverse coherence. Indeed, figure 8 shows that the use of a long, 45 cm, plasma column provides the necessary conditions ($T_e \sim 100$ eV, $N_e \sim 0.5 \times 10^{18}$ cm^{-3}, $r_{pl} \approx 0.25$ mm) with relatively low radial gradients would provide almost refraction-free laser operation in the super-fluorescent mirror-less mode with a gain-length product $GL \approx 16$. The beam angle divergence, $\varphi \sim r_{pl} / L \approx 0.5$ mrad) is determined by the radius $r_{pl} \approx 0.25$ mm and length $L_{pl} \approx 45$ cm of the plasma column rather than the refraction of amplified radiation[11].

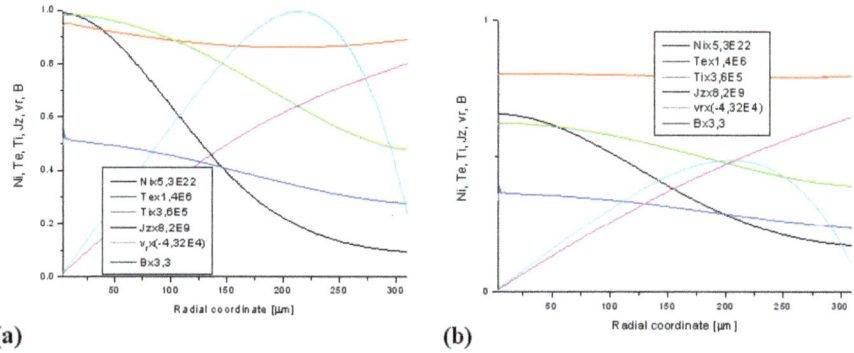

(a)   (b)

Figure 8 *The radial distributions of the parameters N_e, T_e, N_i, T_i, j_z, v_r and B of the lasing plasma calculated using the 1-D, 1-fluid, 2-temperature MHD model: (a) I = 24 kA and (b) I = 19 kA.*

Analysis of figure 8 also indicates that a very low electric current (less than about half of the total current) flows through the central high-density hot region of the plasma. The relatively low current in this region is a favourable condition, because MHD instabilities (typical of ordinary pinches) cannot easily occur in such cases.[18,38]

3 EXPERIMENTAL VALIDATIONS

This section describes the aims, methods, techniques, experimental setup and diagnostics of the Ar^{+8} laser excited by capillary z-pinch discharges. The experimental results can readily be understood by a simple analysis using the theoretical background presented in Section 2.

3.1 Aims and Methods

The main aim of the experiments was to reduce the size of Ar^{+8} lasers from laboratory size to table top. The experiments concentrated on research and development of practical table-top Ar^{+8} lasers excited by relatively low current (< 22 kA) and voltage (< 200 kV) capillary z-pinch discharges. The technical realization of the laser was based on the use of a long ($L \approx 0.5$ m) capillary. The MHD codes showed that the long plasma column would provide the gain-length product $GL \sim 16$, as required for saturated laser operation, at a relatively low gain, $G \approx 0.3$ cm^{-1}. Such a low gain could be produced at relatively low plasma temperature and density, with correspondingly low voltage and current. The advantage of a low density, long plasma column is the possibility to decrease the laser beam divergence thereby increasing the transverse coherence of the beam. The use of a long plasma column of radius $r_{pl} \approx 0.25$ mm with low radial and electron density gradients could provide almost refraction free laser operation in the super-fluorescent mirrorless; the beam angular divergence ($\varphi \sim r_{pl}/L \approx 0.5$ mrad) is determined by the radius and length of the plasma column rather than the refraction of amplified rays (see section 2). This is the physical mechanism behind the method used in the experiments.

Another experimental aim was to demonstrate laser operation without using any external low current pre-ionization circuit. The role of the circuit that generated the pre-ionization current with amplitude about 20 A and duration about 5 µs has been stated to be crucial in previous studies.[15,16,45,46] For the circuit-less operation, the argon gas automatically pre-ionized by a so-called gliding discharge on the internal surface of the capillary driven by the main excitation circuit.[18,19,20] Such a technique considerably simplifies the laser device. The use of the relatively low voltage (≈ 200 kV) of the main excitation pulse allows the use of a gas (SF$_6$ or air at atmospheric pressure) as an electrical insulator in the Marx generator instead of the impractical biodegradable transformer oil. The reduction of the voltage and current of the main excitation pulse should also reduce the ablation of the electrodes and capillary walls, increasing their lifetimes.[21,22,23,24,25,26]

3.2 Experimental Arrangement and Diagnostics

In the experiments describe here, the arrangement shown schematically in figure 9 was used. The laser active medium was generated by discharging a 6 nF water dielectric capacitor, initially charged to high voltage by a custom built six stage Marx voltage ($U \approx 200$kV) through a low inductance circuit which contained a water insulated spark gap and the capillary channel. The energy stored by the discharge capacitor was about 0.2 kJ. The lasing was obtained in an aluminum oxide ceramic capillary of length 0.45 m. The excitation current pulse, monitored by a custom built Rogowsky coil, had a peak value of 17–20 kA and half cycle duration of about 150 ns. In laser schemes using an external low current pre-ionization circuit the main discharge pulse is preceded by a 3–4 µs long current pulse with amplitude of about 20A, which pre-ionizes the argon and assures uniform initial conditions for the z-pinch plasma compression. The lasing is obtained using 3.1 mm diameter alumina capillary channels, filled with continuously flowing argon gas at a pressure in the region of 0.1–0.8 mbar. Figure 10 shows photographs of the Ar^{+8}-laser at the University of Pecs.

The electrical discharges in alumina capillaries are characterized by low wall material ablation, which is important for uniform compression and efficient heating of the plasma in the capillary z-pinch. The spectra were recorded in single shots either using a Jobin–Yvon spectrometer coupled to a MCP–CCD detection system or directly on a phosphor film without using the MCP. The system was also used to analyze the spatial intensity distribution in the laser beam. The output energy and time characteristics of the laser pulse were measured by a calibrated vacuum photodiode (XRD). To attenuate the laser intensity radiation several aluminium foils with calibrated thicknesses were used. A 250 MHz digital oscilloscope monitored the signal from the photodiode. The far-field intensity distributions of the laser pulses were recorded by a two dimensional imaging detector consisting of a phosphor film coupled to a CCD camera. In the saturation detection regime the phosphor detector was screened by aluminium foils.

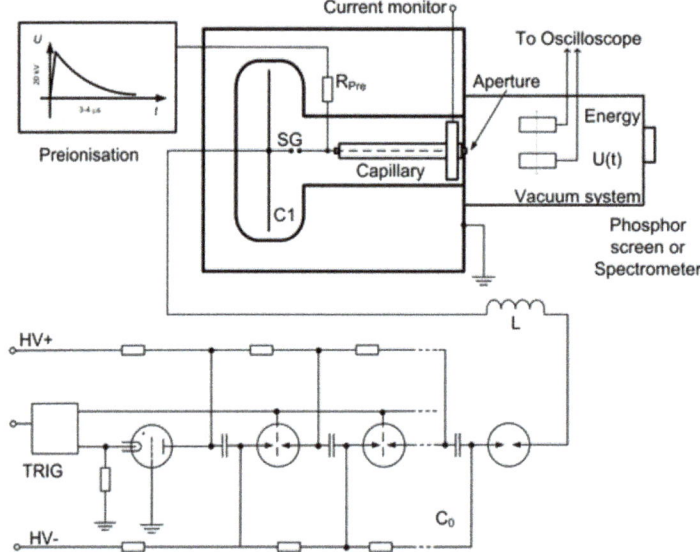

Figure 9 *Schematic of the table-top Ar^{+8} laser at the University of Pecs.*

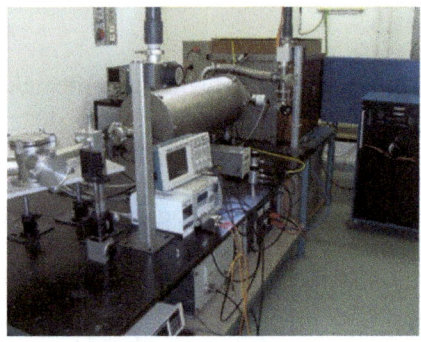 

Figure 10 *Photographs of the table-top Ar^{+8}-laser at the University of Pecs. Left, the laser with the Marx generator filled with SF$_6$; right, the laser with the Marx generator filled with air.*

3.3 Experimental Results

The experiments, which were primarily focused on the reduction in size of Ar^{+8} lasers from laboratory to table top, resulted in the development of a practical table-top laser excited by relatively low current ($I < 22$ kA) and voltage ($U < 200$ kV) capillary z-pinch discharges (figure 10). As an example, figure 11(a) shows lasing with $U \approx 185$ kV and $I \approx 17$ kA. For comparison, figure 11(b) shows the spectrally integrated non-laser radiation observed with non-optimal experimental conditions. The duration of the laser pulse and the non-laser radiation are about 1.5 ns and 30 ns, respectively — see also figures 3 and 6. Figure 12 shows the time integrated spectrum corresponding to the laser pulse of figure 11(a). The 46.9 nm laser line of 2p^{5}3p($J = 0$)–2p^{5}3s($J = 1$) transition of neon-like argon completely dominates the spectrum. In accordance with the results of the MHD codes, the experiments showed that the laser operation depends strongly on the voltage and peak current, and the initial argon pressure. Figure 13 shows the temporal evolution of the laser pulse for different initial pressures. At non-optimal pressures, lower than about 0.2 mbar and higher than about 0.6 mbar, the plasma generates non-lasing (spontaneous) radiation.

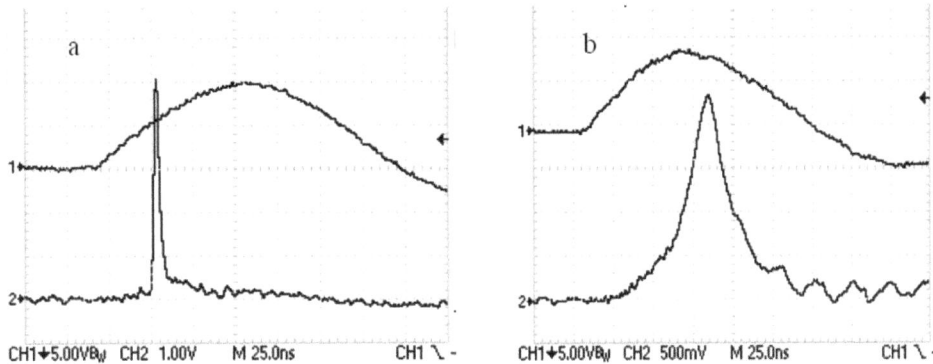

Figure 11 *(a) Lasing at U ≈ 185 kV and I≈17kA. (b) Non-lasing radiation observed at non-optimal experimental conditions.*

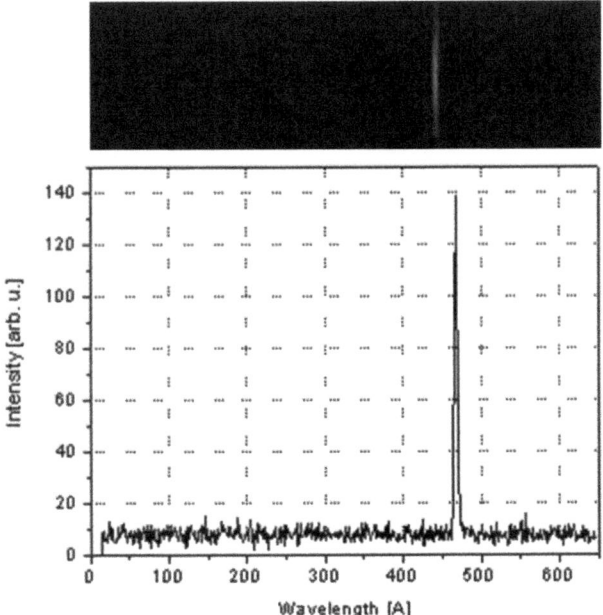

Figure 12 *The time integrated spectrum corresponding to the laser pulse of figure 11(a).* [47]

Figure 14 shows photographs of the laser beams for different experimental conditions. Figures 14(a)–(c) were recorded for a laser using the external low current pre-ionization circuit. Laser operation without using the external low-current pre-ionization circuit, considerably simplifying the laser arrangement, is demonstrated in figures 12(d)–(f). Here, the excitation parameters are $U ≈ 200$ kV, $I ≈ 20$ kA. The capillary wall was composed from the materials: Al_2O_3 (2 mm), epoxy (8 mm) and plastic zx100 (15 mm). That corresponds to the radial distance $Δ ≈ 25$ mm between the internal surface of the discharge channel and the triggering electrode. The argon initial pressures are given by (a,d) – 0.35 mbar, (b,e) – 0.3 mbar and (c,f) – 0.25 mbar.

The use of relatively low voltage for the main excitation pulse allowed atmospheric pressure gas (SF_6 or air) as the electrical insulator in the Marx generator instead of the impractical biodegradable transformer oil. The reduction of the voltage and current of the main excitation pulse also reduced the ablation of the capillary walls and electrodes and increased their lifetimes up to about 10^4 discharge shots.

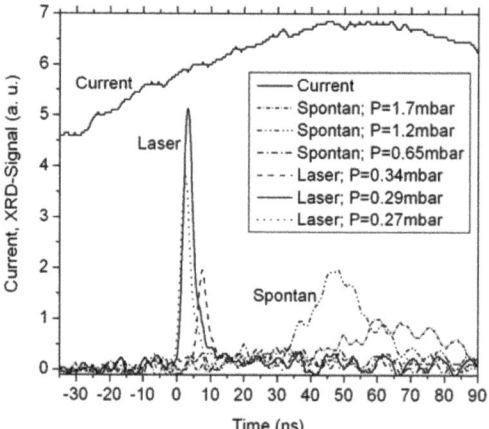

Figure 13 *The temporal evolution of the laser pulse for different initial argon pressures. At non-optimal pressures, lower than about 0.2 mbar and higher than about 0.6 mbar, the plasma generates non-laser (spontaneous) radiation.*[47]

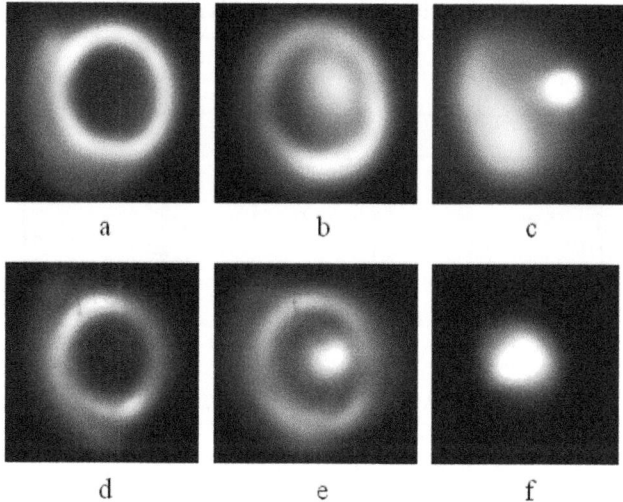

Figure 14 *Photographs of the laser beams for the argon initial pressures: (a,d) – 0.35 mbar, (b,e) – 0.3 mbar and (c,f) – 0.25 mbar.* Laser using the external low current pre-ionization circuit ((a)–(c)). Laser operation without using the external low-current pre-ionization circuit ((d)–(f)).[20]

4 CONCLUSIONS

The short (46.9 nm) laser wavelength perfectly matches the requirements for many potential applications; examples include nanoscience and nanotechnology associated with photolithography, materials processing (ablation), holography, nonlinear optics and laser-induced plasma spectroscopy. The aim of future research is the further clarification and optimization of the physical processes in the laser, as well as decrease of the system dimensions, shorter laser wavelength, increased energy, better coherence and higher pulse repetition rate. The implementation of wave guided light, double pulse excitation in capillary plasmas – i.e., capillary discharge plus laser pulse or two capillary discharges –, colliding hot and cold plasmas, and metal vapours in capillaries are envisaged.[48,49,50]

ACKNOWLEDGEMENTS

This study was supported in part by the Framework for European Cooperation in the field of Scientific and Technical Research (COST, Contract No MP0601). SVK acknowledges all his former students and colleagues for their cooperation in the research field.

References

1 R.C. Elton, *X-ray Lasers*, 1990 (Boston, Academic Press).
2 D. Atwood, *Soft X-ray and Extreme Ultraviolet Radiation*, 1999 (Cambridge, Cambridge University Press).
3 H. Daido, *Rep. Prog. Phys.*, 2002, **65**, 1513.
4 S. Suckewer and P. Jaegle, *Laser Phys. Lett.*, 2009, **6**, 411.
5 J.J. Rocca, V. Shlyaptsev, F.G. Tomasel, O.D. Cortàzar, D. Hartshorn and J.L.A. Chilla, *Phys. Rev. Lett.*, 1994, **73**, 2192.
6 J.J. Rocca, M. Frati, B. Benware, M. Seminario, J. Filevich, M. Marconi, K. Kanizy, A. Ozols, L.A. Artyukov, A. Vinogradov and Y.A. Uspensky, *C.R. Acad. Paris,* 2000, **1**, 1065.
7 B. Kish, M. Shuker, R.A. Nemirowsky, A. Fisher, A. Ron and J.L. Schwob, *Phys. Rev. Lett.*, 2001, **87**, 015002.
8 G. Niimi, Y. Hayashi, M. Nakajima, M. Watanabe, A. Okino, K. Horioka and E. Hotta, *J. Phys. D: Appl. Phys.*, 2001, **34**, 2123.
9 G. Tomassetti, A. Ritucci, A. Reale, L. Palladino, L. Reale, S.V. Kukhlevsky, F. Flora, L. Mezi, J. Kaiser, A. Faenov and T. Pikuz, *Eur. Phys. J. D*, 2002, **19**, 73.
10 A.V. Vinogradov, J.J. Rocca, *Kvant. Electron.*, 2003, **33**, 7.
11 A. Ritucci, G. Tomassetti, A. Reale, L. Palladino, L. Reale, F. Flora, L. Mezi, S.V. Kukhlevsky, A. Faenov and T. Pikuz, *Appl. Phys. B: Lasers and Optics*, 2004, **78**, 965.
12 Y. Zhao, Y. Cheng, B. Luan, Y. Wu and Q. Wang, *J. Phys. D: Appl. Phys.*, 2006, **39**, 342.
13 V.I. Ostashev, A.M. Gafarov, V.Yu. Politov, A.N. Shushlebin, L.V. Antonova, O.N. Gilev, A. Safronov and A.V. Komissarov, *Plasma Physics Reports*, 2006, **32**, 489.
14 K. Kolacek, J. Schmidt, V. Prukner, J. Straus, O. Frolov and M. Martinkova, *Czechoslovak Journal of Physics*, 2006, **56**, 259.
15 S.V. Kukhlevsky, J. Szasz, G. Almasi, J. Hebling, I. Santa, I. Foldes and S. Szatmari, *23rd Summer School and International Symposium on the Physics of Ionized Gases*, 2006, (Kopaonik, Serbia and Montenegro, August 28 – September 1).
16 C.A. Tan and K.H. Kwek, *Phys. Rev. A*, 2007, **75**, 043808.
17 S. Heinbuch, M. Grisham, D. Martz and J.J. Rocca, *Opt. Express*, 2005, **13**, 4050.
18 S.V. Kukhlevsky, *Europhys. Lett.*, 2001, **55**, 660.
19 http://www.patentstorm.us/patents/6408052.html.

20 J. Szasz, M. Kiss, I. Santa, S. Szatmari and S.V. Kukhlevsky, *Phys. Rev. Lett.*, 2013, **110**, 183902.

21 S.V. Kukhlevsky and L. Kozma, *Appl. Phys. B*, 1993, **57**, 213.

22 S.V. Kukhlevsky, L. Kozma, L. Palladino, A. Reale, F. Flora and G. Giordano, *Proc. SPIE*, 1997, **3156**, 174.

23 S.V. Kukhlevsky, Cs. Ver, L. Kozma, L. Palladino, A. Reala, G. Tomassetti, F. Flora and G. Giordano, *Appl. Phys. Lett.*, 1999, **74**, 2779.

24 S.V. Kukhlevsky, J. Kaiser, L. Palladino, A. Reale, G. Tomassetti, F. Flora, G. Giordano, L. Kozma, M. Liska and O. Samek, *Phys. Lett. A*, 1999, **258**, 335.

25 A. Ritucci, G. Tomassetti, A. Reale, L. Palladino, L. Reale, T. Limongi, F. Flora, L. Mezi, S.V. Kukhlevsky, A. Faenov, T. Pikuz and J. Kaiser, *Contr. Plasma Phys.*, 2003, **43**, 88.

26 Y. Li, Z.X. Liu and Y.D. Wei, *Chinese Laser*, 2004, **31**, 538.

27 S.V. Kukhlevsky, A. Ritucci, I.Z. Kozma, J. Kaiser, A. Shlyaptseva, G. Tomassetti and O. Samek, *Contr. Plasma Phys.*, 2002, **42**, 109.

28 C. Breton, C. Michelis and M. Mattioli, *J. Quant. Spectrosc. Radiat. Transfer*, 1978, **19**, 367.

29 D.A. Verner, E.M. Verner and G.J. Ferland, *Atomic Data Nucl. Data Tables*, 1996, **64**, 1.

30 Y.S. Kim and R.H. Pratt, *Phys. Rev. A*, 1983, **27**, 2913.

31 Y. Hahn, *Theory of Dielectronic Recombination*, in: *Advances in Atomic and Molecular Physics*, Vol. 21, 1985 (New York, Academic Press).

32 K.B. Fournier, M. Cohen and W.H. Goldstein, *Phys. Rev. A*, 1997, **56**, 4715.

33 K.W. Hill, M. Bitter, D. Eames, S. Goeler, N.R. Sauthhoff and E. Silver, in: AIP Conference Proceedings No. **75**, *Low Energy X-Ray Diagnostics*, 1981 (New York, American Institute of Physics).

34 K.B. Fournier, M. Cohen, M.J. May and W.H. Goldstein, *Atomic Data and Nucl. Data Tables*, 1998, **70**, 231.

35 S.I. Braginskii, *Rev. Plasma Phys.*, 1965, **1**, 216.

36 V.N. Shlyaptsev, A.V. Gerusov, A.V. Vinogradov, J.J. Rocca, O.D. Cortazar, F. Tomasel and B. Szapiro, *Proc. SPIE Int. Soc. Opt. Eng.*, 1993, **2012**, 99.

37 V.N. Shlyaptsev, J.J. Rocca and A.L. Osterheld, *Proc. SPIE Int. Soc. Opt. Eng.*, 1995, **2520**, 365.

38 N.A. Bobrova, S.V. Bulanov, T.L. Razinkova and P.V. Sasorov, *Plasma Phys. Rep.* 1996, **22**, 349.

39 S.H. Kim, K.T. Li, D.E. Kim and T.N. Lee, *Phys. Plasmas*, 1997, **4**, 730.

40 A. Nemirovsky, A. Ben-Kish, M. Shuker and A. Ron, *Phys. Rev. Lett.*, 1999, **82**, 3436.

41 A.W. Trivelpiece and N.A. Krall, *Principles of plasma physics*, 1986 (San Francisco, Press Inc.).

42 V.V. Vikhrev, *JETP Lett.*, 1978, **27**, 95.

43 S.V. Kukhlevsky, J. Kaiser, A. Ritucci, G. Tomassetti, A. Reale, L. Palladino, I.Zs. Kozma, F. Flora, L. Mezi, O. Samek and M. Liska, *Plasma Sources Sci. Technol.*, 2001, **10**, 567.

44 S.V. Kukhlevsky, J. Kaiser, A. Reale, G. Tomassetti, L. Palladino, A. Ritucci, T. Limongi, F. Flora and L. Mezi, *J. de Physique France IV*, 2001, **11**, 583.

45 Y.L. Cheng, B.H. Luan, Y.C. Wu, Y.P. Zhao, O. Wang, W.D. Zheng, H.M. Peng and D.W. Yang, *Acta Physica Sinica*, 2005, **54**, 4979.

46 M. Shuker, A. Ben-kish, R.A. Nemirovsky, A. Fisher and A. Ron, *Physics of Plasmas*, 2006, **13**, 013102.

47 J. Szasz, M. Kiss, I. Santa, S. Szatmari and S.V. Kukhlevsky, *Contrib. Plasma Phys.*, 2012, **52**, 770.

48 S.V. Kukhlevsky, L. Kozma, L. Palladino, A. Reale, F. Flora, L. Mezi and G. Giordano, *Proc. SPIE*, 1997, **3156**, 180.

49 M. Shuker, A. Ben-kish, A. Fisher and A. Ron, *Appl. Phys. Lett.*, 2006, **88**, 061501.

50 A. Ritucci, G. Tomassetti, A. Reale, L. Palladino, L. Reale, F. Flora, L. Mezi, S.V. Kukhlevsky, A. Faenov, T. Pikuz, J. Kaiser and O. Consorte, *Europhys. Lett.*, 2003, **63**, 694.

NANOMETRE SCALE TAPERED PLANAR WAVEGUIDES FOR FOCUSING X-RAY FEMTOSECOND PULSES

S.V. Kukhlevsky

Department of Physics, University of Pecs, Hungary

1 INTRODUCTION

Straight and tapered capillaries are attracting significant interest for the transmission, focusing and collimation of X-rays, as well as in the production of plasmas for wave-guiding and the generation of incoherent and coherent X-rays. Considerable effort has been devoted to the development of so-called X-ray near-field capillary nano-optics, which typically use continuous or long pulse X-ray nanoscale beams formed by a straight or tapered capillary waveguide (nanometre-scale manipulator) in the near-field zone of the guide output.[1,2]

Recently, new applications which are relevant to the guiding and nanoscale confinement of ultrashort (femtosecond) X-ray pulses have been proposed.[3,4,5,6] Such pulses are produced by laser-induced plasmas using long wavelength femtosecond lasers or directly by 4th-generation synchrotron radiation facilities and free electron lasers (FELs). The new applications require knowledge of the evolution of the detailed structure of the femtosecond pulses as they propagate through the guide. The pulses, which satisfy the wave equation and boundary conditions of the straight multimode guide, are usually described by the conventional theory of modal dispersion, in which the femtosecond pulse can be represented in the form of a Fourier integral. Such an approach is usually used in the study of femtosecond pulses passing through apertures or lenses.[7,8] For long wavelengths (a few hundred nanometres), the eigenmode interpretation works very well to explain most experimental results to date, at least for pulse widths greater than 5 fs.[9,10,11,12] Unfortunately, the formation of input pulses with mode-matched profiles and flat wavefronts is a serious problem, because of the spatial and temporal distortion of the femtosecond pulses by most optical elements.[13,14,15] For tapered waveguides the eigenmode solution approach is extremely complicated, but the geometrical optics approach provides a satisfactory approximation.[16]

In the present chapter, the detailed structures of the field distributions in arbitrary X-ray femtosecond pulses passing through straight and tapered capillary waveguides are determined using scalar diffraction theory and the method of images.[17] At the guide entrance the femtosecond pulses are represented in the form of Fourier integrals. The numerical analysis showed that the ultrashort X-ray wave propagates down the guide as the superposition of many pulses (transient modes) that diffract in the off-axis direction and interfere with each other. In opposition to the eigenmode theory of waveguides (solving a boundary condition problem), the field at the guide entrance can satisfy neither the guide wave equation nor the boundary conditions. That gives the possibility to consider not only the mode-matched waves, but also the pulses having more complicated properties at the guide entrance. For the straight multimode guide the usual result is recovered, namely that the spatial profile of the wave is undistorted along its propagation path so long as the input pulse is mode matched. For non mode-matched waves or tapered guides, the transverse intensity distribution of the pulse depends on the propagation path and on the parameters of the guide and the input wave. In addition, the computations show that changes of the temporal profile and duration of the pulse during propagation are negligibly small for a wide range of guide parameters, at least down to pulse widths of few femtoseconds. Finally, for tapered guides having very small guide exits (a few tens of nanometres), a very interesting behaviour of the femtosecond pulses was found; the numerical analysis showed that the output pulses have uniform transverse intensity distribution with very little dependence on the parameters of the guide and the input wave.

2 THEORY AND COMPUTER SIMULATIONS

First consider the propagation of X-ray pulses through a planar waveguide, consisting of a vacuum core and a cladding, with a taper angle γ (figure 1). The frequency dependent complex refractive index, $n = n(\omega)$, changes abruptly from $n_1 = 1$ to n_2 at the guide boundary. Considering, initially, harmonic fields a wave $E_0(x, z, t)$ arriving at the guide entrance can be represented by $E_0(x, 0, t) = E_0(x)\exp[i\{-\omega t + \phi_0(x)\}]$. Here, $\omega = kc/n$ and $\phi_0(x)$ are the wave frequency and phase. In accordance with the Huygens-Fresnel principle, which is usually regarded as a form of the Helmholtz-Kirchhoff integral theorem, every point $P(x, z = 0)$ of the wave $E_0(P, t)$ can be considered as the centre of a secondary spherical wave. When the secondary wave reaches the core-cladding boundary, it is split into two waves: a leaky wave proceeding into the second medium and a reflected wave propagating back into the first medium. According to the method of images, the reflected wave can be represented as a wave emerging from the corresponding imaginary point $P_1(x_1, z_1)$ of the free-space equivalent virtual source (figure 1). The amplitude and phase of the imaginary wave $E_1(x_1, t)$ are determined by the Fresnel field reflectivity $R_1(\Phi_1, n_1, n_2)$ and the phase change $\Delta\phi_1(\Phi_1, n_1, n_2)$ associated with reflection at an angle Φ_1.[18,19] The field $E_0(P, t)$, after m reflections, can be represented as the field $E'_m(P', t)$ emerging from the mth zone of the virtual source having the field distribution $E_m(P_m, t)$. The field $E'_m(P', t)$ at a general point P' of the guide core is given by the Fresnel-Kirchhoff integral,[17,20]

$$E'_m(P',t) = \frac{1}{2i\lambda} \int_{x_m^{\min}}^{x_m^{\max}} E_m(P_m,t) \frac{\exp\left[in_1c^{-1}\omega r(P'_m, P_m)\right]}{r(P'_m, P_m)} \chi(\Theta_m) \, dx_m \tag{1}$$

$-E_m(P_m,t) = R_m E_0(x_m)\exp\left[i(\phi_0(x_m) + \Delta\phi_m - \omega t)\right], P_m = P(x_m, z_m)$ and $P'_m = P'(x'_m, z'_m)$.

Here (x_m, z_m) and (x'_m, z'_m) are the coordinates of the point P and P' in the coordinate system (X_m, Z_m). The points $P_m = P(x_m, z_m)$ are found by reflection of the points $P = P(x, z)$ at the guide boundary, and Θ_m is the angle that the line (P'_m, P_m) makes with the unit normal $\vec{e}_m$ to the line $(x_m^{\min} x_m^{\max})$.

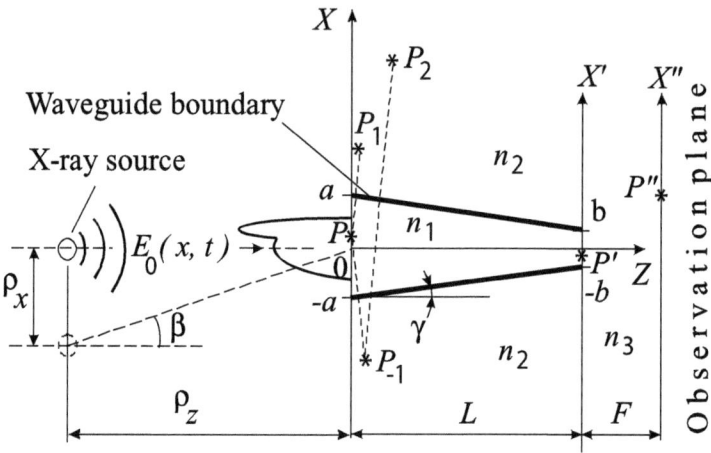

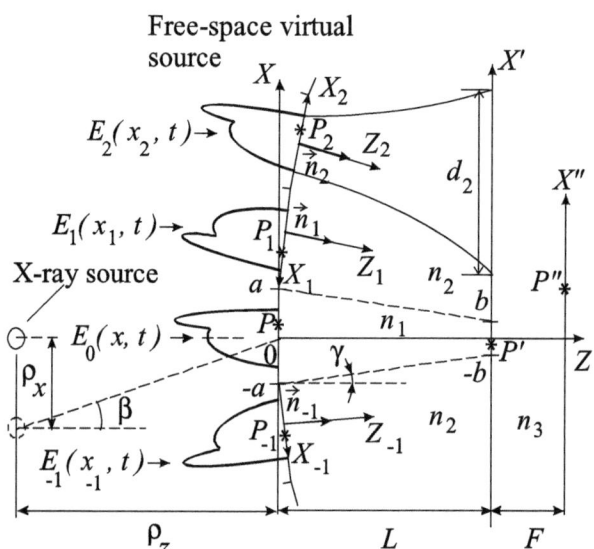

Figure 1 *Top: propagation of an X-ray femtosecond pulse through a tapered optical waveguide. Bottom: schematic diagram of the free-space virtual source (Fresnel waveguide or Fresnel-like lens).*

The transformation can be presented as the rotation, translation and inversion of the coordinate system (X, Z),

$$\begin{pmatrix} x_m \\ z_m \end{pmatrix} = \begin{pmatrix} \cos(2m\gamma), \pm(-1)^m \sin(2m\gamma) \\ \mp(-1)^m \sin(2m\gamma), \cos(2m\gamma) \end{pmatrix} \begin{pmatrix} x_m^{T,I} \\ z_m^T \end{pmatrix} \tag{2}$$

$$x_m^{T,I} = \left[x \mp a \left\{ 1 + \cos(2m\gamma) - f(m) \sum_{j=1}^{m-1} \cos(2j\gamma) \right\} \right] (-1)^m \tag{3}$$

$$z_m^T = z - a \left| \sin\left(2m\gamma\right) - f(m) \sum_{j=1}^{m-1} \sin\left(2j\gamma\right) \right|, \tag{4}$$

where $r(P_m', P_m)$ is the distance between points P_m and P_m', R_m and $\Delta\phi_m$ are the Fresnel reflectivity and phase change for the m reflections, $x_m^{\min}$ and $x_m^{\max}$ define the mth zone of the equivalent source, $2a$ and $2b$ are the dimensions of the guide entrance and exit, $f(m) = 2(\delta_{1m} - 1)$, and δ_{1m} is the Kronecker delta. The $\pm$ signs indicate $x > 0$ (+) and $x < 0$ (−). The total field $E'(P', t)$ at the point P' is found by summing the contributions from the $2M+1$ zones of the virtual source,

$$E'\left(P',t\right) = \sum_{m=-M}^{M} E_m'\left(P',t\right), \tag{5}$$

where $2M+1$ is the number of zones (beams) contributing energy to the field $E'(P', t)$. The value depends on the critical full reflection angle Θ_c ($\approx \Theta_M$), the taper angle ψ and the transverse dimension $d_M(\lambda, a, z)$ of the beam $E_M'\left(P',t\right)$, which can be considered as a transient mode. The field $E'(P',t)$ due to diffraction at the output aperture produce the field $E''(P'', t)$ in the observation plane,

$$E''\left(P'',t\right) = \frac{1}{2i\lambda} \int_{-b}^{b} E'\left(P',t\right) \frac{\exp\left[i n_3 c^{-1} \omega r\left(P'',P'\right)\right]}{r'\left(P'',P'\right)} \chi\left(\Theta'\right) dx'. \tag{6}$$

Next consider the guided propagation of an X-ray pulse, represented by a wide frequency bandwidth field. The input pulse $E_0(x, 0, t)$ at the guide entrance can be represented by a Fourier integral

$$E_0\left(P,t\right) = \int_{-\infty}^{\infty} E_0\left(P,\omega\right)\exp\left(-i\omega t\right) d\omega. \tag{7}$$

Using equations (1)–(6) for the input harmonic field $E_0(P,\omega)\exp(-i\omega t)$ and substituting the results into equation provides the fields in the guide and in the far field zone of the guide output. A simple analysis of equations (1)–(7) shows that the ultrashort wave propagates down the guide as the superposition of many pulses – transient modes $E_m'\left(P',t\right)$ – that diffract in the off-axis direction and interfere with each other.

Examples of computer simulations of the evolution of 150 fs pulses inside the straight and tapered waveguides are now presented, first for a straight hollow guide. The guide parameters used were $2a = 5$ µm, $L = 10$ cm and $R_m = 1$. At the entrance of the multimode guide the Gaussian-shaped pulse, produced by a light source as shown in figure 1, matched the profile of the TE$_2$ eigenmode and had a flat wave front,

$$E\left(x,0,t\right) = E_0\left(x\right)\exp\left[-2\ln\left(2\right)\left(t / \tau_0\right)^2 \right] \exp\left(i\left[-\omega_0 t + \phi_0\left(x\right)\right]\right), \tag{8}$$

where $E_0(x) = TE_2(x)$, $\tau_0 = 150$ fs, $\lambda_0 = 2\pi c/\omega_0 = 10$ nm, $\phi_0(x)$ $=$ const, $\gamma = 0$ mrad and $\rho_z = \infty$ (see, figure 1). The intensity distribution of the pulse was calculated for different distances from the guide entrance, as shown in figure 2, where the normalized intensities at $z = 0$, 5 and 10 cm from the entrance of the straight guide; the three results are essentially identical. Note that the usual result of the modal dispersion theory is recovered, namely that the spatial profile of the wave is undistorted along its propagation path provided the input pulse is mode matched. In the case of non mode matched waves or tapered guides, the transverse intensity distribution of the pulse depends on the propagation path and the parameters of the guide and input wave, equations (1)–(7).

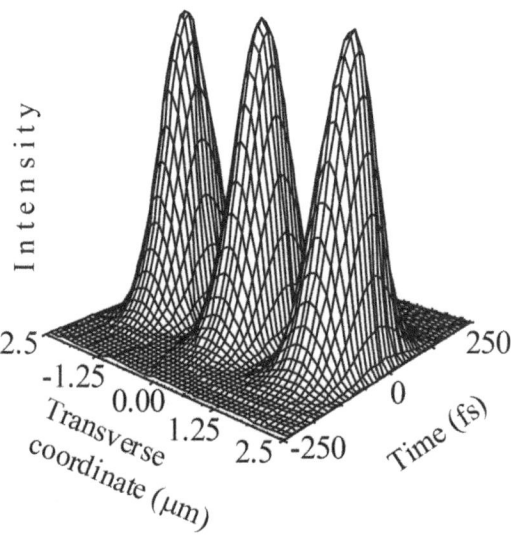

Figure 2 *Normalized intensity distributions of 150 fs pulses at distances z = 0, 5 and 10 cm from the entrance of a straight multimode guide. The guide has a critical full reflection angle Θ_c = 5 mrad, corresponding to M = 100.*[17]

Figure 3 presents the evolution of 150 fs pulses computed for a tapered guide with $2a = 5$ μm, $L = 10$ cm and $2b = 20$ nm. The pulses arrive at the guide entrance in the off-axis direction ($\rho_x \neq 0$, $\beta \neq 0$, see figure 1) with parabolic field distributions and spherical wavefronts; $E_0(x) = (x/a)^2$, $\phi_0(x) = k[\beta(x+a)+(x^2-a^2)/2\rho_z]$, $\beta = 0.1$ mrad and $\rho_z = 1$ mm. The pulses satisfy neither the guide wave equation nor the boundary conditions. Such pulses can be formatted, for example, in tilted pulse schemes by spatial and temporal distortion of the Gaussian pulses with flat wavefronts. Figures 3(a), (b) and (c) show the intensity distribution of the pulse at $z = 0$, 5 and 10 cm. Note that the transverse intensity distribution of the pulse depends on the propagation path. At the guide exit the transverse distribution is uniform as shown in figure 3(c). In addition, for a wide range of guide parameters the computations show that changes of the temporal profile and duration of the pulse under propagation are negligibly small, at least for pulse widths larger than a few femtoseconds. A very interesting behaviour of femtosecond pulses in tapered guides with small exits (a few tens of nanometres) is that the output pulses have uniform transverse intensity distribution, independent of the guide parameters and the input wave. The distributions are similar to that shown in figure 3(c).

4 CONCLUSIONS

In conclusion, the detailed structures of field distributions in arbitrary femtosecond X-ray pulses passing through straight and tapered capillary waveguides have been found using scalar diffraction theory and the method of images. The numerical analysis showed that the ultrashort wave propagates down the guide as the superposition of many pulses (transient modes) that diffract in the off-axis direction and interfere with each other.

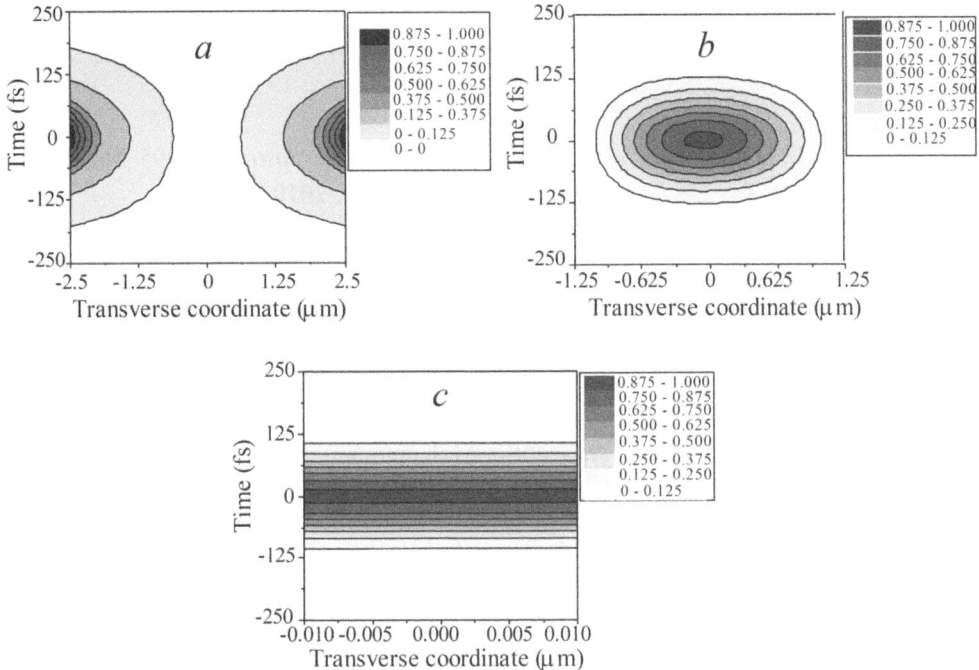

Figure 3 *Normalized intensity distributions of 150 fs pulses at different distances z from the entrance of the tapered guide: (a) 0 cm, (b) 5 cm and (c) 10 cm. The guide has a critical full reflection angle Θ_c = 5 mrad, corresponding to M = 100.*[20]

In contrast to the eigenmode theory of waveguides (solving a boundary condition problem), the field at the guide entrance need satisfy neither the guide wave equation nor the boundary conditions. This provides the possibility to consider not only mode matched waves but also pulses having more complicated properties at the guide entrance. For the straight multimode guide the usual result was recovered, i.e. that the spatial profile of the wave is undistorted along its propagation path provided that the input pulse is mode matched. For non modematched waves or tapered guides, the transverse intensity distribution of the pulse depends on the propagation path and the parameters of the guide and the input wave. In addition, the computations show that changes of the temporal profile and duration of the pulse under propagation are negligibly small for a wide range of guide parameters, at least for pulse widths larger than a few femtoseconds. Finally, for tapered guides small exit dimensions (a few tens of nanometres), the output pulses have uniform transverse intensity distributions, independent of the parameters of the guide and the input wave. Such uniform intensity distributions of the output pulses along with the negligibly small change of temporal profile and duration could play a key role in the development of ultra fast near field, nanometre scale, X-ray optics.

ACKNOWLEDGEMENTS

This study was supported in part by the Framework for European Cooperation in the field of Scientific and Technical Research (COST, Contract No MP0601). The author acknowledges all his former students and colleagues for their cooperation in the research field.

References

1 J.B. Hastings, S.L. Hulbert and G.P. Williams (eds.), *Rev. Sci. Instrum.*, 1995, **66**, 1271.
2 C.A. MacDonald, *X-Ray Optics and Instrumentation,* 2010, **2010**, 867049.
3 S.P. Hau-Riege, H.N. Chapman, J. Krzywinski, R. Sobierajski, S. Bajt, R. A. London, M. Bergh, C. Caleman, R. Nietubyc, L. Juha, J. Kuba, E. Spiller, S. Baker, R. Bionta, K. Sokolowski Tinten, N. Stojanovic, B. Kjornrattanawanich, E. Gullikson, E. Plönjes, S. Toleikis and T. Tschentscher, *Phys. Rev. Lett.*, 2007, **98**, 145502.
4 K.J. Gaffney and H.N. Chapman, *Science*, 2007, **316**, 1444.
5 H.N. Chapman, S. P. Hau-Riege, M. J. Bogan, S. Bajt, A. Barty, S. Boutet, S. Marchesini, M. Frank, B. W. Woods, W.H. Benner, R. A. London, U. Rohner, A. Szöke, E. Spiller, T. Möller, C. Bostedt, D. A. Shapiro, M. Kuhlmann, R. Treusch, E. Plönjes, F. Burmeister, M. Bergh, C. Caleman, G. Huldt, M.M. Seibert and J. Hajdu, *Nature* , 2007, **448**, 676.
6 A. Barty, S. Boutet, M.J. Bogan, S. Hau-Riege, S. Marchesini, K. Sokolowski-Tinten, N. Stojanovic, R. Tobey, H. Ehrke, A. Cavalleri, S. Düsterer, M. Frank, S. Bajt, B.W. Woods, M.M. Seibert, J.Hajdu, R. Treusch and H.N. Chapman, *Nature Photonics*, 2008, **2**, 415.
7 Z. Bor and Z.L. Horvath, *Opt. Commun.*, 1992, **94**, 249.
8 M. Bertolotti, A. Ferrari and L. Sereda, *J. Opt. Soc. Am. B*, 1995, **12**, 1519.
9 S. Jackel, R. Burris, J. Grun, A. Ting, C. Manaka, K. Evans and J. Kosakowskii, *Opt. Lett.*, 1995, **20**, 1086.
10 M. Nisoli, S. De Silvestri, O. Svelto, R. Szipöcs, K. Ferencz, Ch. Spielmann, S. Sartania and F. Krausz, *Opt. Lett.*, 1997, **22**, 522.
11 S. Sartania, Z. Cheng, M. Lenzner, G. Tempea, Ch. Spielmann, F. Krausz and K. Ferencz, *Opt. Lett.*, 1997, **22**, 1562.
12 G. Tempea and T. Brabec, *Opt. Lett.*, 1998, **23**, 762.
13 M.E. Fermann, *Opt. Lett.*, 1998, **23**, 52.
14 M. Kempe, U. Stamm and B. Wilhelmi, *Opt. Commun.*, 1986, **59**, 119.
15 O.E. Martinez, *Opt. Commun.*, 1986, **59**, 229.
16 S.V. Kukhlevsky, G. Lupkovics, K. Negrea and L. Kozma, *Pure and Appl. Opt.*, 1997, **6**, 102.
17 S.V. Kukhlevsky, *X-Ray Spectr.*, 2003, **32**, 223.
18 M. Born and E. Wolf, *Principles of Optics*, 1980, (Oxford, Pergamon Press).
19 J.W. Goodman, *Statistical Optics*, 1985, (New York, Wiley-Interscience).
20 S.V. Kukhlevsky and G. Nyitray, *Phys. Lett. A*, 2001, **291**, 459.

EXTREME ULTRAVIOLET EMISSION FROM MULTI-CHARGED STATE IONS IN POTASSIUM PLASMAS

T. Higashiguchi,[1] B. Li,[2] R D'Arcy,[2] P. Dunne,[2] and Gerry O'Sullivan[2]

[1]Department of Advanced Interdisciplinary Sciences, Utsunomiya University, Yoto 7-1-2, Utsunomiya, Tochigi 321-8585 Japan
[2] School of Physics, University College Dublin, Belfield, Dublin 4, Ireland

1 INTRODUCTION

Recent progress in the development of CO_2 or solid-state lasers facilitates the production of highly-ionized plasma sources for the generation of extreme ultraviolet (XUV) radiation and soft X-rays for applications such as photo-absorption spectroscopy, semiconductor lithography, and biological imaging.[1,2,3,4,5,6,7,8,9,10] The use of laser-produced plasmas (LPPs) to provide highly charged ions has a number of advantages, for example the laser intensity and/or wavelength can be readily altered in order to control the electron temperature and critical electron density. This implies that plasmas with a certain charge state distribution or electron temperature can be produced by selecting the appropriate laser intensity and/or wavelength.

The physics of neutral alkalis is attractively simple because of the presence of only a single valence electron and for this reason their spectra were widely studied in the past. However, due to their higher ionization potentials, emission from transitions involving valence electrons in the ions only occurs in the vacuum ultraviolet (VUV) spectral region. The current knowledge of XUV emission from multiply charge state alkali metal ions of potassium has been summarized by Sugar and Corliss[11] and Sansonetti.[12] XUV photo-absorption spectra due to sub-valence excitation in neutral potassium have been extensively investigated and this work has now been extended to include potassium ions.[13,14]

This chapter discusses the observation of intense XUV emission generated from a hollow cathode micro-discharge source.[15] The spectrum of this capillary discharge produced potassium plasma suffers from some contamination that gives rise to strong line emission at wavelengths around 40 nm from carbon, due to ablation of the capillary material - polytetrafluoroethylene (PTFE). The parameter window for the optimization of this XUV discharge source is restricted by the energy coupling between the electric circuit of the power supply and the plasma impedance in the capillary. The emission spectra of multiply charged potassium ions in the 20-60 nm spectral region are reported here. The spectral behaviour, which is related to plasma parameters such as electron density and temperature, is discussed in the context of the experimental conditions.

The emission spectra were characterised at a time-averaged electron temperature of 12 eV and an electron density of 10^{20} cm^{-3}.[16] Existing experimental data plus Hartree-Fock with configuration interaction (HFCI) approximation calculations using the Cowan suite of codes[17] to provide weighted oscillator strengths (transition probabilities or gf values, where g is the statistical weight of the emitting level and f is its oscillation strength) as functions of the wavelength (λ) was used to generate theoretical spectra for a range of potassium ions, from K^+ to K^{7+}. From these data, it was seen that emission from K^{3+} to K^{5+} combines to produce a strong emission band near 40 nm. Further information was obtained by comparison of these experimental and calculated data with discharge- and laser-produced ions in order to explore the effects of self absorption (opacity) at different electron densities.

2 EXPERIMENTAL ARRANGEMENT

The capillary, which was housed in a vacuum chamber, was 1000 μm long with internal and external diameters of 500 μm and 1000 μm, respectively, made in a PTFE block. The capillary discharge produced the electrons required to initiate an efficient hollow cathode discharge at the potassium electrode. The maximum discharge voltage and current were 30 kV and 200 A using a pulsed power supply connected to a 2 nF capacitor at a repetition rate of 10 Hz. The discharge voltage and current were monitored by use of a high-voltage probe and a Rogowski coil connected to a 2 GHz sampling digital oscilloscope. All equipment was connected to a personal computer for data acquisition and processing. The maximum discharge current had a pulse duration of 150 ns with the light source as the load. During the operation of the light source, the pressure in the chamber was maintained below 3×10^{-3} Pa.

The axial emission of the discharge in the XUV spectral region was analyzed with a normal incidence vacuum spectrograph using an iridium coated grating with 1200 lines/mm (Acton VM502). The distance between the exit of the capillary and the entrance slit of the spectrometer was 1 m. Time-integrated spectra were obtained by a thermoelectrically cooled back-illuminated X-ray charge coupled device (CCD) camera. The typical spectral resolution was better than 0.05 nm. The pressure in the spectrometer was maintained at less than 3×10^{-5} Pa.

A Q-switched Nd:YAG laser operating at 1064 nm with a maximum pulse energy of 2 J and a duration of 10 ns (FWHM) was focused, using a 12 cm focal length lens, onto a planar potassium target of thickness 1 mm inside a vacuum chamber. The focused spot size was monitored using a telescope and a CCD camera placed on the laser axis. The spot size, measured to be 500 μm (FWHM), was chosen to minimize the plasma hydrodynamic expansion loss.[18] The pulse energy, at the fixed spot diameter of 500 μm, was adjusted to vary the focused laser intensity. The emission from the plasma in the XUV spectral region was viewed at 45° with respect to the incident laser axis. The temporal behaviour of the XUV emission at 39 nm was measured using a fast photomultiplier tube connected to a sampling digital photon counter, which was coupled to the spectrometer.

3 CALCULATIONS

Calculations were performed for the transitions $3s^2 3p^n \rightarrow 3s^2 3p^{n-1} 3d^1 + 3s 3p^{n+1}$ and $3s^2 3p^{n-1} 3d^1 \rightarrow 3s^2 3p^{n-1} 4f^1$ ($1 \leq n \leq 5$) in two to six times ionized potassium. The corresponding oscillator strength, g_f, distributions calculated with the Cowan suite of codes[17] are presented in figure 1 to show the resonant emission in ions ranging from neutral potassium to K^{7+}. The scaling factors applied to the Slater Condon parameters[17] were chosen to optimize agreement between the calculations and the available spectral data. In particular, the results show that emission from K^{3+} to K^{5+} can combine to produce a strong emission band near 40 nm. The ion population of the potassium plasma as a function of electron temperature was calculated using a time-dependent collisional radiative (CR) model[19,20] at an electron density of 10^{20} cm^{-3} in a plasma that was assumed to be optically thin for XUV radiation.[17] Figure 2 shows the variation of ion stages with electron temperature, calculated by this steady-state CR model. This shows that 3s-3p, 3p-3d and 3d-4f transitions in the ions K^{3+} and K^{5+} are expected to dominate the spectra around 40 nm for time-averaged electron temperatures of 8–14 eV at an electron density of 10^{20} cm^{-3}.

In the case of discharge-produced potassium plasmas, the temporally and spatially averaged electron temperature, evaluated using the intensity ratios of the line emissions of oxygen ions which occur as impurities in the time-averaged XUV spectra, increased linearly with an increase in the discharge current. At a discharge current of 200 A, the electron temperature was observed to be about 10 eV.

For laser-produced plasmas, the laser intensity was set to be 2×10^{10} W/cm^2 in order to produce a time-averaged electron temperature of 12 eV. The temporally and spatially averaged electron temperature, again evaluated from intensity ratios of line emissions of oxygen impurity ions in the time-averaged XUV spectra, increased with increasing laser intensity.

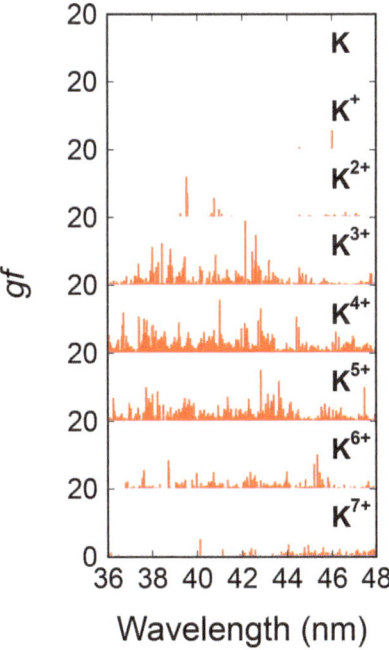

Figure 1 *The weighted oscillator strength (gf) spectra of the resonant lines from K to K^{7+}.*

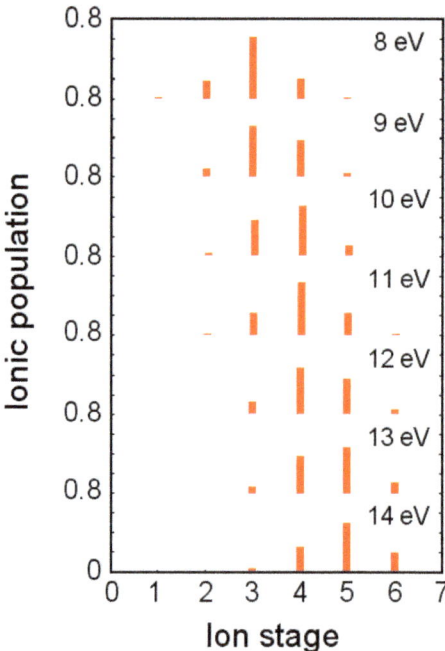

Figure 2 *Ionic populations of each ion charge state at electron temperatures of 8–14 eV.*

4 SPECTRA AND ANALYSIS

The time-averaged spectra around 40 nm from the capillary-discharge and the laser-produced potassium plasmas,[15,16] originate from potassium plasmas at an electron temperature close to 12 eV. The highest ion charge stage attainable was predicted to be K^{4+}. The spectral bandwidth of the emission was approximately 8 nm (FWHM) at 39 nm, mainly due to 3p-3d transitions of K^{3+} and K^{4+}. The discharge-produced spectrum[15] was quite similar to the laser-produced spectrum.[16] The positions of known lines in the potassium spectrum with heights proportional to their measured gA values are included for comparison;[14] g and A are, respectively, the statistical weight and the Einstein spontaneous emission coefficient. Note that the values were not weighted by relative ion populations. The weighted spectra were shown in figure 3 assumed at an electron temperature of 12 eV. The dashed line in figures 3(a) and 3(b) were also experimental spectra for the discharge-produced plasma (a) and for the laser-produced plasma (b), respectively. The similarity between laser and discharge spectra demonstrates that the emission from the discharge-produced plasma occurs in a region of high electron density of the order of 10^{20} cm^{-3}.

To explore the spectral emission further, the theoretical potassium plasma emission was calculated for a time-averaged electron temperature of 12 eV by combining the ion population weightings with the theoretical spectral information given in figure 1. The results are shown in figures 4(a) and 4(b); the former is the spectrum due to emission only while the latter includes opacity effects. Note the different vertical axis scales in Figs. 4(a) and 4(b). The differences in overall profile between these spectra are due to absorption broadening around 39 nm, which was also included in the numerically calculated transmission coefficient – the effective emissivity (the effective ionic plasma) thickness at 39 nm was evaluated to be 5 μm. The 39 nm emission originates from strong resonant lines and this emission is also self-absorbed.

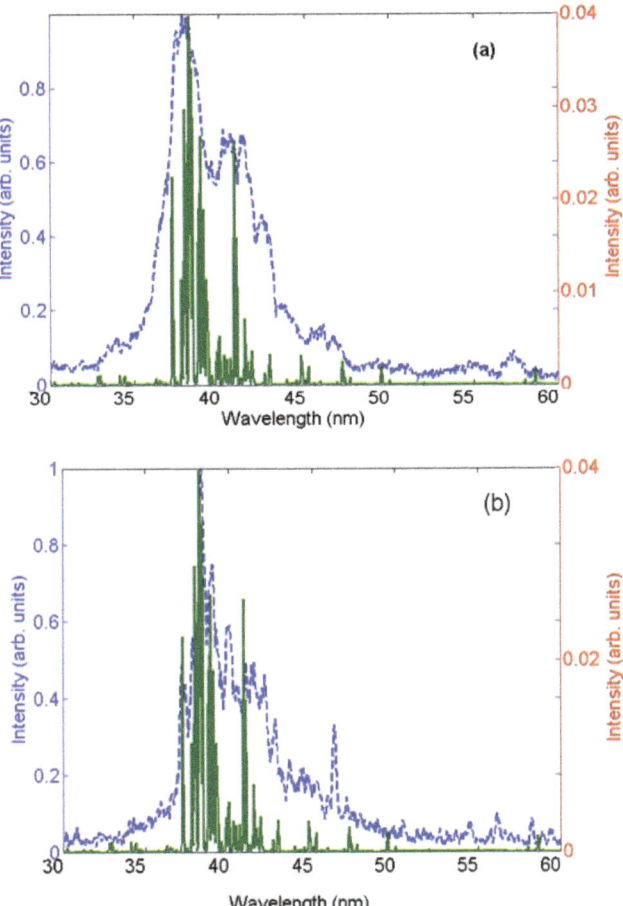

Figure 3 *The emission weighted by ion stage assuming a plasma electron temperature of 12 eV for (a) a discharge plasma and (b) a laser plasma.*

Figure 4(b) shows the numerically calculated spectrum at a time-averaged electron temperature of 12 eV and an electron density of 10^{20} cm^{-3}. The calculations are in reasonably good agreement with experimentally observed spectra,[15,16] and so the numerical calculations provide the insights needed to experimentally optimize the XUV emission.

In order to further understand the spectral behaviour for the emission centred around 40 nm, additional numerical calculations were performed using a hydrodynamic model. The calculations were done using a CR spectral analysis code designed to simulate the atomic and radiative properties of local thermodynamic equilibrium (LTE) and non-LTE plasmas spanning a wide range of conditions. Spectra at different electron temperatures, but at the same electron density of 10^{20} cm^{-3}, are shown in figure 5. When the electron temperature is low, as in figure 5(a), 50-55 nm emission, originating from K^{2+} and K^{3+} ions, is dominant. According to the Cowan code calculation, the weighted oscillator strength of the transitions responsible is weak which agrees with the available Einstein coefficient data.[17] As the electron temperature increases, the main peaks move to the shorter wavelength spectral region around 39 nm with an increase in emission intensity, which can be attributed to the presence of K^{3+}, K^{4+} and K^{5+} ions; the weighted oscillator strength from the Cowan code calculation and measured gA values[12] is predicted to be strong. In addition, the ion stage is increased due to higher ionization attainable at higher electron temperatures, as shown in figure 2.

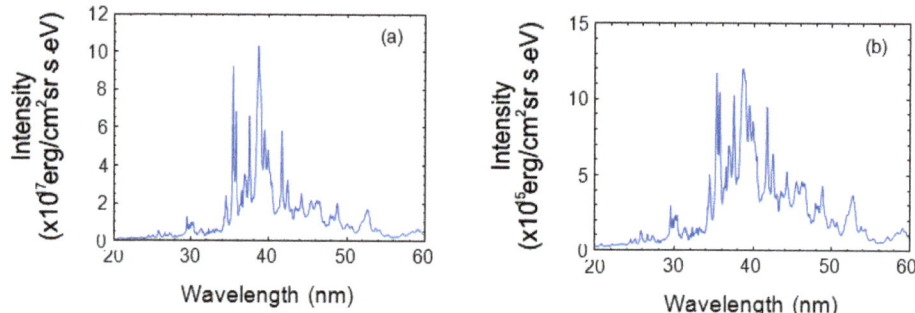

Figure 4 *The numerically calculated spectra of potassium plasmas (a) without and (b) with self-absorption (opacity) effects at an electron temperature of 12 eV based on the time-dependent CR model combined with a Cowan code simulation.*

The K^{3+}-K^{5+} ion emission is dominant at electron temperatures higher than 10 eV, as can be seen from figures 5(b) and 5(c). The observed spectra[15,16] closely resemble the numerically calculated one in figure 5(c) and confirm that $n = 3$ to $n = 3$ transitions in potassium ions ranging from K^{3+} to K^{5+} overlap to form a relatively narrow emission band.

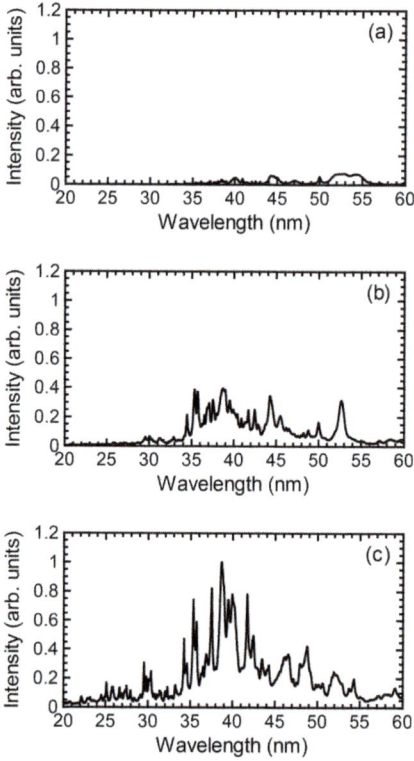

Figure 5 *Calculated spectra at electron temperatures of (a) 6 eV, (b) 10 eV and (c) 14 eV for an electron density of 10^{20} cm^{-3}.*

5 EFFECT OF IRRADIATING LASER WAVELENGTH

To explore self-absorption effects spectra were also observed at different laser wavelengths, because shorter wavelength irradiation produces plasmas with higher ion densities.[9,18,21,22,23] The spectra were seen to be almost identical for laser wavelengths of 1064 nm and 532 nm at the same laser intensity of 2×10^{10} W/cm^2, as shown in figure 6. At first sight, this result indicates that the potassium plasma may be only weakly self-absorbing though the earlier analysis[15,16] would seem to contradict this.

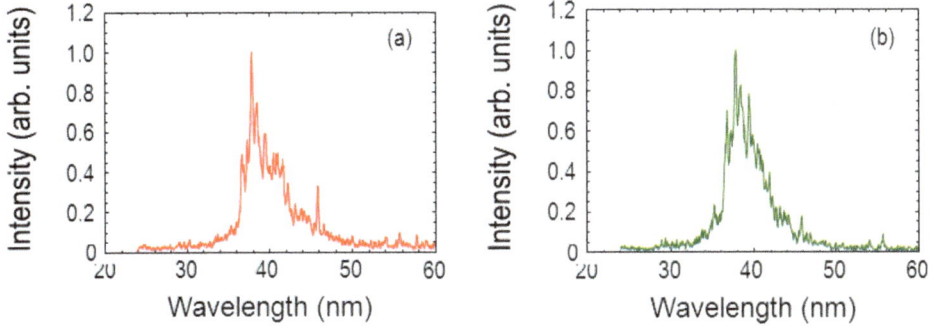

Figure 6 *Normalized time-averaged XUV spectra for laser wavelengths of (a) 1064 nm and (b) 532 nm (b) at the same laser intensity of 2×10^{10} W/cm^2.*

6 TEMPORAL HISTORY OF THE 39 NM EMISSION

In order to confirm the applicability of the time-dependent CR model in explaining the time evolution of the XUV emission, the temporal behaviour of the 39 nm emission from a laser-produced plasma was observed, as shown in figure 7(a). The dots represent the measured emission, and the solid line is a smoothing of the experimental result. Note that the time $t = 0$ ns corresponds to the laser intensity peak over the pulse duration of 10 ns (FWHM). The duration of the emission was 50 ns (FWHM). The dashed (red) line is a numerical simulation of the 39 nm emission. The calculation, shown in figure 7(b), was based on a time-dependent CR model which includes plasma processes such as ionization, recombination, emission, collisional excitation and de-excitation between several levels of the potassium ions. The electron temperature, determined experimentally by the laser pulse profile, was calculated in the hydrodynamic simulation and determines the ion population and therefore the 39 nm emission. The peak and time-averaged electron temperatures were calculated to be 26 eV and 1 eV, respectively, at the chosen laser intensity of 2×10^{10} W/cm2. The dominant 39 nm emission was shown to be generated both spatially and temporally when the electron temperature is close to 15 eV. The highest ion stages attained initially are K^{5+} to K^{7+}; thereafter the plasma recombines to lower ionization states, and starts emitting the 36-42 nm band in this recombining phase, as shown by figure 7(b).

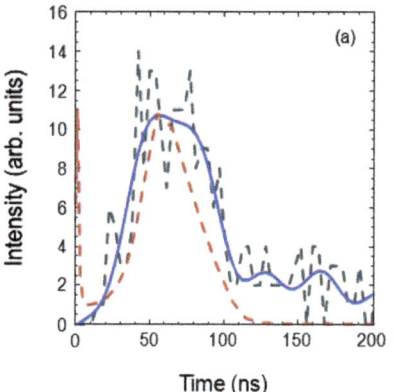 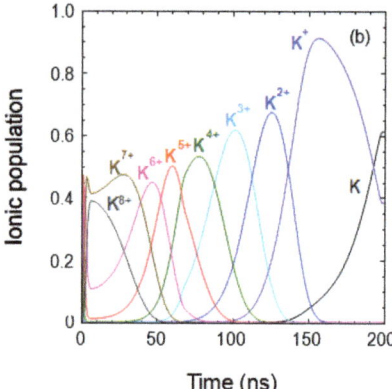

Figure 7 *(a) Temporal behaviour of the emission at 39 nm at a laser intensity of 2×10^{10} W/cm². The dots represent the experimentally measured data and the solid line (blue) is a smoothing of the experimental result. The dashed line (red) is the calculated temporal waveform of the 39 nm emission. (b) Numerical calculation of the temporal change of the ionic population based on a time-dependent CR model.*

7 SUMMARY

In summary, the emission in the 40 nm XUV spectral region has been characterized from capillary discharge and laser-produced plasmas containing potassium ions with time-averaged electron temperatures of 12 eV and electron densities of 10^{20} cm^{-3}. Atomic structure calculations with the Cowan code together with available spectral data from K^{3+} to K^{5+} ions were used to generate theoretical spectra to compare with the observed emission. The spectral behaviour was similar in both cases, showing that the plasma conditions are comparable in both plasma types. Emission at 39 nm was seen to occur during the recombining phase.

ACKNOWLEDGEMENTS

The authors are grateful to Professor Noboru Yugami, Dr Hiromitsu Terauchi, Ms Mami Yamaguchi, Mr Takamitsu Otsuka and Mr. Keisuke Kikuchi for their unparalleled technical support. Part of this work was performed under the auspices of MEXT (Ministry of Education, Culture, Science and Technology, Japan) and "Utsunomiya University Distinguished Research Projects." One of the authors (TH) also acknowledges support from the Canon Foundation. The University College Dublin group acknowledges support from Science Foundation Ireland under Principal Investigator Research Grant No. 07/IN.1/I1771.

References

1 G. O'Sullivan and P.K. Carroll, *J. Opt. Soc. Am.* 1981, **71**, 227.
2 P. Dunne, G. O'Sullivan, and V.K. Ivanov, *Phys. Rev. A*, 1993, **48**, 4358.

3 H. Tanaka, A. Matsumoto, K. Akinaga, A. Takahashi and T. Okada, *Appl. Phys. Lett.*, 2005, **87**, 041503.

4 T. Higashiguchi, N. Dojyo, M. Hamada, W. Sasaki and S. Kubodera, *Appl. Phys. Lett.*, 2006, **88**, 201503.

5 Y. Ueno, G. Soumagne, A. Sumitani, A. Endo and T. Higashiguchi, *Appl. Phys. Lett.*, 2007, **91**, 231501.

6 N. Murphy, A. Cummings, P. Dunne and G. O'Sullivan, *Phys. Rev. A*, 2007, **75**, 032509.

7 K. Fahy, E. Sokell, G. O'Sullivan, A. Aguilar, J.M. Pomeroy, J.N. Tan and J.D. Gillaspy, *Phys. Rev. A*, 2007, **75**, 032520.

8 T. Otsuka, D. Kilbane, J. White, T. Higashiguchi, N. Yugami, T. Yatagai, W. Jiang, A. Endo, P. Dunne and G. O'Sullivan, *Appl. Phys. Lett.*, 2010, **97**, 111503.

9 T. Otsuka, D. Kilbane, T. Higashiguchi, N. Yugami, T. Yatagai, W. Jiang, A. Endo, P. Dunne and G. O'Sullivan, *Appl. Phys. Lett.*, 2010, **97**, 231503.

10 G. Tallents, E. Wagenaars and G. Pert, *Nature Photonics*, 2010, **4**, 809.

11 J. Sugar and C. Corliss, *J. Phys. Chem. Ref. Data*, 1985, **14** (Suppl. 2), 1.

12 J.E. Sansonetti, *J. Phys. Chem. Ref. Data*, 2008, **37**, 7.

13 M. Koide, F. Koike, R. Wehlitz, M.-T. Huang, T. Nagata, J. C. Levin, S. Fritzsche, B.D. DePaola, S. Ohtani and Y. Azuma, *J. Phys. Soc. Japan*, 2002, **71**, 1676.

14 G. O'Sullivan, P.K. Carroll, J. Conway, P. Dunne, R. Faulkner, T. McCormack, C. McGuinness, P. van Kampen and B. Weinmann, *Opt. Eng.*, 1994, **33**, 3993.

15 T. Higashiguchi, H. Terauchi, N. Yugami, T. Yatagai, W. Sasaki, R. D'Arcy, P. Dunne, and G. O'Sullivan, *Appl. Phys. Lett.*, 2010, **96**, 131505.

16 T. Higashiguchi, M. Yamaguchi, T. Otsuka, H. Terauchi, N. Yugami, T. Yatagai, R. D'Arcy, P. Dunne and G. O'Sullivan, *Appl. Phys. Lett.*, 2011, **98**, 091503.

17 R. D. Cowan, *The Theory of Atomic Structure and Spectra* (University of California Press, Berkeley, 1981).

18 R.C. Spitzer, T.J. Orzechowski, D.W. Phillion, R.L. Kauffman and C. Cerjan, *J. Appl. Phys.*, 1996, **79**, 2251.

19 D. Colombant and G.F. Tonon, *J. Appl. Phys.*, 1973, **44**, 3524.

20 T. Fujimoto, *J. Phys. Soc. Jpn.*, 1979, **47**, 265; J. Phys. Soc. Jpn., 1979, **47**, 273.

21 M. Yamaura, S. Uchida, A. Sunahara, Y. Shimada, H. Nishimura, S. Fujioka, T. Okuno, K. Hashimoto, K. Nagai, T. Norimatsu, K. Nishihara, N. Miyanaga, Y. Izawa and C. Yamanaka, *Appl. Phys. Lett.*, 2005, **86**, 181107.

22 S. Miyamoto, A. Shimoura, S. Amano, K. Fukugaki, H. Kinugasa, T. Inoue and T. Mochizuki, *Appl. Phys. Lett.*, 2005, **86**, 261502.

23 J. White, P. Dunne, P. Hayden, F. O'Reilly and G. O'Sullivan, *Appl. Phys. Lett.*, 2007, **90**, 181502.

Source Development

LASER PRODUCED PLASMA X-RAY AND EUV SOURCES FOR LITHOGRAPHY

G.O'Sullivan

University College Dublin, School of Physics, Science Centre, Belfield, Dublin 4, Ireland

1 INTRODUCTION

One of the first major applications of laser produced plasmas was as sources for absorption spectroscopy. Carroll and co-workers discovered that the emission spectra of plasmas of elements with $62 \leq Z \leq 74$ emitted broad regions of line-free continua in the extreme ultraviolet and soft X-ray range.[1] The plasmas were produced by a Q-switched ruby laser and the power density on target was $\sim 10^{11}$ W cm^{-2} giving an average plasma temperature of 50–60 eV. Subsequently the suitability of these plasmas as transfer irradiance standards in the EUV was established.[2,3] In the course of this work comparison was made with the emission from CO_2 laser produced plasmas and the results showed that line emission was noticeably stronger than for the ruby laser, due to the lower plasma density, and also that the intensity could be enhanced by using a double pulse or pre-pulse illumination. In a 1984 study, the application of laser produced plasmas as sources for lithography was first demonstrated using emission from a Nd:YAG irradiated steel target.[4]

In studying the continuum emission, it was noted that in the lower Z elements of the sequence, an intense modulation of the continuum appeared in the 5–6 nm region that moved to shorter wavelength with increasing Z. A similar feature was noted in the spectra of all elements from tin through the lighter lanthanides. Moreover, it was observed that if the concentration of the element was reduced in the target, the feature actually grew in intensity and became significantly narrower.[5] It was therefore concluded that opacity is a limiting factor that could be obviated by reducing the plasma density, either through reducing the concentration of the element of interest in the target, or by using CO_2, rather than solid state, lasers or by using pre-pulses resulting in interaction with lower density target plasmas. It was shown that the modulation arose from $4p^6 4d^n$–$4p^5 4d^{n+1} + 4d^{n-1} 4f$ transitions, which merge to form an unresolved transition array (UTA) consisting of tens of thousands of lines. The peak position is sensitive to atomic number and moves to shorter wavelengths as Z increases; in tin it lies near to 13.5 nm.[6,7] Configuration interaction in the final state is very important and leads to a spectral narrowing and, in the heavier lanthanides, better overlap between the emission from adjacent ion stages than would be obtained by consideration of 4d–4f or 4p–

4d transitions alone. As a result, the UTA can be the brightest emission feature in the spectra of plasmas of these elements. The effects of configuration interaction are to concentrate the emission intensity in the high-energy end of the array and cause the oscillator-strength envelope of the transition array to narrow and overlap at essentially the same energy position in successive ion stages.

Over the last several years the largest international activity in extreme ultraviolet (EUV) science, especially in terms of industrial involvement, has centred on the optimisation and development of sources for EUV lithography (EUVL) and metrology at 13.5 nm. Currently, state of the art lithography is performed using 193 nm argon fluoride (ArF) excimer lasers. The attractiveness of EUVL stems from the fact that it represents the extension of tried and tested optical methods, albeit to another wavelength region. The alternatives of immersion lithography at 193 nm with double exposure and electron-beam lithography are deemed to be too expensive by comparison.

Because of the short wavelength, a suitable source must have a relatively high plasma temperature and so will utilize either discharge or laser produced plasmas. The parameters to be optimized are consequently the plasma composition in terms of elements to be used, the electron and ion densities, the electron temperature and the duration of the emission. Three candidate elements for EUVL which satisfy these requirements are lithium, tin and xenon, all of which have ions with resonance transitions in the 13.5 nm region. Of these xenon was perceived as the most attractive because of its gaseous nature which limits debris issues. However, the conversion efficiency of xenon discharge produced plasma (DPP) sources was found to be at best 0.5% while laser produced plasmas (LPP) generated on solid xenon targets had maximum conversion efficiencies of 1.2% when irradiated with Nd:YAG laser pulses at a wavelength of 1.06 μm.[8] It was shown that the 13.5 nm emission from xenon originated from $4d^8$–$4d^7 5p$ resonance transitions in Xe^{10+}.[9] The population of this ion stage needs to be optimised to attain maximum intensity, but the xenon spectrum is dominated by a $4p^6 4d^n$–$4p^5 4d^{n+1}$+$4d^{n-1} 4f$ UTA near 11 nm and subsequent studies performed at the National Institute of Standards and Technology (NIST) electron beam ion trap (EBIT) source showed a five-fold enhancement in a 2% bandwidth near this wavelength where emission arises from Xe^{11+}–Xe^{17+}.[10]

No suitable mirrors exist at these wavelengths with reflectivities approaching the figure of >70% attainable at 13.5 nm using Mo/Si multilayers, but the analogous transitions in tin lie near to 13.5 nm. In 2005, the semiconductor industry set a power requirement of 120 W in band at intermediate focus, after reflection off the collector optic, as a target for source development. It became immediately obvious that while xenon might have a future in metrology sources for high volume manufacturing where EUV power and a high collection efficiency are essential, tin was a more suitable source. However, opacity is now a significant issue. In a controlled study with tin doped glass targets, summarised in figure 1, it was found that the brightness attainable with solid flat (slab) targets at 5% concentration was some 45% greater than with a pure tin target, and yielded a conversion efficiency close to 3% with a Nd:YAG pulse focused to a power density of 2×10^{11} W cm^{-2}.[11] The transitions responsible for the emission were identified by Churilov and co-workers[12] and shown to originate from Sn^{8+} to Sn^{12+}. Plasma modelling calculations within the collisional-radiative (CR) regime (where collisional excitation is balanced by radiative and three-body recombination) performed to obtain the ion distributions as a function of electron temperature showed that the condition for production of these ion stages for an optically thin plasma corresponds to electron temperatures in the 30–50 eV range.[13]

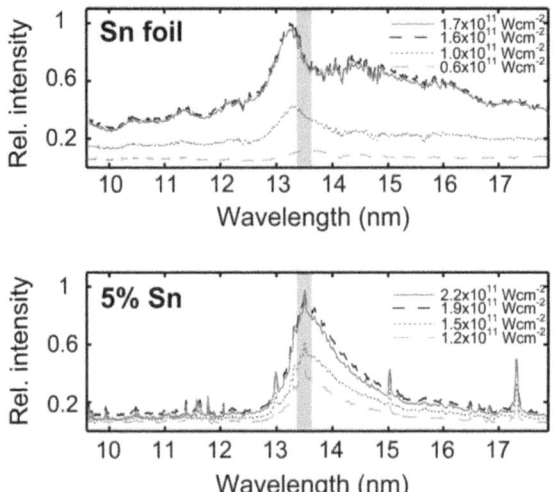

Figure 1 *Comparison of emission from solid and low density tin targets. The in-band intensity is clearly greatest for the plasma containing a tin concentration of 5% by ion number due to reduced opacity.*

2 Nd:YAG BASED SOURCES FOR HIGH VOLUME MANUFACTURING

While tin was already known to be a good emitter at 13.5 nm, early work had concentrated on xenon since, being an inert gas, it posed fewer debris problems than thin. To minimize the problem of debris, mass limited droplets, where the target contains the maximum number of atoms that can be fully ionised were proposed and developed as the optimum EUV source while debris mitigation schemes such as plasma curtains to ionize neutral debris and electrostatic and magnetic deflection setups were developed in attempts to remove debris.[14,15] A number of groups using droplet targets with Nd:YAG irradiation showed that the max conversion efficiency (CE) obtainable was around 2-2.5% and better for shorter laser pulses. The reason that the 6% CE predicted by considering the earlier xenon work[10] was not attained was due to two factors, namely plasma opacity and suboptimal emitting volume. If targets containing a few percent of tin by number were used the CE increased while the use of pre-pulses with 100% tin targets could effectively have the same impact and increase the CE by a factor of up to 80%.[16] With a pre-pulse the plasma has time to expand and the resulting increase is due to the interaction of the laser beam with a larger plasma EUV emitting volume at a lower density than produced by the laser-solid interaction directly, thus reducing opacity effects. Another method of reducing tin concentration was the adoption of foam targets where concentrations of 0.5% gave CEs of around 1.5%.[17] While the results of the Xe EBIT experiments[10] and early calculations[14] pointed to a CE of 5-6% being attainable in the absence of absorption this figure is reduced to 2-3% when opacity effects are included.

In a detailed study using laser irradiation of polymers coated with a 100 nm thick layer of tin, Ando *et al.*[18] measured both the angle of the EUV emission and the CE for pulse lengths of 1.2– 8.5 ns. They found that optical depth increased linearly with pulse length and obtained a maximum CE of 2.2% at a pulse length of 2.3 ns for a power density of 5×10^{10} W cm^{-2}. The power density was varied from 10^{10} W cm^{-2} to 10^{120} W cm^{-2} by using focal spots of 300–900 μm. The results also showed that the maximum EUV emission was obtained in the direction of the target normal and that the optimum pulse duration was determined not only by the optical depth but also by the fraction of the laser energy absorbed in the EUV emitting region, which is located in the lower density corona. Indeed, an earlier theoretical study that involved hydrodynamical modelling found that most of the plasma emission originated from the plasma periphery, in this case the outermost four of the 500 cells used to model a 1-D expansion.[19] The result of the simulation is shown in figure 2. In their paper, Ando *et al.* calculated the fraction of the laser energy absorbed in the EUV emitting region to vary from 25% at 1.2 ns to 85% after 8.5 ns, corresponding to absorption at increasing scale lengths due to plasma expansion. Hence the emission can be controlled by maximizing the scale length while at the same time minimizing the optical depth. The use of pre-pulses provides an efficient method of accomplishing this.

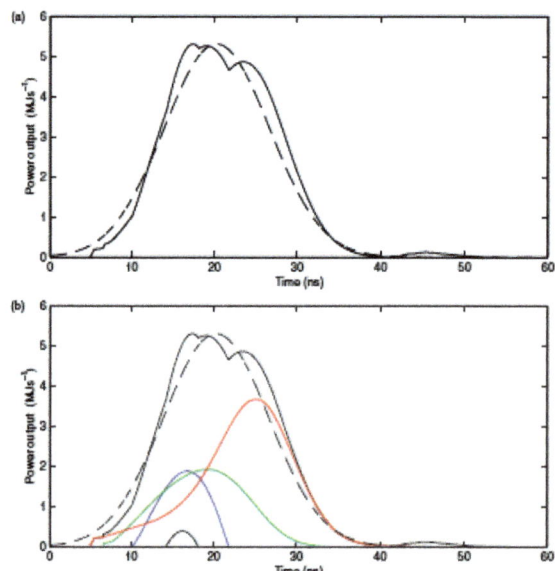

Figure 2 *Time dependence of the net power output for a peak power density of 3×10^{11} W cm^{-2} (top). The dashed curve indicates the laser pulse. The lower figure shows the contribution of the outermost fluid cells, with the red curve indicating the very outermost cell.*

It was also shown that the emission was not isotropic but varied with viewing angle and maximized when the plasma was viewed at normal incidence. Subsequently this effect was studied in more detail.[20,21] Both measurements obtained essentially identical results and the angular distribution of radiation is shown in figure 3.

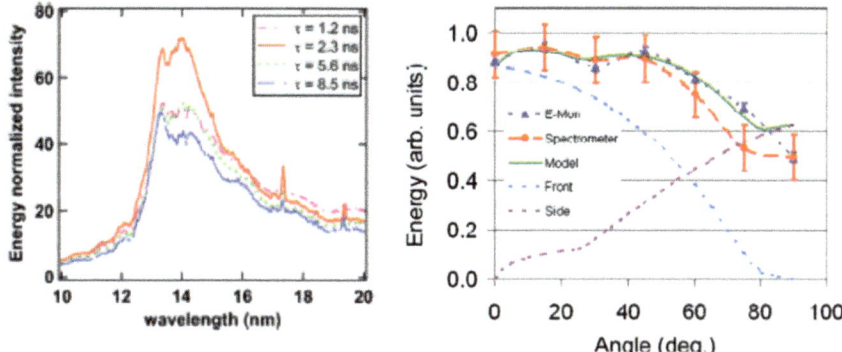

Figure 3 *Angular distribution of the EUV emission and intensities at different pulse lengths for Nd: YAG irradiation of solid tin. (left: Ando et al.[18], right: Sequoia et al.[21]) Reprinted with permission from T. Ando et al. Appl. Phys. Lett., **89**, 151501 and K L Sequoia et al. Appl. Phys. Lett. 92, 221505. Copyright 2006, 2008, AIP Publishing LLC.)*

3 CO$_2$ BASED SOURCES FOR HIGH VOLUME MANUFACTURING

It had already been pointed out[22] that the use of CO$_2$ lasers operating at a wavelength of 10.6 μm would help reduce the opacity problem as the plasma electron and ion densities at this wavelength are a factor of 100 lower than for Nd:YAG produced plasmas since the critical electron density, i.e., the density at which laser light is most efficiently absorbed, is approximately $10^{21}\lambda^{-2}$ cm^3. Calculations by Nishihara and co-workers[23] showed that for CO$_2$ radiation incident on a solid tin target at a power density of 10^{10} W cm^{-2}, a conversion efficiency of 6% was attainable at the critical electron density in a 30 eV plasma (figure 4). Moreover, this figure shows that at lower densities and irradiances even higher CEs are attainable. However the number of ions becomes so small that the absolute emitted intensity is insufficient even at 100 kHz repetition rates to reach the power requirements for high volume manufacturing (HVM). Furthermore the calculations showed that the spectral efficiency reaches a maximum of 35% for an ion density of $(4-5)\times10^{17}$ cm^{-3} and an electron temperature of 45 eV. Under these conditions, the optimum pulse duration is of the order of 10 ns.

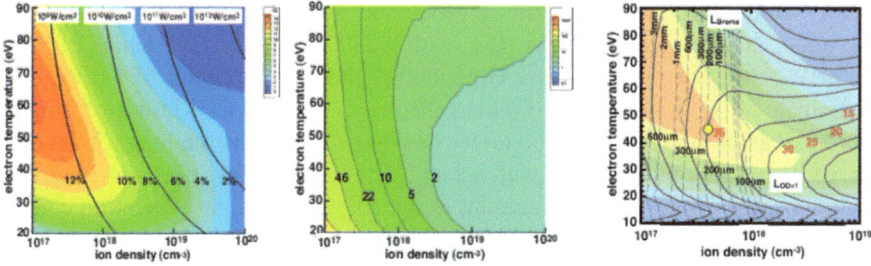

Figure 4 *Optimum conversion efficiency from 1% to 13% in 1% intervals (left), optimum pulse durations (centre) and spectral efficiencies (right) for maximum conversion as a function of ion density and plasma electron temperature. (After Nishihara et al. 2008[23]). (Reprinted with permission from Nishihara et al., Physics of Plasmas, **15**, 056708, Copyright 2008, AIP Publishing LLC)*

Early work with CO_2 lasers and solid tin slab targets indicated that conversion efficiencies approaching 4% could be obtained for 10 ns pulses incident on slab targets though the CE dropped to less than 3% for longer pulses. Ueno *et al.*[24] found that using 100 ns long CO_2 pulses the CE increased from 2% to 4% if plasmas were formed in 200 μm deep cavities at power densities of $(1–3)\times10^{10}$ W cm^{-2} while the CE was unchanged for Nd:YAG produced plasmas for any cavity depth. Subsequently Tao *et al.*,[25] using an identical pulse length and power density found that the CE increased from 2% to close to 5% after twenty laser shots at the same target position while a similar study[26] using 30 ns CO_2 pulses at a focused power density of 6×10^{10} W cm^{-2} found that the CE increased from 2.7% to 5% after around sixty pulses on the same position. In a controlled experiment to study the effects of grove width it was found that the highest CEs were obtained when the grove width was between one and two times the focal spot diameter. Studies showed that the electron density inside the groove reached a maximum of 10^{19} cm^{-3}. Compared to planar targets, where the plasma expands freely and the density decreases rapidly away from the surface, grooved targets control the hydrodynamic expansion by confining the plasma. This in turn leads to a denser region appearing in front of the target with a gentler density gradient. In addition, the geometrical hydrodynamic confinement prevents energy loss from the EUV production zone through lateral expansion thereby increasing radiation losses.

However, slab targets cannot be used for high volume production due to the heating produced at the required laser repetition rates of up to 100 kHz. One solution is to use mass limited droplets, the first successful demonstration of which involved the use of 10 ns Nd:YAG laser pre-pulses incident on 40 μm diameter tin droplets at a power density of 5×10^8 W cm^{-2}.[23] If the main 10 ns CO_2 laser pulse arrived 180 ns after the end of the pre-pulse, a CE of 6.5% was attained in the experiments. Later, using 36 μm droplets, irradiated with Gaussian 1.06 μm pulses at intensities of $5\times10^{10} – 4\times10^{11}$ W cm^{-2} while the main 10.6 μm pulse had a duration of 30–50 ns and a power density of 10^{10} W cm^{-2}, a CE of 4% was measured at an inter-pulse delay close to 1 μs at which time the droplet has expanded to fully fill the focal spot of the CO_2 laser.

The above discussions show that, currently, there are two main contenders for EUV sources at 13.5 nm, namely discharge-produced plasmas and laser-produced plasmas. The collector configuration favours the latter since normal incidence optics exist with collection efficiencies of 30%, three times that for the grazing incidence configurations used with DPP sources. Debris is still an issue and one novel solution that is being tested is the use of liquid metal coated optics using a tin alloy as the liquid metal.[27]

Currently DPP sources are capable of producing 34 W of in-band radiation after the collector optic, at intermediate focus (IF), with a CE of around 2% while LPP sources produce 21 W. However the company Gigaphoton, the main supplier of LPP sources for HVM anticipate 50 W at IF using a 100 kHz droplet source and CO_2 laser operating with a CE of 2.5%. Part of the reason for the lower than expected CE is the longer than optimum duration of the 10.6 μm pulse; novel schemes to extend the plasma lifetime while maintaining the plasma density are being investigated. In their first-generation commercial source Gigaphoton hope to realise 140 W at IF assuming a 4% CE while their second-generation source is expected to deliver 250 W of in-band EUV at IF, again with a CO_2 irradiated tin droplet at a CE of 5%.

4 LITHOGRAPHY SOURCES BEYOND 13.5 NM

In 2009 ASML announced that sources would be needed at 6.x nm for future lithography beyond 13.5 nm predicated on the availability of mirrors, this time La/B$_4$C multilayers which reflect in a 0.06 nm bandwidth near 6.7 nm.[28] The theoretical reflectivity is 80% and recent work has seen the experimental figure double to over 40%.[28] Both gadolinium and terbium have $4p^6 4d^n - 4p^5 4d^{n+1} + 4d^{n-1} 4f$ transitions near this wavelength and will emit intense UTAs. The emission processes is more efficient than in tin because of the more complete overlap of emission in adjacent stages and also due to the fact that the 4f wave function is completely contracted in the relevant ions, thereby maximizing the oscillator strength. In addition, 4d – 4f transitions of the type $4d^{10} 4f^n - 4d^9 4f^{n+1}$, not present in tin, may also contribute to the in-band emission. Thus many ion stages can contribute and strong emission is expected over a wide range of plasma temperatures. In modelling the optimum conditions, the presence of these open 4f sub-shell configurations make calculation of spectra much more complex than in the case of tin because of the greatly increased number of transitions possible and the close proximity of configurations containing variable numbers of 4f, 5p and 5s electrons in lower ion stages.[29]

The spectra of vacuum spark and laser plasma emission from laser produced plasmas of gadolinium and terbium are shown in figure 5.[30] The optimum ion stages should be those containing fewest lines so that the emission is not divided among many transitions. Thus for gadolinium, for example, the strongest lines should result from 4d–4f lines in the spectra of Ag-like Gd XVIII, Pd-like Gd XX and Rh-like Gd XXI; the electron temperatures in a gadolinium plasma should be that which maximises the population of these ions. From calculations assuming CR equilibrium, the ion stage distribution as a function of electron temperature for a Nd;YAG produced plasma is shown in figure 6. It can be seen that the optimum plasma temperature is expected to be in the range 100–130 eV.

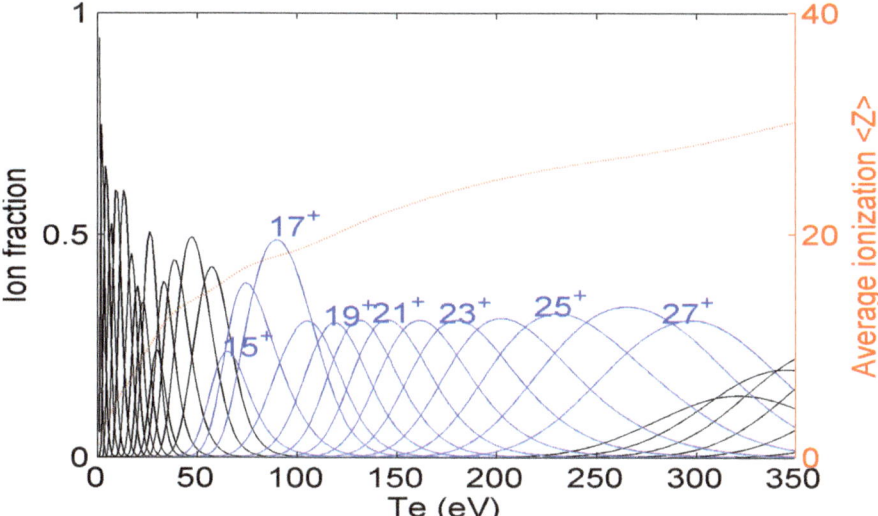

Figure 5 *Calculated ion stage distribution as a function of electron temperature in a Nd:YAG LPP of gadolinium*

Time integrated spectra have been recorded at spot diameters in the range 30–300 μm using a 2J laser with a 10 ns pulse length, giving corresponding power densities 1.5×10^{13}–

1.5×10^{11} W cm^{-2}.[30] The peak emission was observed at a power density of 4×10^{11} W cm^{-2}, a spot diameter of 210 μm. The plasma temperature in this case was 45 eV, which is suboptimal, but the decreased emission at higher temperatures results from the decreased plasma emitting volume and increased lateral expansion at smaller focal spot sizes which effectively transfers energy from radiation to plasma kinetics. Comparison of emission from plasmas produced by fundamental and frequency doubled Nd:YAG lasers clearly showed increased emission by satellite transitions at longer wavelengths in the latter and reduced UTA emission because of increased opacity for higher energy excitation. From their observations, the authors identified the need for shorter pulse irradiation, a low initial density target and low electron density plasmas such as LPPs or DPPs, produced using a CO_2 laser, the same conditions as for tin plasmas. In a subsequent paper,[31] the wavelength effects on opacity were further explored for an irradiance of 1.6×10^{12} W cm^{-2} and a 50 μm focal spot diameter, clearly showing that satellite emission is enhanced at shorter wavelengths. Moreover, a comparison of emission from slab targets of gadolinium and Gd_2O_3 showed that use of the latter led to an increase in both emitted in-band flux and spectral purity; the measured CEs (within a 2% bandwidth) were 1.05% and 1.3%. With targets containing gadolinium concentrations of 30% and 10% the CEs were improved to 1.6% and 1.45% respectively,[32] clearly showing the effects of opacity. Use of a pre-pulse with the 30% target increased the overall CE to 1.8% for inter-pulse time delays of around 500 ns.

5 CONCLUSIONS

For Sn based sources, the path to incorporation in a lithography tool is clear. The source of choice will be a Sn liquid droplet source, ideally operating at a 100kHz repetition rate preionised by a low intensity Nd:YAG prepulse and allowed to expand until it reaches the critical density for CO_2 radiation when it will be reheated by a CO_2 pulse. The challenges here are tailoring the duration of the CO_2 pulse to match the lifetime of the plasma target when it is at the optimum density for absorption of the 10.6 μm radiation and sustaining the high required repetition rate for both stable droplet injection and laser operation. For 6.x nm based sources a major increase in CE needs to be attained in order to demonstrate the feasibility of that technology as well as the experimental realisation of the theoretical mirror reflectivity values for the LaB_4C multilayers. Even then, the design of a droplet injection system will pose a major additional challenge. Should all these difficulties be overcome the power requirement at intermediate focus will probably exceed 1KW because of the increased resolution required on the resist.

References

1 P.K. Carroll, E.T. Kennedy and G. O'Sullivan, *Appl. Opt.*, 1980, **19**, 1454.

2 G. O'Sullivan, J.R. Roberts, W.R. Ott, J.M. Bridges, T.L. Pittman and M.L. Ginter, *Opt. Lett.*, 1982, **7**, 31.

3 G. O'Sullivan, P.K. Carroll, T.J. McIlrath and M.L. Ginter, *Appl. Opt.*, 1981, **20**, 3043.

4 D.J. Nagel, C.M. Brown, M.C. Peckerar, M.L. Ginter, J.A. Robinson, T.J. McIlrath and P.K. Carroll, *Appl. Opt.*, 1984, **23**, 1428.

5 G. O'Sullivan and P.K. Carroll, *J. Opt. Soc. Am.*, 1981, **71**, 227.

6 P.K. Carroll and G. O'Sullivan, Phys. Rev., 1982, **A25**, 275.

7 P. Mandelbaum, M. Finkenthal, J.L. Schwob and M Klapisch, Phys. Rev., 1987, **A35**, 5051.

8 H. Shields, S.W. Fornaca, M.B. Petach, M. Michaelian, R.D. McGregor, R.H.Moyer and R.J. St. Pierre, *Proc. SPIE*, 2002, **4688**, 94.

9 S. Churilov, Y.N. Joshi and J. Reader, *Opt. Lett.*, 2003, **28**, 1478.

10 K. Fahy, P. Dunne, L. McKinney, G. O'Sullivan, E. Sokell, J. White, A. Aguilar, J.M. Pomeroy, J.N. Tan, B. Blagojević, E-O. LeBigot and J.D. Gillaspy, *J. Phys. D: Appl. Phys.*, 2004, **37**, 3225.

11 P. Hayden, J. White, A. Cummings, P. Dunne, M. Lysaght, N. Murphy, P. Sheridan and G. O'Sullivan, *Microelectronic Engineering*, 2006, **83**, 699.

12 S.S. Churilov and A.N. Ryabtsev, *Phys. Scr.*, 2006, **73,** 614.

13 J. White, P. Hayden, P. Dunne, A. Cummings, N. Murphy, P. Sheridan and G. O'Sullivan, *J. Appl. Phys.*, 2005, **98**, 113301.

14 T. Aota and T. Tomie, *Phys. Rev. Lett*, 2005, **94**, 015004.

15 S.S. Harilal, B. O'Shay, Y. Tao and M.S. Tillack, *Appl. Phys.s B: Lasers and Optics*, 2006, **86**, 547.

16 G. O'Sullivan, A. Cummings, G. Duffy, P. Dunne, A. Fitzpatrick, P. Hayden, L. McKinney, N. Murphy, D. O'Reilly, E. Sokell and J. White, *Proc. SPIE*, 2004, **5196**, 273.

17 S.S. Harilal, B. O'Shay, M.S. Tillack, Y. Tao, R. Paguio, A. Nikroo and C.A. Back, *Phys. D: Appl. Phys.*, 2006, **39**, 484.

18 T. Ando, S. Fujioka, H. Nishimura, N. Ueda, Y. Yasuda, K. Nagai, T. Norimatsu, M. Murakami, K. Nishihara, N. Miyanaga, Y. Izawa and K. Mima, *Appl. Phys. Lett.*, 2006, **89**, 151501.

19 A. Cummings, G. O'Sullivan, P. Dunne, E. Sokell, N. Murphy and J. White, *J. Phys. D: Appl. Phys.*, 2005, **38,** 604.

20 O. Morris, F. O'Reilly, P. Dunne and P. Hayden, *Appl. Phys. Lett.*, 2008, **92**, 231503.

21 K L Sequoia, Y Tao, S Yuspeh, R Burdt and M S Tillack 2008 *Appl. Phys. Lett.* 92, 221505

22 G. O'Sullivan and P. Dunne, *International SEMATECH EUV Source Workshop*, 2002 (Dallas, USA). http://www.sematech.org/meetings/archives/litho/euvl/20021014/16-Spectro.pdf

23 K. Nishihara, A. Sunahara, A. Sasaki, M. Nunami, H. Tanuma, S. Fujioka, Y. Shimada, K. Fujima, H. Furukawa, T. Kato, F. Koike, R. More, M. Murakami, T. Nishikawa, V. Zhakhovskii, K. Gamata, A. Takata, H. Ueda, H. Nishimura, Y. Izawa, N. Miyanaga and K. Mima, *Physics of Plasmas*, 2008, **15**, 056708.

24 Y. Ueno, G. Soumagne, A. Sumitani, A. Endo and T. Higashiguchi, *Appl. Phys. Lett.*, 2006, **91**, 231501.

25 Y. Tao, M.S. Tillack, S. Yuspeh, R. Burdt, N. Shaikh, N. Amin and F. Najmabadi, *IEEE Transactions on Plasma*, 2010, **38**, 714.

26 S.S. Harilal, T. Sizyuk, V. Sizyuk and A. Hassanein *Appl. Phys. Lett.*, 2010, **96**, 111503.

27 K. Fahy, F. O'Reilly, E. Scally, I. Kambali and P. Sheridan, *Proc. International Workshop on EUV Sources*, 2010, (University College Dublin, Ireland).

28 A. Makhotkin, E. Louis, E. Zoethout, R.W.E. van de Kruijs, Andrei M. Yakunin, Stephan Müllender and F. Bijkerk *Proc. International Workshop on EUV Sources*, 2010, (University College Dublin, Ireland)

29 D. Kilbane and G. O'Sullivan, *J. Appl. Phys.*, 2010, **108**, 104905.

30 S.S. Churilov, R.R. Kildiyarova, A.N. Ryabtsev and S.V. Sadovsky, *Phys. Scr.*, 2008, **80** 045303.

31 T. Otsuka, D. Kilbane, J. White, T. Higashiguchi, N. Yugami, T. Yatagai, W. Jiang, A. Endo, P. Dunne and G. O'Sullivan, *Appl. Phys. Lett.*, 2010, **97**, 111503.
32 T. Otsuka, D. Kilbane, T. Higashiguchi, N. Yugami, T. Yatagai, W. Jiang, A. Endo, P. Dunne and G. O'Sullivan 2010, *Appl. Phys. Lett.*, 2010, **97**, 231503.

PRACTICAL ASPECTS OF XUV GENERATION BY NON-LINEAR FREQUENCY CONVERSION

Bill Brocklesby

Optoelectronics Research Centre, University of Southampton, Highfield, Southampton, Hampshire. SO17 1BJ, UK

1 INTRODUCTION

The use of nonlinear optics for frequency conversion is now widespread in photonics, from complex optical parametric amplifiers to simple second harmonic generation in green laser pointers. Frequency conversion from visible to soft X-ray radiation is an extreme example of the process, but is possible given the very high intensities available from lab-scale laser sources. X-ray sources based on frequency conversion have many of the useful properties of the pump lasers – they can produce very short pulses (down to less than 100 attoseconds), they can have very high spatial coherence, and produce very well-collimated beams. However, the production process is not efficient, and the necessary pump lasers are both complex and expensive. This review is intended to provide basic information about the frequency conversion process – its physical basis (high-harmonic generation, or HHG), and its present and potential applications. The applications tend to fall into two categories: the science and engineering of attosecond pulses, and the production of soft X-rays for imaging and spectroscopy. The latter area will be covered within this chapter on "practical aspects" – excellent reviews of attosecond science are available elsewhere.[1] Section 1 will review briefly the laser sources necessary for HHG. Section 2 will give an overview of the single atom processes that are responsible for the nonlinearity, and section 3 will look at the complex process of phase matching the nonlinear interaction, which is critical for high efficiency X-ray generation. Section 4 will consider experimental aspects of HHG, comparing the output characteristics of typical HHG sources with other sources in the soft X-ray spectral region, and section 5 will outline a few of the many applications for these sources.

1.1 High-Intensity Pulsed Lasers

The advent of mode locked lasers gave a route to lab-based laser sources of high intensity. A mode locked laser system produces a train of high-intensity pulses via the creation of an interference pattern in time, leading to pulses spaced in time by the cavity round-trip time of the laser, whose time width is determined by the inverse bandwidth of the gain medium. Thus the use of a gain medium with a very broad bandwidth, like titanium-doped sapphire, makes the production of pulses in the sub-50 fs regime straightforward. However, peak powers reached in these systems might typically be ~250 kW, in a pulse of total energy

~10 nJ. In order to reach the peak intensities necessary for HHG, further amplification is needed.

Amplification of femtosecond pulses to high peak powers is made possible using chirped pulse amplification (CPA). Chirped pulse amplification[2] avoids one of the major problems in pulse amplification, which is that the high peak powers in the amplified pulses cause nonlinear effects within the amplifying medium, and distort the pulse, increasing its temporal length in a manner that is difficult to reverse. At very high peak powers, these nonlinear effects can cause damage to the amplifier gain medium.

In order to amplify without appreciable nonlinear distortion, the pulse can be chirped. Pulse is propagated through an optical system with appreciable second order dispersion, producing a phase shift that is linearly proportional to frequency within the pulse. This stretches the pulse in time, changing the relative delay of each of the frequency components making up the pulse. Stretching of the pulse reduces the peak power significantly, reducing nonlinear effects. After stretching, the pulse can be amplified, and then the opposite second order dispersion applied using a different optical system, and the frequency components can be reassembled into a short pulse. In this case, the full peak power is not achieved until the final pulse compression, meaning that few of the optics in the system ever have to tolerate the highest peak powers. Typically stretchers and compressors based on gratings are used, although many other variants are possible.

A typical lab-scale system for CPA might produce pulses of length ~30 fs, with energy of ~1 mJ – a peak power of ~30 GW. This comes at a reduced repetition rate; typical CPA systems using Ti:sapphire amplifiers run at 1 kHz, using a frequency-doubled diode-pumped Q-switched Nd laser as the pump. The repetition rate of the Q-switched pump limits the overall repetition rate to a few kilohertz.

1.2 Comparison of Electric Field Strengths

The 30 GW peak power from a CPA system, when loosely focused to ~100 μm diameter spot, produces a peak irradiance of almost 10^{15} W/cm². It is illuminating to compare the electric field at the focal spot to the Coulomb field at the Bohr radius within a hydrogen atom. The laser field is ~80 GV/m; the field at 1 Bohr radius from the proton is ~500 GV/m. The laser field is a substantial fraction of the Coulomb field, and hence the effect of laser irradiation on the atom will be very significant.

1.3 Fibre Laser Sources

Progress to higher average powers than those available using Ti:sapphire gain media is possible using optical fibre-based CPA systems. High power continuous wave optical fibre lasers are now routinely available[3] with average powers above 1 kW. This high average power is very attractive for nonlinear generation; however, the ability of the fibre source to produce very short pulses is severely limited by nonlinearity in the fibre amplifiers, meaning that very high peak powers are not easy to achieve. Additionally, the bandwidth available from typical fibre laser gain media, often ytterbium-doped silica fibre, is not large enough to support very short pulses – pulse lengths down to ~250 fs are available directly, and further shortening must be produced by using nonlinear effects to generate broader bandwidths.

Despite the inherent difficulties in producing very short pulses from a fibre system, the progress of fibre CPA sources is impressive. Sub-500 fs pulses with energies up to 2 mJ have been demonstrated[4] at average powers ~10 W, and average powers of up to ~800 W have been produced for pulses of ~900 fs and 10 μJ energy.[5] Fibre CPA systems have already been used to generate high-harmonic radiation[6] at high repetition rates. The available pulse parameters, coupled with the ability to perform nonlinear pulse compression to reach much shorter pulse lengths, and the recent demonstration of coherent combination of pulses from multiple fibre CPA systems,[7] make fibre CPA sources a strong candidate for future high-flux HHG sources.

2 SINGLE ATOM NON-LINEAR PROCESSES

2.1 Non-linear Processes: Quantum Model

The effect of irradiating an atom with a laser field of similar magnitude to the binding field is to produce non-linear effects on a different scale to those usually available with lasers. For moderate fields, the polarization P due to the nonlinear response of the atom to a laser field E of frequency ω is usually described by $P = \chi E$, where

$$\chi = \chi_1 + \chi_2 E + \chi_3 E^2 \dots \tag{1}$$

The non-linear polarizabilities χ_i are the coefficients of a Taylor expansion for χ in the field E, and get progressively smaller for higher order terms in the expansion. In the case of high-field effects like HHG, the assumption that the response to the electric field is a perturbation of the atom is no longer valid, and the Taylor expansion in equation (1) cannot be used. Instead, it is necessary to calculate, either analytically or numerically, the response of the atom to the laser field, and from this, the dipole moment produced by the field can be extracted. Prediction of the single atom polarization was first demonstrated analytically by Lewenstein *et al.*[8] who described the form of the spectrum of harmonics, which is very different from the rapid fall-off usual with perturbative non-linear optical effects. The spectrum has an initial fall-off as the harmonic order is increased, but then extends into a 'plateau' region, up to a distinct cut-off, as illustrated schematically in figure 1. The energy of the cut-off is directly related to the input laser intensity.

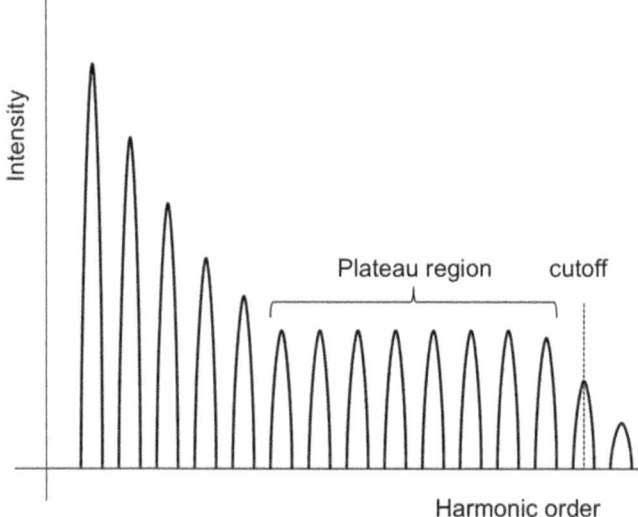

Figure 1 *Intensity variation of different harmonic orders in HHG*

The energy of the cutoff is described by a simple formula[9]

$$E_{\text{cut-off}} = I_P + 3.2U_P \tag{2}$$

where I_P is the ionization potential of the atom or molecule in use, and U_P is the ponderomotive energy of the electron in the laser field, given by

$$U_P = \frac{e^2 E^2}{4m_e \omega^2} \tag{3}$$

where E is the electric field of the laser, m_e is the electron mass, and ω is the laser frequency.

2.2 Semi-Classical '3-Step' Model

Physically the $3.2U_P$ corresponds to the maximum kinetic energy that the electron can have when it returns to the atomic core due to being accelerated in the field of the laser. Its value can be calculated using classical equations of motion for the electron in the laser field, within a model which breaks down the HHG process into three steps: tunnel ionization,[10] followed by classical motion of the electron in the laser field, followed by return of the electron to the atomic core causing light emission via an excited dipole. Many of the parameters important for HHG can be obtained from this relatively simple model. As an example, HHG requires significant probability of ionization during the time of the laser cycle. The necessary intensity for this ionization can be calculated from the Keldysh parameter, γ, which is defined as the ratio of the laser frequency and the tunnelling frequency, and given by $\gamma = \sqrt{I_P/2U_P}$; tunnel ionization dominates when γ < 1. For argon, with an ionization potential of 15.7 eV, this regime corresponds to intensities greater than ~10^{14} W/cm², which sets the intensity scale necessary for HHG using argon atoms.

The consequence of equation (2) is that the maximum energy photons which can be obtained for a given laser intensity is fixed, and increasing the maximum photon energy can be achieved by either increasing intensity or increasing the wavelength of the illuminating radiation. This has consequences for the choice of pulse source, which will be explored in section 3.5.1.

2.3 One-Dimensional Time-Dependent Schrödinger Equation Models

More detailed properties of the single atom nonlinearity can be explored reasonably simply by direct integration of the time-dependent Schrödinger equation (TDSE) for a model one-dimensional atom in the laser field. Figure 2 shows the evolution of the wave function of an electron excited by a 7 fs laser pulse with peak intensity 3×10^{14} W/cm² in a soft Coulomb potential. A portion of the wave function is pulled significantly away from the atom by the high electric field – significant electron density exists several nanometres from the atom, two orders of magnitude further than the Bohr radius. As the field direction reverses, much of this electron density is swept back toward the atomic core, and the interference between two parts of the wave function causes high-frequency oscillation of the overall dipole moment of the atom. These high-frequency oscillations are responsible for X-ray emission.

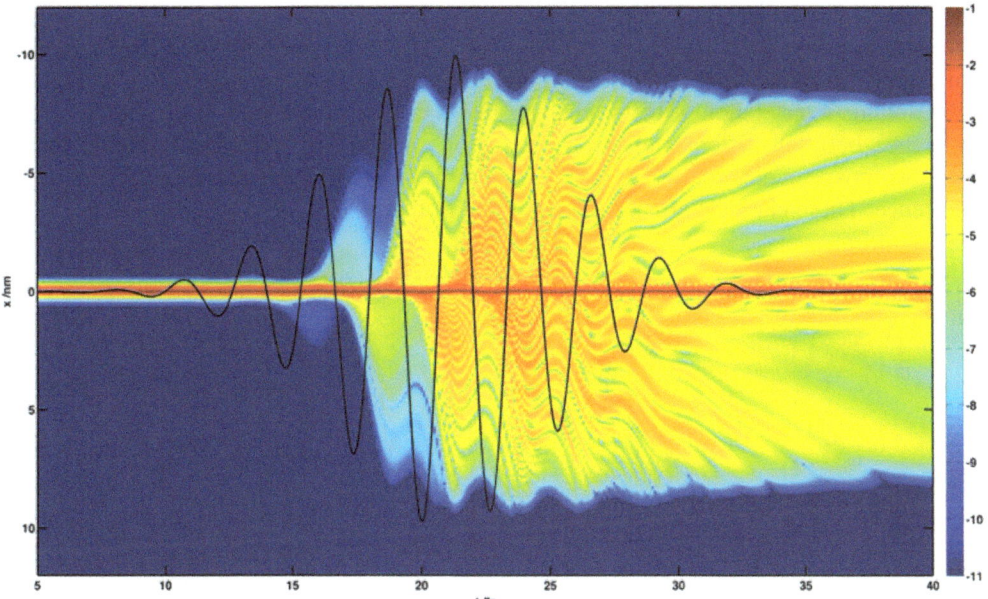

Figure 2 *Electron wave-function evolution with time in a simple 1-dimensional atom driven by a 7 fs laser pulse (black line). Colour scale is base 10 logarithmic.*

If the high-frequency oscillations of the dipole moment are isolated by spectral filtering (Figure 3), it can be seen that bursts of high-frequency oscillation occur every half cycle of the laser field. This corresponds to the semi-classical picture of ionization near each peak of the laser field, followed by acceleration away from and then back toward the atomic core, and recombination. The half-cycle repeat time of the X-ray bursts leads to the harmonic structure of the final X-ray spectrum, with harmonic spacing of twice the laser frequency, and the symmetry of the process determines that only odd harmonics are produced.

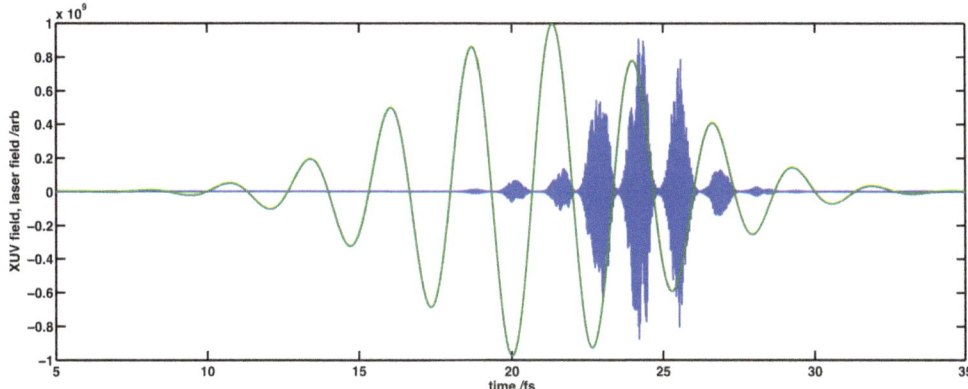

Figure 3 *Dipole moment (electron acceleration) versus time for simple 1D atom driven by a 7 fs laser pulse. Low frequency oscillations have been filtered out to show the high-frequency X-ray bursts.*

Figure 4 shows the unfiltered spectrum of the time-domain X-ray bursts shown in Figure 3. Harmonic structure is only really visible at energies around the high-frequency cutoff, which in the classical picture would be ~72 eV for this laser intensity. Below cut-off, quantum interference effects change the spectral structure. These result from interference between different electron oscillations at the same energy. In the classical model, two possible electron trajectories can contribute to each output energy, resulting in interference. These classical effects are illustrated in Figure 6.

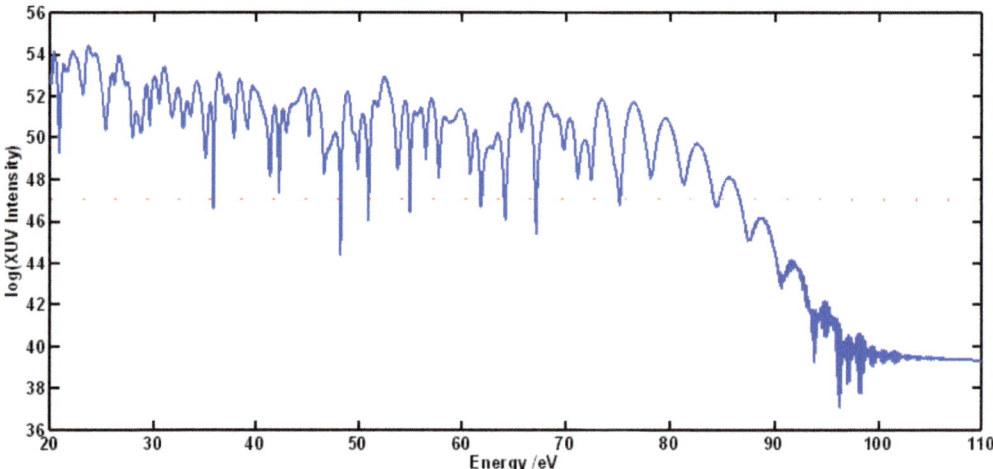

Figure 4 *XUV output spectrum from a 1D atom driven by a 7 fs pulse. Harmonics are clearly visible only near the cut-off, at ~80 eV.*

In a real experiment, where a radial intensity variation occurs due to beam propagation, the quantum interference effects in the single atom spectrum become less marked, because radiation from each of the two trajectories has a different spatial profile. Thus as the X-ray beam diffracts in propagating away from the sources, the harmonics are much more clearly visible.

Figure 5 shows the spectral structure of the radiation produced using a 3 fs, almost single cycle pulse. In this case, only a single intense X-ray burst is produced, as only one half cycle of the laser field is strong enough to cause significant ionization. The actual spectrum in this regime will depend very critically on the phase relationship between the carrier wave and the envelope of the pulse.[11] An odd-parity pulse with a sine carrier wave may have two half cycles that can produce an X-ray burst, resulting in the spectral equivalent of a Young's slits pattern – $\cos^2$ fringes in the spectrum. An even-parity pulse with a cosine carrier wave will only produce a single burst, resulting in no spectral modulation at all; the ideal situation for producing attosecond pulses. Hence effective attosecond pulse production relies on control of the carrier envelope phase relationship, which in an unstabilised femtosecond pulsed laser may vary rapidly with time.

2.4 Polarization

The real HHG process is three dimensional, and the trajectory of the electrons is affected by the polarization of the laser. Linear polarization ensures that the electron trajectories return to the atomic core. For circular polarized light, the electrons do not return to the core, and no harmonic generation is seen. This has been used to 'gate' short sections out of a longer pulse in order to create an effectively single cycle pulse for single attosecond pulse generation.

The simple 1D model also lacks any information about the actual wave functions of the atoms used. One significant example of the effect of the real atomic wave function on the HHG spectrum is the 'Cooper minimum' seen in generation from argon, which prevents efficient generation between ~50 eV and ~100 eV.[12] Generation from molecules is possible, and provides rich information about the interaction between the returning electron and the molecule itself, which can be used to measure the molecule wave function.[13]

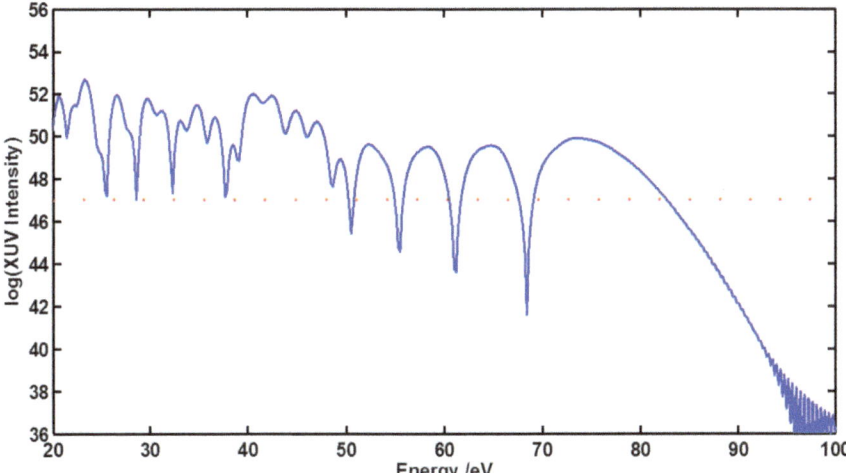

Figure 5 *X-ray spectrum from a 1D atom driven by a 3 fs laser pulse. Note that the harmonic structure has completely disappeared from the spectrum, as only a single X-ray burst is produced.*

2.5 Effective Non-linearity

When considering production of an efficient high-flux source, one important parameter is the effective nonlinear parameter for the target atom in question. Several factors contribute to the strength of the effective non-linearity. The overlap of the returning electron with the atomic core affects the magnitude of the dipole moment created. This has two immediate consequences: in general, stronger generation is seen from larger atoms; xenon will be more effective than neon or helium, for example. Secondly, the wavelength of the radiation affects the effective non-linearity significantly because of the spread of the wave function during the time spent in the continuum – a scaling factor of $\sim\lambda^{-6.5}$ has been observed experimentally.[14] These factors have major consequences for the choice of non-linear medium when considered alongside the problems of phase matching, which will be described in the next section.

2.6 Solid Surfaces

Electrons in solid surfaces can also be used to generate high-order harmonics. In principle, the very high electron densities could produce much more efficient generation. Two regimes exist for HHG at surfaces: below relativistic intensities, where well-collimated but low-order harmonics can be produced by collective oscillations of the electrons in the metal, and a regime at high intensities[15] where a plasma is formed at the surface which acts as a 'moving mirror' because of ponderomotive pressure, and the Doppler shift from this moving mirror causes extreme frequency shifting of the light. Both are potentially interesting as high-flux, well-collimated sources of X-rays, but require laser sources with an order of magnitude higher intensity than those needed for HHG.

2.7 Multiple Wavelengths

The dynamics of the electron motion during the laser pulse, which define the cut-off and also the generation efficiency, can be significantly altered by the use of multiple or shaped pulses rather than the Gaussian, flat phase pulses assumed in simple theory. The use of a second pulse at twice the laser frequency, which is added to the original laser pulse, has been demonstrated by several groups[16,17] to increase harmonic yield significantly. The second pulse can be easily generated from the pump pulse using non-linear crystals, but careful adjustment of the relative delay and amplitude produces the best results. Production of a second pulse using an optical parametric amplifier (OPA) and addition of the two unrelated frequencies has been demonstrated to extend cutoff,[18] and based on this idea, optimised pulses have been theoretically derived[19] to produce cu-toffs extended by a factor of 2.5 while retaining harmonic yield.

2.8 Field Enhancement

The use of a CPA laser for generation of high harmonics limits the applicability of the source – CPA lasers are expensive and complex. To work around this limitation, HHG gas been demonstrated using a Ti:Sapphire oscillator, and reaching the necessary field strengths for HHG by using field enhancement based on plasmonic antennae.[20] In the first experiments in this area, gold 'bow-tie' antennae were produced using lithography on a silicon substrate. The bow-tie structures consist of two gold triangles with a gap between their apices of ~50 nm. In this gap, the electric field can be several orders of magnitude larger than the incident field. By introducing a noble gas (xenon) into this gap, harmonics up to the 17th order were produced using pulses of energy 1.3 nJ. Efficiency was low, but the attraction of using a much simpler laser system is clear, and further work in the area may produce useful sources.

3. PHASE MATCHING

As with all non-linear generation process, the efficient production of radiation depends on phase matching, or conservation of momentum k within the nonlinear process. Phase matching requires that for HHG $\bar{k}_q = q\bar{k}_{in}$ for a collinear process generating the qth harmonic. This presents problems because of the hugely varying physical properties of the generating medium at the two wavelengths, as well as two other factors implicit in the HHG process – the 'atomic phase' shift caused by the single atom generation process, and the presence of plasma due to ionisation by the high field.

3.1. Material Indices

The refractive indices of rare gases through the visible and soft X-ray regions are reasonably well known, and so their contribution to the phase matching is easy to calculate, if the density is known. The effect of ionisation during the pulse has to be taken into account, however, and will be discussed in section 3.3.

3.2. Atomic Phase

The semi-classical description of the HHG process has the electron following a trajectory that goes away from the atom, and then returns after the electric field reverses during the optical cycle. This delay between the initial tunnelling event and the eventual X-ray burst produces a phase shift of the generated X-ray field that is dependent on the intensity of the laser. The classical time delay and kinetic energy of electrons ionised at different times through the optical cycle is shown in figure 6.

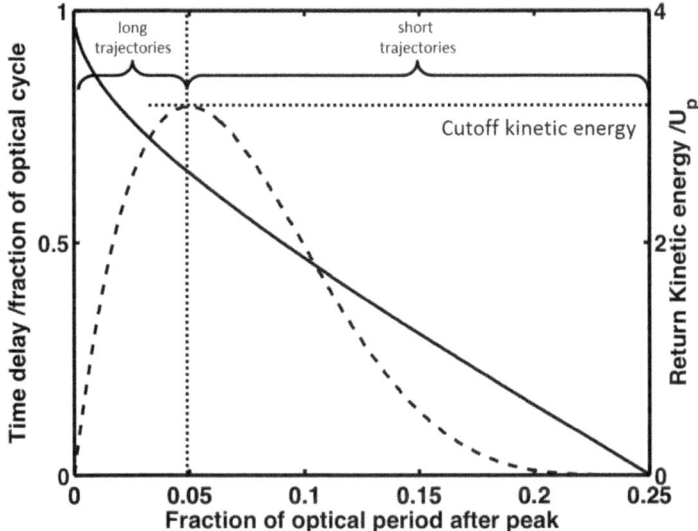

Figure 6 *Time delay and kinetic energy of returning electron as a function of time of ionization through laser cycle. Zero time is at the peak of the laser field.*

Figure 6 shows several important features of the HHG process. Firstly, it shows that the time delay between ionization of the electron and recombination can be a significant fraction of an optical cycle. Secondly, it is clear that different kinetic energies can be associated with different trajectories and time delays. Importantly, it illustrates that the peak kinetic energy available to the electron on recombination is ~3.2U_P and, at kinetic energies below the peak value, there are two possible trajectories which can contribute, and thus two different time delays. This results in quantum interference effects[21] that are visible in the TDSE–calculated spectra shown in figure 4, where clear harmonic structure is seen around cut-off, but below cut-off the regular structure disappears and is replaced by a complex interference pattern. The two trajectories contributing at each output photon energy are usually divided into the 'short trajectory', for electrons ionised after the peak kinetic energy (after $t = $ ~ 0.05 in figure 6) and 'long trajectory' for those electrons ionised before the peak kinetic energy. The atomic phase contribution to phase matching is approximately dependent on the delay shown in figure 6. Quantum calculations of the atomic phase produce very similar results.[22,23]

3.3 Plasma

The major complication in achieving phase matching in HHG comes from the generation of a plasma. Ionization is a necessary part of the HHG process, and so plasma generation is inevitable; any ionization which happens during the first or third quarters of the optical cycle produces electrons whose trajectories never return to the atomic core. The generation of plasma acts as a non-linearity when calculating propagation of the pulse, as the electron density in the plasma varies with laser intensity, producing an intensity dependent refractive index. The degree of non-linear propagation depends on the plasma density, which in turn depends on the initial gas density and the laser intensity. Because the plasma builds up through the pulse, distortion of the temporal and spatial profiles of the pulse will occur. Overcoming the non-linear propagation due to plasma generation is one of the most significant problems in producing high-efficiency, high-flux X-ray sources using HHG, because the conditions for efficient generation – high laser intensity, high gas density – are the conditions under which the non-linear propagation effects are largest.

3.4. Geometrical Effects

In order to offset the effects of the contributions listed above and produce phase matching, geometrical effects are often used. The two most common are the Gouy shift and propagation in a waveguide.

3.4.1 Gouy Shift. The Gouy shift occurs when a beam propagates through a focus. For a Gaussian beam the laser-field envelope $A(r,z)$ through a focus is given by

$$A(r,z) = \frac{A_0}{z_0 + iz} \exp\left(-\frac{kr^2}{2(z_0 + iz)} \right),\tag{4}$$

where A_0 is a constant, k is the wave vector of the radiation, r is the radial distance away from the optical axis, z is the distance from the beam waist and z_0 is the Rayleigh length. The exponential factor is the familiar Gaussian beam solution, but the pre-factor gives an additional phase shift from one side of the focus to the other of π radians. Most of this variation occurs within one Rayleigh length either side of the focus, and contributes an effective phase velocity change, which can be used to phase match the HHG interaction.[24]

3.4.2 Waveguiding. Propagation of the pump light in a waveguide can be used to change the phase velocity of the pump in order to produce phase matching.[25] The waveguides used must be hollow, in order to be filled with the gas which acts as the non-linear medium. Most work in the area has used silica glass capillaries with internal diameters of around 100 μm, but generation has also been observed in gas filled photonic crystal fibres.[26]

The optical modes of hollow capillary fibres[27] have different properties to the more usual modes of solid optical fibres. The modes are lossy, and the guides are always multimode. This in itself is a problem for HHG, as the propagation of intense pulses in gas filled capillaries leads to nonlinear mode coupling, changing the phase velocity of the pump light.

3.5. Common Phase Matching Schemes

The combination of all of the factors described above has led to several common phase matching schemes in popular use; we will concentrate on those aimed at high-efficiency, high-flux sources in this chapter.

3.5.1 Balancing Gouy shift against atomic phase. In the case where the gas density is small, the effect of the medium refractive index and to a lesser extent the plasma generation can be considered small, and phase matching can be achieved by a balance between the atomic phase and the Gouy shift.

This balance is shown in figure 7. The figure shows the intensity contours of the focused laser, and the shading represents the coherence length for nonlinear generation at harmonic $q = 29$, in millim. It can be seen that for short trajectory generation, there is a significant region with a coherence length of 3-4mm about 7-8mm beyond the focus of the laser. Thus if a jet or cell full of gas is placed at this region, well phase-matched generation will occur. For long trajectory generation, there is a short region with long coherence length just before the focus, and predominantly long trajectory generation will occur if the gas cell is placed here.[28] The short coherence length region is often used in experimental generation.

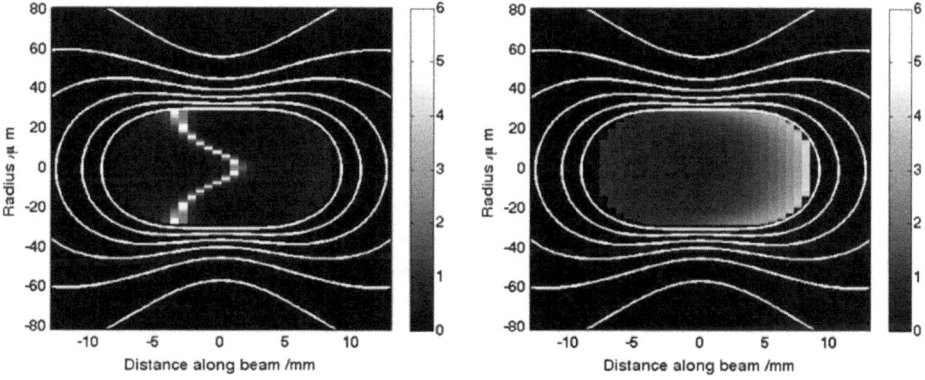

Figure 7 *Coherence length map for long (left) and short (right) trajectories. The colour scale gives the coherence length in millimetres.*

3.5.2 Move to long wavelength sources. In order to scale this phase-matching regime to high pressure and high-laser intensities to produce high-flux beams at short wavelengths, the additional effects of the gas index and non-linear propagation must be addressed. One effective technique is to limit the peak intensity of the laser, limiting the degree of ionization within the target gas – the 'neutral gas' regime. As an example, for a 40 fs pulse in argon gas, an upper limit of ionization of 5% implies a peak intensity of $\sim 1.7 \times 10^{14}$ W/cm². This immediately limits the shortest possible generated wavelength; the direct relation between cut-off and intensity (equation 2) implies that generation with this intensity will only reach the 31st harmonic. In order to increase the maximum generation energy, it is necessary to increase the wavelength of the exciting pulse since cut-off energy is proportional to λ^2. The use of optical parametric chirped pulse amplifiers has become popular in order to reach wavelengths in the near to mid infrared,[29] making extended cut-off generation possible while maintaining the 'neutral gas' regime for phase matching. However, an issue with the use of infrared radiation for generation is the rapid decrease of the single atom non-linearity with wavelength – approximately $\lambda^{-6.5}$. Hence it is necessary to compromise between extended cut-off and reduced non-linearity – wavelengths of 1.4–1.6 µm provide a good compromise. Using this technique, relatively efficient generation into the water window ($\approx$300–500 eV) has been reported.[30]

3.5.3 Phase matching in capillaries. The use of capillaries to provide phase matching relies on offsetting the effects of the gas refractive index against the plasma index produced by ionization. The intensity-dependent atomic phase is assumed in many calculations not to contribute to the phase matching. Typically the capillary can offset the plasma index up to a certain value of ionization – early work in argon gas at pressures of ~100 mBar required an ionization limit of ~4%, again limiting peak intensities and thus high energy cut-offs. Recently the use of longer wavelengths and higher pressures has led to phase matched generation with much higher cut-offs,[31] again into and beyond the water-window region.

3.5.3 Quasi-phase matching. An alternative to the exact phase matching described above is commonly used in other branches of non-linear optics. In quasi-phase matching (QPM), the non-linear material is given a periodic structure, in which the nonlinear coefficient is changed – either reversed, removed, or altered in some way. This affects the build-up of the generated radiation. For non-zero Δk the integrated build-up of generated radiation along the sample shows distinct maxima and zeros. In a QPM sample, the changing non=linear coefficient along the sample is designed to prevent the destructive interference that reduces the integrated output. In traditional non-linear materials such as lithium niobate, this change in non-linear coefficient can be achieved by poling the material to invert the crystal structure, reversing the sign of the non-linear coefficient. In a gas used for HHG, the variation is less effective, but any periodic change in intensity, gas density, etc., can potentially be used for QPM. QPM has been demonstrated by modulating the diameter,[32] and by mode beating between high-order modes in a capillary.[33] It has also been demonstrated by the use of counter-propagating pulses in a capillary,[34] where the counter-propagating pulses affect the atomic phase, effectively changing the phase of the non-linear emission and producing QPM. In free-space propagation, the use of gas density variation by multiple gas jets has been suggested.[35] In many wavelength regimes, the high absorption of HHG materials limits the usefulness of QPM in extending the interaction region.

4 EXPERIMENTAL ASPECTS OF HHG SOURCES

4.1 Free Space Propagation

The experimental configurations used for HHG are determined principally by the need for phase matching. In free space, where the Gouy shift is necessary to phase match the interaction, the HHG medium – most often a noble gas – has to be placed in a particular position relative to the focus of the laser, as illustrated in section 3.5.1. The short coherence lengths created by Gouy shift phase matching mean that the useful propagation length in the gas can only be a fraction of the Rayleigh length. If the laser source is ~1 mJ in a 40 fs pulse, then in order to reach the necessary intensities for ionization, a focal spot of ~100 µm is necessary, implying a Rayleigh length of ~4 cm. Thus a gas cell of a few millimetres is all that is necessary. Since no window materials exist which can tolerate the laser intensities, the gas–filled region is usually in the form of a transverse jet, or a cell with entry and exit holes matched to the laser focal dimensions. Gas cells can produce higher densities, enabling experiments to reach the pressure region where generation is limited by gas absorption[a] rather than by coherence length. Gas loading of the generation vacuum chambers can cause significant losses due to absorption as the beam propagates away from the gas jet/cell, and any significant pressure in the chamber can also cause ionization-induced defocusing of the input beam, reducing intensity by increasing spot size. Hence cells with small transverse holes or pulsed gas jets are necessary.

4.2 Capillary Propagation

Propagation in a capillary to improve phase matching is potentially useful, but the associated experimental problems compared to free space propagation make it challenging. The gas is introduced into the central section of a capillary, typically of diameter ~150 µm, via one or more holes through the sidewall. This allows the creation of a region of constant pressure. The laser is coupled into the capillary modes via a lens or mirror. Good coupling into the fundamental EH_{11} mode of the capillary requires both good matching of the spot sizes, and also good beam quality. Coupling efficiency should in in theory be better than 90% for a matched Gaussian beam into the lowest order EH_{11} mode of a hollow capillary.[36] In practice, coupling with efficiency of 50–70% is easily achieved, but in this regime the actual fraction of power coupled into EH_{11} is often relatively low, and many higher order modes are excited. Monte Carlo studies of coupling into capillaries for various alignment errors show that many combinations of alignment parameters will produce coupling up to 70%, but for most alignment parameters less than half of the coupled energy goes into the EH_{11} mode. Above 80% coupling efficiency, most of the possible alignment parameters give a very high fraction of coupled power in the fundamental EH_{11} mode. Since phase matching relies on propagation in a single mode, efficient coupling into EH11 is critical, and overall coupling well above 80% is necessary to ensure it. Practically, beam stability is also critically important to maintain good coupling.

A significant issue with capillary usage is the multimode nature of the guide. Unlike the more common optical-fibre geometry, no wavelength cut off exists for the modes of a capillary. Hence all propagation is multimode. The significance of this for HHG is that under conditions where ionization is occurring, non-linear mode coupling will always occur, distorting the pulse both spatially and temporally. Hence phase matching based on propagation in the fundamental mode is limited to low gas pressures and low ionization fractions, which is not ideal for high-flux X-ray sources.

The X-ray output of the capillary is effectively unguided, because the effective size of the capillary at the X-ray wavelength is so large that the bean never reaches the capillary walls (at least for short trajectory generation). The capillaries could be used to guide at the X-ray wavelengths, because they exhibit total external reflection, but the guiding is lossy, particularly at very short wavelengths.

4.3 OUTPUT CHARACTERISTICS OF HHG SOURCES

The usefulness of HHG sources for real experiments will depend on how well their output parameters compare to other sources. Many of the output characteristics of HHG sources are significantly different from synchrotrons or laser-plasma X-ray sources and give HHG sources unique potential for experiments.

---

[a] The absorption length for XUV and soft X-rays is quite short in many of the gases used for HHG. As an example, a 1 cm cell of argon gas at 0.1 atmosphere has a transmission of only ~10% at 27 nm, with a peak transmission in the range 10–40 nm .

4.3.1 Pulse Length

The most obviously unique parameter of the output from HHG is the length of the pulses produced in time. Each of the X-ray bursts produced in HHG is less than a femtosecond long. If single-cycle pulses are used for generation, then it is possible to produce an isolated pulse of sub-100 attosecond (as) duration. The opportunities for new experiments opened by the existence of a laboratory source of attosecond pulses are enormous, and has opened up a whole new branch of physics. A review of this area has been given by Krausz.[1]

4.3.2 Coherence

The spatial coherence of radiation from a non-linear generation processes typically reflects the degree of spatial coherence of the source. In the case of HHG, the pulse source has high spatial coherence, and it has been demonstrated experimentally that it is possible for HHG sources to show very high spatial coherence.[37] The temporal coherence of an HHG source is primarily influenced by its bandwidth. An isolated 100 as pulse clearly has very low temporal coherence; however, if a single harmonic is obtained by filtering, then its bandwidth is determined by the pulse length of the original laser source. Thus temporal coherence times of tens of femtoseconds are possible, equating to lengths of ~10 μm.

The combination of full spatial coherence and 10 μm temporal coherence means that applications in imaging and holography of small objects are possible with very high resolution. The spatial coherence is not determined primarily by source size, as it might be in a laser-plasma or synchrotron source, so long distances are not necessary to regain spatial coherence.

One particular variety of imaging which is well suited to HHG is coherent diffractive imaging (CDI). CDI uses computer algorithms to replace the traditional objective lens in an imaging system. This is ideal for the soft X-ray region of the spectrum, where objective lenses are hard to manufacture. CDI will be described in more detail in section 5.1.1.

4.3.3 Flux

The absolute flux available from HHG sources is very dependent on the wavelength range of choice. HHG has been demonstrated at energies beyond 1 keV; however, the flux available at very high energies is extremely low. In the most common spectral region used for HHG experiments, at around 27 nm (harmonic number $q = 29$), using argon gas as a target, efficiencies up to ~10^{-5} have been demonstrated – i.e., for a 1 mJ pulse, 10 nJ of XUV radiation is generated. This corresponds to ~10^9 photons/pulse.

Calculation of brightness in the units usually used for synchrotron radiation is possible with some approximations. For a typical generation setup at the 29th harmonic using argon gas, the source spot size w_0 is ≈ 30 μm. The angular divergence is given by Gaussian beam theory to be $M^2 \times \approx 0.3$ mrad, where M^2 is the usual beam propagation factor. This is measured to be around 2-3, giving a divergence of ~1 mrad. The calculated brightness is then ~10^{14} photons/sec/mm^2/mrad^2 in a bandwidth of 0.1%, making sensible assumptions for the harmonic bandwidth. This compares well with the average brightness of, for example, synchrotron bending magnet radiation in the same wavelength region. Peak brightness is much higher than for synchrotron radiation because of the very short pulse length. Compared to laser-plasma sources, HHG sources are also several orders of magnitude brighter in this spectral region.

Although HHG sources can be of high brightness in the XUV, efficiency of generation typically drops rapidly as the wavelength is reduced. Efficiencies of $\approx 2 \times 10^{-7}$ at 100 eV and 5×10^{-8} at 300 eV have been reported using the 'neutral gas' phase-matching regime.[30] Hence one of the biggest challenges for HHG sources is to improve efficiency of generation at higher X-ray energies.

5. APPLICATIONS OF HIGH HARMONICS

HHG forms the basis of much of the very broad topic area of attoseconds physics. This review is not intended to cover this area, but this section will cover a selection of the possible applications of HHG which are not directly related to attosecond pulses. The principal areas covered will be examples of HHG in imaging and spectroscopy., where the spectral and spatial properties of the HHG source make it well-suited.

5.1 Imaging

The use of the high harmonic process itself for imaging of molecular wave functions has been demonstrated by several laboratories.[38] This is an elegant demonstration of the new physics enabled by strong-field lasers. However, this review is intended to look at imaging in which the HHG source acts purely as a source, rather than being part of the experiment.

5.1.1 Coherent Diffractive Imaging. The coherence properties of the output of an HHG source make possible new forms of soft X-ray microscopy. In particular, the use of coherent diffractive imaging (CDI) for microscopy has been demonstrated to enable soft X-ray imaging with resolutions in the 50-10–nm range, close to the diffraction limit for the 20-30 nm radiation used.

Coherent diffractive imaging is a technique that has its origins in X-ray crystallography, where the lack of information about the phase of a diffraction pattern recorded on film or a CCD limits the information available. The phase information required to reconstruct the original object from the intensity of its scattered light can be obtained using iterative algorithms in certain circumstances.[39] One particular circumstance is when the object is known to be on a blank support – this constraint in the object plane, together with known modulus of the diffraction pattern (the 'modulus constraint'), allows an iterative algorithm to reconstruct the phase of the light in the diffraction plane. By simple propagation the field at the object, and thus the optical properties of the object itself, can be obtained.

In order for the CDI technique to work, good spatial coherence of the illuminating radiation is necessary. CDI has been demonstrated with electromagnetic radiation from visible light to X-rays, and for electrons. It is particularly applicable in the soft X-ray regime because of the comparative difficulty in preparing optics in this spectral region. CDI with radiation from synchrotrons is possible, but the production of good spatial coherence usually requires significant reduction in the flux from the source – spatial coherence is obtained by using small pinhole and large propagation distances. HHG sources produce highly spatially coherent light to start with, and so CDI with HHG sources is potentially an attractive technique. Imaging at around 29 nm[40] and around 13 nm has been demonstrated, with almost diffraction-limited resolution.

Although their spatial coherence makes HHG sources suited to CDI, the spectral width of the output does not. The CDI process is straightforward for monochromatic radiation; for polychromatic radiation, the object forms several overlapping diffraction patterns at different wavelengths, each different because of the change in wavelength and change in absorption and refractive index of the object itself. These patterns add incoherently on the detector, and cannot be easily separated. This makes the use of simple algorithms inappropriate. More complex algorithms which attempt to separate the different diffraction patterns are presently being developed;[41] these work on specific objects, but generally the area is still under development. As HHG almost always produces multiple wavelengths, the other option is to monochromate the radiation before use – this has been done in most of the imaging experiments so far demonstrated. However, this wastes a large amount of the generated radiation, and so improvements in this area are a priority for the use of HHG in imaging.

Demonstration of CDI from a single shot of HHG radiation has been reported[42] using very high energy laser pulses for generation; 50 mJ laser pulses were used to produce XUV pulses with ~10^{11} photons/pulse at 32 nm. At this signal level, CDI of test objects was possible with resolution of ≈113 nm.

5.1.2 Holography. The same spatial coherence that makes HHG sources ideal for CDI also makes them appropriate for holography. In-line holograms of small objects have been created, but little work has been done so far in the area.

5.2 Spectroscopy

The HHG process itself is affected by the spectral properties of the atoms or molecules used for generation; an example is the previously-noted effect of the 'Cooper minimum' in ionisation of argon, which manifests itself as a reduction of HHG efficiency[12] in the spectral region 20-25 nm. The vibration of molecules involved in the generation process has also been measured via the effect on HHG efficiency.[43] The focus here is on the use of HHG as a source for experiments on other systems, rather than study of aspects of the HHG process itself.

The broad spectral range of HHG sources makes them attractive for spectroscopy in the EUV and soft X-ray regions, which can be a large fraction of the spectral range of the radiation – 20-30% is not uncommon. Also, the pulse envelope in time is typically ~10 fs, making transient absorption over large spectral ranges a possibility. Examples of spectroscopic measurements to data that have used HHG include EXAFS, and the determination of refractive indices.

5.2.1 Extended X-Ray Absorption Fine Spectra.

With appropriate phase matching, in this case the use of very short pulses to prevent plasma build-up during generation, HHG has been used to measure absorption spectra[44] of several metals and gases in the spectral region from 500 eV up to almost 3.5 keV. Samples of metals in the form of thin foils were used for transmission measurements, and the resulting absorption spectra show not only the typical elemental edges, but also fine structure at the edges, corresponding to extended X-ray absorption fine spectra (EXAFS). The typical nearest-neighbour inter-atomic distances could be extracted from the Fourier transform of the absorption data.

5.2.2 Refractive Index Measurement. Detailed measurements have been made of the complex indices of materials in the XUV region by studies of diffraction from regular arrays.[45] Nanoscale spheres of PMMA were self-assembled into a regular lattice. A diffraction pattern is produced by the lattice formed, but is modulated by the form factor of scattered light from a single sphere. Mie scattering from a sphere is determined by its refractive index, and hence the index can be determined. The regular lattice creates diffraction patterns from each harmonic which are spatially separated, and so the refractive index can be separately determined at each incident harmonic frequency from its individual form factor. Comparison of the index measured with values from standard databases, which in this spectral region are often extrapolated from other spectral regions, shows significant differences, indicating that HHG can provide new information using this technique.

5.1 Seeding for FEL and X-ray Lasers

The spatial and temporal properties of HHG pulses make them ideal seed pulses for amplification. Two significant examples of this are the seeding of laser-pumped X-ray amplifiers[46] and seeding of free electron laser systems (FELs).[47] In both cases, the high-quality spatial and temporal structure of the HHG beam were used to improve the amplifier output.

6. SUMMARY

In addition to its high profile role as the source of attosecond pulses, high-harmonic generation has moved on from being an atomic physics experiment to providing a viable laboratory-scale source of soft X-rays and XUV radiation, with unique properties suitable for many applications. Its physical basis as an extreme non-linear optical effect is well understood, and advances in the ability to produce more flux and shorter wavelengths are proceeding as phase matching becomes better controlled. The potential applications are widespread, and will become more generally used as laser sources providing the appropriate pulses become simpler and cheaper.

References

1 F. Krausz, *Rev. Mod Phys.*, 2009, **81**, 163.

2 D. Strickland and G. Mourou, *Opt. Comm.*, 1985, **56**, 219.

3 J. Nilsson and D.N. Payne, *Science,* 2011, **332**, 921.

4 T. Eidam, J. Rothhardt, F. Stutzki, F. Jansen, S. Hädrich, H. Carstens, C. Jauregui, J. Limpert, A. Tünnermann, *Opt. Express*, 2010, **19**, 255.

5 A. Tünnermann, T. Schreiber and J. Limpert, *Appl. Opt.*, 2010, **49**, F71.

6 J. Boullet, Y. Zaouter, J. Limpert, S. Petit, Y. Mairesse, B. Fabre, J. Higuet, E. Mével, E. Constant and E. Cormier, *Opt, Lett.*, 2009, **34**, 1489.

7 E. Seise, A. Klenke, J. Limpert and A. Tünnermann, *Opt. Express*, 2010, **18**, 27827.

8 M. Lewenstein, P. Balcou, M. Yu. Ivanov, A. L'Huillier and P.B. Corkum, *Phys. Rev. A*, 1994, **49**, 2117.

9 J.L. Krause, K.J. Schafer and K.C. Kulander, *Phys. Rev. Lett*, 1992, **68**, 3535.

10 V.S. Popov, *Physics-Uspekhi*, 2004, **47**, 855.

11 T. Brabec and F. Krausz,. *Rev. Mod. Phys.*, 2000, **72**, 545.

12 H.J. Wörner, H. Niikura, J.B. Bertrand, P.B. Corkum and D.M. Villeneuve *Phy. Rev. Lett.*, 2009, **102**, 103901.

13 C. Altucci1, R. Velotta, J.P. Marangos, E. Heesel, E. Springate, M. Pascolini, L. Poletto, P. Villoresi, C. Vozzi, G. Sansone, M. Anscombe, J-P. Caumes, S. Stagira and M. Nisoli, *Phys. Rev. A*, 2005, **71**, 013409.

14 A.D. Shiner, C. Trallero-Herrero, N. Kajumba1, H.-C. Bandulet, D. Comtois, F. Légaré, M. Giguère, J-C. Kieffer, P.B. Corkum and D. M. Villeneuve, *Phys. Rev. Lett.*, 2009, **103**, 073902.

15 U. Teubner1, G. Pretzler, Th. Schlegel, K. Eidmann, E. Förster and K. Witte *Phys. Rev. A*, 1003, **67**, 013816.

16 I.J. Kim, C. Kim, H.T. Kim, G.H. Lee, Y.S. Lee, J.Y. Park, D. J. Cho and C.H. Nam, *Phys. Rev. Lett.*, 2005, **94**, 243901.

17 N. Dudovich, O. Smirnova, J. Levesque, Y. Mairesse, M. Yu. Ivanov, D.M. Villeneuve1 and P. B. Corkum *Nature Physics*, 2006, **2**, 781.

18 T. Siegel, R. Torres, D.J. Hoffmann, L. Brugnera, I. Procino, A. Zaïr, J.G. Underwood, E. Springate, I.C.E. Turcu, L.E. Chipperfield and J.P. Marangos, *Opt. Express*, 2010, **18**, 6853.

19 L.E. Chipperfield, J.S. Robinson, J.W.G. Tisch and J.P. Marangos *Phys. Rev. Lett.*, 2009, **102**, 063003.

20 S. Kim, J. Jin, Y-J. Kim, I-Y. Park, Y. Kim and S-W. Kim, *Nature*, 2008, **453**, 757.

21 M.B. Gaarde, F. Salin, E. Constant, P. Balcou, K.J. Schafer, K.C. Kulander and A. L'Huillier *Phys. Rev. A*, 1999, **59**, 1367.

22 M. Lewenstein, P. Salières and A. L'Huillier, *Phys. Rev. A*, 1995, **52**, 4747.

23 F. Lindner1, W. Stremme1, M.G. Schätzel, F. Grasbon, G.G. Paulus, H. Walther, R. Hartmann and L. Strüder *Phys. Rev. A*, 2003, **68**, 013814.

24 P. Salières, A. L'Huillier and M. Lewenstein, *Phys. Rev. Lett.*, 1995, **74**, 3776.

25 A. Rundquist, C.G. Durfee III, Z. Chang, C. Herne, S. Backus, M.M. Murnane and H.C. Kapteyn, *Science*, 1998, **280**, 1412.

26 O.H. Heckl, C.R. Baer, C. Kränkel, S.V. Marchese, F. Schapper, M. Holler, T. Südmeyer, U. Keller, J.S. Robinson, J.W. Tisch, F. Couny, P. Light, F. Benabid and P.S. Russel, *Lasers and Electro-Optics/International Quantum Electronics Conference*, 2009, paper JThH1.

27 E.A.J. Marcatili and R.A. Schmeltzer, *Bell System Technical Journal*, 1964 **43**, 1783.

28 P. Balcou, P. Salières, A. L'Huillier and M. Lewenstein *Phys. Rev. A*, 1997, **55**, 3204..

29 E.J. Takahashi, T. Kanai, Y. Nabekawa, K. Midorikawa, *Appl. Phys. Lett.*, 2008, **93**, 041111.

30 E.J. Takahashi, T. Kanai, K. Ishikawa and K. Midorikawa *Lasers and Electro-Optics/International Quantum Electronics Conference*, 2009, paper JThG2.

31 T. Popmintchev, M-C Chen, A. Bahabad, M. Gerrity, P. Sidorenko, Oren Cohen, I.P. Christov, M.M. Murnane and H.C. Kapteyn, *PNAS*, 2009, **106**, 10516.

32 A. Paul, R.A. Bartels, R. Tobey, H. Green, S. Weiman, I.P. Christov, M.M. Murnane, H.C. Kapteyn and S. Backus, *Nature*, 2003, **421**, 51.

33 M. Zepf, B. Dromey, M. Landreman, P. Foster and S.M. Hooker, *Phys. Rev. Lett.*, 2007, **99**, 143901.

34 A.L. Lytle A, X. Zhang, P. Arpin, O. Cohen, M.M. Murnane and H.C. Kapteyn, *Opt. Lett.*, 2008, **33**, 174.

35 T. Auguste, B. Carré and P Salières, *Phys. Rev. A*, 2007, **76**, 011802(R).

36 R.K. Nubling and J.A. Harrington, *Opt. Eng.*, 1998, **37**, 2454.

37 Y. Tamaki, J. Itatani, K. Midorikawa, M. Obara, *Jpn. J. Appl. Phys.*, 2001, **40**, 1154.

38 T. Kanai, S. Minemoto and H. Sakai, *Nature*, 2005, **435**, 470.

39 J.R. Fienup, *Opt. Lett.*, 1978, **3**, 27.

40 R.L. Sandberg, A. Paul, D.A. Raymondson, S. Hädrich, D.M. Gaudiosi, J. Holtsnider, R.I. Tobey, O. Cohen, M.M. Murnane and H.C. Kapteyn *Phys. Rev. Lett.*, 2007, **99**(9), 098103.

41 R.A. Dilanian, B Chen, G.J. Williams, H.M. Quiney, K.A. Nugent, S. Teichmann, P. Hannaford, V. Dao and A.G. Peele, *J. Appl. Phys.*, 2009, **106**, 023110.

42 A. Ravasio, D. Gauthier, F.R.N.C. Maia, M. Billon, J-P. Caumes, D. Garzella, M. Géléoc, O. Gobert, J-F. Hergott, A-M. Pena, H. Perez, B. Carré, E. Bourhis, J. Gierak, A. Madouri, D. Mailly, B. Schiedt, M. Fajardo, J. Gautier, P. Zeitoun, P. H. Bucksbaum, J. Hajdu, and H. Merdji, *Phys. Rev. Lett.*, 2009, **103**, 028104.

43 N.L. Wagner, A. Wüest, I.P. Christov, T. Popmintchev, X. Zhou, M.M. Murnane and H.C. Kapteyn, *PNAS*, 2006, **103**, 13279.

44 E. Seres, J. Seres and C. Spielmann, *Appl. Phys. Lett.*, 2006, **89**, 181919.

45 B. Mills, C.F. Chau, E.T.F. Rogers, J. Grant-Jacob, S.L. Stebbings, M. Praeger, A. M. de Paula, C.A. Froud, R.T. Chapman, T.J. Butcher, J.J. Baumberg, W.S. Brocklesby and J.G. Frey, *Appl. Phys. Lett.*, 2008, **93**, 231103.

46 P. Zeitoun, G. Faivre, S. Sebban, T. Mocek, A. Hallou, M. Fajardo, D. Aubert, P. Balcou, F. Burgy, D. Douillet, S. Kazamias, G. de Lachèze-Murel, T. Lefrou, S. le Pape, P. Mercère, H. Merdji, A.S. Morlens, J.P. Rousseau and C. Valentin, *Nature*, 2004, **431**, 426.

47 A. Azima, J. Bödewadt, M. Drescher, H. Delsim-Hashemi, S. Khan, T. Maltezopoulos, V. Miltchev, M. Mittenzwey, J. Rossbach, R. Tarkeshian, M. Wieland, H. Schlarb, S. Düsterer, J. Feldhaus, T. Laarmann, *Proc. 11th European Particle Accelerator Conference*, 2008, paper MOPC028.

ELECTRON TRAJECTORIES IN HIGH HARMONIC GENERATION

I.B. Földes[1] and K. Varjú [2]

[1]Department of Plasma Physics, Association EURATOM, Wigner Research Centre for Physics of the Hungarian Academy of Sciences, H-1121 Budapest, Konkoly-Thege u. 29-33, Hungary
[2]Department of Optics and Quantum Electronics, University of Szeged, H-6720 Szeged, Dóm tér 9, Hungary

1 INTRODUCTION

The most flexible tool for generating coherent EUV radiation by an intense laser pulse is the generation of its high harmonics. Harmonics of ultrashort laser pulses in gases cover practically the full spectral range from the visible to the kilo-electronvolt X-rays. In most cases odd harmonics are generated in gases up to a limit defined by the laser intensity and the ionisation potential of the gas giving an upper limit of the generated photon energies. The other method for high harmonic generation (HHG) is that from the steep density surface of laser plasmas, for which – depending on the intensity – several mechanisms are responsible. Harmonics generated in this way can be both odd and even order, and in principle there is no ionization limit for the highest available harmonic order.

The statement that in gases only odd harmonics and in plasma gradients odd and even harmonics are generated is however an oversimplified statement. This chapter will compare the models of generation of gas and plasma harmonics, emphasizing symmetry considerations. It will be shown that the picture which is based on the classical considerations of Bloembergen based on the symmetries of the high order susceptibilities,[1,2] is a simple model and – especially in the case of laser pulses of just a few cycles – both the accurate pulse shape and the geometrical symmetry of the interaction must be taken into account. The aim is to show similarities and differences between harmonic generation in gases and from plasma gradients. In the latter case the range of moderately relativistic intensities will be emphasized in more detail, when the coherent wake emission (CWE) model can be applied which illustrates these symmetry considerations more clearly.

2 GAS HARMONICS

2.1 Model

In the interaction of an intense laser pulse with gas particles (atoms, molecules), optical ionisation may occur as the result of the distortion of the Coulomb potential by the intense electric field. The freed electron is driven by the laser electric field, and may recombine with its parent ion, emitting the excess energy in the form of a high energy photon, as illustrated in figure 1.[3,4] This process is repeated every half-cycle, and the temporal periodicity of the process leads to the appearance of discrete spectral lines at harmonics of the laser frequency.

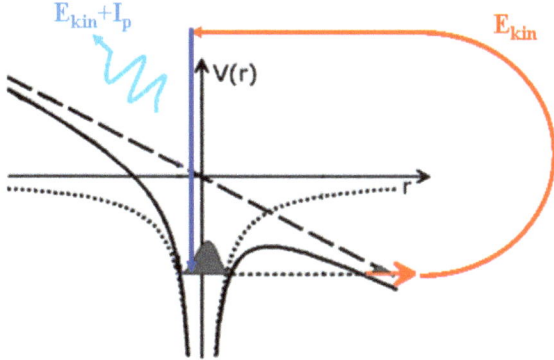

Figure 1 *Illustration of the three steps in the HHG process in a gas*

The emission of a photon is only one possible route after the return of the electron to the parent ion. If the electron is re-scattered instead of being reabsorbed, above threshold ionization (ATI) is observed. The photon spectrum produced in HHG and the electron spectrum observed in ATI are analogous, with a slight difference since an electron emitted during the positive values of the electric field travels in the opposite direction to one emitted in the next half cycle and these distinguishable processes lead to a full cycle periodicity of the process. Thus ATI peaks are separated by a photon energy corresponding to adjacent harmonics whereas the HHG spectrum contains only odd harmonics.

The third path for the returning electron, when its kinetic energy is higher than the ionisation potential, is to induce the emission of another electron, and hence non-sequential double ionisation (NSDI) takes place.

2.2 Symmetry of the Process

The HHG process is not sensitive to the sign of the electric field; the same ionisation/acceleration/recombination processes can take place when the electron is freed at positive or negative values of the laser electric field. This is why the process is repeated every half cycle of the laser field. The half-cycle periodicity translates to the appearance of harmonics separated by double the laser frequency (see figure 2).

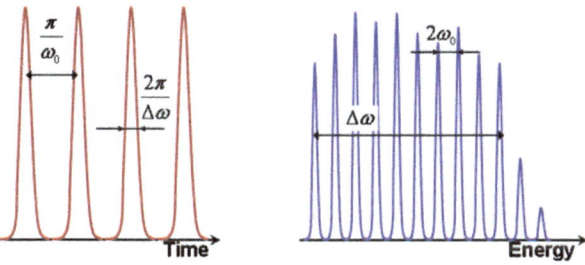

Figure 2 *The temporal periodicity of the process is manifested in the spectrum. The duration and separation of the bursts produced in every half cycle determines the widths and separations of the spectral peaks*

Considering the harmonic burst emitted during two consecutive half cycles, it may be assumed in general that the amplitude A and phase of the bursts are functions of the harmonic order, but for simplicity here we take $A_1(q) = A_2(q) = 1$, where q is the harmonic order. The half-cycle time delay between the two bursts lead to a phase term of $i\pi q$.

The harmonic spectra produced by these two bursts, illustrated in figure 3, can be written as

$$S(q) = A_1(q)e^{i\varphi_1(q)} - A_2(q)e^{i\varphi_2(q)}e^{i\pi q} = e^{i\varphi_1(q)} - (-1)^q e^{i\varphi_2(q)} \tag{1}$$

$$S(q) = e^{i\varphi_1(q)}\left[1 - (-1)^q e^{-i\left[\varphi_2(q) - \varphi_1(q)\right]}\right], \tag{2}$$

where $\varphi_1(q)$ and $\varphi_2(q)$ are the dipole phases that the electron accumulates during its travel in the laser field.

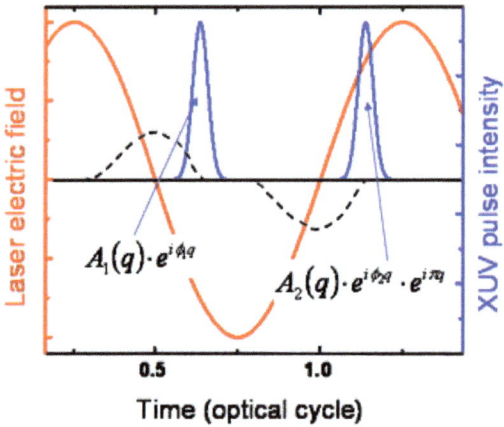

Figure 3 *Harmonic bursts emitted in consecutive half-cycles. The red line depicts the electric field of the generating laser pulse and the dashed black lines illustrate the electron trajectories leading to the emission of the XUV bursts in blue*

2.3 Odd Harmonics Generated by a Long Driver Laser Pulse

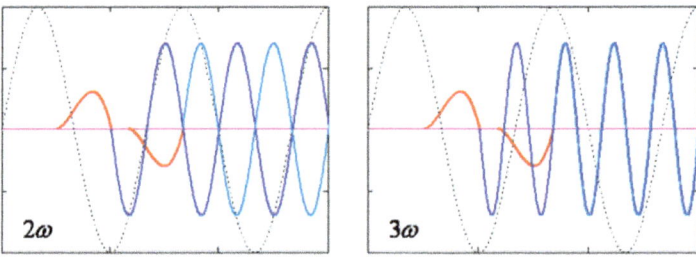

Figure 4 *Constructive interference for odd and destructive interference for even harmonics is responsible for the production of the harmonic spectrum. In the figures the dotted lines represent the laser electric field, while the red lines illustrate the electron trajectories in two consecutive half-cycles. The radiation emitted in the two half-cycles is opposite in phase.*

When harmonics are generated by long laser pulses consecutive half cycles steer the electron wave packet identically – apart from the obvious sign difference -, and the dipole phases are the same. i.e, $\varphi_2(q) - \varphi_1(q) = 0$, and equation (2) gives $S(q) = 0$ for all even values of q, but nonzero for the odd harmonics. This is equivalent to stating that there is constructive interference for odd harmonics, and destructive interference for even, as shown in figure 4.

Figure 5 shows the difference of the calculated dipole phase in two consecutive half cycles $(\varphi_2(q) - \varphi_1(q))$ of a 50 fs, 800 nm, 4×10^{14} W/cm^2 laser. The figures were calculated via the non-adiabatic saddlepoint method where the extrema of the phase are found to determine the relevant electron trajectories.[5,6] The two most important electron trajectories are plotted that contribute to harmonic emission, labelled short and long referring to the time the electron spends in the continuum. In the case of short trajectories the electron return time to the atom is close to the half period of the laser oscillation, whereas the long path ones have return time near to one period. It can be seen that the phase difference of the radiation emitted in consecutive half cycles is very close to zero, and that only odd harmonics are observed, both for the short (red) and long (blue) trajectory components, as expected from equation (2).

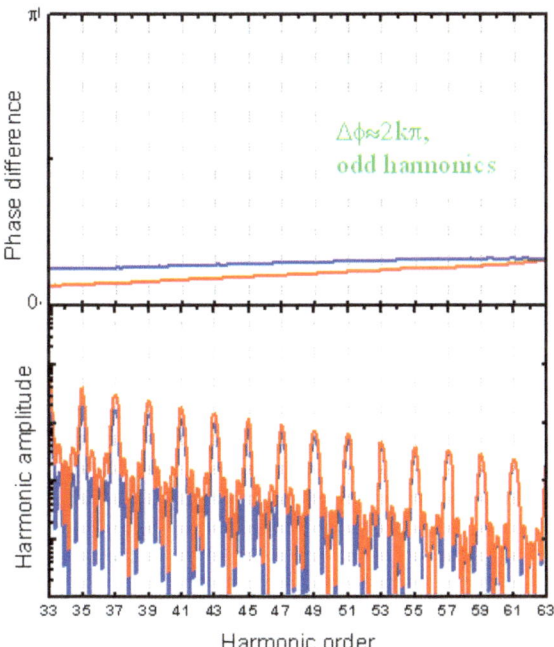

Figure 5 *Harmonic phase differences and the produced spectrum for a long driving pulse. The red (blue) curves illustrate short (long) trajectory components. The vertical dotted lines guide the eye for the positions of the odd harmonics.*

A consequence of the presence of odd harmonics and the half-cycle periodicity is the phase jump of π radians in the attosecond pulses in the produced train, leading to varying carrier-envelope phase (CEP). Figure 6 shows how the consecutive attosecond pulses switch phase by π.

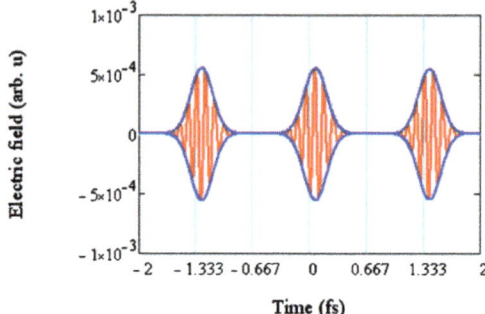

Figure 6 *The π radians phase jump in the electric field of the driving laser pulse leads to a π phase jump in the produced harmonic bursts.*

2.4 CEP Tunable Harmonics using Short Driver Laser Pulses

If a short, few-cycle laser pulse is used for harmonic generation, the consecutive half cycles are not equivalent. Then, if the condition $\varphi_2(q) - \varphi_1(q) = \pi$ is satisfied, this leads to the appearance of even harmonics in the spectrum.

    Here, calculations are presented for generating harmonics in argon using a 5.2 fs laser pulse with a central wavelength of 800 nm focused to an intensity of 4×10^{14} W/cm^2. For such a short driving pulse the electric field varies strongly from half cycle to half cycle, and the phase differences between consecutive half cycles vary between zero and 6π. Consequently odd and even harmonics are seen in parts of the spectrum. Figure 7 indicates phase differences of odd and even multiples of π, corresponding to the appearance of even and odd harmonics, respectively.

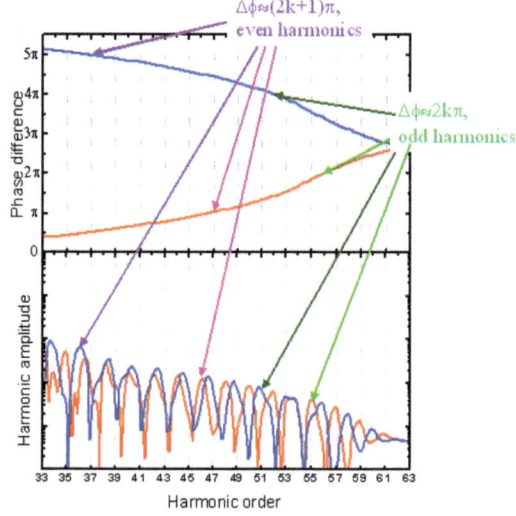

Figure 7 *Harmonic phase differences and the produced spectrum for a short driving pulse. Red curves illustrate short and blue curves long trajectory components. The vertical dotted lines guide the eye for the positions of the odd harmonics.*

With the variation of the CEP the electric field and hence the freed electron trajectories are altered, and thus there is a spectral variation of the phase difference with CEP. This affects the spectrum, the harmonic peaks shifting with CEP. Such an effect was observed experimentally by Calegari *et al.*[7]

3 HARMONICS FROM LASER PLASMAS

Early results on the generation of harmonics from laser-plasma interactions[8] in the non-relativistic regime were explained through interactions with plasmons generated near the critical surfaces. Thus, similarly to gas harmonics, an upper limit for the generated photon energy was foreseen at a plasma frequency corresponding to the electron density in solids.[9] The rapid development of high-power short-pulse laser systems using chirped-pulse amplification (CPA) allowed several orders of magnitude increase in laser intensities. When the intensity is so high that the electrons oscillate with relativistic velocities in the intense laser field, harmonic generation is described by the so-called relativistic oscillating mirror (ROM) model.[10] In this case the available harmonic orders and thus the obtainable highest frequency of the available coherent radiation is no longer limited since the Doppler shift of light reflected from the collective oscillating plasma produces this effect.[11] The theory of high harmonic generation has been well elaborated for the weakly relativistic case defined by the normalized vector potential $a_0 \approx 1$, where $a_0 = eE / m\omega_L c = \left[I\lambda_L^2 / \left(1.37 \times 10^{18} \right) \right]^{1/2}$, the intensity I is in W/cm^2 and the laser wavelength λ_L is in micrometres. In this case the so-called coherent wake emission (CWE)[12] is responsible for harmonic generation whereas in the strongly relativistic case the ROM mechanism is dominant, for which a more detailed theory has also been given.[13] Some recent excellent reviews of harmonic generation using plasma mirrors[14,15] give detailed descriptions of the main theoretical and experimental backgrounds. The following paragraphs provide some comparisons with harmonic generation in gases based on symmetry considerations.

A summary of a detailed discussion[11] of the effect of symmetry on harmonic generation is as follows. Using cold relativistic plasma fluid equations, i.e. relativistic hydrodynamics plus Maxwell equations, a wave equation containing both CWE and ROM harmonics was derived. This indicated three factors that lead to harmonic generation: the electron density n, including the critical surface and its motion, the longitudinal Lorentz factor, containing the longitudinal velocity, and a term containing the vector potential and the angle of incidence.

For normal incidence of linearly polarized light there is no difference between the s and p polarizations since the electric field of the beam oscillates parallel to the target surface. Hence there is no directionality of the generated harmonics, and similarly to gas harmonics there is a half-cycle periodicity leading to the generation of only odd harmonics. This is therefore analogous to the case of gas harmonics and for pulses of a few cycles a model similar to that described in section 2.4 can be applied. No harmonics are generated by circularly polarized light.

However, oblique incidence is the one which is basically different from the previously discussed cases. In the case of p-polarized incoming radiation the component of the electric field perpendicular to the target surface has a full-cycle periodicity because the direction of it is either inward or outward. This electric field component therefore leads to the generation of even and odd harmonics according to the simple considerations of section 2. It can be shown that the generated harmonics will be only p-polarized for p-polarized incident radiation, but for s-polarized radiation it is more complicated. This is because s-polarized odd harmonics are generated in a similar way as for normal incidence, and p-polarized even harmonics are generated with intensities weaker than those of odd harmonics, especially for small angles of incidence and disappearing for normal incidence. The description of Lichters *et al.*[11] is in agreement with earlier simulations[16] as it gives harmonics as high as $50\omega_0$ for the mildly relativistic case with $a_0 = 3$.

Before going into the details of the two main mechanisms of harmonic generation it is to be noted that the half-cycle and full-cycle periodicities have an important consequence for the generation of attosecond pulses. It has long been known that phase locking of the generated harmonics results in a train of attosecond pulses;[17] the simplest method to select a single attosecond pulse is to use filters which transmit only the highest observable harmonic order[18] generated by a few-cycle laser. In the case of full-cycle periodicity this selection will work with somewhat longer pulse durations than for the half-cycle periodicity of gas harmonics or normally incident light pulses, and thus requires less stringent conditions on the laser beam. This means that no additional methods such as polarization gating need to be applied.

3.1 Coherent Wake Emission Harmonics for Moderate Laser Intensities

The coherent wake emission model[12] has several similarities with the theory of gas harmonics. The original model consists of three steps.[15] Initially, electrons from the plasma surface are pulled out into the vacuum by the electric field of the laser beam and then pushed back into the dense plasma after gaining energy from the laser field. This step is analogous to the ionization of electrons by the laser field in the case of gases, with the difference that this process is directional as the electric field has a component inward or outward perpendicular to the solid target. In the second step the electrons propagate in the dense plasma, forming ultrashort bunches, thus generating Langmuir waves in their wake. The third step is in the inhomogeneous part of the plasma density gradient where an integer multiple of the laser frequency coincides with the local plasma frequency. These collective electron plasma oscillations radiate at different local frequencies as the excited plasma waves undergo linear mode conversion into electromagnetic waves via inverse resonance absorption.

CWE is the dominant harmonic generating mechanism from plasma mirrors for the moderately relativistic regime with $a_0 \approx 1$. This type of process occurs once in every laser optical cycle, and thus the spectrum will contain even and odd harmonics up to the maximum plasma frequency ω_p. This frequency corresponds to the total electron density of the cold solid material leading to a maximally available harmonic order of 15-30 for a 800 nm laser field depending on the initial electron density in the solid target with the maximum available frequency corresponding to the total electron density of the solid material. Resonance[19] and Brunel absorption[20] are responsible for the pulling out the electrons and their energy gain from the laser field. Calculated trajectories of the electrons show that those returning to the plasma will have a crossing point in the over-dense material, thus leading to bunching. This finding led to an alternative three step model[21] in which it is assumed that the plasma Langmuir-waves are generated at a given depth of the plasma gradient. There is only a slight difference between the two methods in the description of the phase properties of harmonics. The first model[15] describes more accurately the phase difference which is caused by the depth of the plasma, because it takes into account the exact time needed for the waves to reach the critical surface and so describes better the so-called attosecond chirp, i.e. the finite duration of a single bunch of harmonics of attosecond duration. The second model[21] gives an elegant formulation of the so-called femtosecond chirp, which is caused by the phase difference of harmonics in the bursts following their generation (i.e. once in every full cycle), whereas the first model has to use an additional free parameter for it.

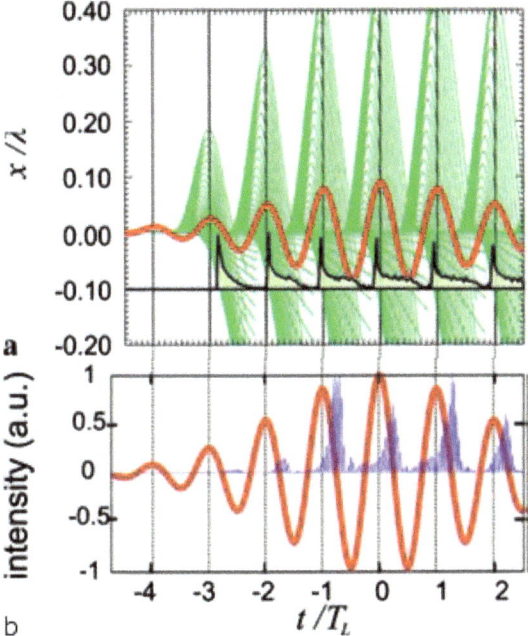

a

b

Figure 8 *Results of 1D calculations based on the 3-step model.[21] (a) The trajectories of the electrons released into the vacuum and returning to the plasma surface are shown in green and the number of electrons per unit time at a depth of x=-0.1 is given by the black line. The driving electric field is plotted by the red line. (b) The generated attosecond pulses (blue) are emitted at times coinciding with the return of the expelled electrons.*

Figure 8 illustrates the trajectories of the electrons and the generated attosecond pulses for a three-cycle laser pulse based on a one-dimensional model[21] for an angle of incidence of 45° and an intensity corresponding to $a_0 = 1.5$. The instant of release of the electron from the plasma–vacuum interface is denoted as t_0. The electrons at the vacuum–plasma interface are initially at rest and uniformly distributed, i.e. t_0 is varied in equidistant intervals. After re-entry into the plasma region they are assumed to travel with a constant velocity, acquired during the excursion into the vacuum. Only the orbits of the electrons returning to the plasma region are plotted in figure 8. The histogram of the electrons per unit time crossing a line located at $x = -0.1$ behind the interface clearly shows a temporal bunching of the returning electrons within a time window of a fraction of a cycle. The reason for this bunching is merely the electron dynamics for different phases of the sinusoidal electric field of the laser. A careful examination of the orbits indicates that at the beginning of each cycle (positive-to-negative field crossing) some electrons are pulled out but due to the weak field their excursion is short. As the field continues to grow, the subsequent electrons move further from the surface but due to their higher velocity they return approximately at the same time as the previous ones. This results in a crossover of the orbits inside the plasma and the formation of electron bunches. Since the harmonics are generated via these bunches of energetic electrons (e-bunches), the emission time of the individual attosecond bursts will follow the occurrence of the electron temporal localization. It is to be noted that simulations carried out with a 1D-PIC code confirm these results, the bunching of electrons near the *x=-0.1* line as well as the emission time shown in figure 8b.

The experimental results using laser pulses of a few cycles for harmonic generation show the effects of the carrier-envelope phase.[21] The spectrum of the harmonics is broadened as would be expected from the attosecond bunch durations. However, a substructure of harmonics can also be observed, and this, according to the three-step model, can be attributed to the shot-to-shot variation of the carrier envelope phase.

3.2 Relativistic Oscillating Mirror Harmonics for High Intensities

The basic idea of the relativistic oscillating mirror (ROM) originates from Bulanov *et al.*[10] and was described in detail by Lichters *et al.*[11] It is assumed that due to the $\mathbf{v}\times\mathbf{B}$ component of the Lorentz force the reflection occurs from an oscillating source which can be described by a step function. More precisely it is the ponderomotive force which drives the harmonic oscillation of the critical surface that is proportional to the second harmonic of the laser field. Therefore the position of the mirror can be described by $x(t') = A_{\mathrm{m}} \sin(2\omega_{\mathrm{L}} t' + \varphi)$ where A_{m} is the amplitude of the mirror oscillation and φ is the relative phase between the incident wave and the mirror position.[18] The incident laser field $E_{\mathrm{inc}}(t) = \sin(\omega_{\mathrm{L}} t)$ gives rise to a reflected field $E_{\mathrm{refl}}(t) = \sin(\omega_{\mathrm{L}} t + 2k_{\mathrm{L}} x(t'))$ on this mirror, neglecting time-independent phase shifts. The reflected field is thus clearly a source of harmonic generation. Note that as this is a relativistic effect there is a retardation between the observer and the mirror which gives the relation between the retarded time t' and the observer's time t as

$$t' = t + x(t')/c. \qquad (3)$$

For the mildly relativistic case the retardation effect can be neglected and – for normal incidence – odd harmonics were derived in terms of Bessel functions by Lichters *et al.*[11] who pointed out that full agreement with the relativistic particle in cell (PIC) codes can only be obtained if the retardation effects are taken into account. Again, it must be emphasized that due to the $2\omega_{\mathrm{L}}$ modulation the density is modulated with a half period of the laser field resulting in the generation of only odd harmonics. For oblique incidence it was difficult to obtain agreement between the analytical model of Lichters *et al.* and the PIC simulations.[11] Summarizing these results, the generated harmonics for s-polarized radiation originate from the ponderomotive force only, giving – similarly to normal incidence – only odd harmonics, whereas for p-polarized radiation stronger harmonic generation is expected with both odd and even harmonics caused by the surface modulation at a frequency ω_{L}, i.e. with a periodicity of $\tau=2\pi/\omega_{\mathrm{L}}$.

Although PIC codes include highly relativistic effects, an alternative analytical theory based on self-similar solutions of the relativistic Vlasov equation was introduced by Gordienko *et al.*[22,23,24] and further developed by Baeva *et al.*[13] – this is called the γ-spiking model. In a collisionless plasma - which is a good approximation for electrons moving with relativistic velocities -, neglecting ion motion the electron distribution function $f(t,r,\mathbf{p})$ is governed by the Vlasov equation, i.e. only by the electromagnetic field,

$$\left\lfloor \partial_t + \mathbf{v}\partial_r - e(\mathbf{E} + \mathbf{v}\times\mathbf{B}/c)\partial_{\mathbf{p}} \right\rfloor f(t,r,\mathbf{p}) = 0, \qquad (4)$$

with the self-consistent **E** and **B** fields satisfying Maxwell's equations

$$\nabla_r \mathbf{E} = 4\pi e(n_e + \rho), \qquad \nabla_r \mathbf{B} = 0$$
$$c\nabla_r \times \mathbf{B} = 4\pi \mathbf{j} + 4\pi \partial_t \mathbf{E}, \qquad c\nabla_r \times \mathbf{E} = -\partial_t \mathbf{B}, \tag{5}$$

where n_e is the background electron density, $\rho = -\int f d p$, $\mathbf{j} = -e\mathbf{v}\int f d\mathbf{p}$ and $\gamma = \left[1 - (v/c)^2\right]^{-1/2}$ is the relativistic factor. The eventual collisions may be taken into account by a collisional term at the right hand side of Eq. 4 (Boltzmann-equation) but in this relativistic treatment for the self-similar solution they are generally neglected. For the strongly relativistic regime, i.e. for $a_0 = eA/m_e c^2 \gg 1$, the number of dimensionless parameters reduces to three, namely $k_0 R$, $\omega_L \tau$ and a new similarity parameter, $S = n_e / (a_0 n_c)$.[13,22,23] Here, R is the focal spot size, τ is the pulse duration and n_c is the critical density. Thus the similarity parameter S combines the dimensionless intensity and the dimensionless density. Its relevance is that if the plasma density and the laser amplitude change simultaneously with S constant, then the laser-plasma interactions remain similar. The S parameter separates relativistically over-dense plasmas ($S \gg 1$) from underdense ones ($S \ll 1$). Clearly, for constant S the plasma electrons move along similar trajectories with momenta proportional to a_0. A direct consequence of this proportionality is that both the electron momentum component perpendicular to the plasma surface (p_n) and that parallel with it (p_t) will be proportional to a_0. This results in the electron velocities in the skin layers approaching the speed of light. However, the relativistic γ factor of the plasma surface, $\gamma_s(t')$ has a different behaviour. Whereas the γ factor of the free electrons becomes large and their velocity approaches the speed of light, $\gamma_s(t')$ for the boundary motion does not depend directly on a_0 and γ_s factor is generally of the order of unity. There is one exception when the γ_s factor has an abrupt maximum, a spike. It can be illustrated by simple considerations. The velocity of the plasma surface normalized by c can be written as

$$\beta_s(t) = \frac{p_n(t)}{\sqrt{m_e^2 c^2 + p_n(t)^2 + p_t(t)^2}} \; . \tag{6}$$

The γ_s factor has a sharp maximum when the parallel component of $p_t(t)$ is zero in which case it can be approximated as.

$$\gamma_s = \frac{1}{\sqrt{1 - \beta_s^2}} = \sqrt{\frac{m_e^2 c^2 + p_n(t)^2 + p_t(t)^2}{m_e^2 c^2 + p_t(t)^2}} = \sqrt{\frac{m_e^2 c^2 + p_n(t)^2}{m_e^2 c^2}} \propto a_0. \tag{7}$$

The tangential component of the momentum is proportional to the vector potential of the electric field which disappears here, too. This happens when it changes sign, i.e. at the turning point.

The relativistic γ factor of the plasma boundary thus has a jump from unity to $\sim a_0$ over a time $\Delta t' \propto 1/(a_0 \omega_L)$, and this γ-spike is the source of the generation of high-order harmonics in the form of ultrashort (attosecond) pulses.

Introducing this spiking motion into the mirror position $X(t)$ of Lichters' model[11] gives the intensity spectrum of the harmonics as $I_n \propto 1/n^{8/3}$, which results in a power law decay of the intensity of the harmonics up to a sharp cutoff at the harmonic number n_{cutoff},

$$n_{cutoff} = \sqrt{8\alpha}\gamma_{max}^3,\qquad\qquad(8)$$

with the numerical factor α being of the order of unity.[13] Due to the strict periodicity of γ-spiking these relativistic harmonics are phase locked, and they can be used to generate attosecond or even zeptosecond pulses.

3.3 Experimental results

With the fast improving availability of high-power short-pulse laser systems a large number of experimental results have confirmed the theoretical predictions. CWE harmonics were found to have a weak intensity dependence, agreeing with the expectation of a linear conversion process since harmonic emission is basically inverse resonant absorption,[12] which transforms the plasma wake generated by the returning Brunel electrons to optical radiation. The phase properties of CWE harmonics have contributed to the understanding of coherent sub-laser cycle dynamics of plasma electrons.[25] These investigations were recently extended to high-power few-cycle laser pulses in which case sub-cycle dynamics of electrons showed a similar strong sensitivity to the carrier-envelope phase of the laser pulse.[21] Since CWE harmonics – as discussed in section 3.1 – result from chirping, it was questionable whether they could be used for generating attosecond pulses. However, recent results[26] have shown that even CWE harmonics are phase-locked, i.e. the XUV harmonics are synchronized, and therefore they enable attosecond temporal bunching. A pulse-duration $\tau \approx 0.9\pm0.4$ fs was derived, providing the first experimental evidence of attosecond surface harmonics. It was clearly shown that the periodicity of the attosecond bunches corresponded to the full laser period in contrast to the half period in the case of gas harmonics. Although the quasi-linearity of CWE effects makes single harmonic selection questionable, the full period between the bunches makes it easier for a laser pulse of a few cycles.[21]

The transition from the CWE mechanism to the relativistic moving mirror model has been demonstrated by several groups.[27,28] The most exciting observation was, however, the demonstration of multi kilo-electronvolt harmonics for the highly relativistic regime up to 2.5×10^{20} W/cm^2.[29] This observation demonstrates the scaling law obtained by the γ-spiking model.[13] Harmonics were observed up to the 3200th order (~3.8 keV).[29] The experimentally observed signals were consistent with harmonics energies of ~17 μJ at 1.4 keV and ~5 μJ at 3.1 keV photon energies in a bandwidth of 1%. The observed narrow beaming of the harmonics[29] occurring within a cone angle of 4° is a strong evidence for the coherent nature of the harmonic production process. Based on the experimental evidence and current theory[22] it will soon be possible to generate 10 keV zeptosecond pulses with a conversion efficiency of ~10^{-7}.

4 CONCLUSIONS

Summarizing the results it can be stated that harmonics will soon serve as available coherent radiation sources from 10 eV to 10 keV photon energies and with pulse durations in the attosecond and even in the zeptosecond range.

It was shown that similarly to high harmonic generation in gases single electron trajectories determine the generation of high harmonics from the surfaces of solid targets in case of weekly relativistic laser fields, when CWE mechanism of high harmonic generation is dominant. In the case of strongly relativistic laser field collective behavior of electrons will determine the behavior of harmonics as described by the ROM mechanism and the γ-spiking model.

References

1 N. Bloembergen, *Nonlinear Optics*, 1965, (New York, Benjamin).
2 Y.R. Shen, *The Principles of Nonlinear Optics*, 1984, (New York, John Wiley).
3 K.J. Schafer, B. Yang, L.F. DiMauro and K.C. Kulander, *Phys. Rev. Lett.*, 1993, **70**, 1599.
4 P. Corkum, *Phys. Rev. Lett.*, 1993, **71**, 1994.
5 M. Lewenstein, P. Balcou, M.Yu. Ivanov, A.L'Huillier and P.B. Corkum, *Phys. Rev. A*, 1994, **49**, 2117.
6 G. Sansone, C. Vozzi, S. Stagira and M. Nisoli, *Phys. Rev. A*, **2004, 70**, 013411.
7 F. Calegary, M. Lucchini, K.S. Kim, F. Ferrari, C. Vozzi, S. Stagira, G. Sansone and M. Nisoli, *Phys. Rev. A*, 2011, **84**, 041802(R).
8 R.L. Carman, D.W. Forslund and J.M. Kindel, *Phys. Rev. Lett.*, 1981, **46**, 29.
9 B. Bezzerides, R.D. Jones and D.W. Forslund, *Phys. Rev. Lett.*, 1982, **49**, 202.
10 S.V. Bulanov, N.M. Naumova and F. Pegoraro, *Phys. Plasmas*, 1994, **1**, 745.
11 R. Lichters, J. Meyer-ter-Vehn and A. Pukhov, *Phys. Plasmas*, 1996, **3**, 3425.
12 F. Quéré, C. Thaury, P. Monot, S. Dobosz, P. Martin, J-P. Geindre and P. Audebert, *Phys. Rev. Lett.*, 2006, **96**, 185001.
13 T. Baeva, S. Gordienko and A. Pukhov, *Phys. Rev. E*, 2006, **74**, 046404.
14 U. Teubner and P. Gibbon, *Rev. Mod. Phys.*, 2009, **81**, 445.
15 C. Thaury and F. Quéré, *J. Phys. B*, 2010, **43**, 213001.
16 P. Gibbon, *Phys. Rev. Lett.*, 1996, **76**, 50.
17 G. Farkas and C. Tóth, *Phys. Lett. A*, 1992, **168**, 447.
18 G.D. Tsakiris, K. Eidmann, J. Meyer-ter-Vehn and F. Krausz, *New J. Phys.*, 2006, **8**, 19.
19 V.L. Ginzburg, *The Propagation of Electromagnetic Waves in Plasmas*, 1970, (Oxford, Pergamon).
20 F. Brunel, *Phys. Rev. Lett.*, 1987, **59**, 52.
21 P. Heissler, R. Hörlein, M. Stafe, J.M. Mikhailova, Y. Nomura, D. Herrmann, R.Tautz, R.G. Rykovanov, I.B. Földes, K. Varjú, F. Tavella, A. Marcinkevicius, F. Krausz, L. Veisz and G.D. Tsakiris, *Appl. Phys. B*, 2010, **101**, 511.
22 S. Gordienko, A. Pukhov, O. Shorokhov and T. Baeva, *Phys. Rev. Lett.*, 2004, **93**, 115002.
23 S. Gordienko, A. Pukhov, O. Shorokhov and T. Baeva, *Phys. Rev. Lett.*, 2005, **95**, 103903.
24 S. Gordienko and A. Pukhov, *Phys. Plasmas*, 2005, **12**, 043109.
25 F. Quéré, C. Thaury, J-P. Geindre, G. Bonnaud, P. Monot and P. Martin, *Phys. Rev. Lett.*, 2008, **100**, 095004.
26 Y. Nomura, R. Hörlein, P. Tzallas, B. Dromey, S. Rykovanov, Z. Major, J. Osterhoff, S. Karsch, L. Veisz, M. Zepf, D. Charalambidis, F. Krausz and G.D. Tsakiris, *Nature Physics*, 2009, **5**, 124.
27 A. Tarasevitch, K. Lobov, C. Wünsche and D. von der Linde, *Phys. Rev. Lett.*, 2007, **98**, 103902.

28 C. Thaury, F. Queré, J-P. Geindre, A. Levy, T. Ceccotti, P. Monot, M. Bougeard, F. Reau, P. D'Oliveira, P. Audebert, R. Marjoribanks and P. Martin, *Nature Physics*, 2007, **3**, 424.

29 B. Dromey, S. Kar, C. Bellei, D.C. Carroll, R.J. Clarke, J.S. Green, S. Kneip, K. Markey, S.R. Nagel, P.T. Simpson, L. Willingale, P. McKenna, D. Neely, Z. Najmudin, K. Krushelnick, P. Norreys and M. Zepf, *Phys. Rev. Lett.*, 2007, **99**, 085001.

MODIFIED CATHODE TUBE: X-RAY AND XUV RADIATION FOR NANO-INSPECTION

U. Hinze and B. Chichkov

Laser Zentrum Hannover e.V., Hollerithallee 8, D-30419, Hannover, Germany

1 INTRODUCTION

Investigations with extreme ultraviolet (EUV) radiation and X-rays have to be accompanied by corresponding metrology. This is required for the characterization and quality control of optical components such as mirrors, filters and sensors, and to examine the properties of novel materials at a defined wavelength or in a certain spectral range. The spectral range that is covered by EUV and X-rays is quite large, starting below 100 eV and ending above 100 keV. One of the spectral bands of special interest within this range is at 92 eV (13.5 nm). This wavelength was identified by the international computer chip industry as a target wavelength for the next generation of optical lithography and manifested in joint specifications of the International Sematech for EUV lithography,[1] since when there has been an on-going need for technical solutions for EUV. One central subject is to establish a fast and compact in-house metrology for the characterization of complex optics developed for EUV lithography.

For metrology in the EUV spectral range synchrotrons and plasma sources are often considered. It is not widely appreciated that such metrology can in fact be done with modified X-ray tubes, and that this type of radiation source provides essential advantages. Due to their operation principles EUV and X-ray tubes are free of debris and exhibit long term stability. At the same time the technology allows very compact setups with industrially proven devices.[2]

There are two basic ways in which a modified X-ray tube can be operated for metrology. The first, continuous operation, is done by accelerating electrons emitted by a continuous electron source in a high-voltage field and focusing them onto a target. The second, pulsed operation, is effected by generating bunches of electrons by the photoelectric effect and accelerating these bunches in a high-voltage field towards a target; such pulses may be as short as 500 fs. While the first method is very good for investigations of static samples and elements, the second method is of interest for the investigation of time-resolved fast dynamic processes.

This chapter gives a short overview of sources for metrology in the EUV and X-ray spectral range, with a discussion of their properties. The physics and operation principles of a modified X-ray tube for EUV radiation are presented. Different wavelength ranges that can be accessed and investigated with this tool are identified and discussed. Continuous and pulsed tube operation are described and characterized.

For the EUV spectral range two metrological applications are presented, namely a simple and compact setup for the characterization of optics (mirrors, filters, sensors) and a complex reflectometer for the characterization of large collector optics with high numerical aperture for EUV lithography.

2 METROLOGY SOURCES

Ideally, a metrology source should be simple in operation, always readily available, have low running costs, provide a stable and calibrated output continuously or at a high repetition rate, have a long lifetime and be free of effects such as debris that could damage the component being inspected. Several different sources are available for metrology in the EUV and soft X-ray range, and the advantages and disadvantages of each of them have to be considered for a given application.

Investigations can be done at synchrotron facilities, which are high-quality sources that allow excellent control of the radiation. This is perfect for scientific research, but synchrotrons have drawbacks in application. Experiments cannot be performed in-house, instead the components to be tested must be transported to the synchrotron, where measurements can only be done according to the time management of the facility. Often in-house laboratory sources are preferable for metrology as they allow direct access at any time. Such sources can be plasma sources, laser based sources or EUV tubes.

Different laser based processes have been suggested and investigated as potential EUV metrology sources, e.g. generation of high harmonics with femtosecond lasers,[3] the operation of lasers at 13.2 nm in very highly ionized cadmium atoms[4] and free electron lasers.[5] However, such sources are, in practice, rarely used in metrological applications. Many metrological laboratory sources for the EUV are based on discharged produced plasmas (DPP) or laser produced plasmas (LPP), with first applications demonstrated in the 1990s.[6,7] Plasma sources are often based on the emission of ionized xenon,[8–10] however special techniques are required to provide stable emission and to reduce contamination by debris.

It is usually overlooked that characteristic EUV radiation can be generated directly by electrons incident on solid targets, avoiding many plasma-related drawbacks, and that this type of radiation source provides essential advantages. An EUV source operated like this is basically an X-ray tube which has been modified for operation in the 5-20kV high-voltage range.

3 EUV TUBES

3.1 Operation Principles

In EUV and X-ray tubes electrons are emitted by a filament. A high-voltage field is used to accelerate these electrons to the anode, where characteristic emission is generated by electron-impact ionization (excitation) of atomic inner shells followed by radiative decay. In particular, electron-induced emission from silicon ($Z = 14$) is of interest for applications for EUV lithography (EUVL), as the silicon $L_{2,3}$ emission covers the spectral range from 12.5-15 nm with a maximum at 13.5 nm. Note that the emission spectrum fits remarkably well with the reflection characteristics of Mo/Si multilayer mirrors.[11] Other materials suitable for this spectral region are beryllium ($Z = 4$), with a K_α emission peak at 11.4 nm, and tungsten which provides broadband emission (see figure 1).

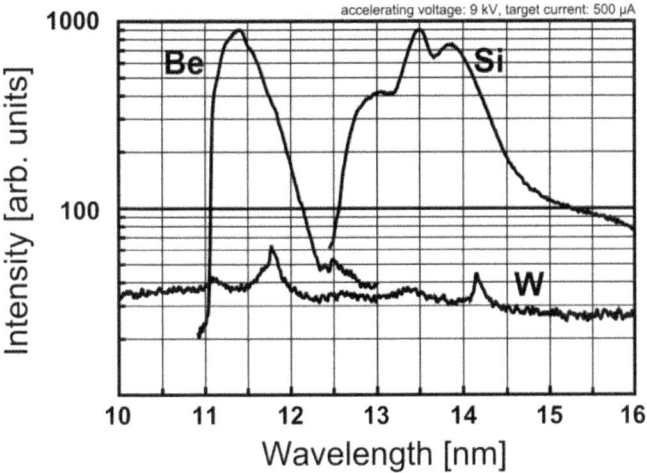

Figure 1 *Emission spectra of silicon, beryllium and tungsten.*

In addition to the desired emission (e.g. 13.5 nm for silicon L in EUVL) there are additional emissions and radiative contributions from an X-ray source, including emission from other X–ray transitions such as 1.7 keV emission for silicon K, a weak broadband Bremsstrahlung background, visible and infrared light from the filament of the X-ray tube and scattered and secondary electrons. These undesirable emissions have to be considered and suppressed in measurement setups, however, this is easily done. Unwanted X-rays can in total be up to 20 times more intense than the EUV radiation, they can be removed by a single reflection from a multilayer mirror. Visible and infrared light can be stopped by a zirconium filter and electrons can be removed by a magnetic field.

The maximum EUV tube output power is limited by the heat load that can be applied to the target. The damage threshold of silicon, for example, is about 1 kW/mm^2. At higher powers the target material starts to melt and evaporate and so becomes damaged. To study this, a silicon target was exposed to electron beams with varying powers, and the target was subsequently inspected in an electron microscope. The results are shown in figure 2.

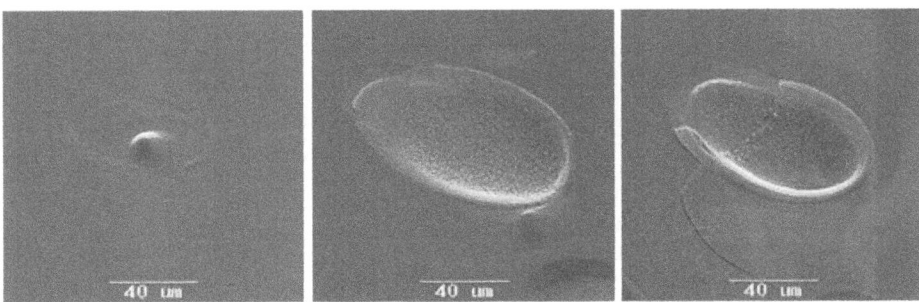

Figure 2 *Electron microscope images of a silicon target operated above the damage threshold: left, 2.4 kW/mm^2, middle 3.9 kW/mm^2 and, left, 19.2 kW/mm^2.*

The power load capacity of EUV targets can be increased by using rotating anodes or localized cooling, as confirmed by numerical modelling and experimental investigation. However, the results indicate that the cost-benefit ratio does not justify the effort for such implementations.[12]

For the work described here a modified X-ray tube was used (GE/phoenix X-ray sem|20, figure 3) equipped with a silicon target for the generation of EUV radiation. The device has an electromagnetic focusing unit, which allows control of the electron spot on the target. The acceleration voltage can be varied from 5 kV to 20 kV, and the maximum electrical power loading on the target is 50 W. The tube must be operated at a vacuum of 10^{-4} mbar or better. An electromagnetic unit removes scattered electrons from the output.

Figure 3 *The GE/phoenix sem|20EUV X-ray tube.*

The conversion efficiency of EUV tubes changes with the acceleration voltage. The maximum output power is achieved for 10 kV, which is thus typically used, giving a conversion efficiency of 2.5×10^{-4} in a solid angle of $2\pi sr$ and a bandwidth of 2%. The total EUV power in is 80 µW/mA and the in-band EUV power at 13.5 nm is 23 µW/mA, both over 2π sr and in 2% bandwidth.

An electromagnetic lens in the EUV tube allows control of the size and the position of the electron spot on the solid target. The EUV source spot size was characterized by a series of pinhole camera images. The typical source size was found to be some tens of micrometers (figure 4), which is much smaller than other EUV sources. This is of advantage in imaging applications as the spatial resolution is governed by the size of the source. In addition, control of the electron spot provides a very stable spatial stability – in a series of measurements over three hours the maximum detected drift was less than 10 µm.

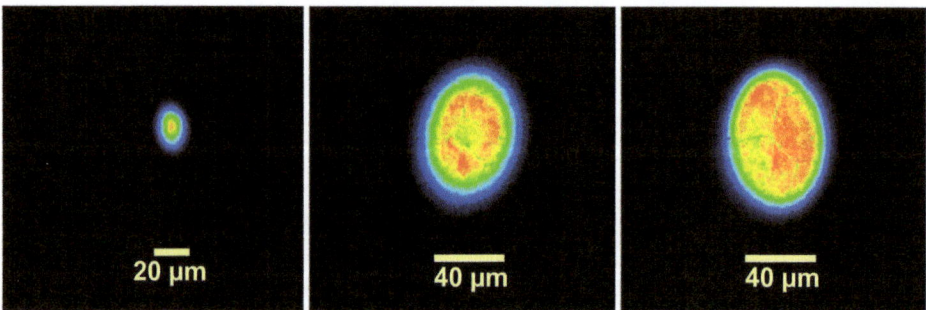

Figure 4 *Pinhole camera images of the EUV emission spot for tube currents of (left) 20 µA, (middle) 400 µA and (right) 900 µA (acceleration voltage 10 kV, pinhole size 5 µm).*

As well as spatial stability, two more factors are important for EUVL metrology sources, namely long-term power stability and mitigation of debris emission. The influence of these parameters was investigated by putting a Mo/Si multilayer mirror 10 cm from the target. The reflected EUV emission was detected by an AXUV-100 photodiode with

Ti/Zr/C coating. The source was operated at 7 kV and 700 μA in this configuration for three days continuously, and the photodiode current was recorded. The measured signal stability was found to be better than 0.5%, limited by the accuracy of the photodiode and picoampere current measurement. This indicates that the stability of the output power was better than 0.5%. There was no indication of debris on the multilayer mirror or degradation of its reflectivity.

4 METROLOGICAL APPLICATIONS

4.1 Optics Characterization

The EUV tube allows very compact setups for metrology. For example, multilayer mirrors can be characterized by illuminating them with an appropriate part of the EUV radiation cone and detecting the reflected intensity. This was demonstrated by measuring the EUV in-band transmission curve of a double multilayer mirror configuration. The same set of mirrors was tested at the Bessy II synchrotron and by using the EUV tube. For both experiments an EUV spectrograph (Jenoptik Mikrotechnik) was used to analyze the reflected radiation. The results, shown in figure 5, exhibit excellent agreement. This would be very difficult to achieve with a plasma source, as the emitted spectral lines result in a fluctuating signal-to-noise ratio.

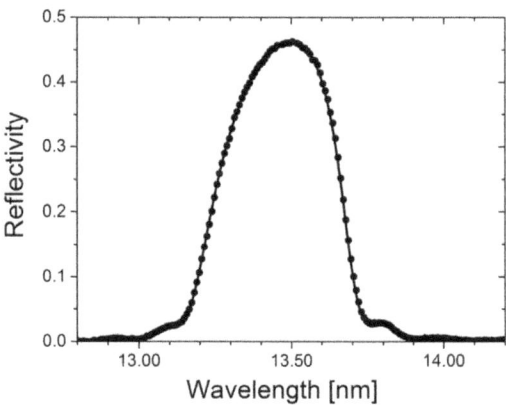

Figure 5 *Multilayer reflectivity measured with an EUV tube (line) and with a synchrotron (dots).*

Based on these investigations a setup was designed to allow computer assisted characterization of debris samples (figure 6). The input requirement was to characterize curved samples with variable unknown curvature at grazing incidence before and after they were exposed to debris. A gold coated toroidal mirror was chosen to image the small emission spot of the EUV tube onto the sample surface. By this the angle of incidence could be well controlled and set for any spot on the surface of the curved sample, and measurements could be performed with well defined parameters. The reflected radiation was measured with an AXUV-100 photodiode; system control and data acquisition were done automatically by computer.[13]

Figure 6 *Compact reflectometer for debris samples.*

To test the relative accuracy of this approach a sample was characterized, removed from the setup, reinserted and characterized again. A comparison of these results reflects any deviation between the measurements, including such parameters as the accuracy of the photodiode current measurement, the vacuum level and the precision of the mechanical mounting of the sample. Experimentally the relative measurement accuracy was found to be better than 0.5%, which is again attributed to the limited precision of the photodiode current reading. The results are shown in fig. 7(a). Although the setup was not designed to be used for measurements with absolute accuracy it was characterized in this way by measuring the same curved sample at grazing incidence angles at the Bessy II synchrotron. Very good agreement of the measurement results was found as shown in see figure 7(b); the differences are mostly less than 1%. The largest difference found is 3%, which is attributed to the fact that the EUV tube reflectometer was operated with 2% in-band radiation while the synchrotron measurement was performed with monochromatic radiation.

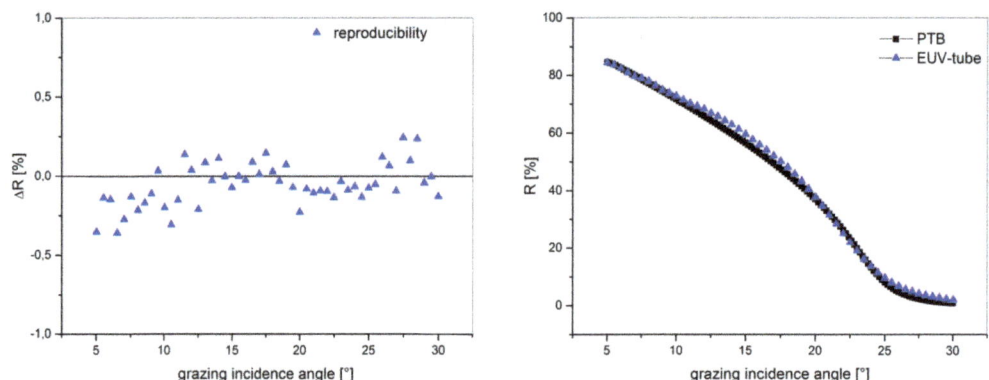

Figure 7 *Repeatability of EUV reflectivity measurements (left) and comparison of synchrotron (squares) and EUV tube (triangles) measurement (right).*

In another experiment the EUV tube was combined with a CCD camera to allow investigations of the quality of optical surfaces. For this experiment a silicon wafer was coated with a Mo/Si multilayer system as an optical test object. The coating was illuminated with the EUV radiation and the reflection was imaged with the EUV CCD camera. Single-shot images were taken to characterize the quality of the multilayer coating. The result is shown in figure 8. The black grid is a shadow of the supporting mesh of the zirconium filter, with a period of about 360 μm. While the mirror seemed to be perfect visually, the image showed several defects. Moreover, the intensity variations indicate that the roughness of the silicon wafer substrate significantly degrades the homogeneity of the reflected EUV radiation.

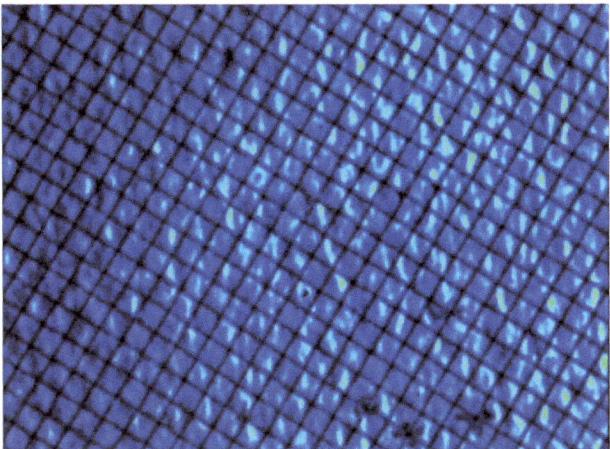

Figure 8 *Single-shot reflected image of a multilayer coating on a rough substrate.*

4.2 EUV Collector Reflectometer

One of the most exciting applications of the EUV tube is the characterization of EUV collector optics. The efficiency and lifetime of EUV source collectors have long been identified as critical parameters in EUVL.[1] This is since both parameters have direct impact on the throughput and the cost of ownership of semiconductor production. Characterizing the large and sensitive collectors at wavelength is a sophisticated task, which is usually performed in large scanning reflectometers at synchrotrons. However, from the viewpoint of the manufacturers of EUV lithography sources and collectors there is strong demand for in-house at-wavelength metrology for the collectors. This is required for shorter development cycles and quality assurance. For this an EUV collector reflectometer was designed in cooperation with Xtreme technologies,[14] specifically for Wolter shell type 1. It is based on an EUV tube, in order to utilize the point-like, debris-free and stability properties. It is thought that this is the first full size at-wavelength reflectometer for EUV DPP collectors.

The Wolter shell collectors consisted of sets of nested reflecting mirrors mounted in a supporting housing – see figure 9(a). The mirrors were designed to reflect incident radiation at grazing incidence and to collect at solid angles up to 2 sr (up to NA $0.73 \equiv 47°$) on the source side. At the intermediate focus side radiation was emitted at solid angles up to 0.24 sr (up to NA $0.28 \equiv 16°$).

To characterize a Wolter shell collector it should be illuminated from the source side with the reflected EUV radiation detected in the intermediate focus, figure 9(b). However, there are several technical constraints, which led Xtreme technologies to the conclusion that an inverse geometry would be preferable. Difficulties arise since the output cone of the EUV tube is significantly smaller than the acceptance angle of the collector on the source side. This means that it would be technically complex to expand the EUV tube's output to 2π sr and to filter and shape the radiation at the same time. This can be avoided in an inverse geometry where the metrology source is placed at the intermediate focus and the detector is placed at the original source position, as in figure 10.

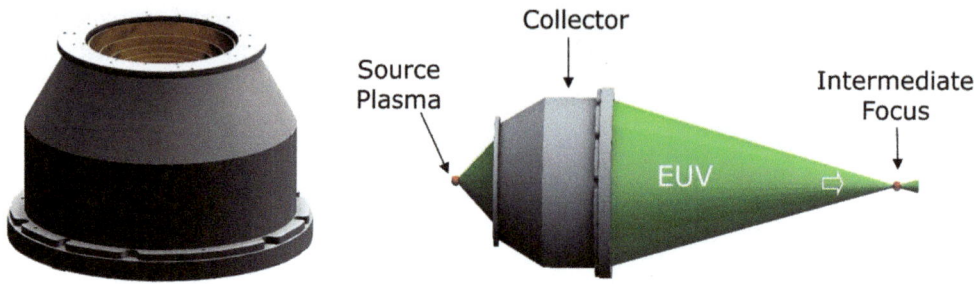

Figure 9 *Wolter shell type 1 source collector (left) configured for EUVL with a discharge produced plasma source (right).*

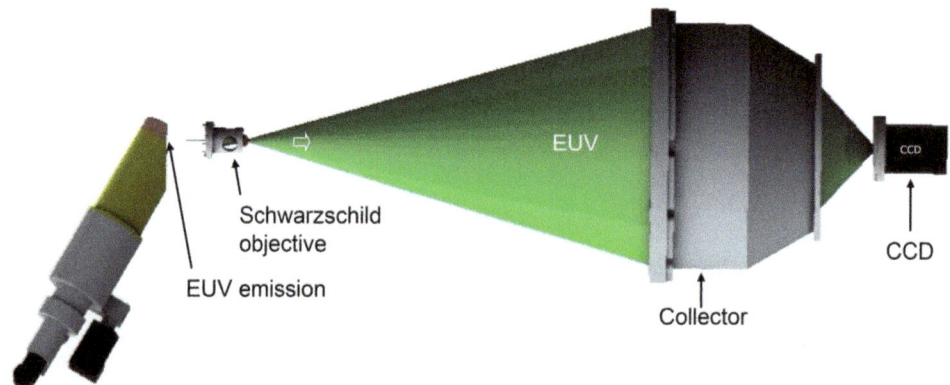

Figure 10 *Configuration of the EUV collector reflectometer in inverse geometry.*

The EUV radiation cone used for probing the Wolter shell collector must be tailored in a way to fit with the numerical aperture of the collector, and unwanted radiative contributions (X-rays) should be filtered. For this purpose a Schwarzschild objective was designed and built by Fraunhofer-Institut für Angewandte Optik und Feinmechanik IOF. The objective contained two Mo/Si multilayer mirrors with centred transmission holes, providing an internal focus just in front of the surface of the second mirror. This, second, mirror had a conical micro hole which served as the output aperture. By this device an input numerical aperture of 0.01 (0.57°, 3×10^{-4} sr) was expanded to 0.27 (15.7°, 0.23 sr) at the output corresponding to a demagnification of 28×. Note that in this configuration only a very small part (NA 0.01) of the EUV tube output was used due to constraints of the objective design.

A characterization of the EUV output behind the Schwarzschild objective recorded with a CCD camera (thinned for EUV) is shown in figure 11(a). The CCD camera was placed about 20 mm behind the internal focus of the Schwarzschild objective. The total power in this image taken behind the objective is about 30-50 pW in in-band EUV radiation. The black circle with the attached bars in the centre of the image is the shadow of a 300 μm diameter beam dump, used to stop that part of the emission from the EUV tube which passes directly the mirror holes of the Schwarzschild objective.

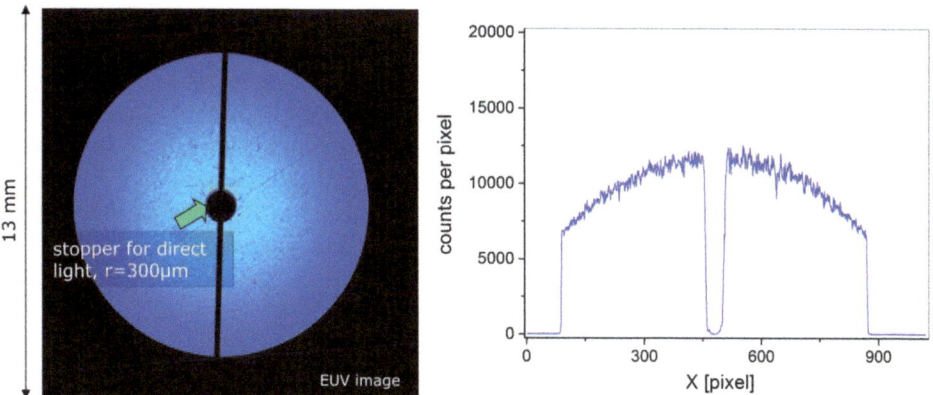

Figure 11 *Characterization of the Schwarzschild objective output: (left) image of the Schwarzschild objective EUV output and (right) trace of the intensity distribution.*

A cross section of the intensity distribution in this CCD image is shown in figure 11(b). The intensity profile of this objective turned out not to be a perfect flat top profile. This was caused by a small mismatch of the multilayer coating gradient to the extreme curvature of the mirrors in the Schwarzschild objective. This was the first coating prepared for this objective, and it could be improved to achieve a real flat top profile by iteratively correcting the coating parameters according to the intensity profile.

In addition, the angular sensitivity of the CCD camera had to be considered, as it would be expected to decrease with increasing radiation incidence angle, usually caused by thin EUV absorbing silicon oxide layer on top of the CCD sensor. The EUV absorption increases when the incidence angle is increased, as the optical path through the absorbing layer becomes longer. For this reason the EUV CCD camera was characterized at the Bessy II synchrotron using the facilities of the German National Standards organization, the Physikalisch-Technische Bundesanstalt (PTB). The results (figure 12) show that the sensitivity of the camera drops from about 19.6 counts/photon at normal incidence to 18.8 counts/photon at an angle of 45°.

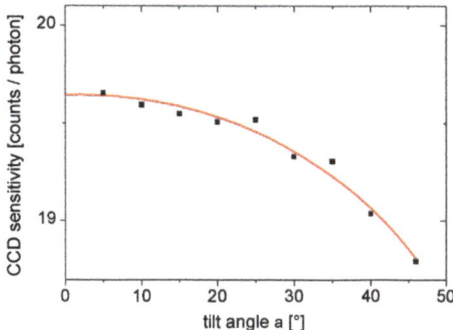

Figure 12 *Angular sensitivity of the EUV CCD camera.*

The Wolter shell collector was mounted in a vacuum chamber behind the Schwarzschild objective. The EUV CCD camera was mounted directly behind the collector as indicated in figure 10. A photograph of the complete hardware is shown in figure 13; the EUV tube can be seen in the lower part of the supporting frame, the vacuum chamber containing the collector is in the upper half. The CCD camera is located on the very top of the vacuum chamber. This vertical configuration of the hardware was chosen to allow easy insertion of collectors with a crane from the top and to avoid asymmetrical forces on the nested mirror shells during measurement and characterization. The long optical paths in this setup provided a good opportunity for precise optical alignment of the system, which was done by moving the source module relative to the collector. When the configuration was well aligned the collector focus coincided with radiation that passes the collector through a small on-axis aperture; a well-aligned configuration is shown in figure 14.

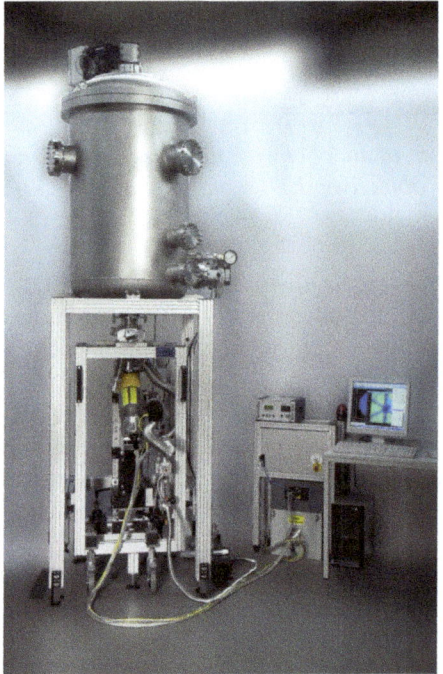

Figure 13 *The EUV collector reflectometer.*

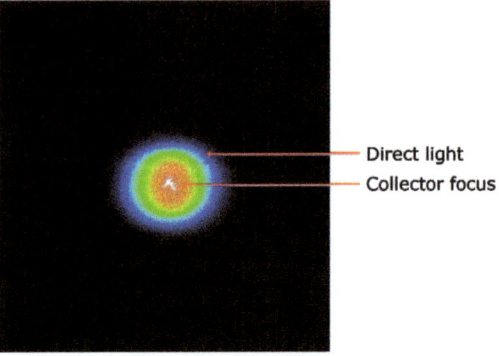

Figure 14 *Collector focus and direct light coincide well in a well-aligned configuration.*

Two EUVL collector properties are of special interest, namely the size of the focal spot and the collection efficiency. The focal spot can be observed by placing the CCD camera behind the collector. The distance between the camera and the collector focus can be varied in this setup by moving the CCD, which allows the structure of the focus to be measured. Because of the large numerical aperture of the collector this had to be done at a resolution of a few tens of micrometres. Figure 15 shows the experimental results for a typical Wolter shell type collector.

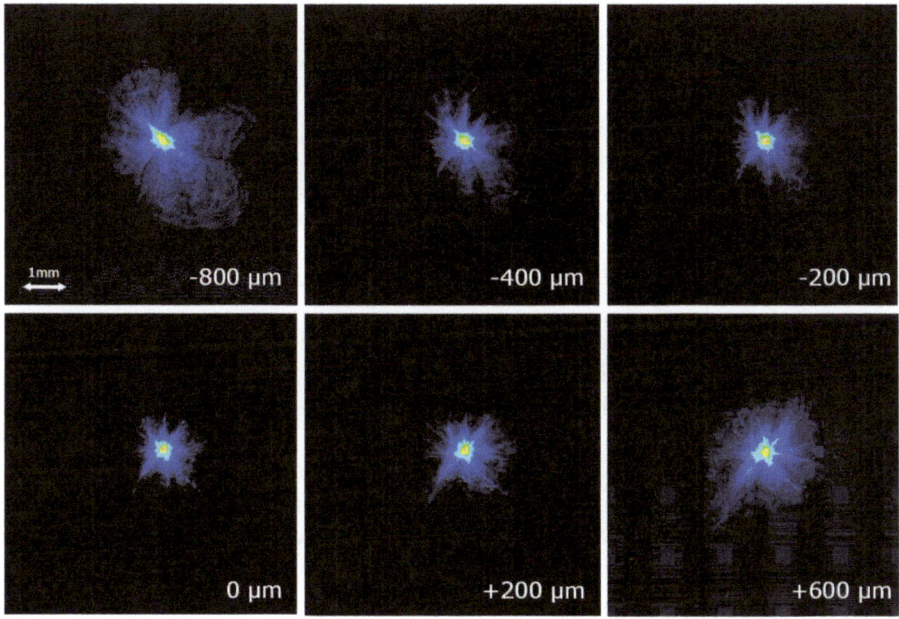

Figure 15 *CCD images recorded close to the collector focus.*

A series of images was recorded between 800 μm in front of and 600 μm behind the focus. A focal size of significantly less than 1 mm was observed, and the spatial structure of the focus the divergence of the radiation can be seen. These images were taken with a point-like illumination source; such highly resolved features can only be observed with this illumination. With the larger emission spot of a high power DPP source for EUVL most of the features seen in the images would be washed out, but these features would still affect the performance of the collector. The excellent resolution of this collector reflectometer will help manufacturers in improving of their tools.

The EUV collector reflectometer can also be used to investigate the transition from focal position to extra focal images up to a distance of about 24 mm out of focus. This distance is basically defined by the numerical aperture of the collector and the size of the CCD chip. Figure 16 shows three images of the radiation profile behind the Wolter shell collector. The first (on the left) shows a small and symmetric focal spot. The image in the centre shows the structure of the radiation field 3.5 mm out of focus; the eight black spokes meeting in the middle of the image are shadows of the support structure of the collector. It can be seen that the radiation traces of the nested mirrors develop in a quite complex, slightly distorted and non symmetrical way. This is because the intensity distribution was not homogeneous over different regions of the mirror shells. The image on the right shows the radiative distribution 23 mm out of focus. The intensity distribution shows clear traces of the mirror shells. Regions of different radiation intensity and collection efficiency can be identified and this information provides information to manufacturers and users about contamination, distortion or defects of the collector optics.

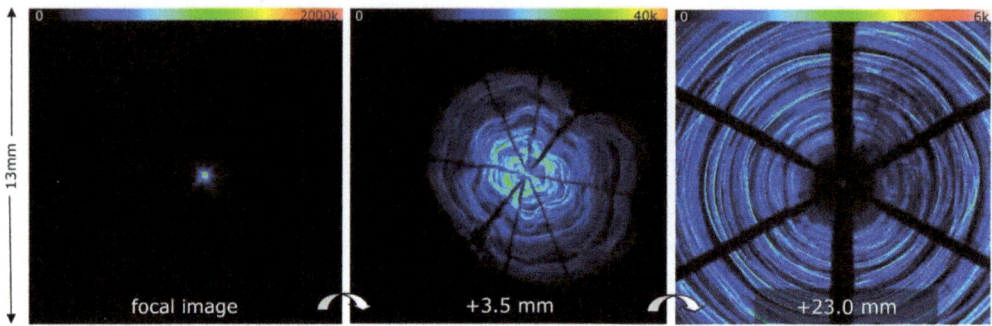

Figure 16 *CCD images recorded close to the collector focus.*

To calculate the collection efficiency the radiation behind the Schwarzschild objective (I_0) was recorded and compared with the output (I) behind the Wolter shell collector. The resulting collection efficiency is $\eta = C\ I/I_0$, where C is a correction factor to take into account geometrical and physical features of the EUV setup, including the angular distribution of the input radiation – see figure 11(b), the numerical aperture, the inverse geometry (based on ray tracing calculations) and the angular sensitivity of the CCD camera (see figure 12). The reflectometer measurement accuracy was estimated to be better than 1.5 % in total.

The reflectometer was used to characterize typical Wolter shell type 1 collector configurations, e.g. a single-shell collector coated with gold. The calculated collection efficiency was 2.34% which agrees well with the measured value of 2.37%. For a dual-shell configuration with physical vapour deposited (PVD) ruthenium the calculated value was 9%, in good agreement with the measured value of 7.8–8.1%. Even for a full configuration with PVD ruthenium the calculated value of 17.1% was confirmed by a measurement of 16.4–17%.

Many different collectors have been studied with this device, to characterize their focal spot size, collection efficiency, optical quality, lifetime and aging.

5 PULSED EUV AND X-RAY SOURCES

EUV and X-ray sources based on generating electron-induced characteristic radiation from solids also have the potential to be used in pulsed mode. This has become of significant interest for experimental investigations of ultrafast processes such as chemical reactions and lattice dynamics.

The pulsed source discussed here, and shown schematically in figure 17, is based on the photoemission of electrons from metallic surfaces by low-power femtosecond laser pulses.[15] After photoemission the electrons are accelerated by a high-voltage field towards a solid target. This configuration provides a pulse train of ultrashort electron bunches. When an electron bunch hits the target EUV radiation and X-rays are produced by inner shell emission and Bremsstrahlung. This work follows from earlier studies of -ray sources driven by pico- and nanosecond laser pulses.[16,17]

Electrons are generated by photoemission from the surface of the photocathode, and thus it is advantageous to use a highly efficient cathode material. However, the photocathode cannot be too sensitive since otherwise it would quickly degrade, compromising long term stability. Two materials were tested in this work, namely a standard copper cathode and the high efficiency material S20, containing sodium, potassium, antimony and caesium.[18]

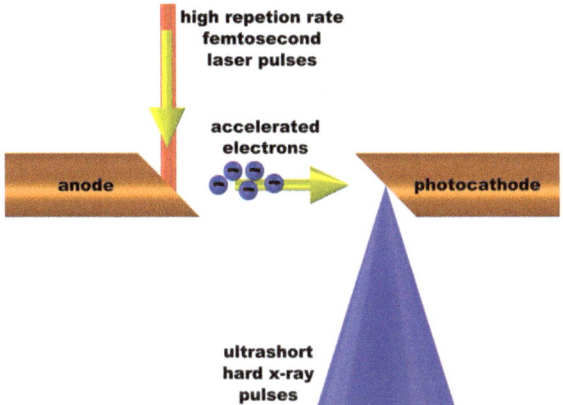

Figure 17 *Principle of a pulsed EUV/X-ray source.*

The quantum efficiency for electron generation from a copper cathode was characterized with a pulsed femtosecond laser (780 nm, 2.5 µJ, 160 fs, 250 kHz). The quantum efficiency achieved for this target material at an acceleration voltage of 60 kV was 1.2×10^{-6}. The charge in a single electron bunch was found to be 1.8 pC/pulse, corresponding to an electron current of 450 nA.

Similar measurements were made for electron generation from a S20 cathode using a femtosecond laser (780 nm, 1–10 pJ, 120 fs, 82 MHz). The quantum efficiency achieved for this target material was 10^{-2}, with a charge of 0.06 pC/pulse in a single electron bunch that results in an electron current of 4.7 µA. However, high efficiency photocathode materials such as S20 are chemically unstable and have to be permanently kept under high vacuum, which cannot be maintained in a typical laboratory setup. For this reason the experimental investigations discussed here were performed with a copper cathode.

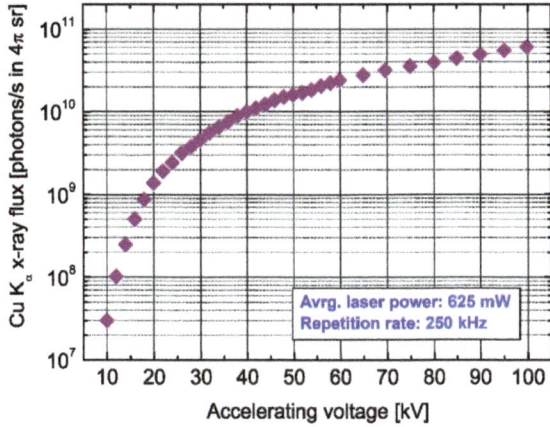

Figure 18 *X-ray flux generated from copper electrodes as a function of acceleration voltage.*

The anode material determines the emission spectrum of the X-ray source with characteristic emission and bremsstrahlung. Many anode materials and their emission characteristics are well known from X-ray tubes; for the experiments described here copper anodes were used with characteristic K emission at about 8 keV. The copper K_α flux generated depends on the acceleration voltage, as shown in figure 18 for a femtosecond laser with an average power of 625 mW, pulse duration of 160 fs and repetition rate of 250 kHz.

The characterization of the X-ray pulse duration was performed with an ultrafast X-ray streak camera provided by the Max Planck Institute of Quantum Optics. This camera was specifically designed for detection of X-ray pulses and so allowed time resolved analysis of the copper anode emission at 8 keV. The experimental setup included a femtosecond laser operated at 780 nm, 150 fs, 300 µJ and 1 kHz. The output beam of the laser was split in two parts, the first of which was used to drive the X-ray source by focusing the laser light onto the photocathode, where electrons were emitted subsequently. The electrons were accelerated to the copper anode to produce electron induced X-ray emission. The X-rays were directed through a beryllium window towards the input of the X-ray streak camera, which was equipped with a potassium iodide photocathode.

The detection on the CCD unit of the streak camera was triggered by the second part of the laser beam, which was sent through an optical delay line. After this the beam hit a gallium Auston switch at the streak camera, which was used as a trigger allowing the trace of the X-ray pulse to be moved into the streak camera's observation window of about 300 µs. This meant that many shots could be accumulated in one image for signal analysis, since the temporal resolution provided by this camera was previously determined to be 1.3 ps.[19]

A streak camera image is shown in figure 19; the strong vertical line in the centre is caused by hard X-ray photons which directly hit the phosphor screen of the CCD camera without being converted into electrons, creating a shadow image of the streak camera entrance slit. The streak signal itself is the thin line to the right-hand side of the image.

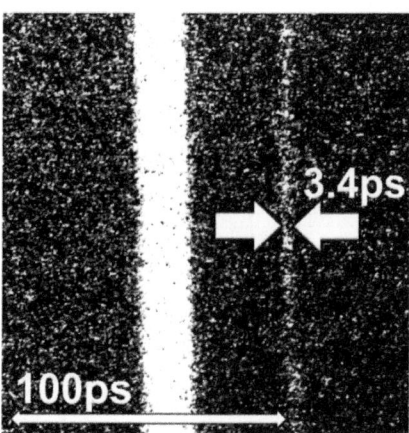

Figure 19 *Streak camera image of a 3.4 ps X-ray pulse, acceleration voltage 90 kV.*

From the measured series of streak images a relation between acceleration voltage and laser power could be derived (figure 20). For larger distances between the electrodes and increased electron density the electron bunches expand during their travel between the electrodes, resulting in broader X-ray pulses. On the other hand, increasing the acceleration voltage or the focused electron spot reduces the pulse broadening due to smaller space charge effects. The shortest pulse of 3ps (corresponding to 2.7 ps taking into account the streak camera time resolution of 1.3 ps) observed in these experiments was generated at an acceleration voltage of 95 kV at a laser pulse energy of 3 µJ. This corresponds to 10^7 electrons in each bunch and 10^5 X-ray photons per pulse. Note that a reduction of the laser pulse energy would reduce the number of electrons by the same factor[20] and so could provide sub-500 fs X-ray pulses.

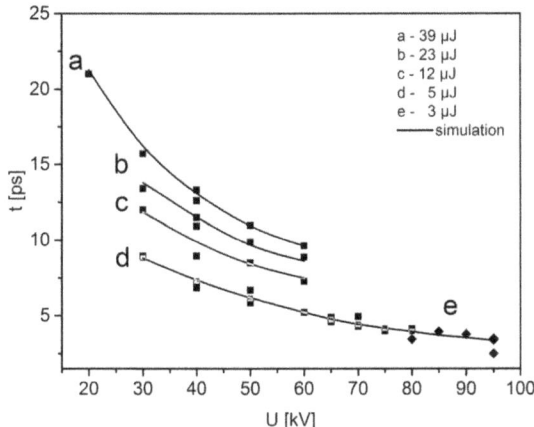

Figure 20 *X-ray pulse duration as a function of acceleration voltage and laser power.*

6 CONCLUSIONS

EUV tubes are remarkable sources for metrology. Based on industry proven X-ray technology they provide an outstanding stability in terms of power and spectrum. Point-like emission spots of only a few tens of micrometres in diameter create a superior spatial resolution for in-house metrology, but it should be noted that the technology can be transferred to X-ray tubes with sub-micrometre focal spots that recently became available as industrial sources. Metrology setups based on such nano-focus tubes will provide unique spatial resolution, opening up new dimensions in EUV optics characterization not only in Wolter shell type collectors, but also in EUV collectors for laser produced plasma sources.

ACKNOWLEDGEMENTS

The authors would like to thank André Egbert, Klaus Eidmann, Torsten Feigl, Michel Fokoua, Thomas Mißalla, Antonio Ritucci, Max-Christian Schürmann, Guido Schriever, Uwe Stamm, Boris Tkachenko and Biofabrication for Nife.

References

1 K. Suzuki, A. Miyake, N. Harned, 2009 International Symposium on Extreme Ultraviolet Lithography, http://www.sematech.org/meetings/archives/litho/8653/index.htm
2 A. Egbert, S. Becker, B. Tkachenko and B.N. Chichkov, Proc. SPIE, 2004, **5448**, 693.
3 Y. Wang, M.A. Larotonda, B.M. Luther, D. Alessi, M. Berrill, V.N. Shlyaptsev and J.J. Rocca, *Phys.Rev. A*, 2005, **72**, 053807.
4 X. Zhang, A.R. Libertun, A. Paul, E. Gagnon, S. Backus, I.P. Christov, M.M. Murnane, H.C. Kapteyn, R.A. Bartels, Y. Liu and D.T. Attwood, *Opt. Lett.*, 2004, **29**, 1357.

5 W. Ackermann, G. Asova, V. Ayvazyan, A. Azima, N. Baboi, J. Bähr, V. Balandin, B. Beutner, A. Brandt, A. Bolzmann, R. Brinkmann, O.I. Brovko, M. Castellano, P. Castro, L. Catani, E. Chiadroni, S. Choroba, A. Cianchi, J.T. Costello, D. Cubaynes, J. Dardis, W. Decking, H. Delsim-Hashemi, A. Delserieys, G. Di Pirro, M. Dohlus, S. Düsterer, A. Eckhardt, H.T. Edwards, B. Faatz, J. Feldhaus, K. Flöttmann, J. Frisch, L. Fröhlich, T. Garvey, U. Gensch, C. Gerth, M. Görler, N. Golubeva, H.-J. Grabosch, M. Grecki, O. Grimm, K. Hacker3, U. Hahn, J.H. Han, K. Honkavaara, T. Hott, M. Hüning, Y. Ivanisenko, E. Jaeschke, W. Jalmuzna, T. Jezynski, R. Kammering, V. Katalev, K. Kavanagh, E.T. Kennedy, S. Khodyachykh, K. Klose, V. Kocharyan, M. Körfer, M. Kollewe, W. Koprek, S. Korepanov, D. Kostin, M. Krassilnikov, G. Kube, M. Kuhlmann, C.L.S. Lewis, L. Lilje, T. Limberg, D. Lipka, F. Löhl, H. Luna, M. Luong, M. Martins, M. Meyer, P. Michelato, V. Miltchev, W.D. Möller, L. Monaco, W. F.O. Müller, O. Napieralski, O. Napoly, P. Nicolosi, D. Nölle, T. Nuñez, A. Oppelt, C. Pagani, R. Paparella, N. Pchalek, J. Pedregosa-Gutierrez, B. Petersen, B. Petrosyan, G. Petrosyan, L. Petrosyan, J. Pflüger, E. Plönjes, L. Poletto, K. Pozniak, E. Prat, D. Proch, P. Pucyk, P. Radcliffe, H. Redlin, K. Rehlich, M. Richter, M. Roehrs, J. Roensch, R. Romaniuk, M. Ross, J. Rossbach, V. Rybnikov, M. Sachwitz, E.L. Saldin, W. Sandner, H. Schlarb, B. Schmidt, M. Schmitz, P. Schmüser, J.R. Schneider, E.A. Schneidmiller, S. Schnepp, S. Schreiber, M. Seidel, D. Sertore, A.V. Shabunov, C. Simon, S. Simrock, E. Sombrowski, A.A. Sorokin, P. Spanknebel, R. Spesyvtsev, L. Staykov, B. Steffen, F. Stephan, F. Stulle, H. Thom, K. Tiedtke, M. Tischer, S. Toleikis, R. Treusch, D. Trines, I. Tsakov, E. Vogel, T. Weiland, H. Weise, M. Wellhöfer, M. Wendt, I. Will, A. Winter, K. Wittenburg, W. Wurth, P. Yeates, M.V. Yurkov, I. Zagorodnov and K. Zapfe, *Nature Photonics*, 2007, **1**, 336.

6 E. Gullikson, J. Underwood, P. Batson and V. Nikitin, *J. X-Ray Sci. Technol.*, 1992, **3**, 283.

7 R. Stulen, *IEEE Journal of Selected Topics in Quantum Electronics*, 1995, **1**, 970.

8 R. Lebert, K. Bergmann, G. Schriever and W. Neff, *Microelectronic Engineering*, 1999, **46**, 449.

9 S. Kranzusch, C. Peth and K. Mann, *Rev. Sci. Instrum.*, 2003, **74**, 969.

10 S.F. Horne, M.M. Besen, D.K. Smith, P.A. Blackborow and R. D'Agostino, *Proc. SPIE*, 2006, **6151**, 61510P.

11 T.Feigl, S. Yulin, N. Benoit, N. Kaiser, Microelectronic Engineering, 2006, **83**, 703

12 A. Egbert, S. Becker, B. Tkachenko and B.N. Chichkov, *Proc. SPIE*, 2004, **5448**, 693.

13 U. Hinze, M. Fokoua and B. Chichkov, *Proc. SPIE*, 2007, **6586**, 65860Q.

14 U. Hinze, B.N. Chichkov, T. Feigl, U.D. Zeitner, C. Damm, D. Bolushkin, J. Kleinschmidt, G. Schriever and M-C. Schürmann, *Proc. SPIE*, 2008, **6921**, 69213A.

15 U. Hinze, A. Egbert, B. Chichkov and K. Eidmann, *Opt. Lett.*, 2004, **29**, 2079.

16 P. Chen, I.V. Tomov and P.M. Rentzepis, *J. Chem. Phys.*, 1996. **104**, 10001.

17 J-P. Girardeau-Montaut, B. Kiraly, C. Girardeau-Montaut and H. Leboutet, *Nucl. Instrum. Methods Phys. Res. A* 2000. **452**, 361.

18 A.H. Sommer, *Rev. Sci. Instrum.*, 1955, **26**, 725.

19 F. Pisani, U. Andiel, K. Eidmann, K. Witte, I. Uschmann, A. Morak, E. Förster and R. Sauerbrey, *Appl. Phys. Lett.*, 2004, **84**, 2772.

20 B-L Qian and H.E. Elsayed-Ali, *Appl. Phys.*, 2002, **91**, 462.

CHARACTERISTICS OF A SUB-PICOSECOND TITANIUM Kα SOURCE USING RELATIVISTICALLY INTENSE LASERS

U. Zastrau,[1,2] I. Uschmann,[1,3] E. Förster,[1,3] A. Sengebusch,[4] H. Reinholz,[4] G. Röpke,[4] E. Kroupp,[5] E. Stambulchik[5] and Y. Maron[5]

[1] Institute for Optics and Quantum Electronics, Friedrich-Schiller-University of Jena, 07743 Jena, Germany
[2] Matter in Extreme Conditions MEC, Linac Coherent Light Source LCLS, SLAC National Accelerator Laboratory, Menlo Park, California 94025, USA
[3] Helmholtz Institute Jena, 07743 Jena, Germany
[4] Institute for Physics, University of Rostock, 18051 Rostock, Germany
[5] Faculty of Physics, Weizmann Institute of Science, Rehovot 76100, Israel

1   INTRODUCTION

In this chapter, the microscopic characteristics of a bright, short-pulsed source of Ti Kα radiation are studied. This x-ray emission is generated from fast electrons that are generated when a relativistically intense laser pulse interacts with a solid metal surface. The electrons have average energies significantly exceeding the ionization threshold of the K-shell (5 keV) and give rise to K-radiation when the K-shell recombines with a lifetime of a few femtoseconds only. Hence the duration of the Kα emission is dominantly determined by the time these fast electrons are present. But at the same time, the electrons also generate a solid-density plasma state at several tens of electronvolts temperature (e.g., several 100 000 K). This alters the emission probabilities of the Kα source, potentially effecting the brightness of the x-ray source. These mechanisms and possible optimizations are subject of this chapter.

The term Warm Dense Matter (WDM) refers to materials at temperatures of few electronvolts (eV) in temperature at solid-like densities. These states are of paramount importance in modelling astrophysical objects. The physics of WDM[1] is of increasing interest because of the location in the transition region from cold condensed materials to hot dense plasmas. They further occur as transient non-equilibrium states in experiments to generate high energy densities, most notably in studies of inertial confinement fusion.

The laboratory generation of such high energy density state can only be realised for extremely short times: a very large amount of energy has to be deposited in a sample almost instantaneously (< 1 ps $= 10^{-12}$ s), so that no hydrodynamic expansion occurs during the interaction and the sample stays at its initial density by inertia of the nuclei. With the implementation of lasers with pulse durations in the order of a few hundred femtoseconds, the deposition of energy into a sample became possible in a time shorter than the hydrodynamic motion scale, allowing information on such exotic states to be inferred. In the last 30 years these lasers have become well established, reaching petawatt peak powers.

The physical mechanism of coupling optical radiation into solid-density matter is highly non-linear and complex. First, the photon energy of a 1-2 electronvolts is not sufficient to photo-ionize the bound electrons. Second, most materials are not transparent or even reflecting. Third, due to the tight focusing necessary to reach the high intensities, the excited volumes are limited to a few cubic micrometres. Hence, the hot plasma occupies a very small spot at the sample surface, and the escaping fast electrons collide with the cold solid and heat it to warm dense conditions.

In this chapter, the creation of WDM by irradiating thin foils by an intense short pulse optical laser is described. Using X-ray spectroscopy, the properties of the solid-density plasma can be inferred, including a radial map of the plasma temperature and the corresponding characteristic X-ray yield. This is achieved with a spatial resolution comparable to the laser focal spot size. The influence of the target foil thickness is also investigated, and the properties of the laser-generated hot electrons and their interaction with the target are derived.

2 WARM DENSE MATTER

The creation and investigation of WDM under controlled conditions in the laboratory is difficult. Using short pulse optical lasers, non-linear absorption leads to rapid temporal variations, steep spatial gradients, and a broad spectrum of plasma physical processes, including non-uniform heating,[2] transient changes of the dielectric function,[3] optical properties,[4] conductivity[5] and shock compression.[6] Pioneering techniques such as shock heating,[6,7] X-ray heating,[8,9,10] ion heating techniques[11,12] and XUV free electron laser irradiation[13,14] have been developed in order to improve the plasma heating mechanism.

In WDM the mean electron energy $k_B T$, where k_B is the Boltzmann constant and T is the temperature, is comparable to the Fermi energy E_F. Furthermore, the inter-particle Coulomb correlation energy is equal to or exceeds the thermal energy. Thus both electrons and ions exhibit temporal and spatial correlations which depend strongly on the plasma temperature and density. The theories of ideal plasmas and condensed matter both fail in this regime. Knowledge of such strongly coupled plasmas also enables continued improvement of key applications including laser-driven sources of X-rays,[15] for example to serve as backlighters[16,17] or particle accelerators,[18,19,20] providing alternative radiation sources for scientific and medical applications.

With the advent of the current generation of high-power lasers, a high-energy density (HED) environment, similar to that in the cores of stars, can be achieved in the laboratory. Typically, pressures exceeding 1 Gbar are referred to as HED, and the central process under investigation is thermonuclear fusion[21,22] in a fuel such as a mixture of deuterium and tritium (DT). Instead of compressing the capsule completely down to self-ignition with the driving lasers, the concept of fast ignition[23] uses an intense short-pulse laser to ignite pre-compressed fuel. This process is analogous to a petrol engine in which the fuel is ignited by a spark plug, unlike diesel engines which compress the fuel to initiate self-ignition; the spark plug plays the role of a beam of laser accelerated electrons. An advantage is that this process works at lower fuel density so that more of the capsule mass can be ignited, leading to higher gain.

A possible way to create WDM is to focus an ultra-intense optical laser onto a solid sample, generating a relativistic electron beam which heats the cold solid to typical WDM conditions. The process is accompanied by X-ray line emission from K-shell vacancies created by the electron impact. Hence these plasmas are bright, small, and sources of short X-ray bursts; a very efficient X-ray plasma source can be provided at moderate laser intensities,[15,24,25] for example using a short pulse laser of a few millijoules (mJ) at kilohertz (kHz) repetition rates and a continuously replenished target design to provide a sub-picosecond table-top source, suitable for university laboratories interested in time-resolved X-ray diffraction.[26,27] Using multi-terawatt high-energy lasers allows the creation of much brighter, single-shot X-ray sources. Furthermore, short-pulse X-ray sources are suitable for scattering experiments at solid density, so long as the wavelength is short enough. The important plasma parameters — of ion and electron temperature and free electron density — are accessible via X-ray Thomson scattering.[28,29]

3 RELATIVISTIC OPTICAL LASER INTERACTIONS WITH METALS

3.1 Generation of fast electrons

Any target placed at an intense laser focus undergoes rapid ionization even before the pulse has reached its peak intensity; the processes that take place are shown schematically in figure 1 and described in the following. All types of non-linear ionizations take place within the skin depth $l = c/\omega_p$ of the target, where the optical radiation asymptotically penetrates. The plasma formed in this manner comprises the usual fluid-like mixture of electrons and ions, but many of its basic properties are essentially controlled by the laser field, rather than by the density and temperature. While plasmas at lower electron density are transparent to the laser radiation, they reflect the radiation when the laser frequency is equal to the plasma frequency ω_p, corresponding to the critical electron density $n_c = 1.1 \cdot 10^{21} \lambda^{-2}$ cm^{-3}, where λ is the wavelength in micrometres.

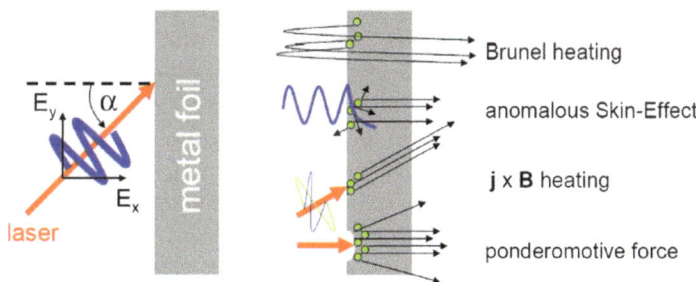

Figure 1 *Physics of femtosecond-laser absorption at a sharp vacuum-target boundary and intensities $> 10^{18}$ W/cm².*

In order to distinguish different interaction regimes at the critical density surface, the plasma scale length D is compared to the Debye screening length λ_D (which is the distance at which the potential energy between particle in the plasma is comparable to the thermal energy). For a rather smooth plasma boundary with $D > \lambda_D$, hydrodynamics and the under-dense (electron density $n_e < n_c$) regions of the plasma have to be considered. If the laser beam is p-polarized, the electric field enters the target and resonance absorption dominates.[30] Plasma waves are driven by the laser and the plasma is heated by collisions. On the other hand, when the plasma gradient is steep, $D < \lambda_D$, and the laser pulse is short and intense, it easily reaches the overdense ($n_e > n_c$) plasma.

In the electric field $\mathbf{E}(\mathbf{r},t)$ of the laser pulse, a free electron experiences a force. Its average kinetic energy is given by the so-called ponderomotive potential U_p. At moderate intensities, $I \sim 10^{13-14}$ W/cm², and for a typical laser wavelength, $\lambda \approx 1\mu$m, U_p is a few electronvolts. At higher intensities, $I > 10^{15}$ W/cm^2, electrons are accelerated to relativistic velocities v_{0s} and, therefore, the laser magnetic field starts to play a role via the Lorentz force. The amplitude, a_0, of the normalized vector potential is defined to be unity at an intensity of 10^{18} W/cm^2 which defines the transition to the relativistic regime where the electric laser field starts to exceed the inneratomic Coulomb forces. If $a_0 \geq 1$, a fully relativistic description is necessary.

Now, the so-called collisionless absorption mechanisms[31,32,33,34] come into play. These are (figure 1):

(i) Brunel heating. If the laser beam is p-polarized, the electric field may accelerate free electrons away from the target. When the field changes its sign, these electrons are injected into the target material with high kinetic energy and their mean free path may exceed the target thickness of a few μm. Inside the target, the laser electric field is screened, and the electrons propagate further.

(ii) Anomalous skin effect. The free electrons within the skin depth $l = c/\omega_p$ receive kinetic energy from the penetrating laser electric field. If their mean free path exceeds the skin depth significantly, these electrons will diffuse into the target and deliver their energy to a larger volume.

(iii) **jxB** heating: The magnetic term of the Lorentz force ($e\mathbf{v}\mathbf{x}\mathbf{B}$) becomes comparable to the electric force ($e\mathbf{E}$). In this case, the so-called **jxB** heating[35] accelerates bunches of electrons into the target in the direction of the laser pulse at twice the laser frequency.[36]

(iv) Ponderomotive force: When the light pressure is much greater than the thermal pressure in the focal spot, the plasma will be pushed in the laser beam direction. In addition, a radial ponderomotive force due to the transverse intensity gradient pushes electrons radially away from the centre of the beam.

Many of these effects depend on the laser irradiance $I\lambda^2$, which means that the threshold intensity for a given phenomenon can vary depending upon the laser wavelength. Which process dominates depends on the experimental conditions, including the preplasma scale length, the intensity, the laser wavelength, its temporal profile and the initial free electron density. Experimental results for the conversion of laser energy into energy carried by the hot electrons differ by around 50%,[37] reflecting the diversity of possible absorption mechanisms in the individual experiments.

Once a fraction of the laser pulse is absorbed, the electrons have a Maxwellian temperature distribution, most often assumed as one or several Maxwell-Boltzmann distributions with temperature T_{hot}. If relativistic effects are important, a relativistic Maxwell-Jüttner distribution function for a relativistic electron gas should be used instead. When $a_0 > 1$ the temperature of the electron distribution scales with the ponderomotive potential.[36,38,39] Several measurements of the electron energy distribution in laser-solid interactions, either directly using electron spectrometers or, earlier, indirectly via methods including Bremsstrahlung, have indicated that T_{hot} scales approximately as $(I\lambda^2)^{1/3}$.[40]

It should be noted that the assignment of a single temperature is only valid in describing the falling wing of the electron distribution at energies well above 100 keV; electrons of lower energies do not follow the same distribution function. The fastest (collisionless) electrons leave at the rear side of the foil into the vacuum,[41] but those of lower energy deposit their energy inside the foil.

When a hot electron distribution with a temperature of around 1 MeV propagates into a solid metal, a space-charge electric field is set up to which the plasma reacts with a return current almost immediately on a timescale of less than a femtosecond (fs, 10^{-15} s).[42] The Coulomb cross section decays with increasing temperature, and so only the cold electrons in the target experience the low-temperature resistivity (which increases with temperature) and couple to the material efficiently, while the hot electrons couple weakly.

Typically fast, ~ 1 MeV, electrons have mean free paths of hundreds of micrometres and collisional energy loss times of a few picoseconds[43] in a solid. Consequently, fast electrons can transport the absorbed energy to parts of the target well away from the laser spot over time scales comparable to the laser pulse length.

For each 1 J of laser energy, the average current $I = eN/\Delta t$ carried by N electrons over the laser pulse duration Δt amounts to several tens of mega amperes (MA).[44] The question as to whether a net current of this order of magnitude can flow is strongly connected with the arise of self-induced magnetic fields; there are two main arguments against it. First, assuming a parallel electron beam, the self-induced magnetic field force points inwards. If the beam current reaches the Alfvén limit,[45] $I_A = 17\ \gamma\beta$ kA, the electrons are reflected through 180° and stop; Here, γ is the Lorentz parameter, which can be expressed as the electrons kinetic energy E_{kin} divided by its rest mass of 511 keV, $\gamma = E_{kin} / 511\ keV - 1$. Further, $\gamma\beta$, the dimensionless electron momentum, is equal to 2 at $I = 10^{19}$ W/cm^2; where β is the electron speed relative to the speed of light. Although the concept of the Alfvén limit is derived for a constant current, the present case relates to a transient short pulse current. This implies that the care has to be taken to translate this concept without considering ultrafast electron kinetics.

The second argument for a limited net current is that the energy content of a self-induced magnetic field from a laser-induced electron pulse would be about three orders of magnitude larger than the initial laser energy, which is impossible due to energy conservation. This implies that a return current has to flow.[46] The net current is limited to negligible values.

According to computational models, the self-induced magnetic field possibly exceeds several hundred Tesla.[47,48] This can cause a strong electron beam to break up into filaments and lead to refocusing inside the target. The return currents can then couple to the material, and an electric field $\mathbf{E} = \eta\mathbf{j}$ is set up. The result is ohmic inhibition of the electron flow. To couple electrons into matter efficiently the return current must flow freely; hence the ohmic inhibition must be small.

3.2 Strongly coupled plasmas

The highly non-linear electron creation and acceleration processes lead to an electron distribution with energies ranging over several orders of magnitude.[49] Solid targets heated by collisions with these electrons are far from being homogeneous, resulting rather in non-equilibrium states with steep spatial gradients.[50] The transfer of energy between these "hot" electrons and the target is important for many potential applications such as isochoric heating of solid targets to high temperatures or fast ignition.[2,23] Thus, it is of primary importance to obtain precise knowledge of physical properties such as the plasma temperature at solid density with high spatial resolution.[2]

The initial plasma state created by relativistic laser-solid interactions is still at solid density, and bulk electron temperatures are in the order of a few tens of electronvolts, comparable to the Fermi energy E_F. For metallic titanium the Fermi energy E_F is equal to 14 eV, which means that in such plasmas the bulk electrons are still dominantly degenerate; $k_B T_e/E_F \sim 1$. Further, the ion coupling parameter Γ is greater than or equal to unity, i.e., the interparticle Coulomb correlation energy is equal to or exceeds the thermal energy. Thus, electrons and ions exhibit strong temporal and spatial correlations.

The electrons, both from the beam and the return currents, are in principle capable of impact ionizing the atomic shells when they carry sufficient energy. In particular, the cross-section σ_K for the production of a vacancy in the atomic K shell has been investigated both theoretically and experimentally. Amongst others, Casnati *et al.*[51] developed an empirical model, which was extended to non-negligible relativistic effects by Quarles *et al.*[51,52]

In 2003, Santos *et al.*[53] performed an extensive review of different models and their comparison to available experimental data. The relativistic approach by Casnati and Quarles (red curve as shown in figure 2) is in good agreement with the experimental data published by Santos within the error bars (not shown here).

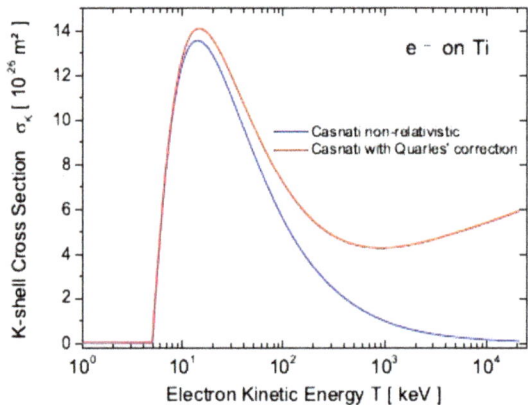

Figure 2   *Left: K-shell ionization cross section for titanium. Right: Comparison of experimental data (symbols) with models (curves) for copper.*[53] *The non-relativistic model*[51] *is indicated by the blue curves, and the red curves have the relativistic correction*[52] *applied.*

The present discussion is restricted to solid density titanium plasmas. The electron configuration of atomic titanium ($Z = 22$) is $[Ar]3d^2 4s^2$. In the bulk state, delocalized quasi-free electrons have to be taken into account. The four outer electrons (in the 3d and 4s orbitals) form the conduction band and so, the lowest charge state is argon-like Ti V with an electron density, in the low-temperature limit, $n_e \approx 4n_i$ where n_i is the ion particle density in solid titanium, $n_i \approx 5.66 \cdot 10^{22}$ cm^{-3}. Further details are given in figure 3.

A so-called non-diagram line is a line not emitted by a cold solid material, but by an excited state. These lines are often close to strong diagram lines and hence are frequently called satellites. Their origin, briefly, is recombination when there is a vacancy in a higher shell. In the particular case of titanium, blue Kα satellites are observed when a vacancy in the M-shell is present. The line shift is 2–3 eV per vacancy[54,55] towards higher photon energies. Since the M-shell is the first bound shell in the metallic state, its electrons are ionized thermally as a function of the bulk temperature. In the case of L-shell vacancies in titanium, the shift is much larger, about 25 eV per missing electron.[56]

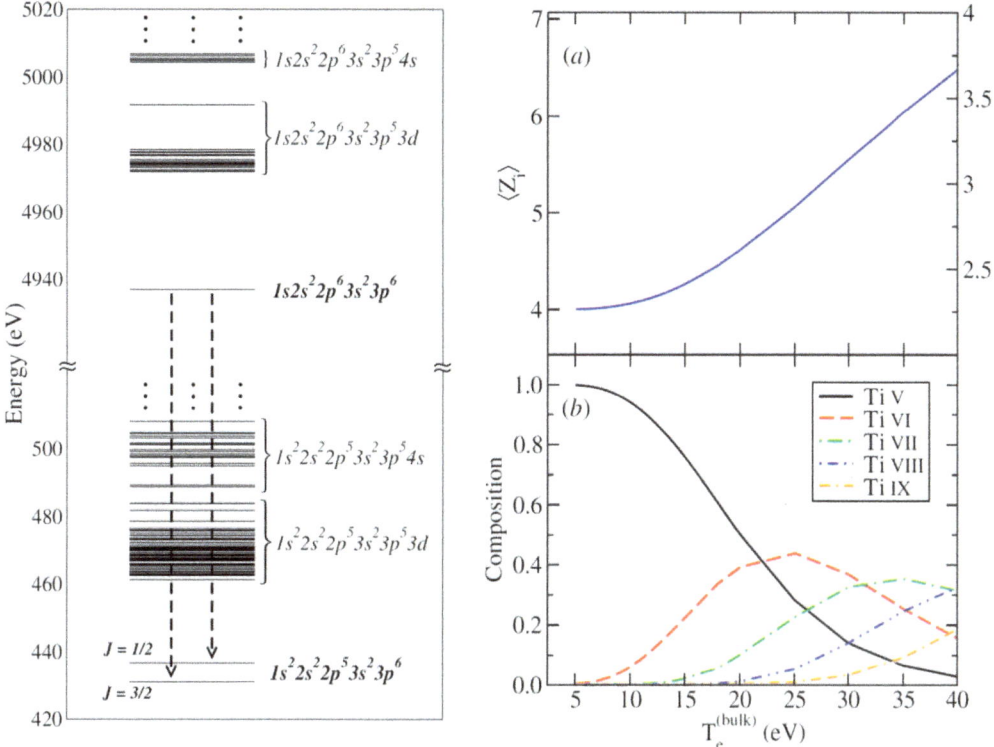

Figure 3 *Left: Ti VI Grotrian diagram showing the K transition and its satellites. Right: (a) Electron density (right axis resembles the corresponding free electron density n_e in units of $10^{23}/cm^3$) and (b) mean ion charge Z_i and charge state distribution in titanium plasmas as functions of the bulk electron temperature.*[57]

Recent progress in line-shape modelling of titanium inner-shell emission lines has been achieved by two different approaches. Thorough, self-consistent modelling requires calculations encompassing many different processes and effects, such as satellite formation and blend-in, plasma polarization, Stark broadening, solid-density quantum effects and self-absorption.

In the first approach,[58] a perturbative Ansatz is chosen to calculate line profiles emitted from WDM. A chemical *ab-initio* code using a simplified atomic structure with respect to the fine structure components yields unperturbed emission energies and orbitals of various ionization stages. The perturbing plasma potential is calculated within a self-consistent ion sphere model, and first-order perturbation theory yields shifted line positions and continuum lowering. The line width is treated as a free parameter in this approach. Modelling of the complete emission profiles is based on a self-consistent calculation of the plasma composition under local thermodynamic equilibrium (LTE) conditions. For this coupled Saha equations are solved applying a Planck-Larkin renormalization to the partition function.

In the second approach,[57] collisional-radiative (CR) calculations treating thousands of atomic transitions from the Flexible Atomic Code (FAC)[59] were performed. The presence of the two electron distributions (bulk and hot), differing in energies by several orders of magnitude, results essentially in a non-equilibrium state. In practical terms these two very different electron distributions complicate the CR calculations significantly, since all atomic processes involving free electrons have to be tabulated over a very wide range of electron energies. In the calculation, a hot electron (E = 150 keV) concentration of 0.1% was assumed.

Spectra of the titanium Kα doublet can be modelled using both approaches, yielding approximately the same results. The bulk electron temperature T_e is the only free parameter in the modelling of the Kα shape, assuming that the fast electrons represent a small fraction of the total electron density.[57] Thus, by varying T_e, the total electron density, the charge-state distribution and the populations of different levels are calculated self-consistently.

4 Kα SPECTROSCOPY OF SOLID-DENSITY PLASMAS

4.1 Diagnostics for Solid-Density Plasmas

Measurements of the characteristic Kα emission have proven to be reliable diagnostics for solid-density plasmas.[60]. First, the plasma parameters strongly affect the structure of the emitted lines. In particular, the emergence of the blue satellites due to the formation of M-shell holes is an indicator of the bulk electron temperature[61] as described in section 3.2. Second, solid-density plasma is transparent to the emitted X-rays which, therefore, yield unique information on the target interior. This is a large advantage over measurements of XUV or visible radiation. Moreover, the x-ray emission occurs within a few picoseconds, well below the typical hydrodynamic expansion time.[24,62,63]

Processes such as ionization, Doppler and pressure broadening, opacity and electric fields in the plasma may modify the shapes of individual emission lines. Examination of emission line profiles therefore reveal detailed information on plasma conditions. High-resolution X-ray spectroscopy has thus made significant contributions to the diagnostics of solid-density plasmas.

For mid-atomic number elements, K-shell emission spectra occur in the X-ray region and spectrometers that can resolve wavelengths to the tens of picometre level are required. The use of one- and two-dimensionally curved crystals, respectively, is common in state of the art X-ray spectrometers.[64,65] Bending the crystal to a radius 2R (R being the Rowland radius), as proposed by Johann,[66] leads to monochromatic focusing of the source onto the detector plane. The great advantage of any geometry that uses the Johann setup is that no slit is needed for high resolution, so that the full divergence of the source can be used to illuminate the spectrometer crystal. This is the key advantage of large aperture bent crystals. Their much higher luminosity compared to other X-ray optics such as grazing incidence reflection or zone plates primarily arises because of their much larger collection solid angles.

Due to the deviation between the Rowland circle and the crystal surface, a geometrical aberration, the so-called Johann error, is present that grows with the crystal extent ξ in the dispersion direction.

The Johann error is negligible at large bending radius R, large Bragg angle θ_B and for reflections close to the crystal centre. In particular, in the work described here the practical values were $\xi \leq$ 10 mm, R = 450 mm and θ_B = 76.6°, yielding $\Delta\lambda/\lambda \leq 3\cdot10^{-6}$, which is nearly an order of magnitude smaller than the resolution limited by the rocking curve width. Rocking curve widths of around 50 arcsec at θ_B = 76° are typical for the crystals used in the experiments described here, yielding $\Delta\lambda/\lambda \approx 6\cdot10^{-5}$ (from Bragg's equation), or, in other words, a resolving power of potentially 15,000.

4.2 Properties of toroidally bent crystals

Additional focusing of the sagittal rays, and therefore one-dimensional imaging of the plasma, can be obtained with two-dimensionally bent crystals. A toroidally bent crystal forms a surface that has both horizontal (meridional) and vertical (sagittal) radii of curvature; the lattice planes are assumed to be parallel to the crystal surface.[67] The use of perfect single crystals of quartz, silicon, germanium, gallium-arsenide or comparable materials allows spectral resolutions of several thousand due to their narrow rocking curves with widths of some tens of arcseconds. These allow much higher resolution than the use of mosaic highly ordered pyrolytic graphite (HOPG) crystals, which, however, provide higher luminosity.

The crystal bending radii in the horizontal, R_h, and vertical, R_v, directions determine focal distances in each plane. The theoretical respective focal lengths are given by $f_h = (R_h \cdot \sin\theta_B)/2$ and $f_v = R_h/(2\sin\theta_B)$.

If $f_h = f_v$ then monochromatic imaging of the source is achieved with a high two-dimensional spatial resolution and a spectral range of a few tens of picometres.[68] However, if f_h and f_v differ, the crystal works as a one-dimensional spatially resolving spectrometer.[69,70]

As shown in figure 4, the detector is placed onto the Rowland circle at $\ell_b = R_h\sin\theta_B$ from the crystal, so that highest spectral resolution is obtained. Hence, the plasma source S has to positioned at the vertical focal point ℓ_a, so that spatial resolution occurs in the y direction, which is dictated by the lens equation in vertical y-direction: $1/f_v = 1/\ell_a + 1/\ell_b$. The spatial magnification factor, $M = \ell_b/\ell_a$, is given by the ratio of the crystal-detector distance ℓ_b to the plasma-crystal distance ℓ_a. It relates y' on the detector plane through $y' = -My$. The spectrometer geometry in figure 5 has $M > 1$, resulting in the vertical extent of the source being magnified at the detector plane. Accurate detector alignment is required to avoid the effects of defocusing in order to maintain high spatial resolution.

However, limitations are given by the width of the rocking curve, by the wavelength of the photons of interest and by the availability of suitable crystals.

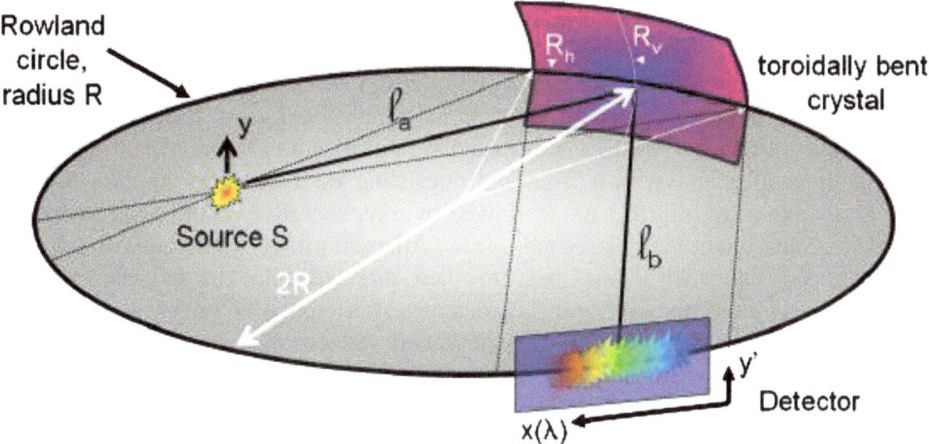

Figure 4   *Schematic of a toroidally bent crystal spectrometer. The source S is located inside the Rowland circle with radius R.*

4.3   Description of the X-ray Spectrometer

To investigate the K-shell emission of titanium at $\lambda = 0.27$ nm a GaAs crystal was chosen in (400) reflection with a Bragg angle of $\theta_B = 76.6°$. The perfection of the crystal should be as high as possible in order to minimize broadening of the rocking curve of a flat GaAs crystal. From a double-crystal setup,[69] the rocking curve was measured to be 55 arcsec FWHM, agreeing with calculations of the reflection properties which give 53.6 arcsec. The calculations also show that due to the large bending radii, $R_h = 450$ mm and $R_v = 306$ mm, the rocking curve FWHM increases by less than 2.5 %.

The size of the effective crystal surface is 30 mm in the dispersion plane and $\Delta h = 10$ mm in the perpendicular direction. The geometry for this particular crystal means a distance to the plasma of $\ell_a = 245$ mm and a distance to the detector (located on the Rowland circle) of $\ell_b = 440$ mm; the magnification in the spatially resolving plane is therefore $M = 1.8$. The 1D spatial resolution of the particular crystal used was determined to be 4 μm by imaging a gold grid,[64] which is of course only valid at perfect alignment.

The efficiency of the spectrometer setup can be calculated with high accuracy if all input parameters are known,[65] including the transmission through foils that protect the crystal from debris and the film detector from visible light. In the present case, the ratio of photons incident on the detector to Kα photons emitted by the plasma source is $N_{det}/N_0 = 2 \cdot 10^{-5}$. X-ray film was used as the detector because it is comparatively insensitive to the noise, present in typical high energy density experiments due to fast electrons, high-energy x-rays, and strong electromagnetic shocks (EMP). EMP affects CCD cameras by inducing pulsed currents in electric circuits and connection cables. Also, fast electrons hit both the detector chip and any shielding (mostly high-Z elements such as lead) and the chamber walls (steel, aluminium), where they produce cascades of secondary hard X-rays. The main drawback of film is the time needed to extract it from the target chamber and to develop it. Thus venting of the chamber is necessary, but this was not a limiting factor since single-shot targets were used.

5   RADIAL BULK TEMPERATURE AND LINE PROFILES

The experiments reported here[74] were performed at the LULI laser facility[71,72,73] at Ecole Polytechnique, in Palaiseau, France. Operational since 1997, this facility is based on chirped pulse amplification (CPA)[75] and can deliver pulses with powers of 100 TW.

The laser chain starts with a front-end consisting of a titanium-sapphire oscillator working at a wavelength of 1057 nm, followed by a regenerative amplifier at a repetition rate of 10 Hz. Subsequently, the laser pulses are further amplified by a chain of multi-glass amplifiers doped with Nd^{3+} ions and are then directed into the experimental hall, containing a vacuum compressor and the vacuum experimental chamber. A fast Pockels cell with a rise time of about half a nanosecond reduces the amplified spontaneous emission (ASE) and prepulses.

The spatial beam profile is surveyed by wavefront sensors and optimized using an adaptive mirror between the two passes in the last disc amplifier. This provides an almost Gaussian beam profile at a final rate of one shot per 20 minutes, although the amplifiers do not cool to thermal equilibrium in this period and thus thermal lensing cannot be avoided but is compensated for.

After compression the FWHM of the pulse is (330±30) fs. The beam is then focused by a dielectric off-axis f/3 parabolic mirror to a spot diameter of about 8 μm. The maximum energy on target was about 14 J at the fundamental laser wavelength, providing intensities of $(5\pm1)\cdot10^{19}$ W/cm^2. ASE and prepulses within the first nanosecond before the main pulse had a contrast ratio of $C_I = 10^{-8}$. To maximize the contrast of the main pulse compared to any plateau, the laser pulse can be frequency doubled ($\lambda = 529$ nm), which allows C_I to be reduced to better than 10^{-10}.[76]

To achieve the highest intensities the laser pulses need to be focused to diameters of a few micrometres, where lateral gradients of the generated plasma parameters are expected to be of a similar scale. Thus, in order to achieve precise knowledge on the physical properties, such as the plasma temperature at solid density, high spatial resolution is needed.[2,50] For example, it has been reported that the plasma temperature is a function of the irradiated foil thickness,[61,77] however no evidence of radial variations of the plasma properties has yet been experimentally obtained at solid density.

The laser incidence angle to the target normal was 11°, as shown in figure 5. As targets, 25 μm, 10 μm and 5 μm thick titanium foils, as well as bulk titanium, were used. The 5 μm foil was coated with 250 nm of copper to eliminate the Kα radiation from the hot plasma at the surface of the foil. Time integrated single pulse spectra of the Kα doublet emission (4490-4530 eV) were detected using the spectrometer described in section 4.3, in the manner shown in figure 4.

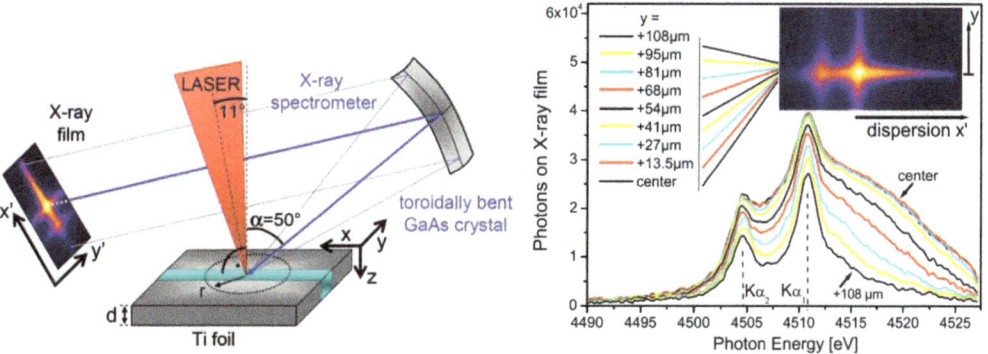

Figure 5 *Left: Schematic of the experimental setup. The energy dispersion axis x' on the X-ray film is perpendicular to the spatial resolution, y, direction. Right: Single-pulse X-ray spectra of the laser-irradiated 10 μm thick titanium foil at a laser intensity of $5\cdot10^{19}$ W/cm^2 and wavelength 1057 nm, registered at different target positions y. The X-ray film scan (false colours) is given in the inset.[74]*

The measured spectrum $I_M(x',y)$ as a function of the film coordinates is the local emissivity $I(E,r,z)$ integrated along both the x axis and the target depth z, as indicated by the highlighted slice in figure 5; the conversion from x' to photon energy E is obtained by applying a non-linear dispersion relation.[65]

All spectra were recorded using absolutely calibrated Kodak Industrex X-ray film. Together with the known crystal efficiency it was thus possible to infer absolute photon numbers. The films were scanned along the spatially resolved axis using a Zeiss densitometer with steps of 13.5 μm to give an optimal signal-to-noise ratio.

The bent crystal spectrometer provides 1D resolved spectra of the plasma emission in y-direction, as shown in figure 4 and 5. In this single exposure, a $5\cdot10^{19}$ W/cm^2 laser pulse at the fundamental wavelength hit a 10 μm thick titanium foil. From the centre of emission up to $y = \pm27$ μm, i.e., very close to the laser focus, the same profiles and emissivities were observed. With every further step of 13.5 μm the profiles change. All profiles show a significant high energy wing with rather smooth profiles and no additional peaks, and the positions of the components are not shifted.

If the amount of the resonant Kα self-absorption over a propagation distance of the attenuation length is negligible, then it can be assumed that only the latter contributes and ℓ = ℓ_0 = 20 μm of the solid titanium, and to be practically constant over the narrow spectral range of interest,[78] i.e., w is solely a function of z, $w(z)$ = exp[-z/($\ell_0$cos α)]. For the thin foils, d is comparable to or smaller than $\ell_0$cos α, and the attenuation is fairly small, $w \approx 0.5$ for photons coming from the rear side of a 10 μm foil. On the other hand, spectra obtained from bulk targets represent mostly radiation from the front 20 μm layer.

Applying the inverse Abel transform[79] to the measured spectra leads to recovery of the radial dependence of the plasma emission; the spectrum inferred, however, is averaged – with the proper weight $w(z)$ – over the target depth d.

For each photon energy, the inverse Abel transformation using the onion-peeling method[80] was applied, yielding radially resolved spectra in steps of 13.5 μm. In detail, the choice for the two-dimensional data sets was to apply an iterative code based on geometric calculations of concentric emission areas A^i_j with radii $i \times 13.5$μm and where j is the segment index, as shown in figure 6. The code starts with a spectrum $S(0)$, 100×13.5 μm from the irradiation centre. There was no plasma emission from this area within the spectral range of the X-ray spectrometer. The spectrum $S(0)$ is normalized in intensity to the emitting area and called $R(0)$.

Next, the spectrum from a lateral strip 13.5 μm closer to the centre, $S(1)$, is treated as the sum of the radial spectrum at this position, $R(1)$, and $R(0)$. The weighting ratio is calculated by the contributing geometrical areas. This procedure is repeated iteratively for all radii until reaching the centre.

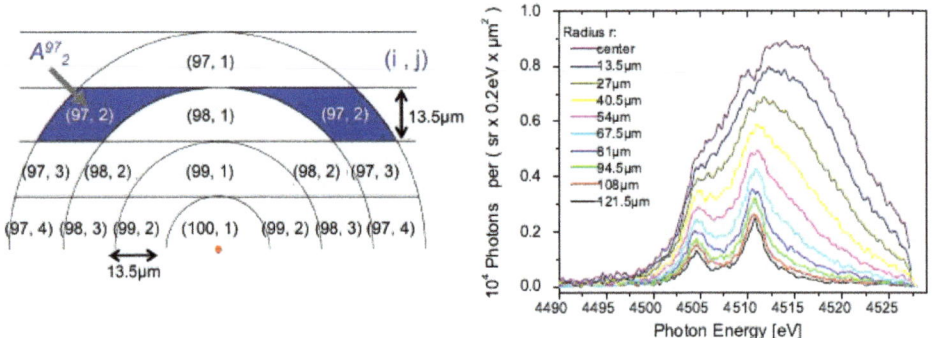

Figure 6   *Left: Illustration of the discrete Abel inversion applied to the measured spectra. The plasma emission is assumed to be cylindrically symmetric and the spectra are divided into segments with radius index i and segment index j. The emission from a lateral strip is then converted into the emission from a certain radius. Right: Radially resolved titanium K spectra obtained by Abel deconvolution from the spectra of figure 5 (10 μm foil). For regions more than 120 μm from the emission centre a narrow doublet structure is observed, while the central emission shows a strongly broadened and blue-shifted shape and 4 times increased peak emission.[74]*

A drawback of this iterative process is possible accumulation of noise, which can make a reasonable reconstruction impossible. Therefore, cautious filtering was applied to the spectroscopic data where the noise level significantly exceeded the K emission.

Figure 7 shows radially resolved titanium Kα spectra obtained by Abel transformation of the data presented in figure 6. For radii ≥ 120 μm a narrow doublet structure is observed, similar to those obtained from X-ray tubes. Close to the emission centre, a high energy wing emerges on both lines. For radii ≤ 30 μm, the fine structure is totally smeared out, resulting in a broad line profile with a maximum at 4515 eV, a FWHM of about 20 eV and an integrated emission 11 times stronger than that at 120 μm.

5.1   Radial profiles

Using published model spectra[57] to fit the deconvolved data the radially resolved electron temperature was determined; the results are shown in figure 7 and 8. For higher temperatures, significant contributions of higher ionization stages and excited state satellites affect the line shape, resulting in the blue shift (a shift towards higher photon energies) and broadening of the Kα spectrum. Stark broadening contributes to the smoothing of the entire line shape. The error values in figure 8 were determined by varying T_e around the best fit value (which gives a minimal χ^2 according to the standard statistical method of least χ^2) up to a 50% increase in χ^2. As can be seen from figure 7, good agreement with the model spectra was achieved for all radii. Applying this procedure for every data set, the radial variation of T_e is determined with a resolution of 13.5 μm. For radii ≥ 200 μm the lineshape of cold titanium is observed. The line shapes do not change significantly below $T_e \approx 5$ eV, resulting in relatively large errors at large radii.

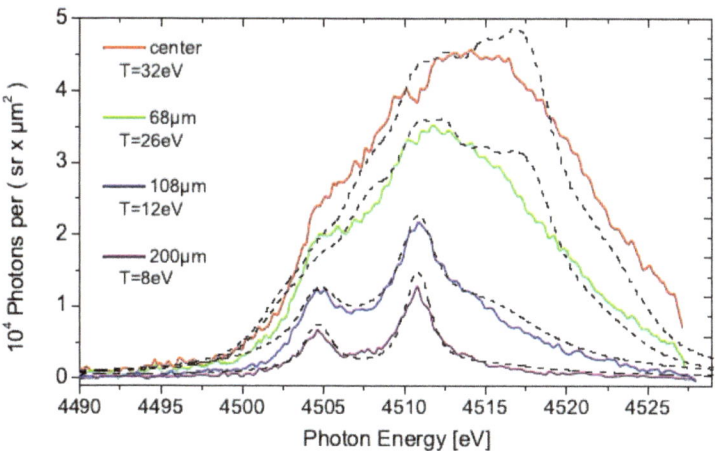

Figure 7   *Comparison of Abel-inverted experimental spectra (solid lines) from irradiation of a 10 μm titanium foil with the best fit modelled spectra (dashed lines).* [74]

The radial temperature profiles and Kα yield for both lower laser contrast (ω) and high laser contrast (2ω) are shown in figure 8. The maximum temperature is constant at around 30-40 eV within the error bars. At ω, this temperature extends over ≈ 50 μm which is an order of magnitude larger than the laser focal spot. This radius increases for thinner foils, especially for the 10 μm foil, although the 5 μm foil has a smaller hot area with a maximum temperature exceeding 40 eV. The 2ω exposures show a more systematic dependence: both the maximum temperature and the radius of the heated area increase when the foil thickness decreases; the temperature increases from 32 eV to 37 eV, and the radius from 30 μm to 60 μm. In addition, the Kα yield decreases with thinner foils. At larger radii temperature gradients of ~ 1 eV/μm are present.

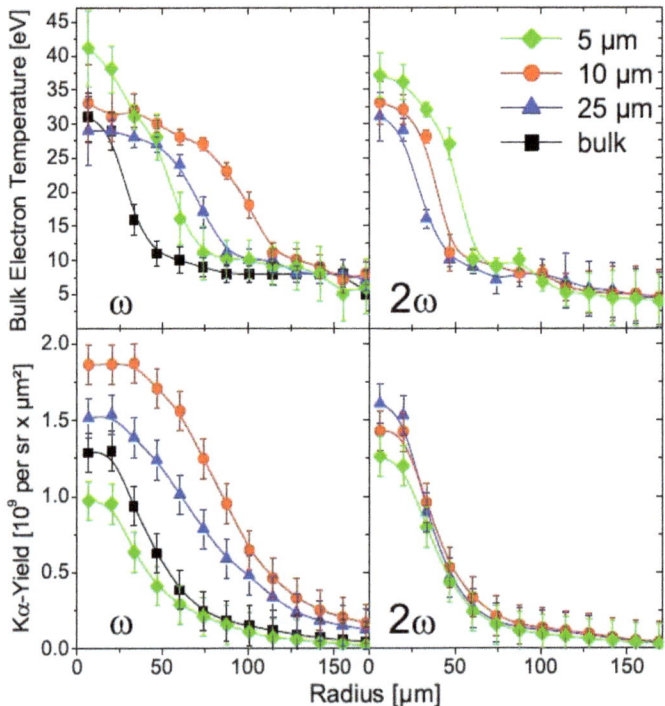

Figure 8   *Radial temperature (top) and integrated Kα yield (bottom) distributions of the different targets used, irradiated with the ω and 2ω laser pulses [74]. No bulk data for 2ω laser pulses was recorded.*

Although some of the dependences can be attributed to pulse to pulse laser parameter variations, a strong correlation between the spatial distributions of T_e (bulk) and Kα yield is observed for all single pulse measurements. This is evident from figure 9, where the FWHM values of the distributions are shown for different target thicknesses, both for ω and 2ω pulses. These results show that there is a strong correlation between the mechanism(s) responsible for the radial distributions of the electrons with $E > 5$keV – giving rise to the Kα radiation – and those of the WDM bulk heating, indicating that for the latter the faster electrons could play an important role.

The differences between the 2ω and ω exposures are most likely related to the significantly suppressed laser pre-plasma interaction due to the higher contrast ratio at 2ω. Hydrodynamic simulations[76] show that for a metal foil the surface of the critical electron density, $n_c \approx 10^{21}$ cm^{-3}, moves about 10 μm away from the foil due to the pre-pulse. At ω, only the 5 μm foil shows a comparatively small heated area as compared to the thicker foils. This may also be related to preplasma formation, since this foil was coated with 250 nm of copper in which pre-plasma formation is likely, and may also be responsible for similar radial distributions for this foil for ω and 2ω (see figure 8), contrary to the other targets.

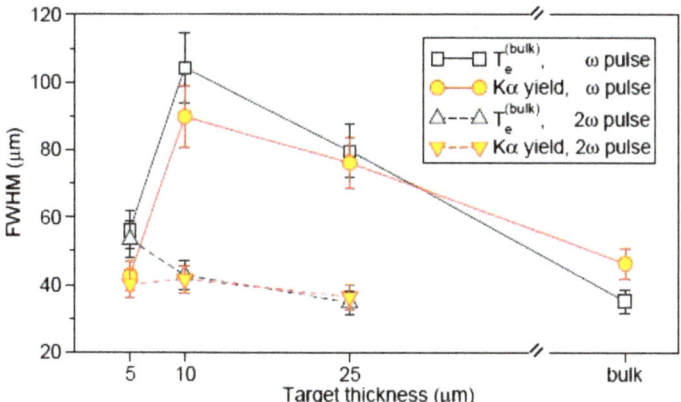

Figure 9 *Radial spatial extend (FWHM) of the bulk electron temperature range compared to the Kα yield radial extend for the various targets and laser pulses. Strong correlations between the FWHM values for the individual exposures are evident.*

5.2 Temperature and spot sizes

The top left graph of figure 10 illustrates the spatial properties of the plasma emission from fundamental laser frequency exposures at two different X-ray energies. The emission FWHM is 6 times larger than the laser focal spot size. The emission of the high energy wing at 4516-4522 eV has also been analyzed; for the 5 μm foil the FWHM is similar to the laser focus diameter, while for the 10 μm foil it is twice as large. The emission area is lower for the bulk sample and the 5 μm foil, with higher maximum temperatures as shown in the bottom left part of figure 10.

5.2.1 Hydrodynamic Motion.

When a thin (2 μm) pure titanium foil was irradiated at the fundamental laser frequency no further increase in temperature was observed, in contradiction to previous measurements.[61] The prepulse at the fundamental laser frequency ω generates an energy density of about 1 kJ/cm^2 at the focal spot within half a nanosecond. Hydrodynamic simulations using the one dimensional Helios code[81] indicate that the first micrometre of the foil is ablated and expands into the vacuum, while the other half is pre-heated. The density in the remaining foil volume increases temporarily by a factor of two due to a shock wave, as can be seen from figure 10. When the main pulse arrives 0.5 ns later the foil already has a complex density gradient. Since the line shape modelling assumes solid density, the error bars are large for foils thinner than 5 μm and so the results for these foils were excluded from the temperature analysis.

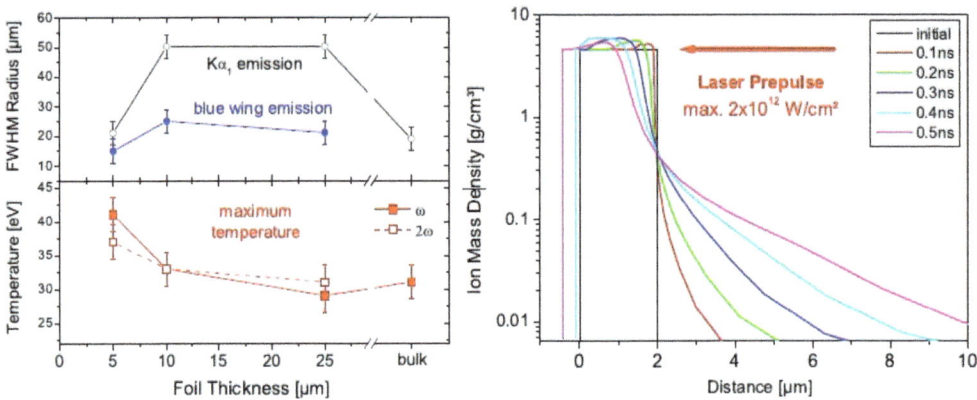

Figure 10   *Left top: The emission FWHM radius of emitted Kα₁ photons (4508–4512 eV) and those emitted in the high energy (blue) wing (4516–4522 eV) as functions of target thicknesses at ω. Left bottom: Maximum bulk temperature as a function of target thickness for ω and 2 ω. Right: Hydrodynamic simulation of the laser prepulse incident on a 2 μm titanium foil using the 1D HELIOS code. The prepulse is assumed to be a linear ramp, reaching 2.5·10¹² W/cm² after 0.5 ns. At this time, the main pulse arrives and interacts with both a lower density preplasma and the about 1.5 times compressed residual foil.* [74]

The overall conclusion[47] is that the high contrast 2ω pulse, known to create negligible preplasma,[76] does not inject a large number of hot electrons into the solid. There are two reasons for this: (i) pre-ionization is suppressed and therefore the laser pulse interacts with a steep density gradient,[82] and (ii) the temperature of the electron distribution scales [36,40] as $T_{hot} \sim (I\lambda^2)^{0.3\text{-}0.5}$; The laser intensity I is about 50% lower for the frequency doubled pulse compared to the fundamental. These effects are counterbalanced by the fact that more laser energy can be coupled into to the target, because the critical density is higher for shorter wavelengths; in addition, preplasma formation is suppressed. Finally, the achievable laser focal radius is theoretically smaller with shorter wavelength for constant beam divergence.

5.2.2 Electron Distribution.

Although the results contain three dimensional details of the characteristic X-ray emission of the generated plasma, there is no direct information about the electron spatial and energy distributions. However, conclusions can be drawn from the X-ray emission via the relativistic cross-section σ_K for titanium K shell ionization.[51] This cross-section reaches a maximum at an electron energy of 15 keV, and is about 60 % lower but almost constant for E > 100 keV. A significant fraction of fast electrons with kinetic energies greater than 500 keV escapes from the target and can be detected.[49] This charge separation is the source of strong electric fields of up to megavolts per micrometre at the foil boundaries.[83] The mean free path of electrons with energy less than 100 keV is in the order of tens of micrometres; they are probably confined to the target foil by the strong electric field and deposit their kinetic energy inside the foil by traversing it several times – called refluxing. Thus, electrons with energies up to a few hundred kiloelectronvolts (keV) are the dominant source of K-hole creation.

The properties of K-shell emission as discussed within this experiment are sensitive to the low energy part of the electron distribution; while electrons with $k_B T \geq 5\,\text{keV}$ are capable of K-shell ionization, the creation of L-shell, and even M-shell, holes is dominated by slower electrons with energies above 500 eV. M-shell holes affect the K line shape by blue shifting, as observed in the presented data, and serve as bulk electron temperature indicator. This general picture is complicated by effects such as secondary electrons, radiation heating and electron-electron interaction.

5.2.3 X-ray yield

The conversion efficiency of optical laser energy to X-ray photons in a narrow bandwidth ΔE is a very important parameter for applications such as time resolved X-ray diffraction, backlighting applications such as shadowgraphy and X-ray Thomson scattering.[84] For laser-driven X-ray sources with femtosecond pulses, the absolute number of photons per Joule of laser energy emitted over 4π sr is also a meaningful quantity for comparison with other experiments.

Figure 11 shows seven measurements of X-ray yields: the data for three different foil thicknesses as well as bulk targets were analysed for ω and 2ω irradiation. The ω exposures show nearly constant yields for the 10 μm and 25 μm foils, and for the bulk sample and the 5 μm foil. The increase by a factor of 2.5 of the former compared to the latter is remarkable. However, the exposures at 2ω do not show this pronounced dependence on foil thickness.

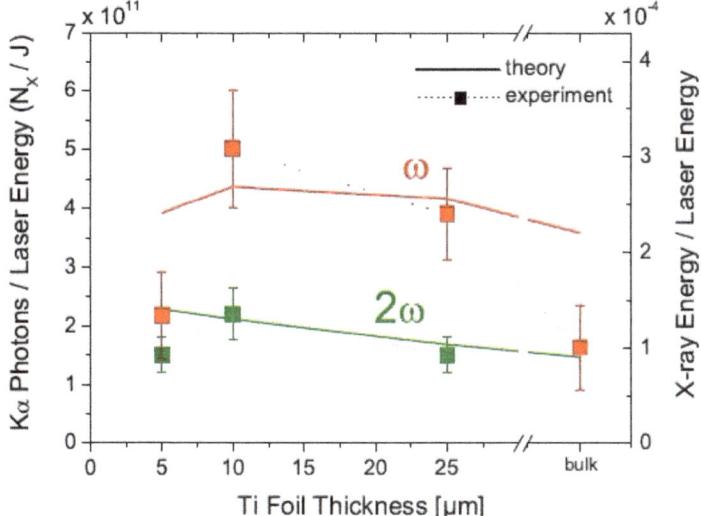

Figure 11 *The 4π yield of emitted K photons in the energy range 4490-4530 eV per Joule of incident laser energy as a function of target thickness. Alternatively, the right ordinate shows the data as ratio of X-ray energy to laser energy. The solid lines are theoretical values deduced from figure 13. To match the experimental results, a laser absorption of 50% was assumed for ω and 10% for 2ω.* [74]

In order to model the measured X-ray yield, a Monte-Carlo (MC) simulation[15] is employed to calculate the conversion efficiency η_{eX} from electrons into K photons. As an initial electron distribution the relativistic Maxwell-Jüttner function was assumed. If the electrons traverse the foil only once, i.e., refluxing is neglected, thicker foils will emit more photons, only limited by reabsorption, as shown by the dashed curves (labeled *"no reflux"*) in figure 12. This is in clear contrast to the observations.

The MC model was therefore extended to account for refluxing: electrons with energies greater than some value E_{crit} are assumed to escape the target space charge, while slower electrons are reflected inside the foil boundaries. The solid curves in figure 12 show that if E_{crit} is assumed to be 100 keV then the overall efficiency is increased. For the 5 μm and 10 μm foils η_{eX} reaches a maximum of 1.3% at T_{hot} = 50 keV, while it is only around 0.4% for $T_{hot} \geq 1$MeV. For thicker foils, the effect of refluxing electrons is less pronounced, and it is negligible for bulk material. Qualitative agreement with the measured target thickness dependence at ω is achieved at electron temperatures around 100 keV. Another model[85] which assumed total refluxing, i.e., all the electrons were confined to the target foil, predicted a general increase of the yield, especially for electrons of a few mega-electronvolts (MeV), but did not reproduce the observed dependence on foil thickness.

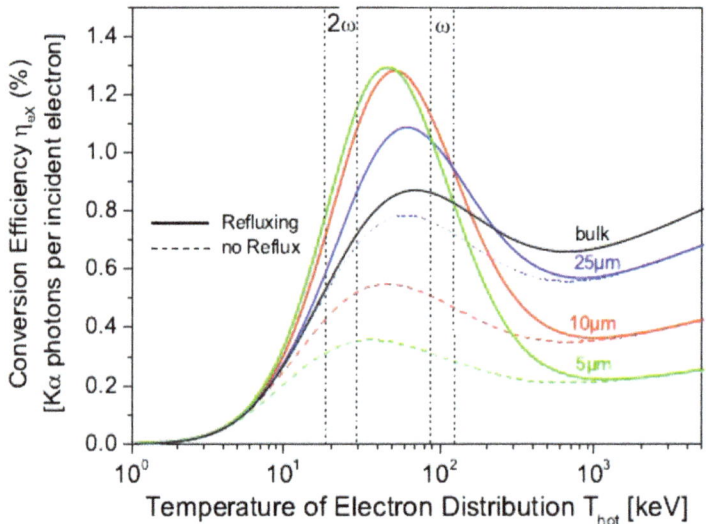

Figure 12   *The theoretical conversion efficiency of electrons into Kα photons, as a function of the electron temperature T_{hot}, into K photons for different titanium foil thicknesses. The solid lines result when refluxing of electrons with E<100 keV, compared to the dashed lines with no refluxing. The marked areas indicate the electron temperature estimate for the ω and 2ω exposures, respectively.* [74]

Therefore an electron distribution with T_{hot} = 100 keV was assumed for all exposures at a laser frequency ω. From the scaling laws discussed above, the hot electron temperature for 2ω is then about 35 keV. An upper limit for the number of electrons is given by assuming that all the incident laser energy generates electrons, independent of target thickness. Applying the conversion efficiency from figure 11 then allow to predict an upper limit for the Kα yield. Comparison with the experimental data shown in figure 11 indicates that 50% and 10% of the laser energy is converted into hot electrons for ω and 2ω. Within typical errors, these values agree with previously published values.[85,86,87]

At the fundamental frequency, the large numbers of free electrons in both the preplasma and the preheated foil result in an increased luminosity compared to the exposures at 2ω. This is probably also the reason for the larger emission area shown in figure 8. In addition, the derived electron temperatures are about one order of magnitude lower than one would expect from the general scaling laws. These were derived from the slopes of the high energy tails of the time integrated distributions, while the production of K shell vacancies strongly depends on electrons of several tens of kiloelectronvolts within the first picosecond after irradiation. Thus the derived electron temperature is valid for the distribution of kiloelectronvolt electrons, which are usually not accessible by electron spectroscopy at the rear of the target.

The intensity of several times 10^{19} W/cm^2, used in the experiments described here is about two orders of magnitude too high for efficient Kα generation in titanium, since the peak of the hot electron distribution does not overlap with the peak of the K shell ionization cross-section.[15,24]

5.2.4 Energy deposition as a function of depth in the target.

Usually the Kα yield and the overall deposition of energy into the target are functions of depth, leading to a depth dependent bulk electron temperature.[88] The results discussed here have high lateral resolution but are averaged over the target depth, weighted by the attenuation factor for Kα photons in titanium. However, the Monte Carlo Code, including refluxing, yields depth dependent information on the energy loss of the hot electrons.

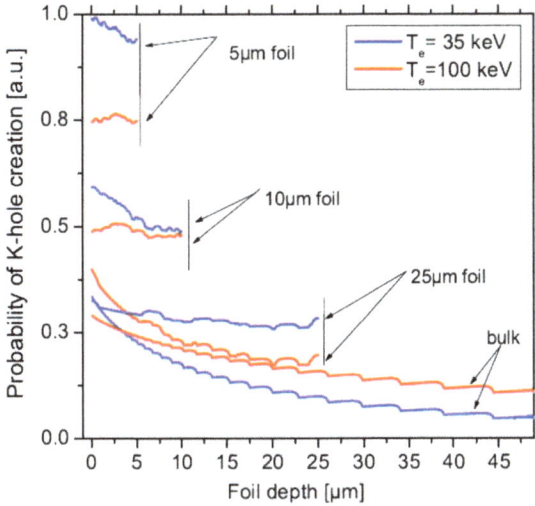

Figure 13   *Results from Monte Carlo simulations of the depth dependent K shell ionization probability for two hot electron temperatures and the various titanium foil thicknesses. Refluxing of the electrons is included for all the foils, but no refluxing is assumed for the bulk target.*

Figure 13 shows the probability of K shell ionization for electron distributions with temperatures of 100 keV (ω) and 35 keV (2ω). Due to the refluxing of electrons, the thin foils show a higher concentration of K shell holes than the thick foils. Additionally, the depth dependency is negligible for foil thicknesses up to 10 μm; only the 25 μm foil and the bulk target (without refluxing) show a drop of the K hole concentration. The hole concentration at a target depth of 20 μm, the attenuation length, is not more than about 60% of that at the irradiated surface. The slower electrons from the 35 keV distribution deposit more energy in the first 5 μm of each foil, while the faster electrons with T_e^{bulk} = 100 keV lead to a more homogeneous deposition of energy.

5.2.5 Brilliance limitations.

Laser driven X-ray sources are suitable for probing the dynamics of solid density plasmas.[89] Important examples are time resolved X-ray diffraction and X-ray Thomson scattering. While maximum yield and short X-ray emission duration are crucial in both cases, Thomson scattering also needs a narrow bandwidth of $\Delta E/E$ = 0.01 for non-collective scattering[90] and $\Delta E/E$ = 0.002 for collective scattering detection.[91] The results presented here show that the plasma emits the cold line shape resulting from collective scattering only if its temperature does not significantly exceed 20 eV. At higher temperatures, the emission is broadened and blue-shifted, which also affects the use of X-ray optics in, for example, monochromatic imaging.[65] The use of thicker foils, on the other hand, results in a longer X-ray pulse duration.[15]

The peak brilliance, that is the number of photons per second in a defined bandwidth, emission area and solid angle is plotted in figure 14 as a function of foil thickness d, using the results of the MC calculations. The pulse duration is possibly prolonged by the refluxing electrons. As illustrated in figure 10, the laser ASE prepulse evaporates about 50% of the 2 μm foil, which results in decreased emitted photon numbers. A foil thickness of 5-10 μm allows both high brilliance and sub-picosecond pulses, while the bulk target emits a pulse of about 1.5 ps.

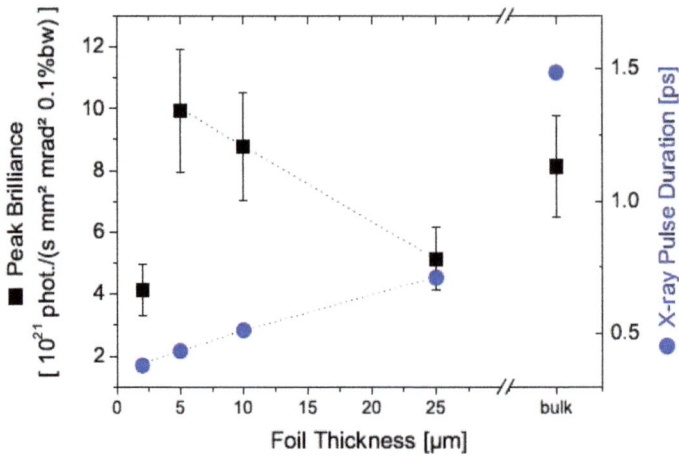

Figure 14   *Peak brilliance (squares) and estimated X-ray pulse duration (circles) as functions of foil thickness for irradiation at the laser fundamental frequency. A foil thickness of 5-10 μm allows both high brilliance and sub-picosecond pulse duration. Note the loss of brilliance for the thinnest foil.* [74]

6 SUMMARY

Although femtosecond lasers are able to deliver several Joules of laser energy into cubic micrometre volumes, the creation of a defined state of warm dense matter is challenging. First, the irradiated spot is only a few micrometers in diameter, resulting in strong gradients. Second, the heating of a solid is indirectly produced by relativistic electrons, since radiation around the visible region cannot penetrate electron densities greater than 10^{21} cm^{-3}. Third, the creation of these electrons by the laser is an extreme non-linear process. This all means that the investigation of these processes with high spectral and spatial resolution is of primary importance. The advantage of such plasmas is the generation of sub-picosecond X-ray spectral lines emitted from small spots, suitable for backlighting. In addition, the generation of electrons for heating high density matter is of major importance for fast ignition.

The key technology in this field is X-ray spectroscopy at the highest possible spectral and spatial resolutions. The electrons traversing the target create vacancies in the inner atomic shells. Recombination to the K shell leads to the emission of characteristic X-rays which are reliable carriers of information, since their line shapes are influenced by the plasma parameters and they penetrate regions of solid density. In section 4 the physical and technological limitations in designing and using a toroidally bent 1D focusing spectrometer for the titanium Kα doublet were discussed. The efficiency of the spectrometer and the X-ray film detector are known, so that absolute photon numbers can be determined.

Laser matter interaction experiments were conducted at the 100 TW laser facility at LULI (section 5). The beam was focused to a few micrometres diameter spot a on thin titanium foils, reaching intensities of several times 10^{19} W/cm^2 for both the fundamental and frequency doubled radiation. Abel inversion allowed to determine radially resolved spectra of the Kα emission. The line profile was compared to temperature dependent calculations, yielding radially resolved bulk electron temperature maps.

In the experiments, the generated plasma showed approximately homogeneously heated cores up to an order of magnitude larger than the laser focal spot, with a strong correlation between the radial distributions of the bulk electron temperature and the Kα yield. The influence of a thin coating on the 5 μm foil resulted in a decrease in both the spot size and the Kα yield. When the laser frequency was doubled and the intensity contrast was increased to 10^{-10}, the inferred distributions become about two times narrower and showed a clear dependence on the foil thickness: thin foils produce hotter plasmas but lower Kα yields.

The line shape at the emission centre changed significantly compared to the cold K doublet observed at large radii: a single broad line with a FWHM of 20 eV was observed, and the central photon energy was blue shifted to 4514 eV with plasma temperatures up to 42 eV. We note that exactly in the central laser focus (the inner few micrometres) much higher temperature and significant ionization is likely. Detection of these small spatial scales lies beyond the current capabilities of curved crystal x-ray optics.

The global K yield was also measured as a function of target thickness and maximum yields were observed for foil thicknesses of 10-25 μm.[71] These results were explained by means of an electron refluxing model. Since the laser generated electrons have energies ranging over several orders of magnitude, the K-shell ionization cross-section was explored using an empirical relativistic model. Monte Carlo simulations based on this cross-section can explain the observations if refluxing of electrons with $E < 100$ keV is assumed. Thus parameters are being derived that allow optimization of the brilliance of such X-ray sources.

These results[71] are of primary importance for the basic understanding of laser matter interaction and benchmarking computer simulations, as well as a range of applications including fast ignition for laser fusion and laser-driven backlighters. Future research in this field will concentrate on how the plasma temperature depends on the target thickness and depth. In addition, time resolved measurement of the X-ray emission using streak cameras need further work to shed light on the complex contribution of refluxing electrons.

7 OUTLOOK

Although we have shown the rich physics of the radial extend of Ka emission, information on the depth dependence (perpendicular to the surface) is the logical next step to arrive at a fully coherent picture of the laser-matter interaction. In the future, we aim to visualize the multi-keV electron distribution in thin foil targets in 3D by employing emission spectro-tomography.

In the experiments described in this chapter, the integration along the z axis (due to accumulation of photons emitted from different depths of the target) results in a weighted averaging of the spectra observed, precluding any depth-resolved conclusions to be made. In order to probe the z-dependence, multi-layered "sandwich" targets are used, with the relatively thin dopant layer placed at a varying depth. However, such specially manufactured targets are relatively expensive, and, more importantly, the multi-layered structure may introduce undesired effects due to the presence of materials with different mechanical, thermodynamic, and electromagnetic properties.

In a proposed experiment,[92] we present an alternative, tomography-like method of resolving the z dimension that requires no specially prepared targets. The proposed method utilizes absorption of the emitted radiation in the target and observed at different angles. We note that the problem of absorption (also termed optical thickness or opacity) arises in the traditional emission tomography methods if the object scanned is not optically thin. Then, the so called attenuation corrections need to be introduced in order to account for the absorption. Although numerical studies suggest that the emission field can be reconstructed with a reasonable accuracy even for optical thickness as high as 6 to 8, the absorption is generally considered to be an obstacle. Contrary to that, in the present method, the absorption within a certain range of opacity is rather an aid, because different lines of sight result in different visible depths of the target.

The simplest type of tomography measurements integrate over the entire foil target as shown in Fig. 25 (without spatial resolution in the y direction) and a moderate spectral resolution that is only sufficient to distinguish $K\alpha$ radiation from the x-ray background. Thus, measured is the total $K\alpha$ radiance as a function of the observation angle. Furthermore, for the sake of simplicity, the $K\alpha$ resonant absorption is assumed to be negligible. The target, made of Ti with an absorption length of 20 μm for its own K radiation, has a thickness of 25 μm. We now consider a series of models where the $K\alpha$ yield in the target decreases with the depth as $\sim exp(z/l)$. This is shown in Fig. 26a for $l = 1$; 20; 10; and 5 μm. The respective radiances as a function of the observation angle, normalized to that at the frontal observation, are shown in Fig. 15.

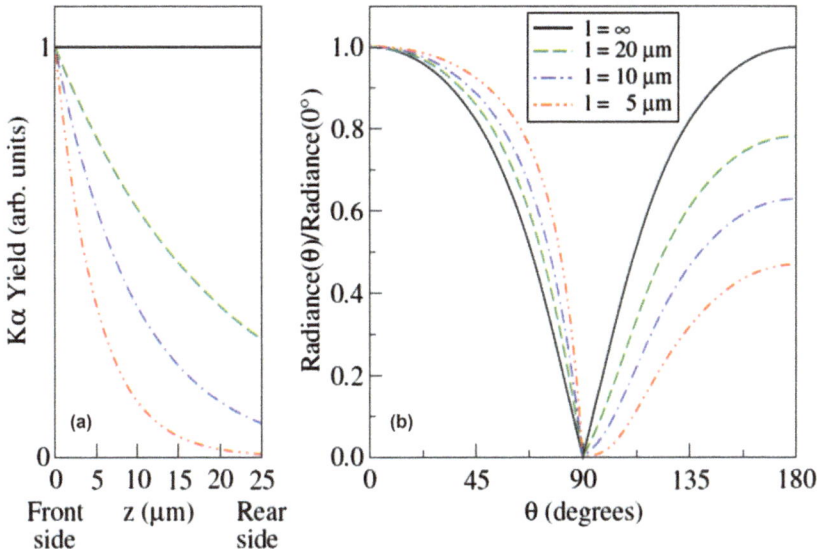

Figure 15   *Demonstration of the method: a) A few models of Kα-yield z-dependence in a 25-μm-thick Ti foil photon free path length of 20 μm; b) Relative (to the zero-degree direction of observation) intensity of the observed radiation as a function of the polar angle θ.* [92]

As is clearly seen, the angular dependencies differ significantly, allowing for discriminating between these simple models using as few as two measurements. Indeed, the ratio of the radiances measured at observation angles close to the rear and front-side normals varies from unity (homogeneous Kα yield, l = 1μm) to more than two for l = 5μm.

Alternatively, with the "reference" directions chosen to be 60 degrees to the front and rear normals, the ratio varies more than fourfold. We note that since only the relative variations of the radiation intensity are used, no absolute calibration of the data acquisition system is required, thus simplifying the experimental setup and reducing instrumental uncertainties. Evidently, performing measurements at only two directions is not sufficient for inferring z distributions that are described by models more complex than the simple one-parameter distribution assumed in this example. In general, the number of observations (at different angles) should exceed the number of parameters in the theoretical model to be tested.

ACKNOWLEDGEMENTS

The authors gratefully acknowledge financial support from the German Research Foundation (DFG), the German-Israeli Foundation for Scientific Research and Development (GIF) via grant I-880-135.7/2005 and the German Federal Ministry for Education and Research (BMBF) via priority programmes FSP 301 and 302. UZ is a Peter-Paul-Ewald Fellow of the Volkswagen Foundation.

References

1   R. W. Lee et al. Plasma-based studies with intense x-ray and particle beam sources. Laser Part. Beams, 20:527, 2002.

2   A. Saemann et al. Isochoric heating of solid aluminum by ultrashort laser pulses focused on a tamped target. Phys. Rev. Lett., 82(24):4843–4846, 1999.

3   Y. Ping et al. Broadband dielectric function of nonequilibrium warm dense gold. Phys. Rev. Lett., 96:255003, 2006.

4   T. Ao et al. Optical properties in nonequilibrium phase transitions. Phys. Rev. Lett., 96:055001, 2006.

5   K. Widmann et al. Single state measurement of electrical conductivity of warm dense gold. Phys. Rev. Lett., 92:125002, 2004.

6   G. W. Collins et al. Temperature measurement of shock compressed liquid deuterium up to 230 GPa. Phys. Rev. Lett.,87:165504, 2001.

7   L. B. Da Silva et al. Absolution equation of state measurements on shocked liquid deuterium up to 200 GPa (2Mbar). Phys. Rev. Lett., 78:483, 1997.

8   R. Sigel et al. X-ray generation in a cavity heated by 1.3- or 0.44-µm laser light. Phys. Rev. A, 38(11):5756–5785, 1988.

9   T. S. Perry et al. Absorption experiments on x-ray-heated mid-z constrained samples. Phys. Rev. E, 54:5617, 1996.

10 J. J. MacFarlane et al. X-ray absorption spectroscopy measurements of foil heating by z-pinch radiation. Phys. Rev. E, 66:046416, 2002.

11 DHH. Hoffmann et al. Present and future perspectives for high energy density physics with intense heavy ion and laser beams. Laser and Particle Beams, 23:47–53, 2005.

12 P.K. Patel et al. Isochoric heating of solid-density matter with an ultrafast proton beam. Phys. Rev. Lett., 91:125004, 2003.

13 U. Zastrau, C. Fortmann, et al. Bremsstrahlung and Line Spectroscopy of warm dense Aluminum heated by XUV Free Electron Laser. Phys. Rev. E, 78:066406, 2008.

14 B. Nagler, U. Zastrau, et al. Turning solid Aluminum Transparent by intense soft x-ray photoionization. Nature Physics, 5:693–696, 2009.

15 C. Reich et al. Yield Optimization and Time Structure of Femtosecond Laser Plasma Kα Sources. Phys. Rev. Lett., 84:4846–4849, 2000.

16 O. L. Landen, D. R. Farley, et al. X-ray backlighting for the National Ignition Facility (invited). Rev. Sci. Instr., 72:627–634, 2001.

17 H.-S. Park, N. Izumi, et al. Characteristics of high energy Kα and Bremsstrahlung sources generated by short pulse petawatt lasers. Rev. Sci. Instr., 75:4048–4050, oct 2004.

18 A. Pukhov and J. Meyer ter Vehn. Relativistic magnetic self-channeling of light in near-critical plasma: Three-dimensional particle-in-cell simulation. Phys. Rev. Lett., 76(21):3975–3978, 1996.

19 W.P. Leemans et al. GeV electron beams from a centimetre-scale accelerator. Nature Physics, 2:696–699, 2006.

20 H. Schwoerer et al. Laser plasma acceleration of quasi-monoenergetic protons from microstructured targets. Nature, 439:445–448, 2006.

21 JD. Lindl et al. The physics basis for ignition using indirect-drive targets in the National Ignition Facility. Phys. Plasmas, 11:339–491, 2004.

22 T. R. Dittrich et al. Review of indirect-drive ignition design options for the nation ignition facility. Phys. Plasmas, 6:2164, 1999.

23 M. Tabak, J. Hammer, et al. Ignition and high gain with ultrapowerful lasers.

Phys. Plasmas, 1:1626–1634, 1994.

24 Ch. Reich et al. Spatial characteristics of k_ x-ray emission from relativistic femtosecond laser plasmas. Phys. Rev. E, 68(5):056408, 2003.

25 E.E. Fill. Relativistic electron beams in conducting solids and dense plasmas: Approximate analytical theory. Physics of Plasmas, 8(4):1441–1444, 2001.

26 M. Silies et al. Table-top kHz hard X-ray source with ultrashort pulse duration for time-resolved X-ray diffraction. Applied Physics A: Materials Science & Processing, 96:59–67, jul 2009.

27 I. Uschmann et al. Investigation of fast processes in condensed matter by time-resolved x-ray diffraction. Applied Physics A: Materials Science & Processing, 96:91–98, 2009.

28 S. H. Glenzer et al. Demonstration of spectrally resolved x-ray scattering in dense plasmas. Phys. Rev. Lett., 90(17):175002, 2003.

29 H.J. Lee, P. Neumayer, et al. X-ray thomson-scattering measurements of density and temperature in shock-compressed beryllium. Physical Review Letters, 102(11):115001, 2009.

30 DW. Forslund, JM. Kindel, et al. Theory and simulation of resonant absorption in a hot plasma. Phys. Rev. A, 11(2):679–683, 1975.

31 F. Brunel. Not-so-resonant, resonant absorption. Phys. Rev. Lett., 59(1):52–55, 1987.

32 P. Gibbon and AR. Bell. Collisionless absorption in sharp-edged plasmas. Phys. Rev. Lett., 68(10):1535–1538, 1992.

33 TY. Brian Yang, L. Kruer, et al. Mechanisms for collisionless absorption of light waves obliquely incident on overdense plasmas with steep density gradients. Phys. Plasmas, 3(7):2702–2709, 1996.

34 LM. Chen, J. Zhang, QL. Dong, et al. Hot electron generation via vacuum heating process in femtosecond laser–solid interactions. Physics of Plasmas, 8(6):2925–2929, 2001.

35 J. Denavit. Absorption of high-intensity subpicosecond lasers on solid density targets. Phys. Rev. Lett., 69(21):3052–3055, 1992.

36 SC. Wilks, WL. Kruer, et al. Absorption of ultra-intense laser pulses. Phys. Rev. Lett., 69:1383–1386, 1992.

37 S.H. Glenzer and R. Redmer. X-ray Thomson Scattering in High Energy Density Plasmas. Rev. Mod. Phys. 81, 1625, 2009.

38 J.P. Gordon. Radiation Forces and Momenta in Dielectric Media. Phys. Rev. A, 8:14–21, 1973.

39 G. Malka and JL. Miquel. Experimental confirmation of ponderomotive-force electrons produced by an ultrarelativistic laser pulse on a solid target. Phys. Rev. Lett., 77(1):75–78, 1996.

40 F.N. Beg, A.R. Bell, et al. A study of picosecond laser–solid interactions up to 10^{19} W/cm². Phys. Plasmas, 4(2):447–457, 1997.

41 B. Hidding et al. Novel method for characterizing relativistic electron beams in a harsh laser-plasma environment. Rev. Sci. Instr., 78(8):083301, 2007.

42 AJ. Kemp, Y. Sentoku, V. Sotnikov, and SC.Wilks. Collisional relaxation of superthermal electrons generated by relativistic laser pulses in dense plasma. Phys. Rev. Lett., 97(23):235001, 2006.

43 P. Gibbon et al. Short-pulse laser-plasma interactions. Plasma Phys. Control. Fusion, 38:769–793, 1996.

44 T. Feurer, W. Theobald, et al. Onset of diffuse reflectivity and fast electron flux inhibition in 528-nm-laser-solid interactions at ultrahigh intensity. Phys. Rev. E, 56(4):4608–4614, 1997.

45   J.D. Huba. NRL plasma formulary. Office of naval research, Washington DC 20375, 2009.

46   A.R. Bell et al. Fast-electron transport in high-intensity short-pulse laser - solid experiments. Plasma Physics and Controlled Fusion, 39:653–659, 1997.

47   R.G. Evans. Modelling short pulse, high intensity laser plasma interactions. HEDP, 1:35–47, 2006.

48   A.P.L. Robinson and M. Sherlock. Magnetic collimation of fast electrons produced by ultraintense laser irradiation by structuring the target composition. Phys. Plasmas, 14(8):083105, 2007.

49   G. Malka, M.M. Aleonard, et al. Relativistic electron generation in interactions of a 30 TW laser pulse with a thin foil target. Phys. Rev. E, 66(6):066402, 2002.

50   G. Malka, Ph. Nicolai, et al. Fast electron transport and induced heating in solid targets from rear-side interferometry imaging. Phys. Rev. E, 77(2):026408, 2008.

51   E. Casnati, A. Tartari, and C. Baraldi. An empirical approach to K-shell ionisation cross section by electrons. J. Phys. B Atom. Mol. Phys., 15:155–167, 1982.

52   CA. Quarles. Semiempirical analysis of electron-induced K-shell ionization. Phys. Rev. A, 13:1278–1280, 1976.

53   JP. Santos, F. Parente, and YK. Kim. Cross sections for K-shell ionization of atoms by electron impact. J. Phys. B Atom. Molec. Phys., 36:4211–4224, 2003.

54   G. Zschornack. Atomdaten für die Röntgenspektral-Analyse. Springer Berlin, 1989.

55   P.H. Mokler and F. Folkmann. X-ray production in heavy ion-atoms colissions in: Structure and colissions of ions and atoms, volume 5. Springer Berlin, New York, 1978.

56   D. Jung et al. Experimental characterization of picosecond laser interaction with solid targets. Phys. Rev. E, 77(5):056403, 2008.

57   E. Stambulchik, V. Bernshtam, et al. Progress in line-shape modeling of K-shell transitions in warm dense titanium plasmas. J. Phys. A: Math. Gen. 42, 214056, 2009.

58   A. Sengebusch, H. Reinholz, et al. K-line emission profiles with focus on the self-consistent calculation of plasma polarization. J. Phys. A: Math. Gen. 42, 214061, 2009.

59   M.F. Gu. The flexible atomic code. Can. J. Phys., 86:675–689, 2008.

60   DC. Eder, G. Pretzler, et al. Spatial characteristics of K_ radiation from weakly relativistic laser plasmas. Applied Physics B: Lasers and Optics, 70:211–217, 2000.

61   S.B. Hansen et al. Temperature determination using $K\alpha$ spectra from M -shell Ti ions. Phys. Rev. E, 72(3):036408, 2005.

62   S.N. Chen et al. Creation of hot dense matter in short-pulse laser-plasma interaction with tamped titanium foils. Phys. Plasmas, 14(10):102701, 2007.

63   P. Audebert, R. Shepherd, et al. Heating of thin foils with a relativistic-intensity short-pulse laser. Phys. Rev. Lett., 89(26):265001, 2002.

64   F. Zamponi. Electron Propagation in solid matter as a result of relativistic laser plasma interactions. PhD thesis, Friedrich-Schiller University Jena, 2007.

65   T. Missalla et al. Monochromatic focusing of subpicosecond x-ray pulses in the keV range. Rev. Sci. Instr., 70(2):1288–1299, 1999.

66   H.H. Johann. Die Erzeugung lichtstarker Röntgenspektren mit konkaven Kristallen. Zeitschrift für Physik, 69:185–206, 1931.

67   I. Uschmann, E. Förster, et al. X-ray reflection properties of elastically bent perfect crystals in Bragg geometry. Journal of Applied Crystallography, 26(3):405–412, 1993.

68   M. Dirksmöller, O. Rancu, et al. Time resolved X-ray monochromatic imaging of a laser-produced plasma at 0.6635 nm wavelength. Optics Communications, 118:379–387, 1995.

69 M. Dirksmöller. Einzel- und Doppelkristallanordnungen zur hochauflösenden röntgenoptischen Anordnung. PhD thesis, Jena University, Jena, 1996.

70 I. Uschmann et al. X-ray emission produced by hot electrons from fs-laser produced plasma- diagnostic and application. Laser Part. Beams, 17:671–679, 1999.

71 J.P. Zou et al. The LULI 100-TW Ti:sapphire/Nd:glass laser: a first step towards a high performance petawatt facility. ICF SPIE, 1998.

72 B. Wattellier et al. Repetition rate increase and diffraction-limited focal spots for a nonthermal-equilibrium 100-tw nd:glass laser chain by use of adaptive optics. Opt. Lett., 29:2494–2496, 2004.

73 B. Wattellier et al. Generation of a single hot spot by use of a deformable mirror and study of its propagation in an underdense plasma. J. Opt. Soc. Am. B, 20:1632–1642, 2003.

74 U. Zastrau, A. Sengebusch, et al. High-resolution radial Kα spectra obtained from a multi-kev electron distribution in solid-density titanium foils generated by relativistic laser-matter interaction. High Energy Density Physics, 7(2):47 – 53, 2011.

75 D. Strickland and G. Mourou. Compression of amplified chirped optical pulses. Optics Communications, 56:219–221, 1985.

76 S. D. Baton, M. Koenig, et al. Inhibition of fast electron energy deposition due to preplasma filling of cone-attached targets. Phys. Plasmas, 15(4):042706, 2008.

77 S.N. Chen et al. X-ray spectroscopy of buried layer foils irradiated at laser intensities in excess of 10^{20} W/cm². Phys. Plasmas, 16(6):062701–8, 2009.

78 B.L. Henke et al. X-ray interactions: photoabsorption, scattering, transmission, and reflection at E=50-30000 ev, Z=1-92. Atomic Data and Nuclear Data Tables, 54:181, 1993.

79 H. R. Griem. Principles of Plasma Spectroscopy. Cambridge University Press, 1997.

80 C.J. Dasch. One-dimensional tomography - A comparison of Abel, onion-peeling, and filtered backprojection methods. Apl. Optics, 31:1146–1152, mar 1992.

81 J.J. MacFarlane et al. Helios-CR – a 1-D radiation-magnetohydrodynamics code with inline atomic kinetics modeling. J. Quant. Spectrosc. Radiat. Transfer, 99(1-3):381, 2006.

82 S. Bastiani, A. Rousse, et al. Experimental study of the interaction of subpicosecond laser pulses with solid targets of varying initial scale lengths. Phys. Rev. E, 56(6):7179–7185, 1997.

83 RA. Snavely, MH. Key, et al. Intense high-energy proton beams from petawatt-laser irradiation of solids. Phys. Rev. Lett., 85(14):2945–2948, 2000.

84 S. H. Glenzer et al. Observations of plasmons in warm dense matter. Phys. Rev. Lett., 98:065002, 2007.

85 W. Theobald, K. Akli, et al. Hot surface ionic line emission and cold K-inner shell emission from petawatt-laser-irradiated Cu foil targets. Phys. Plasmas, 13(4):043102, 2006.

86 F.Y. Khattak, A-M McEvoy, et al. Effects of plastic coating on Kα yield from ultra-short pulse laser irradiated ti foils. J. Phys. D, 36(19):2372–2376, 2003.

87 F. Ewald, H. Schwoerer, and R. Sauerbrey. Kα-radiation from relativistic laser-produced plasmas. Europhys. Lett., 60:710–716, 2002.

88 R.G. Evans et al. Rapid heating of soild density material by a petawatt laser. Appl. Phys. Lett., 86:191505, 2005.

89 C. Rischel, A. Rousse, I. Uschmann, et al. Femtosecond time-resolved x-ray diffraction from laser-heated organic films. Nature, 390:490–492, 1997.

90 O. L. Landen et al. Dense matter characterization by x-ray Thomson scattering.

J. Quant. Spectrosc. Radiat. Transfer, 71:465, 2001.

91   MK. Urry, G. Gregori, et al. X-ray probe development for collective scattering measurements in dense plasmas.
     Journal of Quantitative Spectroscopy and Radiative Transfer, 99:636–648, 2006.

92   E. Stambulchick, et al., Absorption-aided x-ray emission tomography of planar targets
     Physics of Plasmas **21**, 033303, 2014

BREMSSTRAHLUNG X-RAY EMISSION IN ELECTRON-BEAM-PUMPED KRF LASERS

V.D. Zvorykin[1,2] and S.V. Arlantsev[3]

[1]P.N. Lebedev Physical Institute of RAS, 53 Leninskiy Prospekt, Moscow, 119991 Russia. E-mail: zvorykin@sci.lebedev.ru
[2]National Research Nuclear University "MEPhI", Kashirskoe Shosse 31, Moscow, 115409 Russia
[3]Moscow Institute of Physics and Technology (State University), 9 Institutskii Pereulok, Dolgoprudnyi, Moscow Region, 141700 Russia. E-mail: arlantsev@mail.ru

1 INTRODUCTION

The spatial and energy distributions of X-ray emission caused by e-beam deceleration in matter are important issues for any Bremsstrahlung X-ray source.[1] It is also important for the proper design of e-beam pumped large scale KrF laser drivers for Inertial Fusion Energy (IFE) during prolonged operation.[2] Illumination of laser windows by ionising radiation may strongly affect their transmission and durability. It has been demonstrated[3] that both short lived (transient) and long lived (residual) absorption in the windows induced by scattered fast electrons can be almost fully removed by applying a magnetic field. However, induced absorption caused by X-rays, although not significant for single laser pulses, may be important for high repetition rate operation.[4] Hence, it is necessary to characterize the Bremsstrahlung emission from KrF laser cavities and to measure residual X-ray induced absorption of laser radiation in a range of optical materials transparent in the UV spectral domain.[2]

The Bremsstrahlung X-ray emission produced by deceleration of monoenergetic electrons in absorbing media has been discussed by Wyard[5,6] and Evans,[7] who analyzed early experimental and theoretical results for electrons, with initial kinetic energies ε_e up to several megaelectronvolts, that were fully stopped in thick targets. Wyard derived a normalized distribution of photon energy ε

$$z(\varepsilon) = \frac{1}{\varepsilon_e} \left[3.2 \left(1 - \frac{\varepsilon}{\varepsilon_e} \right) + 2.4 \left(\frac{\varepsilon}{\varepsilon_e} \right) \ln \left(\frac{\varepsilon}{\varepsilon_e} \right) \right], \tag{1}$$

and Evans provided a modified version

$$z(\varepsilon) = \frac{2}{\varepsilon_e^2} (\varepsilon_e - \varepsilon). \tag{2}$$

The total Bremsstrahlung energy per electron incident on a thick material integrated over all emission angles is $E = CZ_e^2$ (MeV), where Z is the effective atomic number of the material. The coefficient of proportionality C was 5.77×10^{-4} for equation (1) and

$(7\pm2)\times10^{-4}$ for equation (2). Self absorption of soft quanta was observed in experiments[7,8] in which electrons with ε_e=0.2–2.8 MeV bombarded thick metal targets (Z=4–79) with C=(3.0–3.8)$\times10^{-4}$.

In this chapter a comprehensive investigation is presented of Bremsstrahlung X-ray emission (i) that accompanies e-beam transport from vacuum diodes into the laser chamber through foil windows and (ii) via deceleration of electrons in the working gas mixture of the KrF laser. The energy flux of X-rays incident on laser windows is a key for scaling to larger KrF amplifiers.

2 GARPUN KrF LASER INSTALLATION AND X-RAY MEASUREMENTS

The GARPUN KrF laser[3] has a gas chamber with dimensions 19×22×140 cm and output aperture 16×18 cm (figure 1). Two counter propagating 12×100 cm e-beams with current density 50 A/cm^2 were generated in vacuum diodes and coupled into the chamber from opposite sides perpendicular to the laser axis. The 20 μm thick titanium vacuum foils were supported by a stainless steel hibachi structure with ribs 2 mm wide and 10 mm thick with 8 mm inter-rib gaps. The e-beams were stabilized by a pulsed magnetic field B~0.1 T, generated by a pair of solenoids, parallel to the beams. The ~100 ns, 350 kV pulses were delivered to a pair of cathodes in each vacuum diode by four water filled Blumlein forming lines with wave impedance of 7.6 Ohms. The lines were charged by a 7 stage 14 kJ Marx generator.

2.1 Scintillation Techniques for Measurements of X-ray Emission

A schematic of the GARPUN laser is shown in figure 1. Bremsstrahlung from the laser volume was transmitted through a 4 mm thick polyethylene sheet, used instead of the usual thick CaF$_2$ laser window in order to reduce absorption of the soft X-ray component. The laser chamber was evacuated to 10^{-1} Torr or filled with argon at different pressures. Absorbers of different materials with thickness x, copper (0.05–45 mm), aluminium (0.2–120 mm) and lead (1–6 mm) were placed in front of a scintillation detector to measure attenuation curves $u(x)$ for X-rays. A lead shielding box with 10 mm thick walls and an entrance aperture of diameter 14 mm eliminated scattered X-rays. Another, 25 mm diameter, aperture was inserted at the front end of a 650 mm length tube manufactured of lead to collimate the X-rays in a solid angle of 10^{-3} sr; it could be positioned between the polyethylene laser window and the absorbers. A crystalline NaI(Tl) scintillation detector, 25 mm in diameter and 25 mm thick, was placed 80 cm from the laser output (equivalent to 1 m from the e-beam boundary). Calibrated neutral glass filters were used to attenuate the luminescent flux irradiating a seven stage photomultiplier, which had a rise time of 1.2 ns, an amplification of about 6×10^5 and a linear current up to 2 A. Its maximum sensitivity coincided with the maximum of the scintillation spectrum at 410 nm.

First, it was necessary to show that there were no signal losses due to incomplete X-ray absorption in the scintillation detector in the high-energy region. Losses could result since the detector used was of limited size, comparable to the absorber thicknesses. For this purpose a bigger detector, 70 mm in diameter and 68 mm thick, was tested and demonstrated the same output as the smaller one. X-ray pulses registered by the scintillation detector had rise times of about 30 ns, comparable with that of the e-beam pulse, and a protracted trailing edge. The latter was perhaps determined by the long decay time of the luminescence in NaI(Tl), which is 220–270 ns.[10] The X-ray signals were reproducible with a root mean square error of 2% over a 15 shot run.

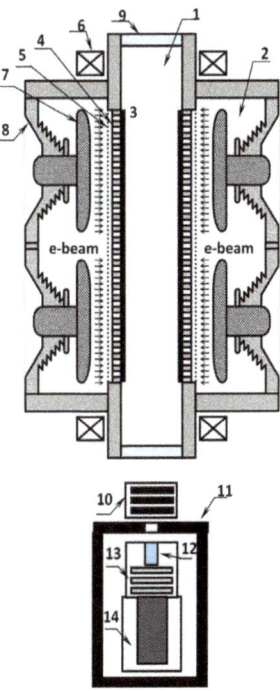

Figure 1   *Cross section of the GARPUN laser arranged for X-ray measurements. 1: gas chamber; 2: vacuum diode; 3: foil; 4: hibachi; 5: anode grid; 6: solenoid; 7: cathode; 8: bushing; 9: output window; 10: X-ray absorbers; 11: lead shielding; 12: NaI(Tl) scintillator; 13: neutral glass filters; 14: photomultiplier.*

Figure 2 shows the transmission of broadband X-rays (presumably 0–350 keV) through different absorbers. Radiation was collected from the whole volume of the laser chamber filled with 1.4 atm of argon. Copper filters seem to be the best for the X-ray range under investigation. Typically a set of copper foils with thicknesses of 0.05 mm, 0.1 mm and 0.5 mm was used, as well as copper plates 5 mm and 10 mm thick. The foils were used for precise measurements in the low energy wing of the X-ray distribution while the plates provided stronger attenuation and were applicable for the high-energy region. By combining different copper filters attenuation curves $u(x)$ were measured in a dynamic range $u_{max}/u_{min} \approx 1000$ with more than 50 data points and total absorber thickness x_{max} up to 50 mm (figure 2a). Aluminium of 0.2 mm thickness was also suitable for precise measurements, but to obtain the maximum attenuation the total thickness absorber needed was too large, over 100 mm (figure 2b). In contrast, 1 mm thick lead filters were too coarse for precise measurements and produced attenuation by three orders of magnitude with a total thickness of 7 mm (figure 2c).

Comparing the attenuation curves $u(x)$ for the three materials shows that they are rather different. For aluminium, the curve is approximately linear over the whole measurement range. For copper filters, the slope of the curve reduces significantly with increasing depth, and becomes approximately linear at greater than 10 mm. A larger slope of $u(x)$ corresponds to higher absorption for the low energy component of the X-ray spectrum. This behaviour is even more pronounced for the lead filters where the slope decreases gradually with increasing thickness. The differences originate from the different dependences of the X-ray absorption coefficients in the filter materials,[11] and so the use of a range of absorbers can give additional information for further X-ray spectral analysis.

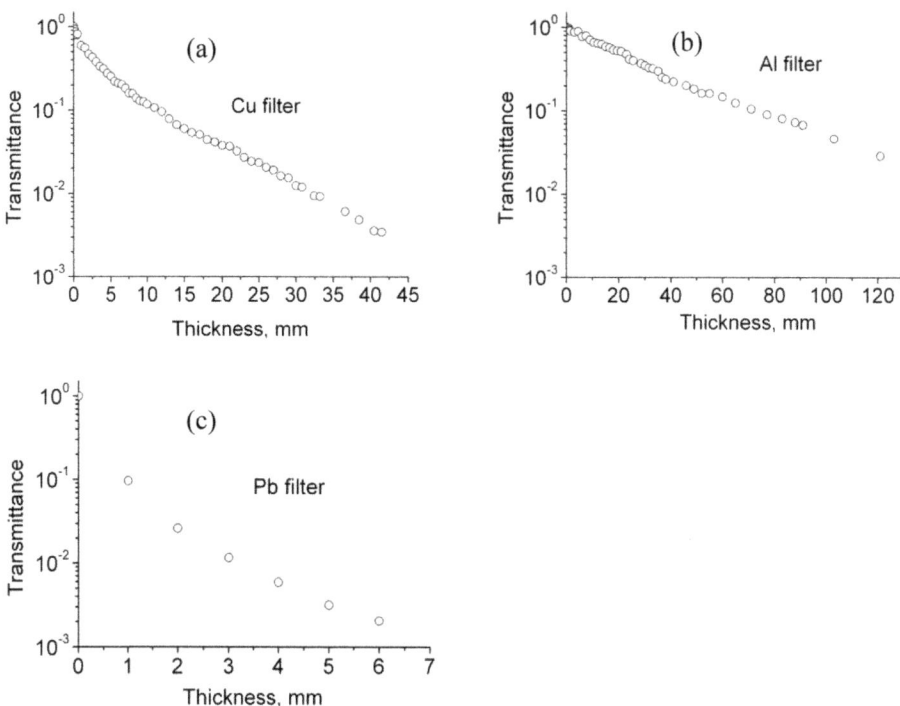

Figure 2   *Transmission of X-rays through (a) copper, (b) aluminium and (c) lead filters. The laser gas chamber was filled with argon at a pressure of 1.4 bar.*

X-rays are emitted because of electron deceleration as they pump the working gas mixture in a laser chamber and are transported from vacuum diodes through pressure foil windows arranged with hibachi support structures. Hence, the X-ray intensity and spectral content depend on the features of the laser module and the e-beam and gas parameters. The relative contributions of the gas and foils to the total X-ray output and radiation spectra were investigated in detail. The X-rays were measured 80 cm from the laser window as a function of argon pressure. Figure 3 shows the amplitudes of the integrated signals from the whole laser volume (without any apertures) and those for X-rays collimated along the axis of the chamber in a solid angle of 10^{-3} sr. The latter data are shown on an expanded scale in figure 4 and compared with measurements obtained when the collimation tube was pointed towards one of the foil windows.

It is clear that at low argon pressures more X-rays are emitted by the foil windows than by the gas. However, even at very low pressure small signals were detected along the axis, apparently caused by X-rays reflected from the rear CaF_2 window. The X-ray signal produced by the foil window reduces slowly with increasing pressure, while the signal connected with the gas increases significantly and for greater than about 0.5 atm exceeds the foil signal. At an argon pressure of 1.4 atm, which is close to optimum for the GARPUN amplifier, the gas contributes a factor of two more to the total X-ray yield than the foil windows. The integrated X-ray output increases by a factor of two when the argon pressure is varied from near vacuum to the maximum value (figure 3).

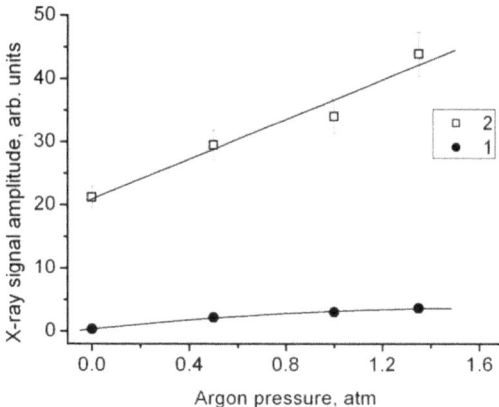

Figure 3 *Amplitude of the X-ray signal for radiation collimated from (1) the centre of the laser chamber and (2) integrated over the volume as functions of argon pressure.*

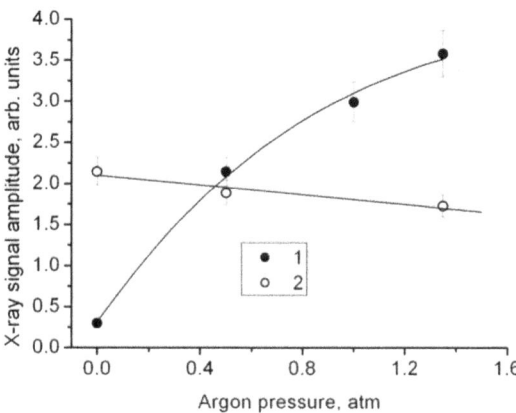

Figure 4 *Amplitude of the X-ray signal for radiation collimated from (1) the centre of the laser chamber volume and (2) from the foil window as functions of argon pressure.*

The evolution of the X-ray spectra with gas pressure has also been investigated for the gas and foil windows using absorption by copper filters. Both the attenuation curves are shown in figure 5 for an argon pressure of 1.4 atm, demonstrating the presence of a soft component in the gas emission by the rapid drop for small thicknesses. This may be explained by the fact that during e-beam deceleration initial fast electrons produce an avalanche of secondary electrons in the gas, which contribute to X-ray emission predominantly in the soft spectral range. As the most of the e-beam energy is deposited in the gas, only a small amount reached the opposite foil window to produce X-ray emission. Electrons incident on this foil window suffered losses mainly due to the massive hibachi, which was manufactured of stainless steel and, therefore, shielded the soft X-ray component. At argon pressures around 0.5 atm both the gas and foil X-ray emission contain similar soft components, since the electron energy is only partially deposited in the gas and the rest reaches the opposite foil window.

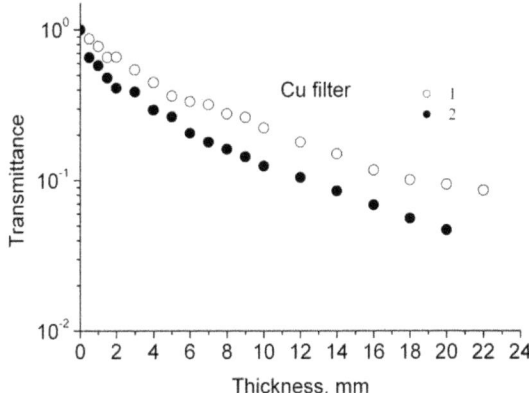

Figure 5   *Transmission of collimated X-rays from (1) the foil window and (2) the centre of laser chamber through copper filters. The argon pressure was 1.4 atm.*

2.2 Absolute Measurements of X-ray Emission and Predictions of Window Irradiation in Large-Scale KrF Lasers

In addition to the time resolved scintillation/photomultiplier technique, absolute measurements of the X-ray emission were carried out using solid state thermo-luminescence Al_2O_3 dosimeters (TLDs) calibrated using a standard ^{137}Cs radioactive source, emitting mono-energetic 0.662 MeV radiation. The TLDs were 1 mm thick disks of diameter 5 mm, and 25 of them were stacked inside an aluminium matrix. Accumulated X-ray doses were measured inside the laser chamber close to the window and outside at 80 cm distance behind the window. The distributions of absorbed X-ray dose in the TLD stacks are shown in figure 6.

It can be seen that all the distributions can be approximated by the same slopes with the exception of a few front stack dosimeters which, since they were inside the laser chamber, detect the soft X-ray component. Integrating the dose distributions over the dosimeter stack allows calculation of the corresponding energy flux F_{X-ray}. Inside the laser chamber, the energy flux incident on the window was $F_{X-ray} \approx 10^{-3}$ J/cm^2. At 80 cm behind the window, the flux was significantly attenuated by window absorption and geometric factor. For an argon pressure of 1.4 atm the flux per shot was $F_{X-ray} = 1.6 \times 10^{-5}$ J/cm^2 and, for the evacuated laser chamber, $F_{X-ray} = 0.7 \times 10^{-5}$ J/cm^2.

Based on the Bremsstrahlung yield, equations (1) and (2), a scaling law can be derived for the X-ray intensity I (W/cm^2) at a distance R (cm) from the centre of the laser chamber of volume V (cm^3) pumped by a specific e-beam power W (W/cm^3),

$$I = \frac{kCZ\varepsilon_e WV}{4\pi R^2}.\tag{3}$$

In a correctly designed laser, the electrons predominantly lose their energy in the working gas inside the laser chamber and the contributions of foil windows and scattered electrons are small. In the experiments described here, with W=0.8 MW/cm^3 in argon (Z=18) and a mean electron energy ε_e=0.3 MeV, the Bremsstrahlung conversion coefficient was found to be C=2.1×10^{-4}, assuming an isotropic X-ray distribution. This is slightly less than for electron interactions with solids measured previously.[8,9] The coefficient k is the ratio of intensities for volumetric and point isotropic sources and can be determined by integration over the volume.

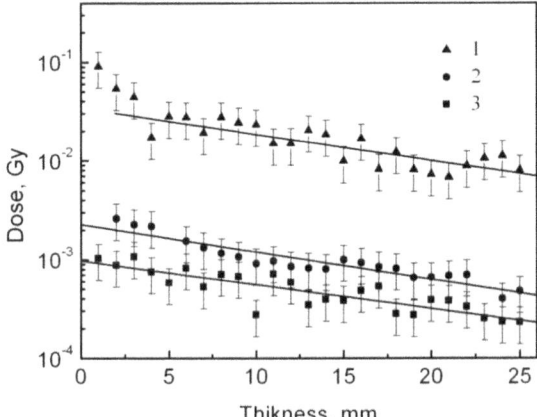

Figure 6 *Distribution of absorbed X-ray dose in TLD stacks (1) inside the laser chamber near the window and (2, 3) 80 cm distance behind the window. For (1, 2) the laser chamber was filled by 1.4atm argon at pressure, while for (3) it was evacuated.*

For Inertial Fusion Energy application[12] a projected large-scale KrF laser driver with ≈100 kJ of energy, an active volume V≈2×10^6 cm^3, pumping power W=0.4 MW/cm^3, electron energy ε_e≈ 0.6 MeV and pulse duration ≈600 ns, the X-ray flux on laser windows being estimated on the base of formula (3) could be as high as F_{X-ray}~10^{-2} J/cm^2. The corresponding accumulated absorbed dose in optics over N = 10^8 shots (i.e. for a 1-year operation cycle) D=$N(\mu_a/\rho)F_{X-ray}$~30 MGy, where the ratio of mass absorption coefficient to density μ_a/ρ ~0.03 cm^2/g for various optical materials and X-ray quanta energy ε =10^2–10^3 keV. According to our trial experiments[1,2] this would reduce significantly the transmittance of laser windows.

3 RECONSTRUCTING X-RAY SPECTRA FROM TRANSMISSION DATA

3.1 Regularization Algorithm

Bremsstrahlung spectra are characterized by the photon number density distribution $z(\varepsilon)$, where ε is the photon energy. By transmitting the radiation through a set of absorbing filters of different thickness x, i.e., of different absorption, and measuring the intensity behind the filters allows attenuation curves, $u(x)$, to be obtained. A functional relation between $z(\varepsilon)$ and $u(x)$ is determined by an integral equation, assuming that measurement arrangement is linear,

$$Az \equiv \int_{\varepsilon_0}^{\varepsilon_{max}} K(x,\varepsilon)z(\varepsilon)\mathrm{d}\varepsilon = u(x), \tag{4}$$

where A is a linear matrix operating on the values of z, a core of integral equation, $K(x,\varepsilon)$, represents a signal induced in a detector by a monochromatic source of unit intensity and energy ε after transmission through an absorber layer of thickness x, and ε_0 and ε_{max} are the boundaries of the radiation spectrum. The problem of determining the radiative spectral composition from a measured attenuation curve $u(x)$ is reduced to solving equation (4) with respect to $z(\varepsilon)$.

Consider a function $z_0(\varepsilon)$ to be an exact solution for the attenuation curve $u_0(x)$. As $u(x)$ is obtained experimentally and thus is an approximation to $u_0(x)$, its use in solving equation (4) would not give $z_0(\varepsilon)$. In addition, a solution $z=A^{-1}u$ (here A^{-1} is the inverse operator) is not stable with respect to small variations in values of $u(x)$, as is typical for inverse problems. If there are a lot of solutions, or no solution at all, for example for under- or over-defined systems of linear equations, respectively, but when a class U of $u(x)$ possibly exists, an approximate solution known as a generalized, quasi- or pseudo-solution may be found. A quasi-solution of equation (4) for a set F is the element $z^* \in F$ at which the distance $\rho_U(Az,u)$ attains its exact lower boundary, i.e.,

$$\rho_U\left(Az^*,u\right) = \inf_{z \in F} \rho_U\left(Az,u\right) \tag{5}$$

– if at $u=u_0$ equation (4) has a solution $z^0 \in F$ it would then coincide with the quasi-solution of the equation $Az=u^0$.

As early as in the 19th century, Gauss and Legendre proposed the least squares method for solving a system of linear algebraic equations (SLAE) which is over defined, generally speaking. Instead of solving the equations $Az=u$ the minimum of a function $\Phi(z)= \|Az-u\|^2$ is found (hereinafter $\|..\|$ designates a norm of function in the linear algebraic space). Such a problem is always solvable. If $\min[\Phi(z)]=\mu=0$, then any value of z at which the minimum is achievable is a solution of $Az=u$. If $\mu>0$, the minimum $\Phi(z)$ corresponds to the quasi-solution. If μ is only slightly greater than 0, then this quasi-solution may be called normal and the problem of finding it can be formulated as

$$\text{find } \min\left[\|z\|\right] \text{ such that } \left[\|Az-u\|^2\right] = \mu = \min\left[\|Az-u\|^2\right]. \tag{6}$$

To solve this problem a Moore-Penrose pseudo-inversion method may be used. A technique of singular value deposition of matrix A gives a so-called pseudo-inverse matrix A^+ which acting on u gives a normal pseudo-solution of the SLAE. If the SLAE has a single solution it is then a normal pseudo-solution, and the pseudo-inverse A^+ is identical to the inverse matrix A^{-1}. This approach is equivalent to

$$\text{find } \min\left[M^\alpha(z)\right] = \min\left[\|Az-u\|^2 + \alpha\|z\|^2\right] \tag{7}$$

where α is a regularizing parameter, introduced to prevent instabilities. Equation (7) has a single solution for any $\alpha > 0$, and it is stable in respect to errors in u. If $\alpha \to 0$ a normal quasi-solution is obtained.

Unfortunately, the pseudo-inversion method is unstable in respect to errors in A matrix if its elements are measured experimentally or calculated with uncertainties. Thus, the problem of finding a normal pseudo-solution, in our case proves to be incorrectly formulated. To illustrate this a model spectrum in the form $z_0(\varepsilon) = 1 - \cos\left[\omega(\varepsilon - \varepsilon_0)\right]$ with $\omega = 4\pi(\varepsilon - \varepsilon_0)/(\varepsilon_{max} - \varepsilon_0)$ and a core of the integral equation $K(x,\varepsilon) \equiv \exp(-\gamma x / \varepsilon)$ with γ=0.5, ε_0=0, ε_{max}=14 were used in test calculations. They were substituted in the equation (4) to compute the exact "attenuation curve" $u_0(x)$. Figure 7 shows solutions obtained using equation (7) as $\alpha \to 0$. The exact values of the operator A and $u_0(x)$ were used as the initial data in calculations. When $\alpha \leq 10^{-14}$ the solution starts to oscillate, becomes unstable as α decreases, and when $\alpha \leq 10^{-15}$ matrix of SLAE approximating the equation (7) becomes degenerate. Note that arithmetic computations were the only source of errors in this case.

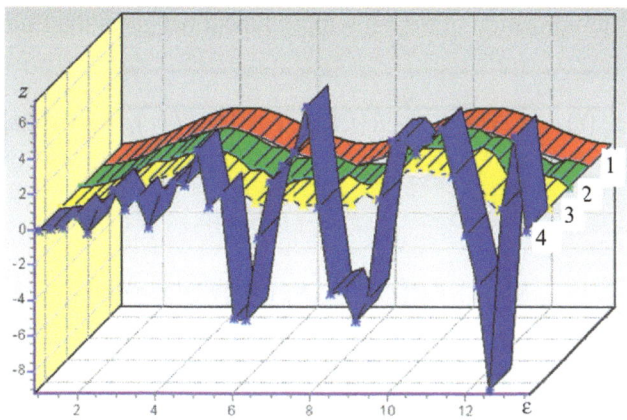

Figure 7 *Exact (1, red) and reconstructed solutions for α, the regularizing parameter, $\approx 10^{-14}$ (2, green), $\approx 4 \times 10^{-15}$ (3, yellow) and $\approx 10^{-15}$ (4, blue).*

An up to date mathematical theory for solving incorrectly formulated problems was developed following the work of Lavrentiev[13] and Tikhonov & Arsenin.[14] Stable methods for solving such problems are based on the idea of under-definiteness. Additional information is needed in order to state criteria for the selection of approximate solutions. This allows the development of stable solution methods called regularizing algorithms. Assuming that for equation (4) $Az = u$, $z \in F$ and $u \in U$, where F are U are Hilbert spaces, and A is a linear continuous operator from F to U, instead of the exact A and u, there are approximations A_h and u_δ such that $\|A_h - A\| \leq \zeta$ and $\|u_\delta - u\|_U \leq \delta$, i.e. the equation $A_h z = u_\delta$ can be solved, with $z \in F$, $u_\delta \in U$. The error of operator ζ is determined both by errors in defining the core of the integral equation, $K(x,\varepsilon)$, and in replacing an integral operator by a discrete operator for a numerical solution of equation (4). It is assumed that $K(x,\varepsilon)$ is a real continuous function in the range $\{c \leq x \leq d; \varepsilon_0 \leq \varepsilon \leq \varepsilon_{max}\}$, $U = L_2$ where L_2 is the normalized integrated space with squared functions, and $F = W_2^1$ where W_2^1 is the normalized space of functions having a square integrated derivative.

The Tikhonov smoothing functional is

$$M^{\alpha}\left[z,u_{\delta}\right] = \left\|A_{h}z - u_{\delta}\right\|_{L_2}^2 + \alpha\Omega\left[z\right], \tag{8a}$$

where the stabilizing functional

$$\Omega\left[z\right] \equiv \left\|z\right\|_{W_2^1}^2 = \int\limits_{\varepsilon_0}^{\varepsilon_{max}}\left\{z^2\left(\varepsilon\right) + q\left[z'\left(\varepsilon\right)\right]^2\right\}d\varepsilon, \tag{8b}$$

q being an adjustable parameter in the range (0, 1). The discrepancy function is

$$\left\|A_{h}z - u_{\delta}\right\|_{L_2}^2 = \int\limits_{c}^{d}\left[A_{h}z - u_{\delta}\left(x\right)\right]^2 dx \text{ with } A_{h}z \equiv \int\limits_{\varepsilon_0}^{\varepsilon_{max}} K_{h}\left(x,\varepsilon\right)z\left(\varepsilon\right)d\varepsilon. \tag{8c}$$

A condition of the minimization of the smoothing functional gives the Euler equation

$$\alpha\left[z_{\alpha}\left(t\right) - qz_{\alpha}''\left(t\right)\right] + \int\limits_{\varepsilon_0}^{\varepsilon_{max}} R\left(t,\varepsilon\right)z_{\alpha}\left(\varepsilon\right)d\varepsilon = f\left(t\right), \quad \varepsilon_0 \leq t \leq \varepsilon_{max}, \tag{9}$$

where

$$R\left(t,\varepsilon\right) = R\left(\varepsilon,t\right) = \int\limits_{c}^{d}K_{h}\left(x,t\right)K_{h}\left(x,\varepsilon\right)dx, \tag{10}$$

$$f\left(t\right) = \int\limits_{c}^{d}K_{h}\left(x,t\right)u_{\delta}\left(x\right)dx \tag{11}$$

and the boundary conditions were chosen to be of the form $z_{\alpha}\left(\varepsilon_0\right) = z_{\alpha}\left(\varepsilon_{max}\right) = 0$. Here z_{α} is a smooth solution at a given regularization parameter α.

As a result, instead of an incorrect Fredholm integral equation of the first kind (4), an integro-differential equation of the second kind (9) must be solved. Because $u_{\delta}(x)$ in equation (9) is not in an explicit form, but is integrated, the effects of fluctuations of $u_{\delta}(x)$ due to measurement errors are considerably smoothed out.

Solutions can be found by minimization of equation (8) itself using gradient methods. However, as shown in the following these methods are inferior, in speed and accuracy, to those in which equation (9) is reduced to SLAE by discretization. In that case the matrix of the system is symmetrical and is positive definite, and to solve such a SLAE efficient methods, such as the Crout (square root) method, can be used.

The discretization of equation (9) can be performed by the finite sums and differences methods. Three variants of squaring formulae were used, namely the formulae of rectangles and trapeziums, and integration by means of cubic splines. In the calculations a non-monotonic model spectrum $z_0\left(\varepsilon\right) = 1 - \cos\left[\omega\left(\varepsilon - \varepsilon_0\right)\right]$ the core of the integral equation (4) $K\left(x,\varepsilon\right) \equiv \exp\left(-\gamma x/\varepsilon\right)$ were used with ω, γ defined above. At discretization uniform grid for energy ε (41 points) and non-uniform grid for filter thickness x (81 points) were used. The given model spectrum allowed a calculation of an exact "attenuation curve" used as $u_0(x)$ for the right hand side of equation (4). Figures 8, 9 and 10 demonstrate the best approaches to an exact solution obtained in each case. The corresponding regularizing parameter has optimum values $\alpha_{opt} \sim 10^{-13}$ for all three approximations, but visually cubic splines provide the best accuracy (figure 10).

For comparison, figure 11 shows the best approach to the exact solution obtained by direct minimization of equation (8) using the conjugate gradients method. The minimization occurred within a number of steps equal to the number of energy points at the calculated spectrum. The accuracy of the solution is worse than for any of the results shown in figures 8-10. Increasing of an iteration number by order of magnitude improves the precision (figure 12), but the solution is still worse than for the other methods. For a uniform grid for x, the obtained solution was much closer to the exact one although still worse than shown in figures 8-10.

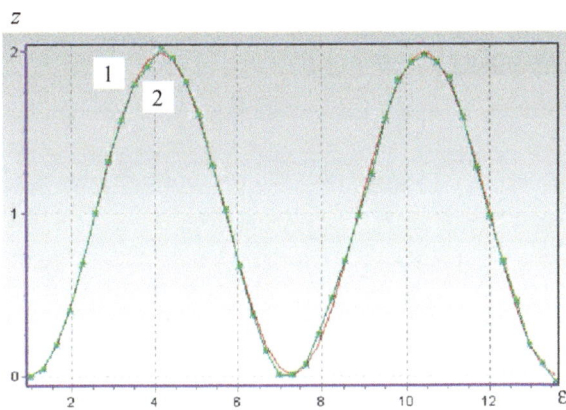

Figure 8 *Exact (1, red) and reconstructed (2, green) solutions obtained using rectangular quadrature for $\alpha \approx 1.5 \times 10^{-13}$.*

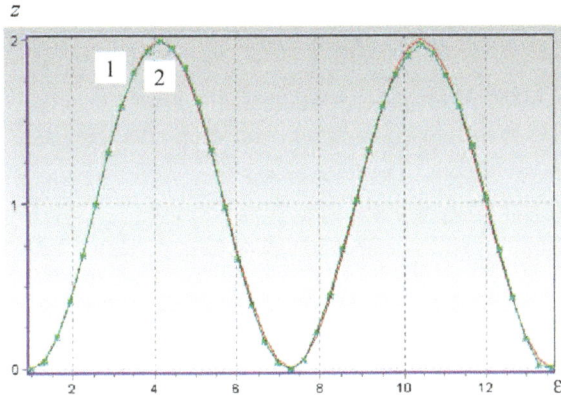

Figure 9 *Exact (1, red) and reconstructed (2, green) solutions obtained using trapezoid quadrature for $\alpha \approx 1.9 \times 10^{-13}$.*

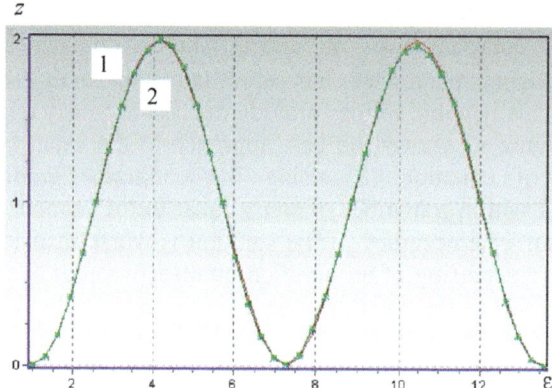

Figure 10 *Exact (1, red) and reconstructed (2, green) solutions obtained using cubic-spline quadrature for $\alpha \approx 1.5 \times 10^{-13}$.*

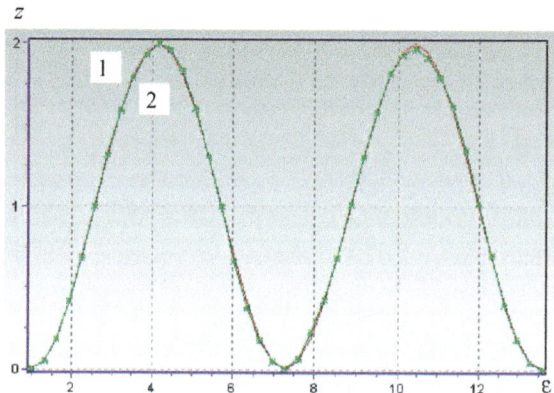

Figure 11 *Exact (1, red) and reconstructed (2, green) solutions obtained using the conjugate gradient method for $\alpha \approx 10^{-15}$.*

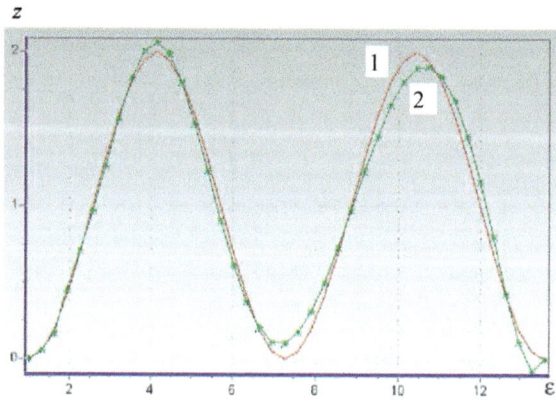

Figure 12 *Exact (1, red) and reconstructed (2, green) solutions obtained using the conjugate gradient method with $10 \times$ more iterations for $\alpha \approx 10^{-15}$.*

It is important to choose correctly the regularizing parameter α, because as shown in figure 7, the problem of minimizing equation (8) becomes unstable when $\alpha \rightarrow 0$. In our test calculations the minimum value of α, down to which rounding errors are insignificant, is about 10^{-14}. There are several criteria for choosing the regularizing parameter. One of the most effective is the method of discrepancy proposed by Phillips,[15] where α is in agreement with the errors of defining A and u. At the extremes of equation (8) a value of α, α_d, for which $\left\| A_h z_\alpha - u_d \right\|_{L_2}^2 = \delta$ should be chosen. The idea of this principle is to choose α_d so that the discrepancy is at the level of sums of all the errors of the method – in defining the operator ζ, of the right-hand part of equation (4) δ, and those arising due to incompatibility of the initial equation μ. At the functional extremes a so-called function of generalized discrepancy $\chi(\alpha)$, which depends on these errors, is formed,

$$\chi(\alpha) = \beta(\alpha) - \mu - \left[\delta + \zeta \sqrt{\gamma(\alpha)} \right]^2 \quad \text{with } \chi(\alpha_d) = 0, \tag{12}$$

where $\beta(\alpha) = \left\| A_h z_\alpha - u_\delta \right\|_{L_2}^2$ is the discrepancy and $\gamma(\alpha) = \Omega(z_\alpha) = \left\| z_\alpha \right\|_{L_2}^2$ is the stabilizing functional meaning at the extremes. This method of choosing α provides $\alpha_d > \alpha_{opt}$ and, as a consequence, an over smoothed solution. In addition, values of δ and ζ are known approximately, and α_d is very sensitive to errors in these parameters.

During the test calculations, "measurement" errors on the attenuation curve were generated using pseudo-random numbers via the formula

$$u(x) = u_0(x)(1 + eps \times \xi), \tag{13}$$

where the values of ξ are random numbers uniformly distributed over $[-1,1]$ and eps is the relative accuracy of the "measurements".

The test calculations, solving equation (9) using SLAE with perturbed $u(x)$, showed that even though α_{opt} is known, since the exact solution of the problem is known, the result obtained may be accepted as satisfactory on provisional grounds. This is illustrated in figure 13 where the functional extremes, equation (8), are represented for different values of the regularization parameter α.

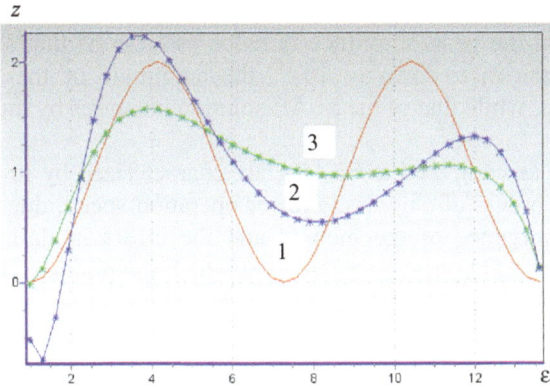

Figure 13 *Exact (1, red) and reconstructed solutions obtained for the perturbed attenuation curves (eps=0.1) for $\alpha=1.2\times10^{-2}$ (2, blue) and 5.6×10^{-2} (3, green).*

The relative error of defining the attenuation curve was *eps*=0.1 and α_{opt} was about 1.2×10^{-2} (blue curve). That curve has the minimum root mean square deviation from the exact solution, but it has a negative peak of no physical sense. To prevent this, a lower limit should be placed on α, leading to a value of 5.6×10^{-2} (green curve). However, this leads to worse agreement with the exact solution. To attempt to overcome this algorithm using the method of conjugate gradient projection (MCGP) has been developed.[16] This differs from the conjugate gradient method,[17] as the functional minimum used is sought within a limited class of possible solutions. Having no additional *a-priori* information about a solution this class is considered to be that of non-negative functions. Figure 14 illustrates the result of this algorithm. It is seen that the solution (2, yellow curve) is closer to the exact solution than that of the conjugate gradient method (3, green curve) with the same value of α (1.5×10^{-2}) even if the negative values are set to zero.

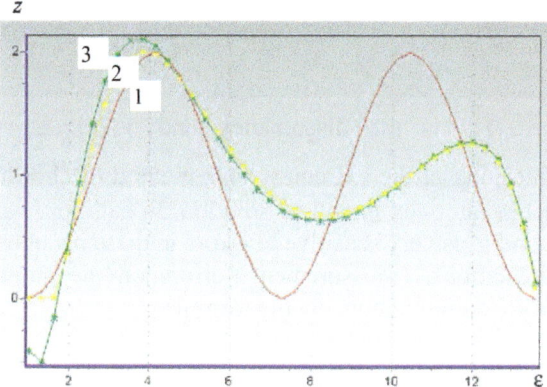

Figure 14 *Exact (1, red) and reconstructed (2, yellow) solutions obtained by MCGP for the perturbed attenuation curve (eps=0.1), compared to that of the conjugate gradient method (3, green) for the same α (1.5×10⁻²).*

As α is decreased the behaviours of solutions obtained by solving SLAE and those obtained using MCGP are essentially different. Figure 15 shows the behaviour of solutions obtained by MCGP as α is decreased from 10^{-2} to 10^{-5}. It is interesting to note that the positions of peaks are close to the maxima of the exact solution. Similar plots obtained from SLAE are depicted in figure 16; the solutions have almost nothing in common with the exact solution, and the peak amplitude is twice as large as that shown in figure 15. With a further decrease in α down to 10^{-14}, the amplitude of the MCGP solution is approximately doubled, while that of the SLAE solution increases by more than four orders of magnitude.

However, MCGPs, as compared to SLAEs, are characterized by lower accuracy in the solutions and about 2 orders of magnitude lower operation speed, due to a comparatively slow (quadratic) convergence of the method and the errors in defining a direction of decline. There is an optimal value ($\alpha \approx 5\times10^{-10}$) at which the gradient is equal to zero, i.e., the minimum is achieved. At lower values the number of iterations increases, and for more than about 1000 iterations rounding errors become evident. At larger values, the accuracy of finding the minimum decreases. As a result, the peaks and gaps in solutions obtained with accurate initial data by this method appear at values of α about orders of magnitude hiher than in SLAE.

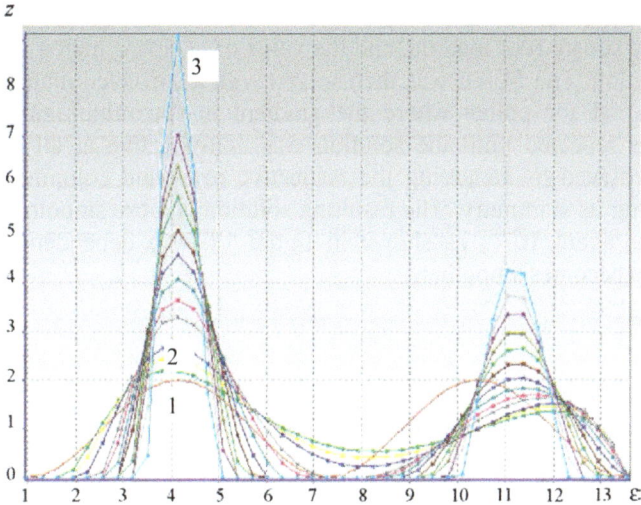

Figure 15 *Exact (1, red) and reconstructed solutions obtained by the MCGP for perturbed attenuation curve (eps=0.1)   with α from 10^{-2} (2, green) to 10^{-5} (3, blue).*

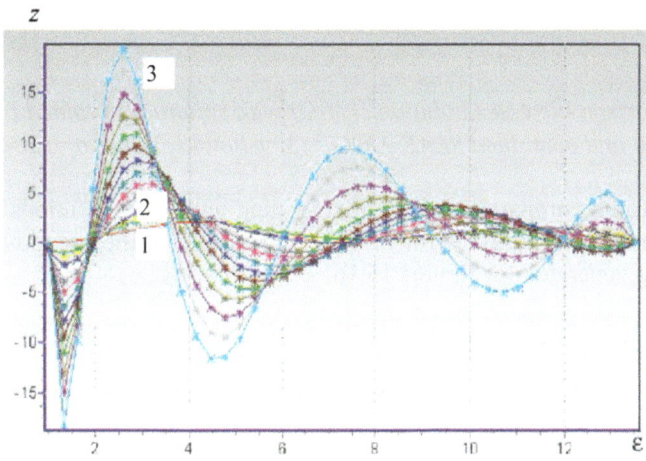

Figure 16 *Exact (1, red) and reconstructed solutions obtained by the SLAE for perturbed attenuation curve (eps=0.1) with α from 10^{-2}(2, green) to 10^{-5} (3, blue).*

It is desirable to combine the advantages of the methods. A simple way to increase the operation speed is to use a predictor (SLAE)-corrector (MCGP) scheme, that is, to use the solution obtained by SLAE, for each α, as an initial approximation for the MCGP. Those solutions, which are positive, should not be corrected by MCGP because the gradient of the minimizing functional will be equal to zero, with high accuracy. At smaller α when the solution has negative parts, the operation speed does not change significantly, it remains at the MCGP level and the accuracy is of intermediate value.

This proved to be successful in obtaining non-negative solutions of a corresponding accuracy with a 2-3× speed decrease compared to MCGP. The diagonal elements of a

SLAE matrix were multiplied by a weight function $g(\varepsilon)$ with, initially, $g(\varepsilon) \equiv 1$. The SLAE was solved by the square root method, and the value of $g(\varepsilon)$ was increased at any negative points of the solution. The SLAE was then solved with a corrected matrix. The procedure was fulfilled at first for points where the gradient of the minimizing functional was positive, and was repeated until the solution was non-negative at all points. The same results can be achieved by removing the respective rows and columns from the SLAE matrix, maintaining its symmetry. The resulting solution is more smooth and non-negative for α as small as about 10^{-15}, as shown in figure 17. The dependence of discrepancy function on α also becomes smoother.

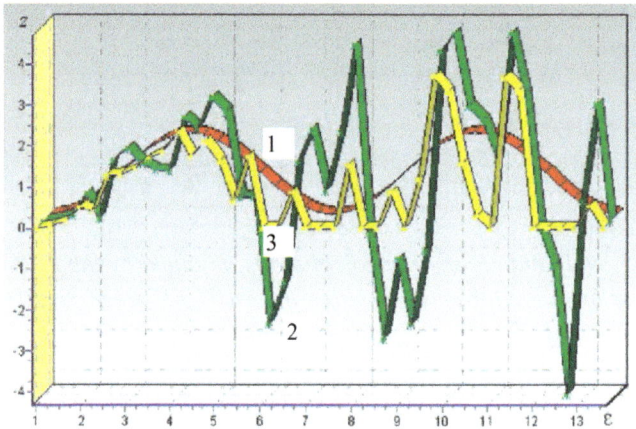

Figure 17 *Comparison the exact solution (1, red) with solutions obtained by the MCGP (2, green) and combined SLAE-MCGP (3, yellow) scheme; α=1.3×10^{-15}.*

Figure 18 shows a comparison of the regularizing and exact solutions for a perturbed attenuation curve, with a relative error of 10% (*eps*=0.1), and the pseudo-random number sequence as being generated for figures 15,16.

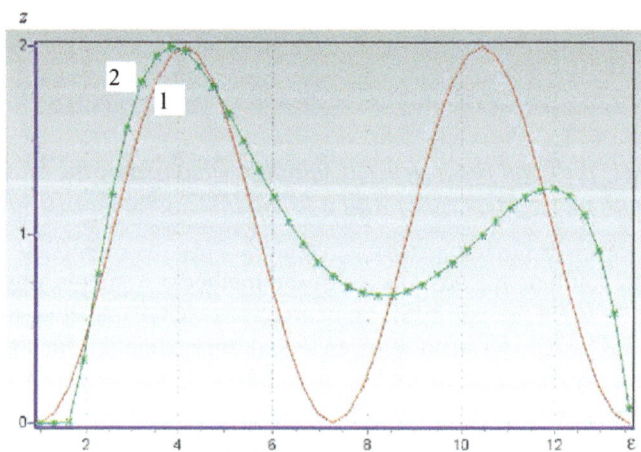

Figure 18 *Comparison of exact (1, red) and regularizing solutions (2, green) for a perturbed attenuation curve (eps=0.1).*

3.2 Calculation of Build-up Factor for X-ray Transfer through Thick Media

The X-ray intensity, or photon flux, incident on detector depends on the radiation source characteristics as well as on the parameters and geometry of components between the source and detector. In the ideal narrow parallel beam geometry photons scattered in the components do not reach the detector, which measures the flux exponentially attenuated by both scattering and absorption, $N(r) = N_0 \exp(-\kappa r)$, where κ is the linear absorption coefficient and r is the distance between the source and detector.

However, the geometry of the experiments to measure the X-ray attenuation curves was very different to the ideal narrow beam. Therefore, the experimental dependencies cannot be used directly to reproduce the energy spectrum by assuming that the attenuation is described by an exponential function. The X-ray detector measured non-scattered and non-absorbed photons together and those which were multiply scattered. This effect of additional photons typically is accounted for by introducing a build-up factor B into the narrow beam attenuation

$$B = 1 + \frac{\varphi_{\text{scat}}}{\varphi_{\text{nonscat}}}, \quad B \geq 1 \qquad (14)$$

where the φ are the relevant fluxes. For a fixed geometry and monochromatic source the build-up factor depends on the initial photon energy and the product κr.

Several papers have been concerned with calculating of build-up factors for X-rays in various materials. They considered, however, the materials and source geometries typical for radiation protection calculations,[11] and these are not relevant to the present discussion. Hence, a code based on the Monte-Carlo method has been developed which allows the simulation of scattering and absorption of X-rays in materials, and it is easy to define the geometry of the absorbing medium and radiation source and their relative position, etc.

Monte Carlo simulations of particle transport mimic physical reality: particles are emitted according to the source distribution, they travel certain distances determined by a probability distribution, which depends on the total interaction cross section, to a collision point. On scattering, they change energy and/or direction according to the differential cross section of the process. They can also generate new particles, which also must be followed. This procedure is continued until all particles are absorbed or leave the volume under consideration. Quantities of interest can be calculated by averaging over a large number of individual particle histories. The resulting quantities are subjected to a statistical uncertainty, which depends on the number N of simulated particle histories, and usually decreases as $N^{-1/2}$. For better statistical accuracy, it may be necessary to carry out many simulations, leading to long calculation times.

Consider the passage of photon in a given material. The considerations are limited to homogeneous random scattering media where the molecules are distributed uniformly. In each interaction with atoms, the photon may lose energy and/or change direction. Each photon starts from the radiation source position with initial direction and energy depending on the characteristics of the source. The distance s between interactions is distributed according to the probability distribution function,

$$p(s) = \frac{\exp(-s/\Lambda)}{\Lambda}, \qquad (15)$$

where Λ is the mean distance between interactions given by $\Lambda = 1/\Sigma_{tot}$, $\Sigma_{tot} = N\sigma_{tot}$ is the so-called macroscopic cross section, N is the number density of atoms and σ_{tot} is the total cross section. The length s to the next interaction is generated by using the sampling formula $s = -\ln(\xi)/\Sigma_{tot}$, where ξ is a uniformly distributed random number between 0 and 1. If the point of the photon interaction is inside the volume under consideration, then the interaction type i is sampled according to the corresponding probability $p_i = \Sigma_i/\Sigma_{tot}$. The energy loss and polar scattering angle are sampled from the differential cross sections of the various processes. The azimuthal scattering angle is generated with a uniform distribution in the interval $(0, 2\pi)$. After sampling the energy loss and scattering angles the energy of the photon is reduced appropriately, and the direction of motion after the interaction is calculated. This is repeated either until the photon leaves the system volume, is absorbed or when the energy becomes lower than a given minimum (stopping) energy.

For the simulation of interactions of X-rays with atoms of the materials elastic coherent (Rayleigh) and inelastic incoherent (Compton) scattering on bound electrons were considered, along with photoelectric absorption. In these collisions, photon energy is transferred to electrons. At lower energies, photoelectric absorption is dominant while at higher energies incoherent scattering is the most important process. The total cross sections of these processes were taken from published data.[18,19] Photon absorption due to electron-positron pair production was not taken into account as the threshold is more than 1 MeV and the maximum energy of quantum in our case was about 0.1 MeV.

For simulating X-ray scattering it is necessary to determine a differential scattering cross-section. In Rayleigh scattering photons are scattered by bound atomic electrons without atom excitation, i.e., without photon energy loss. The differential cross section for Rayleigh scattering has the form[19]

$$\frac{\partial \sigma_{coh}}{\partial \Omega} = \frac{r_0^2}{2}\left(1 + \cos^2\theta\right)\left[F(q,Z)\right]^2, \tag{16}$$

where θ is the photon scattering angle, q is the momentum transferred by a photon to all atomic electrons without energy absorption, and $F(q,Z)$ is the form factor.[20]

In Compton scattering a photon interacts with an atomic electron, causing a secondary (Compton) photon of lower energy to be emitted. The differential cross-section of Compton scattering on bound atomic electrons has the form[21]

$$\frac{\partial \sigma_{inc}}{\partial \Omega} = ZS(q,Z)\frac{\partial \sigma_{KNT}}{\partial \Omega}, \tag{17}$$

where $S(q,Z)$ is the incoherent scattering function, q is the momentum transferred by a photon to an atomic electron, and $\partial\sigma_{KNT}/\partial\Omega$ is the Compton scattering cross section from a free electron at rest, the Klein-Nishina-Tamm cross section[22]

$$\frac{\partial \sigma_{KNT}}{\partial \Omega} = \frac{r_0^2}{2}\left[1 + k(1-\cos\theta)\right]^{-2}\left[1 + \cos^2\theta\frac{k^2(1-\cos\theta)^2}{1+k(1-\cos\theta)}\right], \tag{18}$$

Here k is the photon energy in the units of electron rest mass energy m_ec^2 and r_0 is the classical electron radius.

We assume that the single atom theory can be extended to molecules, i.e., that the molecular cross section for a process may approximated by the sum of the atomic cross sections of all atoms in the molecule.

To test the Monte Carlo code a comparison was made of calculated results with the published data on X-ray transmission. Figure 19 gives a comparison of the calculated transmission, for a normal beam incidence on an aluminium target, with the data[23] for concrete (Z=13.4) at 50–200 keV; the agreement is satisfactory but slightly worse at 100 keV possibly due to different cross sections being used. Figure 20 shows a comparison with published data[24] of the build-up factors for a plane mono-directed 1 MeV source in lead barrier geometry over a range of thicknesses.

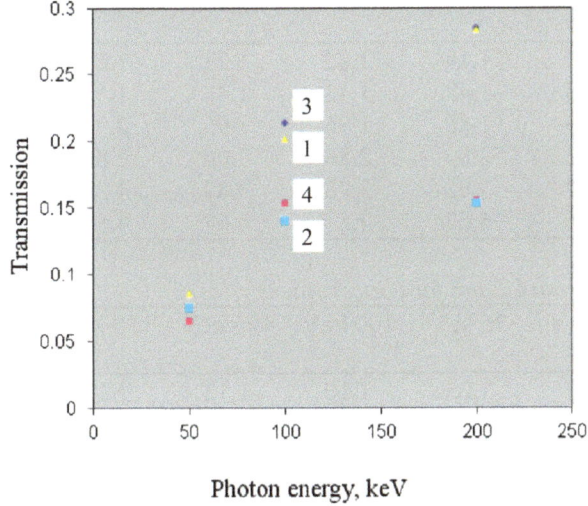

Figure 19 *Comparison of calculated photon (1, yellow) and energy (2, light blue) transmissions for aluminium with experimental data[23] (photon: 3, dark blue, energy: 4, purple).*

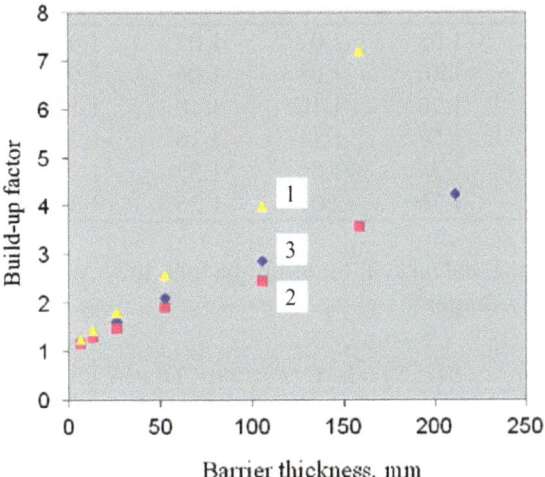

Figure 20 *Comparison of calculated photon (1, yellow) and energy (2, purple) build-up factors for lead with experimental data[24] for energy (3, blue).*

Disagreements in the results would be reduced by 25% if, in the calculations, the atomic electrons were treated as being free, i.e., the non-coherent scattering function $S(q,z) \equiv 1$.

A similar situation was observed in calculations of the build-up factors for aluminium compared to published data.[25]

The results of the build-up factor calculations in the experimental conditions at X-ray energies of 50–250 keV and relative thicknesses κd of the aluminium, copper and lead absorbers in the range 0.5 to 8 mm, are given in tables 1–3.

Table 1 *Energy build-up factors for aluminium filters.*

Relative thickness	50 keV	75 keV	100 keV	125 keV	150 keV	200 keV	250 keV
0.5	1.13	1.18	1.22	1.23	1.23	1.26	1.24
1	1.28	1.42	1.43	1.48	1.51	1.48	1.49
2	1.55	1.88	1.96	1.95	2.00	1.96	1.94
4	2.12	2.84	2.88	3.06	2.87	2.70	2.65
6	3.22	3.62	3.76	3.87	4.23	3.35	3.63
8	4.15	5.02	4.57	4.64	4.58	3.82	4.01

Table 2 *Energy build-up factors for copper filters.*

Relative thickness	50 keV	75 keV	100 keV	125 keV	150 keV	200 keV	250 keV
0.5	1.03	1.05	1.10	1.12	1.16	1.18	1.21
1	1.08	1.12	1.19	1.25	1.30	1.38	1.41
2	1.11	1.24	1.37	1.50	1.59	1.77	1.88
4	1.15	1.38	1.86	2.03	2.21	2.72	2.91
6	1.49	1.69	2.03	2.70	3.45	3.86	4.40
8	1.00	2.28	3.04	3.28	3.46	3.30	4.44

Table 2 *Energy build-up factors for lead filters.*

Relative thickness	50 keV	75 keV	100 keV	125 keV	150 keV	200 keV	250 keV
0.5	1.06	1.05	1.02	1.02	1.04	1.05	1.05
1	1.11	1.10	1.03	1.04	1.05	1.08	1.09
2	1.24	1.20	1.10	1.10	1.12	1.13	1.24
4	1.53	1.57	1.11	1.16	1.31	1.22	1.36
6	1.38	1.56	1.09	1.53	1.21	1.36	1.60
8	1.88	1.76	2.01	1.25	1.51	2.25	3.02

To further utilize the calculated build-up factors they were approximated by the dependence proposed by Berger[24]

$$B(\varepsilon, S) = 1 + C(\varepsilon) S \exp\left[D(\varepsilon) S\right], \qquad (19)$$

where $S = \kappa d$ is the absorber thickness in terms of the mean free path and ε is the photon energy. The coefficients C and D were found by the least squares method, with weights inversely proportional to the calculation errors. For convenience, the C and D coefficients were approximated by dependences of the form $a+1/(b+c\times\varepsilon)$. The total errors of the build-up factor approximation were about 2% within the range of energy ($\varepsilon > 50$ keV) and thickness ($S < 6$) of interest.

To specify and verify a criterion for finding the regularizing parameter a set of calculations simulated a model bell-shaped spectrum $z(\varepsilon) \sim \exp\left[-k(\varepsilon - \varepsilon_0)^2\right]$ with a core of the integral equation (4) taking into account the calculated build-up factors for aluminium absorbers and using cross sections.[18] An exact attenuation curve was calculated using the model spectrum, and based on this an "experimental" curve was derived using random perturbations of points on the exact curve to provide relative errors of 5%. Figure 21 shows plots of the experimental and regularized, or recovered, attenuation curves. A comparison of the recovered spectrum with the model spectrum is shown in figure 22 for different values of the regularizing parameter α. The green curve corresponds to α obtained by the generalized discrepancy method, the yellow curve to a mean weighted value obtained from the analysis of discrepancy and solution, and the blue curve, to the optimal value $\alpha_{opt} \approx 2.5 \times 10^{-8}$. Note that figures 21 and 22 present an example of a quite successful application of the method. In other examples, recovered spectra show peaks at the edges of ranges where perturbations may be stronger. In addition, the quality of a recovered model spectrum, at fixed error, depends on the disposition of perturbations on a plotted curve with simulation of measurement errors using random numbers. This makes the choice of α something ambiguous.

In the present case, however, three independent sets of measurements of attenuation curves (for aluminium, copper, and lead) are available, which allowed comparison of the results of spectrum recovery and α_{opt}. Measurements for each filter material in the same experimental geometry allow at least two attenuation curves to be plotted. The minimum value for α of the discrepancy norm solution should be defined from one set of the experimental data (like as one independent realization), while the right-hand part in equation (4) has to be taken from the remaining set. Such procedure gives α which corresponds to the best approximation to a quasi-solution. The attenuation curve obtained with copper filters has the largest data set with the least dispersion. It is the best for spectrum reconstruction with an estimated $\alpha_{opt} \leq 2.5 \times 10^{-8}$ and total measurement error $\leq 5\%$.

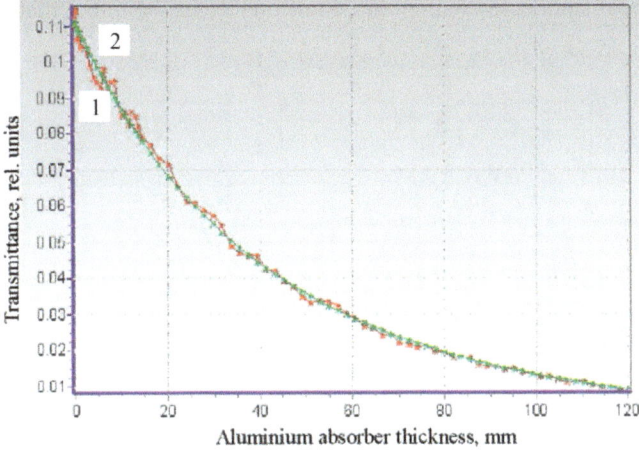

Figure 21 *Attenuation curves for model with random errors (1, red) and regularized spectra (2, green).*

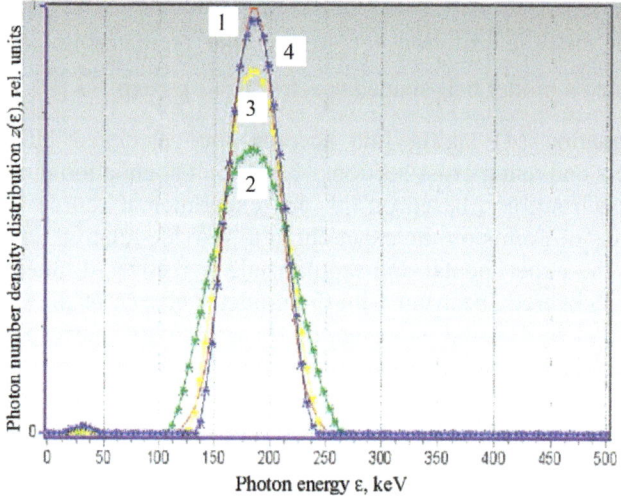

Figure 22 *Comparison of model (1, red) and reconstructed spectra for regularizing parameters: 4.4×10^{-7} (2, green), 7.1×10^{-8} (3, yellow) and 2.5×10^{-8} (4, blue).*

3.3 X-ray Spectral Reconstruction and Analysis

Figure 23 shows spectra recovered from the copper filter attenuation curve, with the regularizing parameter α in the range $\approx 1.5 \times 10^{-7} - 1.5 \times 10^{-9}$. Despite this two orders of magnitude difference, the solutions are similar around the main peak, which is indirect evidence of the correct choice of the range of α. The spectra peak at ≈ 80 keV with FWHM of ≈ 110 keV. Figure 24 shows a comparison of the experimental and recovered attenuation curves. The curves are plotted on a logarithmic scale because on a linear scale the difference between the curves is practically unnoticeable. The maximum difference is observed for larger filter thicknesses, i.e., for high photon energies where the solutions reveals a smooth peak connected with poor conditionality of the initial matrix.

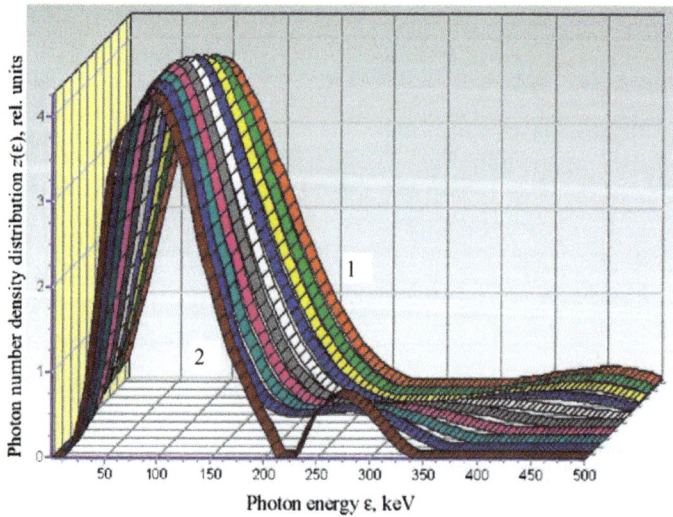

Figure 23 *Results of spectrum reconstruction from the copper attenuation curve, with α ranging from 1.5×10^{-7} (1, red) to 1.5×10^{-9} (2, brown).*

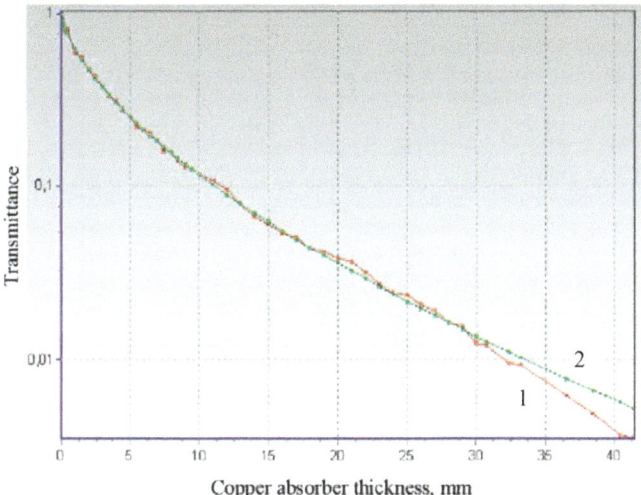

Figure 24 *Comparison of experimental (1, red) and recovered attenuation (2, green) curves for copper.*

Analogous curves for the aluminium filters are given in figures 25 and 26. The α values are larger, in the range $\approx 7.5 \times 10^{-7} - 3 \times 10^{-8}$ with $\alpha_{opt} \approx 2 \times 10^{-7}$, corresponding to lower measurement accuracy. The peak is at ≈ 70 keV and the spectral width is ≈ 70 keV. Figure 27 shows a comparison of the Bremsstrahlung radiation spectra obtained by processing the copper and aluminium attenuation curves, demonstrating a satisfactory agreement between the spectra obtained. For the lead attenuation curve, the reconstructed spectra should be considered as approximate due to a small number (seven) of experimental points. Nevertheless, these spectra do not change in a wide range of α, as shown by figure 28.

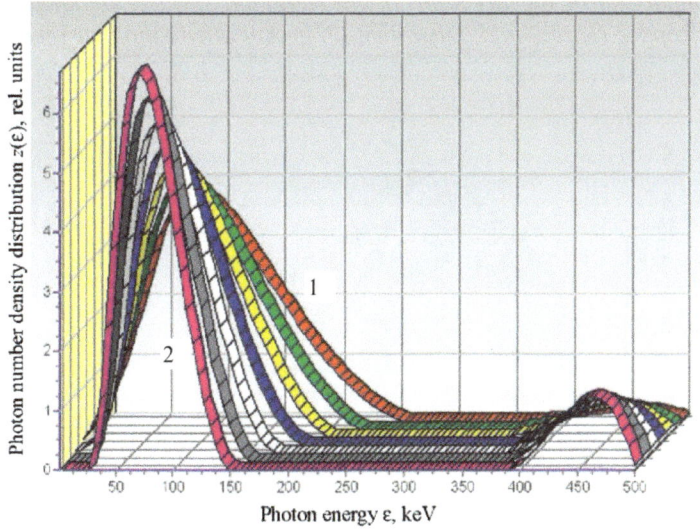

Figure 25 *Results of spectrum reconstruction from the aluminium attenuation curve, with α ranging from 7.5×10^{-7} (1, red) to 3×10^{-8} (2, lilac).*

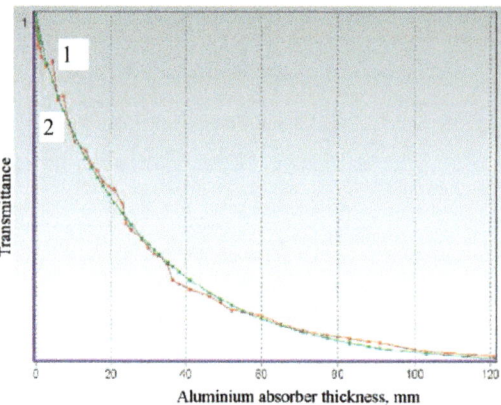

Figure 26 *Comparison of experimental (1, red) and recovered attenuation (2, green) curves for aluminium.*

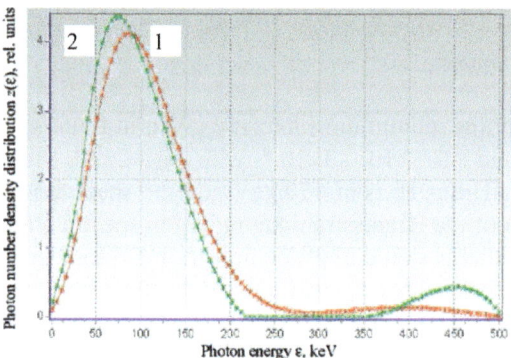

Figure 27 *Comparison of the Bremsstrahlung radiation spectra obtained by processing the copper (1, red) and aluminium (3, green) attenuation curves.*

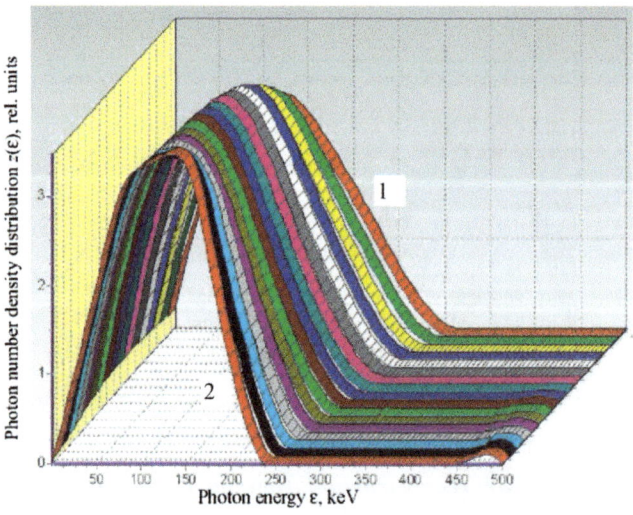

Figure 28 *Results of spectrum reconstruction from the lead attenuation curve, with α ranging from 1.5×10^{-6} (1, red, top) to 3.5×10^{-8} (2, red, bottom).*

The maxima of the reconstructed spectra being in the range of ε=70–100keV is in good agreement with the mean photon energy of Bremsstrahlung radiation $<\varepsilon>$ emitted by an electron beam of mean energy ε_e about 300 keV as in the work described here.[3] The mean photon energy is given by

$$\langle \varepsilon \rangle = \int_0^{\varepsilon_e} \varepsilon z(\varepsilon) d\varepsilon, \qquad (22)$$

and, Wyard's formula, equation (1), gives $<\varepsilon>$= 0.267ε_e=80 keV, while Evan's formula, equation (2), gives $<\varepsilon>$ = 0.333ε_e=100 keV.

It should be noted that if the boundaries of the quantum energy variation are known *a-priori* the accuracy of the solution should be improved. The choice of the range 0–500 keV was deliberate, first, in order to demonstrate the effect of the algorithm, and second, because some electrons observed experimentally had energies which exceeded the voltage applied to the diode due to instability developed in the plasma of a cathode flare.[26] These electrons might be the cause of the high energy peak observed at ≈450 keV in the spectrum reconstructed from the aluminium attenuation curve.

4 CONCLUSION

On the whole, time-resolved scintillation technique and calibrated thermo-luminescence dosimeters (TLDs) have been used to measure Bremsstrahlung X-ray emission in a large-aperture GARPUN KrF laser pumped in a transverse geometry by double-sided electron beams. X-rays were produced by electrons of ~300 keV energy, 60 kA current (50 A/cm^2 current density), and with 100 ns pulse duration when they passed from vacuum diodes through Ti-foil windows into the laser chamber. Spatial distribution of X-rays originated from electron deceleration in foil windows and in working gas in the chamber was measured. Energy flux of X-rays illuminated laser windows was found to be ~10^{-3} J/cm^2. Regularizing algorithm has been developed to reconstruct the X-ray spectrum using experimental data on X-ray transmission through Cu, Al and Pb layers of different thickness. The obtained spectrum had a maximum around 80 keV with FWHM of 110 keV. Based on the measurements it is predicted transmissive optics degradation under X-ray irradiation due to formation of colour centres. This effect should be taken into account for large-size repetition-rate KrF laser drivers designed for a long-time operation in the Inertial Fusion Energy.

ACKNOWLEDGMENTS

We are grateful to N.N. Mogilenetz for the performance of TLD diagnostics, D.A. Zayarnyi for the assistance in experiments. This work was partially supported by the US Naval Research Laboratory.

References

1 A.S. Alimov, B.S. Ishkhanov, V.I. Shvedunov, et al., *Moscow Univ. Phys. Bull*, 2010, **65**, 111.

2 V.D. Zvorykin, A.S. Alimov, S.V. Arlantsev, et al., *Plasma and Fusion Res.*, 2013, **8,** 2405000.

3 V.D. Zvorykin, S.V. Arlantsev, V.G. Bakaev, et al., *Laser and Particle Beams*, 2001, **19**, 609.

4 V.D. Zvorykin, S.V. Arlantsev, V.G. Bakaev, et al., *Proc. SPIE*, 2003, **5120**, 223.

5 S.J. Wyard, *Proc. Phys. Soc.*, 1952, **65 A**, 377.

6 S.J. Wyard, *Nucleonics*, 1955, **13**, 44.

7 R.D. Evans, *The Atomic Nucleus*, , Mc Graw-Hill Book Company, New York, Toronto, London, 1955, pp. 614–617.

8 W.E. Dance, D.H. Rester, B.J. Farmer, et al., *J. Appl. Phys.*, 1968, **39**, 2881.

9 D.H. Rester, W.E. Dance, J.H. Derrickson, *J. Appl. Phys.*, 1970, **41**, 2682.

10 W.J. Price, *Nuclear Radiation Detection*, McGraw-Hill Book Company, Inc., New York, Toronto, London, 1958.

11 B.T. Price, C.C. Horton, K.T. Spinney, *Radiation Shielding*, Pergamon Press, London, New York, Paris, 1957.

12 S.P. Obenschain, J.D. Sethian, and A.J. Schmitt, *Fusion Sci. Tech.*, 2009, **56**, 594.

13 M.M. Lavrentiev, *Some improperly posed problems of mathematical physics,* -Spr. Tracts., New York, 1967.

14 A.N. Tikhonov, V.Ya. Arsenin, *Methods for improper problems solving"*, 2nd. edn., Nauka, Moscow, 1979 (in Russian).

15 D.L. Phillips, *J. Assoc. Comput. Mach.*, 1962, **9**, 84.

16 J.B. Rosen, *SIAM J. Applied Mathematics*, 1960, **8**, 181.

17 M.R. Hestenes and E. Stiefel, *J. Res. NIST*, 1952, **49**, 409.

18 W.J. Veigele, *Atom. Data*, 1973, **5**, 51.

19 E. Storm and H.I. Israel, *Atomic Data and Nuclear Data Tables*, 1970, **7**, 565.

20 J.H. Hubbell and I. Overbo, *J. Phys. Chem. Ref. Data*, 1979, **9**, 69.

21 J.H. Hubbel, W.J. Veigele, E.A. Briggs, et al., *J. Phys. Chem. Ref. Data*, 1975, **4**, 471.

22 O. Klein and Y. Nishina, Z. für Physik, 1929, **52**, 853.

23 M.J. Berger and D.J. Raso, *Radiation Res.*, 1960, **12**, 20.

24 M. Berger and J. Doggett, *J. Res. NIST.*, 1956, **56**, 89.

25 L.R. Kimel' and V.P. Mashkovich, *Shielding of Ionizing Radiation*, Atomizdat, Moscow, 1966 (in Russian).

26 M. Friedman, S. Swanekamp, S. Obenschain, et al., *Appl. Phys. Lett.*, 2000, **77**, 1053.

Integrated Systems

THE BERN ADVANCED GLASS LASER FOR EXPERIMENT (BEAGLE) X-RAY LASER FACILITY

Davide Bleiner and Felix Staub

Institute for Applied Physics, University of Bern, Sidlerstrasse 5, CH-3012, Bern, Switzerland
Present address: EMPA Materials Science & Technology, Überlandstrsse 129, CH-8600 Dübendorf, Switzerland

1 INTRODUCTION

Laboratory-scale short-wavelength lasers have, after decades of research efforts, become reality thanks to a number of breakthroughs that have enabled more efficient light amplification. The generation of a hot and dense plasma column that can serve as an extreme UV or soft X-ray gain medium requires a significant amount of drive power. This becomes especially challenging if one wishes to shrink down the footprint and fit the facility within a laboratory. Figure 1 visualizes the stepwise impact of enabling technologies, such as Q-switching, mode-locking and chirped pulse amplification (CPA), on the improvement of drive beam intensity. Nowadays even much higher intensities for high-energy physics are possible, but this is beyond the focus of this chapter. Indeed for the table-top generation of an X-ray laser, as it is commonly called regardless the fact that the spectrum stretches across the the extreme UV and soft X-ray regions, 10^{12} W/cm^2 can be nowadays sufficient.

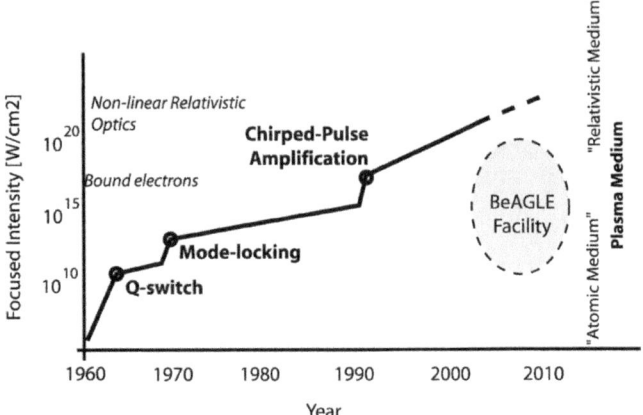

Figure 1 *Focused laser intensity as a function of year, with respect to enabling technologies.*

The BeAGLE facility (Bern Advanced Glass Laser for Experiment) was assembled in the early 1990s, and stepwise upgraded, in a number of laser technology development projects at the Institute of Applied Physics of the University of Bern. The early beginnings of the Uni Bern activity in the field are dating back even to the late 1970s. The system architecture follows the well-established master-oscillator power-amplifier (MOPA) architecture with Nd: glass as the active laser medium. Image relay and spatial filtering are used throughout to keep the nonlinear effects within acceptable values. [1] While an active/passively mode-locked oscillator producing pulses of ~100–500 ps in duration was used in the early phase of x-ray laser experiments, the facility has been upgraded in the 2000s into a CPA system generating high peak power pulses with durations as low as ~1 ps. As the maximum operative output energy is ~20 J, which is limited by the damage threshold of the optics/grating (in principle up to 120 J is possible), the laser thus achieves a maximum pulse power of approximately 20 TW. In the following sections, the various components constituting the facility will be described in some detail. An artist's impression of the layout of the BeAGLE facility is shown in figure 2.

This chapter is organized in such a way that we will follow the optical path of the driver beam from the oscillator down to the target chamber. The laser pulse delivery onto the target, as a line focus, triggers a hot and dense plasma column. The latter sustains the amplification of the spontaneous emission (ASE) which is the "X-ray laser". Here we will put more emphasis on the architecture of the integrated system, which means a module wise description of the drive laser.

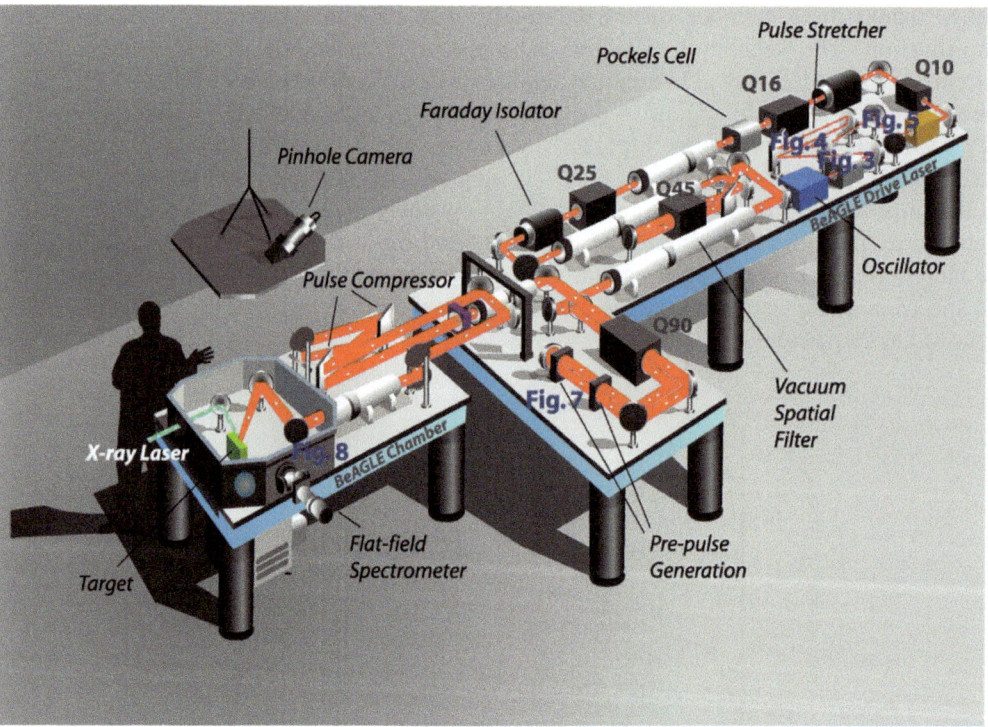

Figure 2 *Artist's impression of the BeAGLE X-ray laser facility.*

2. OSCILLATOR SOURCE

The oscillator of the BeAGLE facility is a commercial mode-locked Nd:glass laser. Nd-doped glass is the laser medium of choice for generating high energy ultrashort pulses, as is the case in current X-ray laser experiments. The bandwidth of 20–30 nm^2 enables pulse durations down to a hundred fs. A diode pumped femtosecond Nd:glass laser oscillator is thus a low-cost (compared to the Ti:sapphire alternative) and compact solution for pumping high-power Nd:glass amplifiers.

Our oscillator emits a train of ~200 fs pulses, generated in a resonator with a nonlinear passive element, i.e. a semiconductor saturable absorber mirror (SESAM). The pulses are emitted at a repetition rate of 75 MHz, having energies of ~1 nJ at the nominal output power of 100 mW. A Pockels cell samples one pulse to the pulse stretcher and amplification stages (figure 3). In the BeAGLE system a beam expander is located at the oscillator exit to allow a well-collimated beam to propagate through the stretcher.

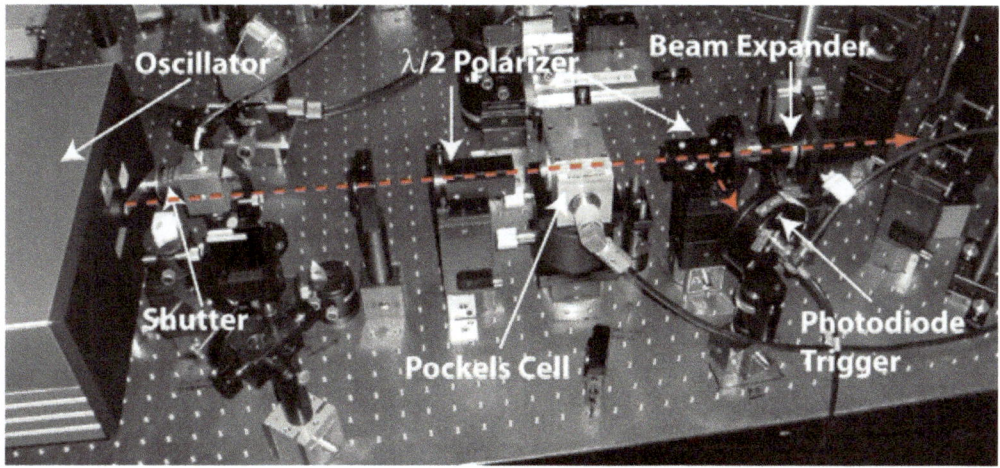

Figure 3 *Pulse sampling setup in front of the oscillator.*

3. PULSE STRETCHER FOR CHIRPED-PULSE AMPLIFICATION

For very high amplification, as required to reach joule level pulses, an amplification approach is needed that can prevent non-linear effects or even damage in the active media. Chirped pulse amplification (CPA)[3] overcomes intensity/damage tradeoffs by stretching in time the pulse duration 10^3–10^4 times prior to amplification. The pulse can thus be amplified in a safe low-intensity and sub-critical regime since the longer pulse gets shallower in peak power.

The amplified pulse is finally compressed back to obtain a high-peak power, short-duration pulse of 1–20 J, prior to delivery onto the target. The stretching is accomplished by means of wavelength-dependent dispersion (chirping) using dispersive optics, such as a grating pair.[4] Compression is the reverse process of un-chirping, also done with dispersive optics. However, the peak power is typically very high at the compressor, henceforth the beam diameter on the gratings has to be large to keep the fluence below the damage specification. For preventing pulse distortion, especially at very high intensity, vacuum is often required. In the BeAGLE system, however, the grating pair is operated at ambient pressure, since a maximum pulse power of ~20 TW is extracted such that nonlinear effects are within negligible range.[5] During chirping the induced wavelength dispersion is not only time dependent (which is desired for stretching the pulse duration), but also subject to spatial dispersion. The latter causes pulse distortion in the sense that an oval cross-section is generated. An oval pulse does not match well with the amplifier modules, whose active media have circular cross-sections. To correct for this, a four-pass approach is adopted in the BeAGLE stretcher.

Figure 4 sketches out the principle of the pulse stretcher and shows its implementation in the BeAGLE system. It follows the design proposed by Lai,[6] essentially being a folded Martinez-type stretcher[7] that provides positive group velocity dispersion (GVD). A single grating having 1740 grooves per millimetre is employed, which, in combination with a 150 cm focal length plano-convex lens, stretches the pulse to a FWHM duration of 1 ns for a bandwidth of 6 nm. In the BeAGLE system the pulse is thus stretched with positive GVD (up-chirp) and then compressed with negative GVD (down-chirp), using the grating-pair approach.[4]

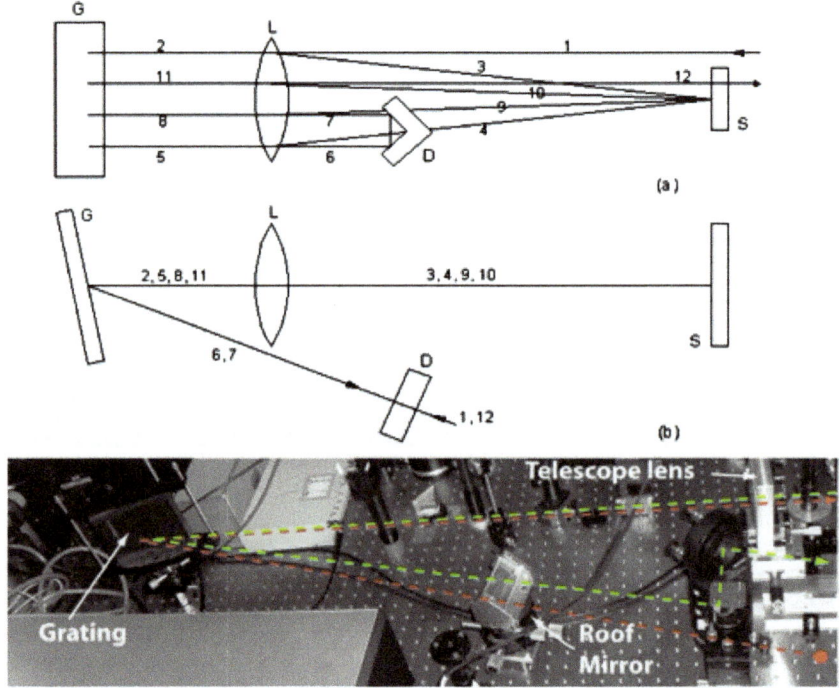

Figure 4 *The pulse stretcher setup. a) Side view, b) top view. The pulse sampled from the oscillator output is wavelength dispersed at the grating and directed as a chirped pulse through a telescope lens to a flat mirror (not shown in the photograph). The sketch shows the optical path up to the regenerative amplifier. The numbers refer to the sequence of beam legs along its optical pathway. Legend: D = Roof-edge mirror, G = Grating, L = Telescope lens, S = Flat mirror.*

3. PULSE AMPLIFICATION

The pre-amplification is done in a diode-pumped Nd:glass regenerative amplifier (RA) with a total of ≈50 round-trips, 7.1 ns round-trip time, and overall gain of approxmately 10^6 with output on the millijoule level. This is a compact and effective way of pre-amplification, with a pulse contrast of approximately 1000:1. A regenerative amplifier is a device that pumps the gain medium across an optical resonator (figure 5) to amplify the input pulse by means of multiple passes (here, ≈50). An optical switch, realized with a

Pockels cell and a polarizer, permits to control the number of round trips, so that a very high overall amplification factor (gain) is achieved. A Faraday rotator at the entry of the RA prevents feedback of the amplified beam due to reflection (isolator) and permits to separate input and output signals by means of rotation of the polarization plane. Regenerative amplification makes it possible to achieve very high gain in a compact setup and in the BeAGLE system it amplifies the nanojoule oscillator signal up to 0.9-1 mJ.

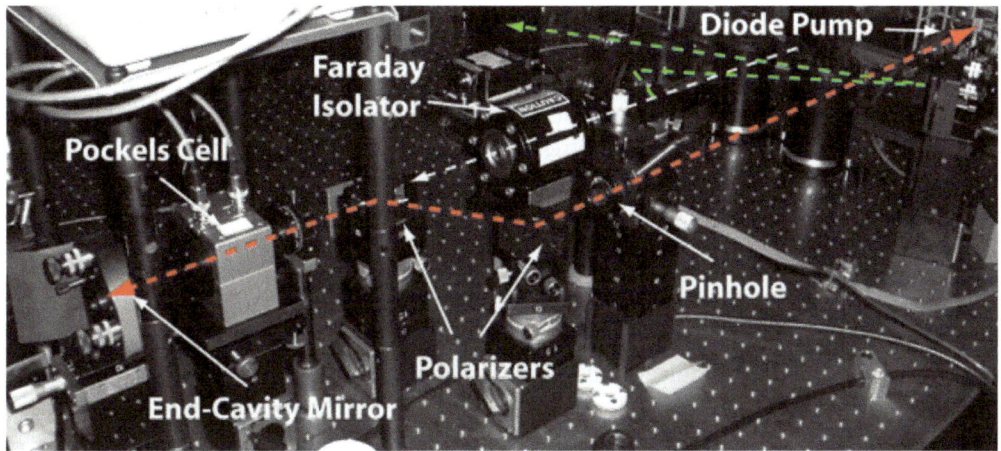

Figure 5 *The regenerative amplifier of the BeAGLE facility.*

The output signal of the regenerative amplifier is then passed to a chain of optical amplifiers (figure. 6). A set of flash lamp pumped Nd: glass rod amplifiers, with diameters of 10, 16, 25, 45 and 90 mm, contributes to the generation of a multi joule level pulse, which is finally compressed with a grating pair. Figure 6 shows that the amplifiers Q16 and Q45 are used in double-pass, in order to achieve higher amplification. Two Faraday isolators and a Pockels cell are inserted between the first three amplification stages as the main elements for pulse isolation. While the role of the Faraday isolators is to prevent backward propagating pulses (feedback) from being further amplified and causing damage to front-end components, the Pockels cell serves to discriminate against amplified spontaneous emission (ASE) and spurious prepulses generated in the regenerative amplifier in the forward direction. Pulse contrast values of typically 10^6 are obtained.

Figure 6 *The flash lamp pumped Nd:glass amplifier chain.*

A variable magnification, three-lens telescope in combination with an apodized aperture in front of amplifier Q16 is used to configure the spatial beam profile such as to provide a super-Gaussian profile at the output of the final amplifier. The profile generated in Q16 is relay imaged to the subsequent amplifiers by a sequence of vacuum spatial filter (VSF) telescopes. Beside their imaging and beam expanding properties, the VSFs serve to suppress high-spatial frequency ripples which otherwise can lead to beam break up and damage to optical components through self focusing.[8]

Finally, the pulse is compressed back by a grating pair that is matched with the stretcher dispersion. In order to be able to achieve transform limited pulse durations, the angles of incidence of the beam on the gratings have to be identical to within a tolerance of much less than a degree,[9,10] with the required accuracy. However, in the BeAGLE setup, one of the gratings is slightly tilted (0.09°) to generate an inhomogeneous pulse with back tilted wavefront.[11] This approach permits to achieve ideal travelling wave irradiation conditions along the line focused beam.

After compression, the pulse is 1.2 ps in duration and has up to 20 J of energy, resulting in a pulse peak power of ~20 TW. The maximum output energy is limited only by the size of the compressor gratings. At this power level, the beam spatial quality is only partly affected by the nonlinear propagation across air to the target chamber. The pulse contrast has been measured by third-order autocorrelation to be better than 10^5:1. The beam spatial quality has been measured with beam-profiling diagnostics that revealed a peak to average intensity ratio of better than 2:1.

4. PRE-PULSE GENERATION

Two pre-pulses, and a main pulse of identical duration, are delivered colinear to the target with an angle of 30°–50° from the surface. The pre-pulses are generated by partial reflection on planar beam splitters with reflectivities in the range 0.5%–2.8% (pre-pulse 1, with 2.8 ns pre-delay), and 8–16% (pre-pulse 2, with 90 ps pre-delay), respectively. These are inserted in the double-pass beam path of amplifier Q90 (figure 7). A full reflectivity mirror at the end reflects the main pulse with the remaining driver energy. The spatial overlap of the pre-pulses and the main pulse on the target (line focus) is controlled by adjusting the orientation of the beam splitters with micrometric screws and acquiring the individual reflections with a CCD camera. The acquisition system can provide the x,y coordinate of each individual pre-pulse and the main pulse. The adjustment is done with a precision of 1 pixel, i.e. < 10 µm. The most critical dimension to adjust is thus the transverse one, since a lateral mismatch can invalidate the pre-pulse effect. A longitudinal pre-pulse/main-pulse mismatch can only partly affect the plasma column at the leading and trailing edges.

5. TARGET CHAMBER AND DIAGNOSTICS

The drive beam with a pulse duration of 1.2 ps is delivered onto the target with an energy of as low as 2.5–3 J for the tin laser with emission at 12 nm. The drive pulses are delivered to the target chamber through a reduction telescope for collimating the beam to a diameter of 80 mm. Figure 8 shows the target chamber and some diagnostics. The pump laser beam is in coming from the right. A 45° planar reflector directs the beam to a parabolic reflector with a focal length of 609.6 mm to generate a line focus onto the target with an incidence angle .

Several X-ray laser lines have been *saturated* in the University of Bern laboratory, including, iron (25.5 nm) and chromium (28.5 nm),[12,13] indium (12.6 nm), palladium (14.7 nm) and silver (13.9 nm),[14,15,16] tin (12 nm),[9,10] antimony (11.4 nm),[17,18] and barium (9.2 nm).[19] X-ray lasers are most efficiently operated in the grazing incidence pumping (GRIP) scheme.[20,21] Here, a preformed plasma is generated in a first step, much as in conventional, normal-incidence pumping, by irradiating the target with a long (~200 ps) laser pulse at normal incidence. The plasma is allowed to expand for a given time to allow the electron density gradients to relax. The short (~picosecond) pump pulse then irradiates the expanding plasma plume at an oblique (grazing) angle, θ, chosen such that the turning point (or apex) electron density of the pump radiation, n_e – given by refraction as $n_e = n \sin^2\theta$, where n is the critical density for the pump laser wavelength – coincides with the density for which maximum gain is predicted for a given X-ray laser transition. Absorption of the pump energy thus occurs directly and very efficiently into this region which is rapidly heated up to the temperatures required to produce strong collisional excitation of the upper laser level. In this sense, the GRIP scheme is an extension of the transient collisional excitation (TCE) scheme to non-normal incidence. Due to the oblique incidence, the pumping pulse beam is inherently a travelling wave with a travelling wave (TW) speed given by $v = c/\cos\theta$ ($v \approx 1.06c$ for $\theta = 20°$). When moving towards shorter wavelengths, higher densities and thus larger GRIP angles are required. For example, for Nd:glass laser pumping ($\lambda = 1054$ nm) of the 11.9 nm tin X-ray laser, $\theta \approx 45°$ and $v \approx 1.41\ c$.

Planar diamond machined targets are used with a new surface at every shot. The target chamber is operated at ~10^{-4} mbar and connected to the diagnostics chamber at 10^{-6} mbar by means of a vacuum gate and with a cold finger to prevent condensation on the CCD detector. The EUV in the application chamber is detected using a 16 bit 1024×1024 pixel back illuminated X-ray CCD with a pixel size of 13 μm. The chip is cooled to –20°C for low background level.

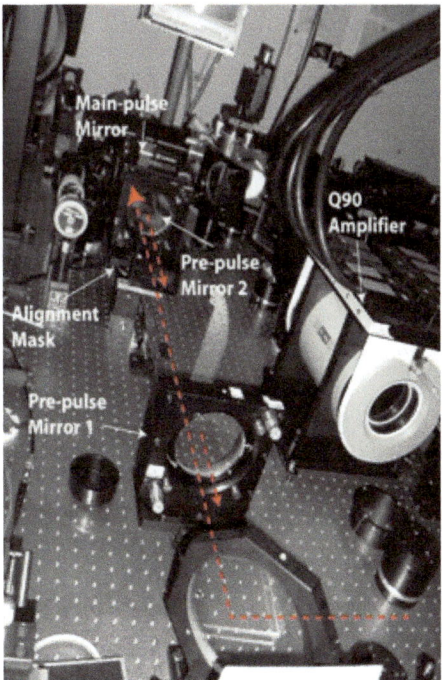

Figure 7 *The pulse train generation system with the planar beam splitters coated for different reflectivities, as explained in the text.*

Two CCD cameras are installed on the target chamber for diagnostics purpose. The flat-field spectrometer system is used for spectral characterisation of the EUV laser by means of an on-axis, time-integrating spectrometer. The latter provides 2D images of angle and wavelength dispersion (5–25 nm). The spectrometer consists of a 1200 line pair per millimetre flat-field Hitachi grating (radius of curvature 5649 mm), working at 3° grazing incidence. The grating disperses the radiation onto a 40 mm diameter P20 phosphor screen, imaged on a cooled CCD camera with a pixel size of 23 μm. The spectral resolution is approximately 0.1 nm, limited by slitless operation.

The pinhole camera system is used for plasma column imaging with a double-slit camera. Images are recorded on a P43 phosphor screen coupled to a CCD camera having a pixel size of 9 μm. The vertical slit (100 μm) is positioned exactly midway between the screen and the target, such that it produces a 1:1 imaging of the plasma column length. The horizontal slit (50 μm) is close to the target to produce a 1:18 magnified image of the plasma column width. This permits to resolve the transverse structure of the plasma.

Figure 8 *Overview of the target chamber.*

6. OUTLOOK: FROM SINGLE-SHOT TO A HIGH REPETITION RATE SYSTEM

The BeAGLE system is a single shot system, where the pulse repetition period is around 10–30 minutes depending on the pulse energy. In order to enhance the repetition rate and consequently both the average power and throughput of the facility, a new beamline is being built. The high repetition rate system will attain amplification using a two stage optical parametric preamplifier (OPA) in place of regenerative preamplification, after having stretched the pulse to attain chirped pulse amplification (CPA). The OPA consists essentially of a commercial 532 nm pump laser and nonlinear optical crystals for amplification. For further amplification to the Joule level, a two-stage Nd:YLF rod amplifier chain using conventional flashlamp pumping technique will be used. The oscillator (signal wave) will be a mode-locked Yb:glass laser, delivering 200 fs pulses at a wavelength of 1053 nm with an average power of 0.2 W and at a repetition rate of 50 MHz. A single pulse is sampled every 0.1–1 s using a Pockels cell and passed to the stretcher. In order to compact the facility footprint to less than ~3 m^2, an all-reflective pulse stretcher design will be adopted as opposed to the refractive design currently in use. The oscillator pulse will undergo a non-linear mixing in the optical parametric stage with the pump wave. OPA is particularly attractive because it provides high gain (10^6–10^7) in a short length. The OPA gain bandwidth is in fact determined by the dispersive properties and the length of the amplifier crystal, rather than by details of a laser transition. OPA has been shown to offer a very attractive alternative to regenerative amplifiers such as the one operated in the existing CPA laser system.

Furthermore, a very important consideration is that an OPA has no energy storage. Therefore it amplifies only while being pumped. This is a key feature that is exploited to attain very high intensity contrast between an amplified pulse and any satellite pulses. The latter is important to prevent plasma pre-ignition and thus having a fully controlled energy delivery at the focal spot. Last, but not least, the absence of heat generation makes OPA ideal for high-power operation at high repetition rate.

Our OPA will consist of a set of lithium triborate (LBO) crystals in a walk-off compensating arrangement. The crystals are mounted in temperature controlled gimbal mounts to mitigate thermal cycling and phase mismatch. The spatial profile of the beam exiting the OP-CPA stage will be transformed into a flat top profile using a commercial Gaussian to flat top converter. Image relay will be used to propagate the flat top profile through the two main amplifiers, which will operate in a double pass geometry. The OP-CPA preamplifier stage, based on a Q-switched, frequency doubled (532 nm) Nd:YAG pump laser delivering 0.2–0.4 J pulses at a repetition rate of 10 Hz, is expected to deliver a pulse energy above 10 mJ in the chirped 1 ns pulse (5 MW). Since optical parametric amplification is a nonlinear process, it is very sensitive to frequency variations of the pump laser.

Further amplification includes two stages of Nd:YLF rod amplifiers having diameters of 6 and 19 mm to produce output pulse energies of approximately 2 J at a repetition rate up to 5 Hz. The pump cavity for the 6 mm rod amplifier is equipped with a 3% doped Nd:YLF rod. The 6 mm amplifier has been pre-evaluated and shown to provide a single-pass gain in excess of 7× at a flash lamp energy of approximately 90 J. Taking saturation effects into account, an energy of 150 mJ should be easily achievable in a double-pass configuration. The final 19 mm amplifier is expected to boost the pulse energy to the projected 2 J in a double pass. Such a system will be suitable for driving the 12 nm EUV laser into saturation with a resulting brilliance up to 3×10^{27} photons s^{-1} cm^{-2} sr^{-1} and a bandwidth of $\Delta\lambda/\lambda = 5 \times 10^{-5}$, which is one and a half orders of magnitude narrower than the 0.1% bandwidth typically used for determining the brilliance.

References

1   J.T. Hunt, P.A. Renard and W.W. Simmons, *Appl. Opt.*, 1977, **16**, 779.
2   D. Kopf, F.X. Kartner, U. Keller and U. Weingartner, *Opt. Lett.*, 1995, **20**, 1169.
3   D. Strickland and G. Mourou, *Opt. Comm.*, 1985, **56**, 219
4   E.B. Treacy, *IEEE J. Quant. Electron.*, 1969, **QE5**, 454
5   S. Seznec, C. Sauteret, S. Gary, E. Béchir, J.L. Bocher and A. Migus, *Opt. Comm.*, 1992, **87**, 331.
6   M. Lai, S.T. Lai and C. Swinger, *Appl. Opt.*, 1993, **33**, 6986
7   O.E. Martinez, *IEEE J. Quant.Electron.*, 1987, **QE23**, 59.
8   J.T. Hunt, P.A. Renard and W.W. Simmons, *Appl. Opt.*, 1977, **16**, 779
9   M. Grünig, *Experimental Study of the Saturated 12 nm Sn X-ray Laser Using Grazing Incidence Pumping*, 2008, (PhD Thesis, University of Bern).

10   M. Grünig, C. Imesch and J.E. Balmer, *Opt. Comm.*, 2009, **282**, 267.

11   J.C. Chanteloup, E. Salmon, C. Sauteret, A. Migus, P. Zeitoun, A. Klisnick, A. Carillon, S. Hubert, D. Ros, P. Nickles and M. Kalachnikov, *J. Opt. Soc. Am. B*, 2000, **17**, 151.

12   F. Löwenthal, *Saturated Neon-like X-ray Laser and Performance Research on a High Power Laser Facility*, 1998, (PhD Thesis, University of Bern).

13   F. Löwenthal, R. Tommasini and J.E. Balmer, *Opt. Comm.*, 1998, **154**, 325

14   R. Tommasini, *X-ray Lasers from Collisionally Pumped Nickel-like Ions*, 1999, (PhD Thesis, University of Bern)

15   R. Tommasini, F. Löwenthal, J.E. Balmerm, *Phys. Rev. A*, 1999, **59**, 1577

16   R. Tommasini, F. Löwenthal, J.E. Balmer, *J. Opt. Soc. Am. B*, 1999, 16, 1664

17   C. Imesch, *Experimental Study of Saturated Nickel-like Soft-X-ray Lasers using Grazing Incidence Pumping*, 1999, (PhD Thesis, University of Bern)

18   C. Imesch, F. Staub, J.E. Balmer, *Opt. Comm.*, 2010, **283**, 66

19   F. Staub, C. Imesch, D. Bleiner, J. Balmer, *Opt. Commun.*, 2012, **285**, 2118

20   V.N. Shlyaptsev, J. Dunn, S. Moon, R. Smith, R. Keenan, J. Nilsen, K.B. Fournier, J. Kuba, A.L. Osterheld, J.J.G. Rocca, B.M. Luther, Y. Wang and M.C. Marconi, *Proc. SPIE*, 2003, **5197**, 221

21   R. Keenan, J. Dunn, V.N. Shlyaptsev, R.F. Smith, P.K. Patel and D.F. Price, *Proc. SPIE*, 2003, **5197**, 213.

ENEA EXTREME ULTRAVIOLET LITHOGRAPHY MICRO-EXPOSURE TOOL: MAIN FEATURES

S. Bollanti, P. Di Lazzaro, F. Flora, L. Mezi, D. Murra, A. Torre

ENEA Frascati Research Center, Technical Unit for Application of Radiation, via E. Fermi 45, 00044 Frascati (Rome), Italy

1 INTRODUCTION

As reported in[1,2], the laboratory-scale Micro-Exposure Tool (MET) for Extreme Ultraviolet projection Lithography (EUVL), realised at the Frascati ENEA Centre within the context of an Italian National Project,[3] was successfully operated in 2008 by achieving a 160-nm resolution imaging of mask patterns onto a polymethylmethacrylate (PMMA) photoresist through 14.4-nm radiation.

EUV ($\lambda \approx$10-15 nm) lithography has come a long way in the last 30 years.[4] The idea of EUVL as a possible extension of conventional optical lithography traces back in fact to 1985,[5] whereas the first experimental results, showing the possibility of resolving 0.5 μm structures using a Schwarzschild objective (SO) as the projection (demagnifying) optics, synchrotron radiation as the source and a reflective-type mask, were reported in 1989.[6,a] Such a result is to be compared with the recent achievement by AMSL, whose NXE:3300B tool offers 22 nm resolution with conventional illumination and 18 nm with off-axis illumination[7].

A survey of more than 130 attendees at the last 2010 SEMATECH Litho Forum[8] confirmed the general trend, that already emerged at previous such forums, that continuation of lithography's current course being considered as the best way to manufacture next-generation chips. Indeed, 193 nm immersion double patterning continues to be considered as the suitable lithographic technology for volume manufacturing at the 22 nm half-pitch node, whilst EUVL is expected to be inserted into semiconductor manufacturing at the 10 and 7nm half-pitch nodes. Several, however, EUV technology

---

[a] The work was presented at the 33rd International Symposium on Electron, Ion, and Photon Beams, Monterey, California, May 30-June 2 (1989). In this regard, in the EUVL historical perspective, addressed to in [4] (Chapter 1, by Kinoshita and Wood), one can read: "At the EIPB symposium banquet..., a Russian scientist, Dr. Tanya Jewell of AT&T, cornered Dr. Kinoshita and proceeded to deluge him with questions. The combination of poor Japanese English and poor Russian English made conversation extremely difficult, so the discussion continued for a long time with Obert Wood of AT&T acting as interpreter. The following year, AT&T announced the printing of 0.05 μm patterns using SXPL. The authors of this chapter regard the discussion that night in Monterey in 1989 as having been *the dawn of EUVL*."

challenges remain; in particular, source power, mask defects, exposure tool throughput, and cost of ownership are rated at the top of the list.

Discharge-produced plasma and laser-produced plasma (LPP) are the leading technologies for generating high-power EUV radiation. In both technologies, a hot plasma of ≈20-50 eV of the chosen fuel material is generated, which produces EUV radiation. Xenon, tin and lithium are the fuel materials of choice for EUV sources.[9] Radiation at 13.5 nm is currently explored for printing. Yet, the ever increasing resolution demanded by the semiconductor industry prescribes that future lithography equipment operates at an even shorter wavelength (6.X nm), i.e. beyond the EUV range (BEUV).[10]

The radiation emitted by the plasma, possibly debris-cleaned by an appropriate debris-mitigating system (DMS), is gathered by the collector optics and focused to the intermediate focus (IF), from where it is relayed to the scanner optics and finally to the wafer. It is at the IF that the source specifications are settled in accord with the high-volume manufacturing requirements, so that the appropriate exposure tool, and particularly its illuminator, does not depend on the EUV source features.

LPP sources seem to offer the most promising technology to reach the power levels needed for high-volume manufacturing. Indeed, the requirements for source power have increased over time as it has become clear that high resist dose is needed to simultaneously meet resolution and linewidth-roughness targets. Thus, it is estimated that for the 22 nm half-pitch, generation an EUV power ≈400 W at the IF within a 2% bandwidth window around the standard exposure wavelength for EUVL is required for 10 mJ/cm^2 photoresist sensitivity to enable > 100 wafer per hour scanner throughput.

Several methods have been proposed and/or implemented to control and/or mitigate the source emitted debris with the intent of protecting the collector mirror, one of the most costly elements in a EUVL setup. DMSs based on the use of ambient gas for moderating the species[11,12] (also combined with a protective covering over the collecting optics[13]), or of foil traps for particle capture[14] as well as on the application of electric[15] and/or magnetic fields[16] for deflection or velocity reduction (also combined with an ambient gas[17]) are well documented in the literature, with the specific DMS structure being devised and optimized depending on the EUV source concerned.[18,19]

The DMS implemented at ENEA MET uses krypton as the ambient gas at a suitable pressure combined with a gas filter and a specifically designed fan. It effectively cleans the Kr-penetrating radiation, which is first collected by one of the two twin ellipsoidal mirrors to the IF and then redirected by the other mirror to the reflective mask through two Mo/Si multilayer mirrors centred at 14.4 nm Finally, a SO is used as the projection (demagnifying) optics, to replicate the patterned mask onto the wafer (figure 1.1).

The main features of the ENEA MET components are synthesized in Sect. 2. Indeed, following the radiation path, we will highlight the characteristics of the source, the DMS, the collector, the mask illuminating mirrors, and the imaging optics. Emphasis will be put on the latter, the SO design, mounting and alignment being described in detail. Section 3 gives an account of the successful operation of the device, reported in,[1] and of the SO transmission measurement, subsequently carried out in order to identify the cause of the unexpectedly poor performance (as regards the overall reflectivity) of the objective. Concluding notes are given in Sect. 4.

2. ENEA MET: SETUP

Figure 2.1 shows the top view of the EUV lithographic setup of the ENEA MET, displaying the source, the collector and the printer modules. These are suitably embedded into two vacuum chambers, which are connected through a flexible duct. One chamber contains the source, the collector and the DMS, whilst the other contains the illumination and projection optics. All the illumination and projection optics lie on a vibration isolated and temperature controlled Invar board.

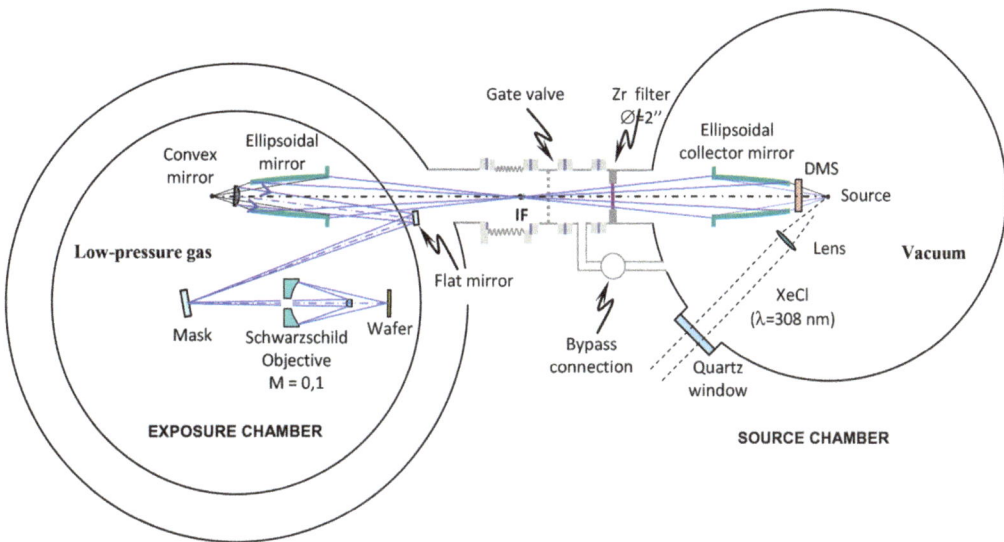

Figure 1.1 *Layout of the ENEA EUVL MET, which uses a laser-produced plasma as a radiation source, a pair of twin ellipsoidal mirrors as collecting optics to gather the 14.4 nm radiation from the source eventually to the mask, an efficient combination of ambient gas and mechanical fan as DMS and finally a low-cost SO as projection optics to image the patterned mask onto the wafer.*

Figure 2.1 *Top-view of the ENEA MET, showing the two vacuum chambers; one chamber (right) contains the plasma source, the collector and the DMS, whilst the other chamber (left) houses the illumination and the projection optics.*

2.1. Source Module

The ENEA MET exploits the LPP solid-tape-target source of the EGERIA (Extreme ultraviolet radiation Generation for Experimental Research and Industrial Applications) facility, which emits over a wide and fairly tunable range from ≈40 eV to ≈2 keV.

Characteristics and performances of the EGERIA source, which presents unique features among similar sources as regards the pulse energy and duration, are detailed in a number of papers.[12,20,21] Here, we briefly recall that the characteristics of the emitted radiation such as spectral range and purity, peak intensity, pulse width, and conversion efficiency can be controlled by the intensity, pulse duration and repetition rate of the drive laser and by the target material.

The source can alternatively be driven by two different high peak-power high repetition-rate XeCl lasers (308 nm); precisely, the laser facility Hercules (developed by ENEA Frascati) whose pulse energy, width and maximum repetition rate are respectively 6 J, 120 ns and 5 Hz and the commercial Lambda-Physik LPX-305 laser with 0.5 J, 30 ns and 50 Hz. Target materials such as Sn, Cu, Ta and In are available for different applications and/or investigations as well as operating wavelength. For instance, Sn and In are appropriate for lithographic processing at, respectively, 13.5 nm and 14.4 nm. The choice of drive laser is dictated by whether a high energy per pulse or a high number of laser shots is required. In any case, emission in the EUV spectral range, which is basically favoured by a long laser pulse, is further optimized by a proper out-of-focus condition of the laser beam. In fact, as can be seen in figure 2.2, by moving the target out of the focus of the 12 cm focal length lens, one can "tune" the emission spectral range as the laser irradiance is made to vary from keVs (at the focal position) to tenths of keVs. Specifically, at about 2 mm away from the focus, the laser irradiance is 10^{10} W/cm^2, by which EGERIA is made to generate EUV radiation from an area having a width of tenths of mm and with a relevant conversion efficiency up to ≈0.7% per eV over the 2π solid angle. Moreover, when defocusing the laser beam, the laser irradiance is much less sensitive to both fluctuations of the target position and laser energy/divergence instabilities, thus yielding a very good shot-to-shot stability.

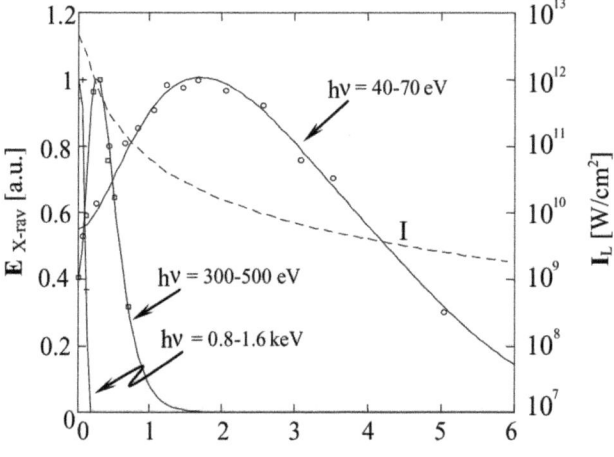

Figure 2.2 *EGERIA-pulse energy in different spectral intervals and laser irradiance on the target (dashed line) vs. target position behind the focal plane for a 100 μm thick Cu tape target irradiated by Hercules laser pulses.*[12,20]

2.2. Collector Module

The radiation pulse, after being debris-cleaned by the patented DMS, sequentially hits a pair of identical ellipsoidal mirrors, which serve to gather the emitted radiation and to transport it from the source to the projection chamber (figure 1.1).

2.2.1 Debris Mitigation System. Debris in a solid-target LPP source includes energetic ions, neutrals and particulates. Of course, atomic and particulate debris behave differently; hence, an efficient DMS must embed different tools to effectively remove both kinds of debris. An accurate speed-size characterization of atomic and particulate debris, emitted by the EGERIA source, has been performed.[22,23] Also, the mitigating efficiency of various tools, i.e. ambient gas (Ar and Kr), mechanical interdicting device (fans with different numbers of blades and speeds) and magnetic field, has been tested by observing the plasma debris contamination of glass slides, exposed - under different environmental conditions and laser shot numbers - to the radiation from a Cu tape, irradiated by the Lambda Physik laser.

As a result of such analyses, the ENEA MET DMS was designed to combine an ambient gas with a fan, further supported by a gas filter for continuous gas cleaning (figure 2.1). The ambient gas, specifically Kr at a pressure of ≈ 1 mbar such to allow for a $\approx 86\%$ transmission of 14.4 nm radiation through the overall target-to-collector distance (≈ 75 mm), serves to slow down and possibly block debris of small/moderate sizes, i.e. with diameters up to 1 μm, which comprise most of the atomic debris (neutrals and ions). The fan is in turn aimed at interdicting the arrival on the collector of larger-size debris(i.e. with diameters ≈ 1-10 μm), which, due to their sizes and formation mechanism, are scarcely moderated by the gas, but have velocities significantly smaller than those of the atomic debris. Specifically, the 5-blade aluminium alloy fan, shown in figure 2.1, was in use during the aforementioned successful operation of the MET. It has a maximum angular velocity of 400π Hz, and is placed at a minimum distance of 4 cm from the source.

The use of Kr (atomic weight 84 amu) is preferred with respect to Ar (atomic weight 40 amu) since, as it emerges from both theoretical estimates and experimental tests,[12,22,24] the mitigating power of Kr is definitely higher than that of Ar under comparable operating conditions, involving also larger ranges of both debris sizes and speeds. However, since Kr inefficiently transmits the EUV radiation up to 13.8 nm, where it becomes transparent, the use of a Kr-based DMS requires a shift of the operating wavelength of the lithographic setup from the standard 13.5 nm to values >13.8 nm. Notably, over the wavelength range 14-14.5 nm the typical Mo/Si multi-layer EUV mirrors maintain the same spectrally integrated reflectivity also after the multiple reflections occurring along the radiation path from the source to the wafer.[22]

Observation of the debris-flux exposed glass plates by both an optical microscope and microdensitometer resulted in very encouraging values of the debris mitigation factor (DMF) for both atomic and particulate debris.[22] In particular, values up to 450 for the DMF for atomic debris were deduced from the optical density measurements by a microdensitometer. As described in detail in[24] and also summarized in,[23,25] the analysis of the exposed glass slides in relation to the particulate debris (i.e. clusters and droplets with diameters $> \approx 10^{-3}$ μm), performed by a code specifically developed for the associated slide-image processing, has confirmed the effectiveness of the DMS. The code, which resorts to some basic image processing routines of the inherent MATLAB toolbox (V5.4),[26] relies on specific procedures, built on the basis of an accurate observation of the various images, for debris recognition, spot fine-structure identification and debris size (both above and below the microscope resolution) evaluation.[24]

Indeed, the rather good performance of the DMS has allowed quite a safe operation of the ENEA MET with the achievement of 160 nm resolution patterns from a multilayered mask on a PMMA photoresist, although a number of laser shots much higher than expected was unexpectedly required.[1]

2.2.2 Collector Optics. The two ellipsoidal grazing-incidence Ru-coated mirrors have been made by Media Lario Technologies (Italy).[27] Their symmetrical configuration is aimed at maintaining the homogeneity of the source angular distribution at the focus of the second mirror. The first ellipsoidal mirror is placed in the source chamber (figure 2.1). It collects the radiation from the source, emitted between 9° and 19° with respect to the target normal, and redirects it to the IF with an efficiency of 8% (over a 2π solid angle) and an integrated reflectivity of ≈80% in the EUV. Actually, before reaching the IF, the radiation is spectrally filtered by a 50% EUV transmittance, 150 nm thick, Ni mesh supported Zr filter (Luxel Corporation, USA), which blocks most of the out-of-band radiation and also isolates the 1 mbar Kr gas in the source chamber from the 10^{-6} mbar vacuum in the projection chamber.

Figure 2.3 shows the spatial distribution at the IF of the 0.8 µm Al foil filtered radiation, as measured by the optical density variation induced on a Gafchromic HD-810 radiochromic dosimetry film. The relationship between EUV intensity and optical density is almost linear in the explored range.[1]

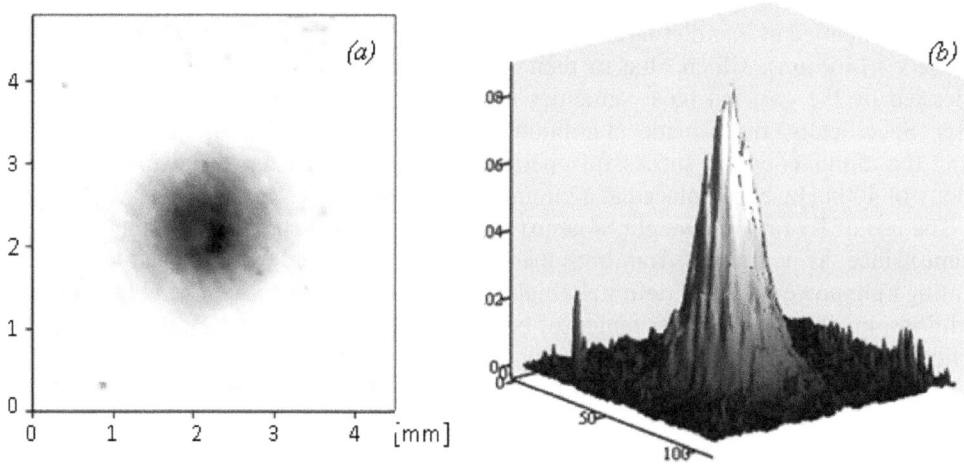

Figure 2.3 *(a) Image and (b) surface plot of the optical density spatial distribution induced by the properly filtered EUV radiation from EGERIA source on a dosimetry film at the IF.*[1]

2.3. Printer Module

The second ellipsoidal mirror is placed in the projection (and exposure) chamber. It conveys the filtered radiation gathered from the IF to the mask illuminating optics (figure 1.1). As such, it can be considered as a component of the printer module, whose central part is the Schwarzschild objective, by which the pattern reflected by the mask is eventually imaged on the wafer.

2.3.1 Mask Illuminating Optics. Two Mo/Si multilayer mirrors, sequentially spherical convex and flat, serve to focus the EUV radiation, conveyed by the second ellipsoidal mirror, on the patterned mask. Both mirror Mo/Si multilayers along with that for the mask were made by the INFN Legnaro Laboratories (Italy)[28] with up to 65% peak reflectivity at the operating conditions, as tested at the BEAR beam-line of ELETTRA in Trieste (Italy),[29] whereas the absorbing patterns on the mask were made by CNR-IFN in Rome.[30] Thanks to the high convex-mirror magnification and to compensation of the convex mirror spherical aberration by a slight longitudinal shift of the second ellipsoidal mirror from the relevant confocal position, the mask is nearly homogeneously illuminated over a 3 mm size area as shown by the image in figure 2.4. This fits well with the high-resolution projection field dimension of the SO, evidenced by the dashed circle in the figure. The image has been obtained exposing at the mask plane an Ilford Q-plate to the incoming radiation; the shadow of the Ni-mesh wires supporting the Zr filter is clearly distinguishable in the figure. The EUV energy density (fluence) on the mask has been measured by an IRD photodiode model AXUV20BNC and is ≈ 10 $\mu J/cm^2/shot$.[1]

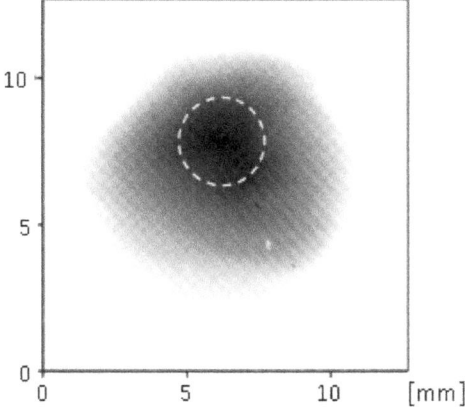

Figure 2.4 *Image of the optical density spatial distribution induced by the properly filtered EUV radiation from EGERIA source on a Q-plate at the mask plane. The dashed circle highlights the high-resolution object field of the SO[1].*

2.3.2 Mask Imaging Optics. The optical system, specifically designated to transfer the desired pattern from the illuminated mask onto the semiconductor wafer, is of pre-eminent relevance within the context of EUVL. SO based configurations have proven to be good candidates for such an issue. The SO is a convex-concave mirror system, in which the mirrors are concentric.[31] Accordingly, it conveys the simplest optical scheme that can be devised as an effective projection system for EUVL, when configured to permit a reduction imaging, i.e. with the illumination field from the object hitting firstly the convex and then the concave mirror, as exemplified in figure 2.5.

The SO is characterized by the object-to-image formula,

$$V(t,\alpha) = t\frac{\sin\alpha}{\sin\alpha'}, \tag{2.1}$$

where V and t denote the axial positions Z_o and Z_i of the on-axis object and image points (in the respective object and image space) scaled to the curvature radius R_2 of the convex

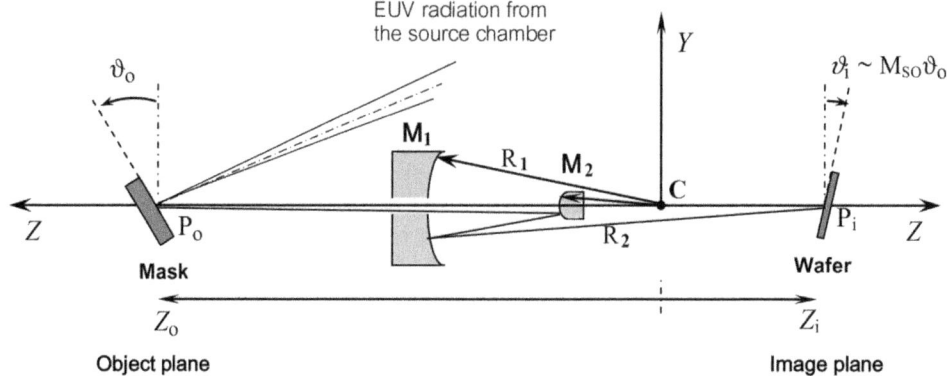

Figure 2.5 *Typical arrangement of a SO type reduction imaging in an EUVL projection setup, showing the mask (i.e. object) plane and the wafer (i.e. image) plane. (C = common centre of curvature of the two SO mirrors).*

mirror, i.e. $V \equiv Z_i/R_2$ and $t \equiv Z_o/R_2$, whereas α and α' are the inclination angles to the optical axis of the incoming and emerging rays. The latter are related by

$$\alpha'(t,\alpha) = 2\sin^{-1}(t\sin\alpha) - 2\sin^{-1}\left(\frac{t}{r}\sin\alpha\right) - \alpha, \qquad (2.2)$$

with r equalling the concave to convex mirror curvature radii ratio, $r = R_1/R_2$.

The standard SO configuration provides an easily feasible system, and so convenient as regards manageability, costs and alignment issues in comparison with multi-element projection optics. Also, it is capable of yielding the high resolution required by EUV lithography, albeit over a rather limited exposure field.

As a basic characterization, the ENEA MET SO has been designed to have an image-space numerical aperture NA = 0.23, a magnification $M \approx 1/9.7$, mirror curvature radii $R_1 = 144.23$ mm and $R_2 = 45.06$ mm, and respective mirror diameters $\Phi_1 = 74$ mm and $\Phi_2 = 12.7$ mm. It has been manufactured by Société Européenne de Systémes Optiques (SESO, France),[32] with a global figure error ≤ 8 nm, and an expected reflectivity curve peak $\geq 65\%$ at 14.4 nm over the whole coated surfaces of both mirrors (by a graded multilayer coating).

More precisely, the configuration implemented in the ENEA MET is the so-called modified SO (MSO) configuration.[33] It preserves the mirror concentricity of the standard configuration[b] while optimizing the SO performance toward a better resolution by placing the mask and the wafer at suitably designed locations, different from those conforming to the standard configuration.[35,36] Specifically, in the ENEA MET SO the mask and wafer axial positions are respectively $Z_o = 340.22$ mm and $Z_i = 36.26$ mm, whereas the values corresponding to the standard configurations would be $Z_o = 350.296$ mm and $Z_i = 36.138$ mm.

---

[b] Let us recall that, in view of extending the good performance characteristics of the SO to a larger exposure field, further schemes have been proposed. For instance, in the SO-based configuration, analysed in [34], the mirrors are slightly displaced from the concentric location while the object and image planes are correspondingly placed at positions such to eliminate, as in the ordinary SO, the third-order longitudinal aberrations.

A detailed analysis of the properties of the MSO configuration, as regards, for instance, the dependence of the image-plane resolution and third-order aberrations on the numerical aperture and on-axis object position has been presented.[35,36] It was shown there that the typical worsening trend of the geometrical resolution in a conventional SO configuration with increased numerical aperture is significantly mitigated (by a factor ≈ 5 for the magnification of concern) in the corresponding modified scheme. In fact, as conveyed by figure 2.6, the resolution *vs.* numerical aperture curve in a MSO lies markedly below that pertaining to the ordinary SO. In conformity to the spirit of the quoted references, the figure reports the results of both a semi-analytical procedure and a numerical simulation. The former resorts to analytical expressions for the geometrical resolution *RES*, defined as the standard deviation of the off-axis distances of the image points over the image plane, which have been numerically elaborated by the Mathcad 13 package, whereas the latter directly conveys the rms radius of the ray distributions around the chief ray, resulting from the sequential ray-tracing performed by the lens-design ZEMAX code.[37] Moreover, the ZEMAX-determined values are synthesized through best fit procedures into compact scaling laws of the system design parameters *vs.* the numerical aperture. Thus, the R_2-scaled geometrical resolution $res \equiv RES/R_2$ in the SO and MSO schemes are given by

$$res_{SO}(NA) = 4.71 \times 10^{-3} NA^{5.07}, \quad res_{MSO}(NA) = 1.17 \times 10^{-3} NA^{5.2}, \quad (2.3)$$

over the examined range of values NA = 0.1–0.3, and M_{SO} = 1/10. Here, M signifies the paraxial magnification relative to an assigned object plane, defined as usual by the ratio of the paraxial image position over the corresponding object position, and hence in a SO it is given by

$$M_{SO}(t,r) = \left.\frac{V(t,\alpha)}{t}\right|_{\alpha \to 0} = \frac{r}{2rt - 2t - r} = \frac{r - 1 + \sqrt{r}}{r - 1 - \sqrt{r}}. \quad (2.4)$$

As said, in the ENEA MET SO, $M \approx 1/9.7$.

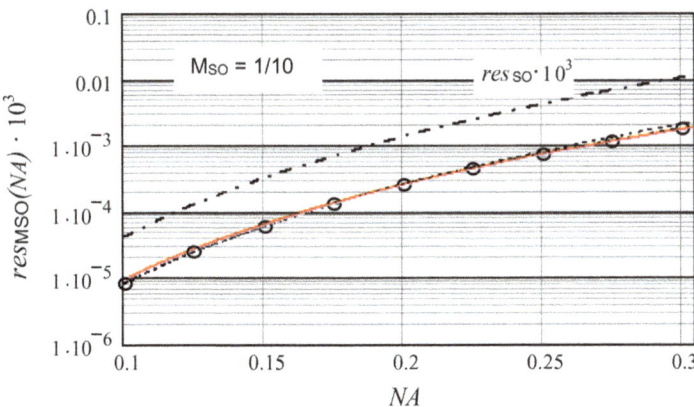

Figure 2.6 *R_2-scaled geometrical resolution vs. NA in a MSO for M_{SO} = 1/10, as conveyed by simulations based on the ZEMAX code (circles) and on specific analytical expressions (dashed line). The solid (red) and dot-dashed lines reproduce the best fits, given in equation (2.3), respectively for a MSO and a SO.*[35]

However, the advantage of the improvement of the geometrical resolution is contrasted by the corresponding reduction of the depth of focus. Therefore, the choice for a specific configuration to implement turns out to be a matter of a convenient compromise.[38]

EUV imaging cannot use refractive optical elements and hence entirely reflective optical components must be used. Therefore, in an EUV lithographic setup the mask is rotated with respect to the optical axis, thus resulting in a degradation of the resolution on the wafer. However, as expected according to the Scheimpflug principle,[39] this effect can effectively be countered by a reciprocal tilt of the wafer. Analysis[36] has in fact confirmed that by a symmetrical suitable tilt of the wafer (figure 2.5) the performance of the untilted mask device can be restored as regards the values and the resolution distribution across the optical axis in the wafer plane. The paraxial magnification roughly rules the ratio between the tilts of the image and object plane, so that, in the ENEA MET SO, the $\vartheta_0 = 10°$ tilt of the mask with respect to the optical axis is compensated by the symmetrical $\vartheta_i \sim M_{SO(MSO)}\vartheta_0 \sim 1°$ tilt of the wafer.

2.3.3 ENEA MET SO mounting and alignment. The above consideration has been concerned only with the geometrical resolution. Actually, the spatial resolution contains also the diffraction limited resolution res_{diff} according to

$$res_{\text{tot}} = \sqrt{res_{\text{SO(MSO)}}^2 + res_{\text{diff}}^2},\qquad(2.5)$$

where the R_2 scaled diffraction limited resolution $res_{\text{diff}} \equiv RES_{\text{diff}}/R_2$ is well known to vary inversely with NA, since

$$RES_{\text{diff}}(NA) = 0.61\frac{\lambda}{NA}.\qquad(2.6)$$

The two contributions to equation (2.5) are plotted in figure 2.7 for both the standard and the modified configurations in correspondence with the specific values $M_{SO} = 1/9.69$, $R_2 = 45.06$ mm and $\lambda = 14.4$ nm pertaining to the ENEA MET SO setup.

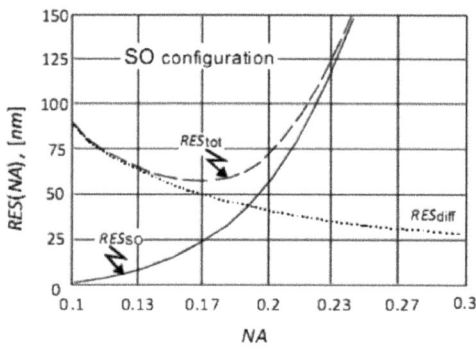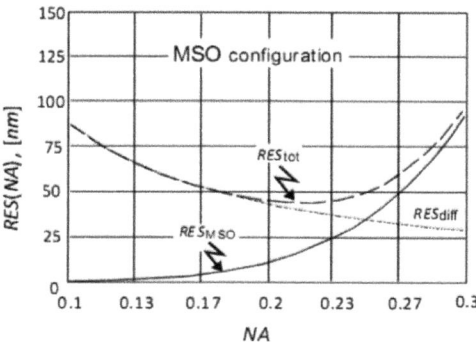

Figure 2.7 *Spatial resolution vs. NA in a standard and a modified SO configuration, in the case of $M_{SO} = 1/9.69$, $R_2 = 45.06$ mm and $\lambda = 14.4$ nm.*[38]

We see that in the standard configuration the geometrical resolution dominates the system performance at relatively small numerical apertures, $NA \geq 0.15$; in contrast, in the modified configuration the geometrical resolution becomes predominant over the diffraction limited resolution only at larger numerical apertures, $NA \geq 0.25$. Accordingly, the choice of the design parameters for the ENEA MET SO setup has been dictated by a reasonable compromise between easy system manageability and resolution request down to 50 nm. Indeed, the geometrical resolution was expected to be $RES_{MSO} \approx 27$ nm for the designed MSO configuration, whilst $RES_{diff} \approx 38$ nm, thus yielding $RES_{tot} \approx 47$ nm as the best resolution one could expect in the designed setup.

In view of approaching such a value as closely as possible, the alignment of the SO mirrors becomes crucial on account of the additional rather severe tolerance on perfect mirror concentricity. In fact, by proper ZEMAX ray-tracing simulations, aimed at analysing the dependence of the geometrical resolution on the two-mirror decentring, the tolerance (defined as the distance between the centres of curvature of the two mirrors which causes a $\sqrt{2}$ worsening in the resolution) has been estimated to be ≈ 10 μm and ≈ 1 μm for decentring respectively along the z-axis and in the transverse plane.

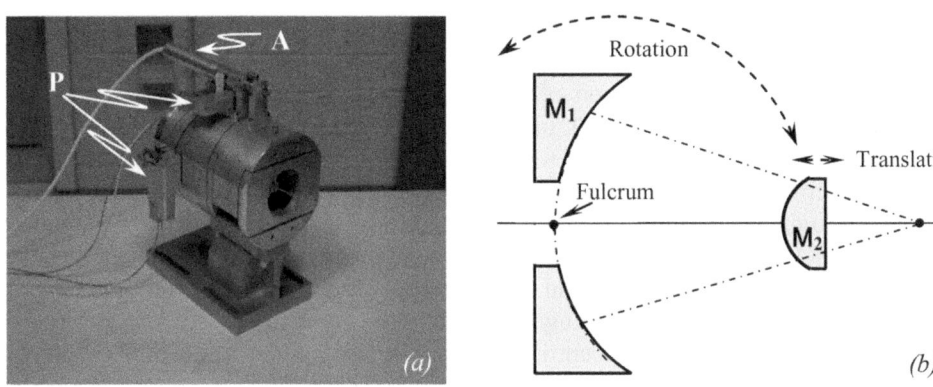

Figure 2.8 *(a) SO mounting-block in the ENEA MET: the two piezoelectric transducers (PT) and the actuator (A) are clearly visible; (b) Schematic of the mirrors control; the coincidence of the concave mirror rotation fulcrum with its vertex enables the steadiness of the two-mirror distance during tilting.*

In this connection, the mechanical mounting of the objective is of specific relevance as to movement precision (repeatability in positioning) and stability. In figure 2.8(a) the SO mounting block is shown. It consists of a cylinder made by a single piece of aluminium alloy with specially thinned sections allowing for three degrees of freedom. The first two movements enabled are rotations of the concave mirror (which is located at the back end of the mount cylinder in the picture) around two mutually orthogonal axes with an angular resolution of ≈ 2 μrad. Both axes are orthogonal to the optical axis, and cross each other at the mirror vertex. The rotations are controlled by piezoelectric transducers (Physik Instrumente, model P-601.3SL), visible in the picture. As a third degree of freedom, the translation of the convex mirror (located at the centre of the parallelogram structure recognizable at the front end of the mount cylinder in the picture) along the optical axis is allowed and is controlled by an actuator (Physik Instrumente, model M-227.10) with a

$\approx 0.1\ \mu m$ longitudinal resolution. The scheme in figure 2.8(b) should clarify the "dynamics" of the SO mirror movements.

The SO alignment has been carried out by exploiting the well-known Foucault technique,[40] whose validity has already been demonstrated in the EUV for a SO.[41-44] However, since the in-band EUV power at the wafer level in the ENEA MET was unexpectedly too low to allow for an at wavelength alignment in a reasonable time, the Foucault test-based SO alignment has been performed by using ultraviolet light and overcoming the inherent limitations (related to the diffraction limit of the alignment wavelength) through a novel procedure, briefly addressed to in[2], while described in detail in.[45] In general, an at wavelength alignment on a low power laboratory plasma source is time expensive and component consuming. Therefore, the proposed alternative technique can conveniently be exploited when EUV/soft X-ray sources with a limited photon flux are involved. It is based on an accurate diagnostics for system aberrations (that evidently indicate mirror misalignments), gained from an experimental characterization and a reliable modelling of the dependence of the SO longitudinal aberrations on the mirror misalignments, so that it is possible to identify individually each aberration source and accordingly adjust the SO alignment parameters.

The setup of the procedure as implemented in the ENEA MET is sketched in figure 2.9. The UV beam of a XeCl laser ($\lambda = 308$ nm) is focused by a 12.5 cm focal length triplet lens in a $\approx 50\ \mu m$ diameter spot on a 5 μm diameter pinhole, which then, being illuminated by the central, quite uniform portion of the Airy disk, provides the on axis point-like object-source for the SO. After the SO, in the "classical" Foucault test the beam would be cut at the focal plane by a knife edge (KE) moving perpendicularly to the optical z-axis; this would generate on the observation plane the typical images, called foucaultgrams, corresponding to different kinds of aberrations, mainly spherical and coma aberrations in the case of the SO. In fact, in the case of the SO, a longitudinal mirror decentring causes a spherical aberration, with marginal and paraxial rays having different focal distances, shorter (longer) for the former according to whether the two mirrors are closer to (further from) each other relative to the concentric condition. In turn, a transverse shift of the two centres of curvature produces a coma aberration, witnessed by a typical comet-like figure on and in proximity to the focal plane. Evidently, on account of the afore-described ENEA MET SO mounting, spherical aberration can be controlled through the longitudinal translation of the convex mirror, whereas coma is controlled by tilting the concave one.

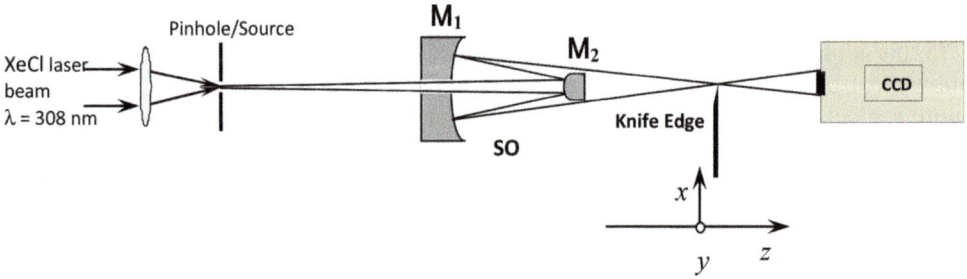

Figure 2.9 *Schematic (not in scale) of the experimental setup for the SO alignment by means of the revised Foucault-test based technique, described in the text.*

The drawbacks of aligning the SO by using a wavelength much larger ($\approx 20\times$, in our case) than the operating one can in some way be overcome by looking at the sequence of foucaultgrams obtained with a longitudinal scanning of the KE kept at 50% of the beam transmitted power. This has two advantages with respect to the transverse KE scanning in the focal plane. Firstly, the longitudinal aberrations extend over a wider spatial scale, increasing the sensitivity to their occurrence; secondly, the transmitted power which reaches the observation plane is almost constant for the different KE positions (that is for each sequence of foucaultgrams).

The foucaultgrams were recorded on the CCD sensor of a 16-bit-dynamics PI-MTE:1300B Princeton Instrument camera, placed at a distance from the SO focal plane such that the 26.8×26.0 mm^2 imaging area was almost completely illuminated. The transverse position of the KE was fixed when the power collected by the CCD was half of the full power (i.e. in absence of the KE) within the range of the KE longitudinal positions of interest. After that, by means of a longitudinal scan of the KE, it was possible to identify and quantify the type of aberration occurring. Examples of camera-recorded foucaultgrams are shown in figure 2.10. They were obtained for a longitudinal KE scan over 20 μm, with the KE transverse position being controlled in order to transmit ≈50% of the power, and clearly reveal the occurrence of spherical and coma aberrations, induced by the controlled decentring of the mirrors, as specified in the figure caption.

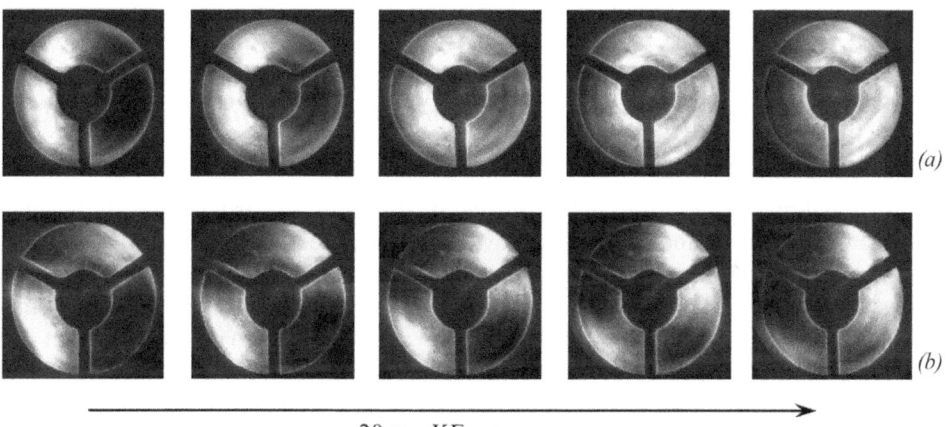

20-μm KE *z*-course

Figure 2.10 *Sequences of foucaultgrams of the ENEA MET SO, obtained for a longitudinal KE scan over 20 μm, corresponding to controlled decentrings of the SO mirrors which induce (a) a spherical aberration with the mirrors nearer to each other with respect to the concentric condition of ≈ 350 μm, and (b) a coma aberration with the concave mirror tilted by ≈ 0.84 mrad, corresponding to a transverse decentring of ≈ 120 μm. In both cases the KE allows the transmission of half the beam power.*[45]

Besides using the ZEMAX code for simulating the SO behaviour, we developed a proprietary Visual C++ ray-tracing program to reproduce the observed foucaultgrams. The program demands as input data the longitudinal distance between the SO mirror centres of curvature and the orthogonal tilts of the concave mirror, along with the KE and CCD positions. For instance, the sequence of camera-recorded foucaultgrams of figure 2.10 is paralleled by that in figure 2.11, showing the foucaultgrams, calculated by the program corresponding to similar mirror decentrings.

As the knife moving along the z-axis crosses a focal plane, the shadow position on the CCD changes from right to left hand. When spherical aberration is present, the focal plane

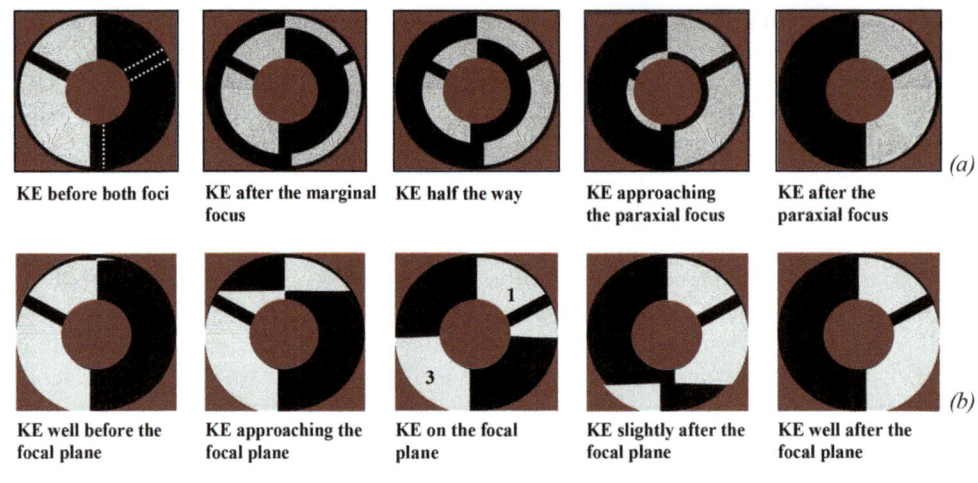

KE before both foci	KE after the marginal focus	KE half the way	KE approaching the paraxial focus	KE after the paraxial focus

(a)

KE well before the focal plane	KE approaching the focal plane	KE on the focal plane	KE slightly after the focal plane	KE well after the focal plane

(b)

20- μm KE *z*-course

Figure 2.11 *Ray-tracing calculated foucaultgrams for the ENEA MET SO in the presence of (a) spherical aberration with the two mirrors closer to each other than in the concentric condition, and (b) coma aberration arising from a tilt of the concave mirror around an axis parallel to the x-axis (see figure 2.9). The program simulates also the three-fin spider mount which supports the convex mirror.*[45]

of the marginal rays differs from that of the paraxial rays. If, in particular, as in the images of part (a) in the figures, the former is at lower *z* values, the knife is between the two foci, and hence the shadow for the marginal rays moves to the left while that for the paraxial rays is still on the right, thus yielding a "double-C" structure in the foucaultgram. In the case of coma, we chose the KE cutting direction in such a way as to cut the right part of the comatic circle, which amounts to obscuring the contributions from the second and fourth quadrants of the beam at the focal plane – see the central image in figure 2.11(b). We can then measure the longitudinal aberration (LA) by looking at sequences like those in figure 2.10, defining the spherical LA as the difference between the KE *z*-positions corresponding to the appearance and disappearance of the double-C structure, and the coma-related LA as the difference between two symmetrical KE *z*-positions with respect to the focal plane corresponding to a defined change in the intensity distribution, which we fixed by the equalization of power in the first and second quadrants and in the third and fourth quadrants, respectively, for decreasing and increasing KE *z*-values from the focal plane.

The modelling of the SO aberrations allows us to correct the mirror relative positions up to the SO alignment. However, the ray-tracing program does not account for diffraction, as is evident when comparing the foucaultgrams in figures 2.10 and 2.11, where it causes the experimental images to be blurred. Because of diffraction, the estimation of the aberration amplitude becomes much more difficult or even impossible when the alignment is progressively improved and the beam dimension at the focal plane becomes comparable to λ / NA.

In order to make the ray-tracing based modelling a reliable diagnostic tool for mirror misalignments (be they externally controlled or not), diffraction is simulated in the C++ code by building, for each KE *z*-value, composite foucaultgrams each made up by the overlapping of single binary contributions from beams whose propagation direction is distributed around the optical axis within the $1.22\lambda/\Phi_1$ diffraction angle (Φ_1 being the concave-mirror diameter), and whose foci are consequently spread in the focal plane within the Airy disk, that is within the diffraction limited focal spot size $1.22\lambda/NA \approx 1.6$ µm. The calculations can be effectively simplified by considering that this overlapping is equivalent to summing up the contributions obtainable by placing the KE at different transverse positions (for each z-value) within the Airy disk. The result is shown in figure 2.12.

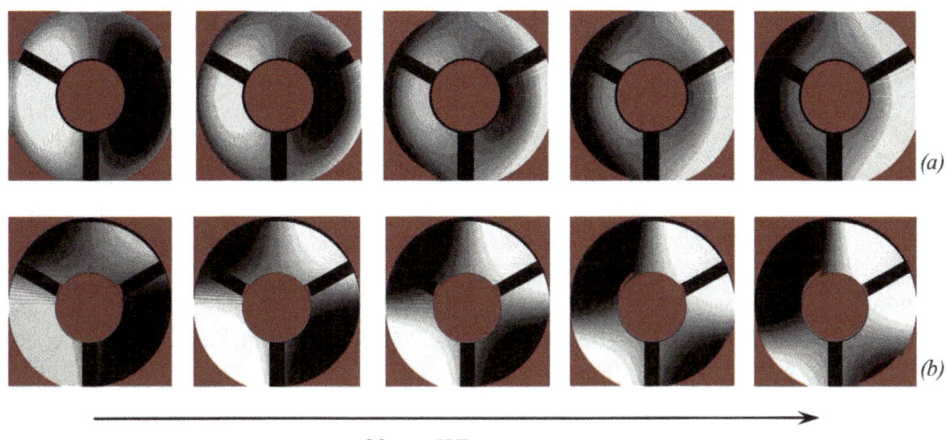

20-µm KE z-course

Figure 2.12 *Foucaultgrams calculated, with the inclusion of the diffraction effects, in the same misalignment conditions as in figure 2.11. Each image has been obtained superimposing 50 beams. By increasing this number the number of grey levels would increase.*[45]

We measured the longitudinal aberrations, as previously defined, by varying one SO free parameter at a time. The procedure was cyclically repeated for the three degrees of freedom until the results were reproducible. The optimum position of the SO mirrors, *i.e.* that making the LA equal to zero for each parameter, could finally be found by means of interpolation. The results in the case of spherical aberration are reported in figure 2.13(a) showing two sets of measurements respectively before and after the coma correction. The improvement in the definition of the best M_2 z-position in the second run is evidenced by both the smaller error bars and the smaller LA values. The M_1 tilt effect on LA for both orthogonal rotation axes is shown in figure 2.13(b). In this case, the error bars are smaller due to the rather more quantitative criteria adopted to evaluate the coma-related LA.

3. ENEA MET: PRINTING PERFORMANCES

As originally reported,[1] the ENEA MET, operating at 14.4 nm with a 30 µm thick Ta tape target, was capable of imaging patterns from a multi-layered photomask onto a PMMA resist with edge response of 90 nm.

3.1 ENEA MET: wafer printing

The exposure was performed using a mask with Cr absorbing patterns in the form of a grating with variable period, starting from 8 µm down to 1µm half-pitch linespace.

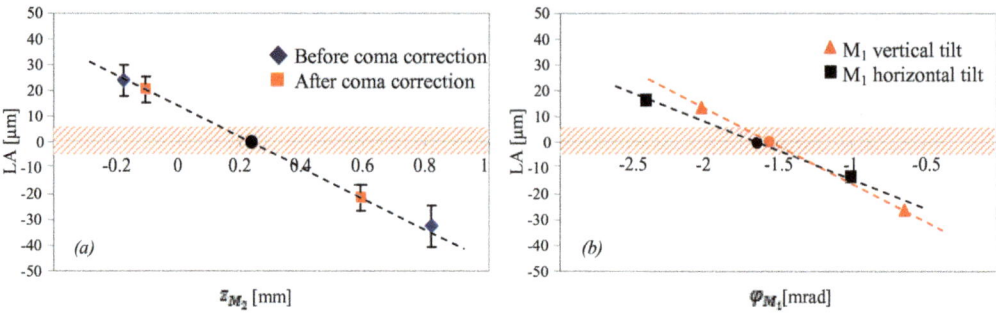

Figure 2.13 *Longitudinal aberration versus SO alignment parameters: (a) LA vs. convex mirror M_2 z-position; (b) LA vs. concave mirror M_1 tilts (error bars are smaller than the symbols). In both cases, the position of the zero value on the abscissa is arbitrary. The dashed area conveys the LA range concealed by diffraction. The circles identify the values chosen as correct alignment parameters.[45]*

The fluence on the wafer can be estimated from that on the mask, taking into account the mask and SO mirror reflectivities, the SO geometrical loss and magnification. Actually, the SO transmission efficiency was revealed to be definitely lower than expected, and yielded an EUV fluence of $\approx 10 \ \mu J/cm^2$/shot on the wafer plane against the expected (and needed) fluence of tens of mJ/cm^2. Therefore, we exposed the 996000 molecular weight, 100 nm thick PMMA resist with 2000 shots, in order to reach the suitable integrated fluence of $\approx 20 \ mJ/cm^2$. The cause of the unexpected SO deficiency was subsequently investigated and the relevant results are presented in the next subsection.

The exposed resist was then developed for 30 s in 30% MIBK-IPA solvent, fixed in IPA and observed by an atomic force microscope. Figure 3.1 shows both the 2D image and the 1D height profile of a 160 nm half-pitch line-space pattern. The profile has been obtained by averaging height values along the lines. The measured edge response (distance to rise from 10% to 90% of the modulation depth) is 90 nm. The pattern is not printed at the full resist height, probably due to a fluence that was still insufficient for the resist used.

The edge response width yields a valuable estimate of the resolution of the device.[46] For an alternative, even though strictly related, estimate, we considered the PMMA modulation amplitude as a function of the spatial frequency of the imaged patterns, which should convey the modulation transfer function (MTF)[c] of the ENEA MET printing line,

---

[c] We recall that the MTF is the spatial frequency response of an imaging system. It is formally given by the absolute value of the optical transfer function, which in turn follows from the 2D Fourier transform of the point spread function (or, impulse response of the imaging system) or the 1D Fourier transform of the line spread function [46, 47]. The latter is in turn the derivative (or first difference) of the edge response. The resolution estimate, based on distinguishable line pairs (lp, i.e. a dark line next to a light line) per unit length (mm or inch), corresponds roughly to spatial frequencies at which the MTF is between 5% and 2%. An MTF of 9% is implied in the definition of the Rayleigh diffraction limit. A more reliable resolution estimate is conveyed by the spatial frequency where the MTF is 50%, when the contrast has dropped by half.

For a diffraction limited circular lens, the following behaviour can empirically be drawn for the MTF as a function of the spatial frequency κ

$$Res_{diff} = 1/\kappa_9$$

where Res_{diff} denotes the lens resolution according to the Rayleigh equation, $Res_{diff} = 0.61/NA$ (see also equation (2.6)) and κ_9 represents the spatial frequency at which, as said, the MTF is 9%. Also, MTF is 80% at a frequency κ_{80} given by $\kappa_{80} \sim (4/17)\kappa_9 \sim 0.23\kappa_9$, thus implying $Res_{diff} = 0.23/\kappa_{80}$, the units being in accord with those used for the spatial frequency (lp per unit length).

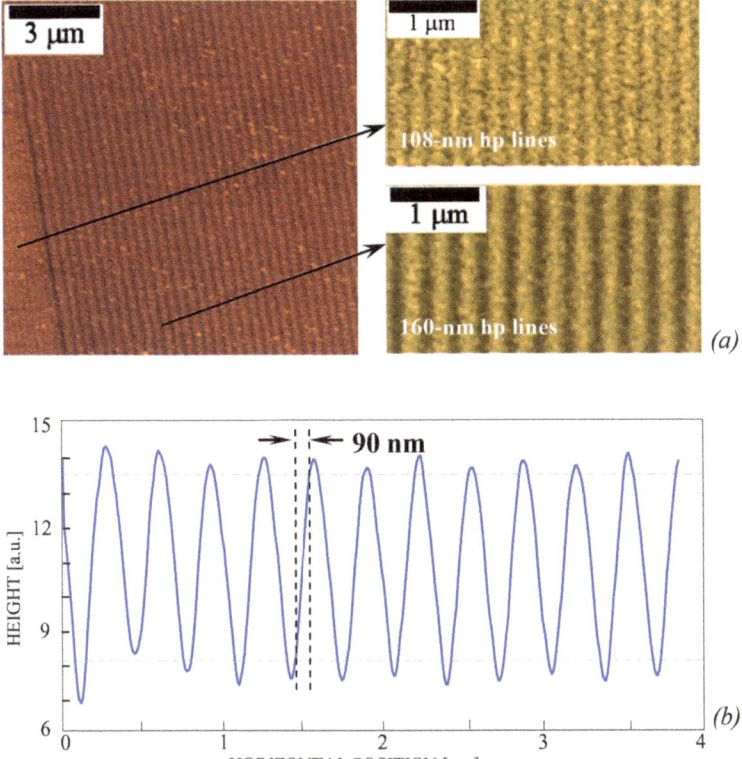

Figure 3.1 *Line-space printing on PMMA: (a) 2D patterns of 160 nm and 108 nm half-pitch (hp) observed by an atomic force microscope. The quality of the 108 nm hp lines seems to be limited mainly by the large granularity of the photoresist. (b) Line profile integrated over a selected portion of the 160 nm hp line PMMA area. The horizontal dashed lines highlight the 10% and 90% modulation levels, conveying an edge response width of 90 nm.*

conforming to the described operation. We observed a reduction of the modulation amplitude of 80% over the 320 nm patterns (figure 3.2), and hence at $\kappa_{80}=1/320$ lp/nm. Therefore, assuming for the MET imaging optics the same behaviour as a diffraction limited circular lens (see footnote c), we may give an estimate for the printing-line optical resolution Res_{opt} according to $Res_{opt} = 0.23/\kappa_{80} \sim 75$ nm.

Moreover, by noting that the lithographic resolution Res_{lith} is determined by the Rayleigh scaling equation[48] $Res_{lith} = k_1 \lambda/NA$, where the dimensionless parameter k_1 is made by recent progress to lie in the range $0.3 \leq k_1 \leq 0.5$, the value extent $32 \text{ nm} \leq Res_{lith} \leq 62 \text{ nm}$ can accordingly be estimated for the ENEA MET lithographic resolution.

We may also note that single-shot operation was addressed by the project.[3] Yet, as already indicated, we were forced to operate the system for 2000 laser shots (i.e. several seconds) in order to reach a sufficient fluence to pattern the photoresist used. It is then reasonable to guess that the result obtained suffered also from thermal and mechanical instabilities that could have been influential over the actual operation time.

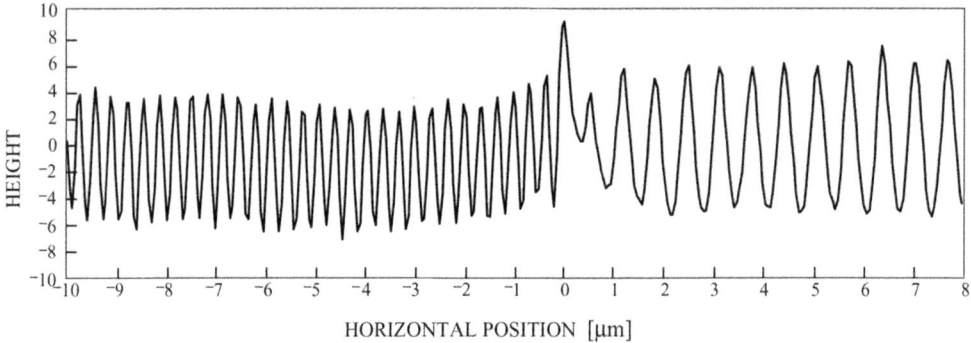

Figure 3.2 *Line profile integrated over a selected portion of the PMMA area comprising contiguous regions with 320 nm (to the left) and 640 nm (to the right) full period lines.*

3.2 ENEA MET SO: checking the transmission

The reported result confirms the very good figure error of both the SO mirrors, measured to be ≤ 8 nm (FWHM). In contrast, as said, the SO transmission at the operating wavelength is markedly smaller (roughly estimable to ≈ 7 times) than expected. In order to investigate the origin of this deficiency, we performed an overall SO characterization at the BEAR beam line of ELETTRA,[29] by measuring the relevant transmission over the wavelength range 13–17 nm.

Technical difficulties forced us to measure the overall transmission of the objective rather than the reflectivity of each mirror, about which however reasonable conjectures can be drawn from the measured transmission, as detailed below.

The BEAR beam line of ELETTRA is very suitable for the characterisation of optics in the EUV range. It is based on a bending magnet source and a high performance monochromator able to cover the full EUV range with an absolute accuracy of photon energy measurement of 0.1 eV, thus implying that at $\lambda = 14.4$ nm the absolute accuracy of a wavelength measurement is better than 0.02 nm.

Due to the rather small divergence of the BEAR EUV beam (1° after the last focal point), it was not convenient to place the SO at the same distance from the last focal point of BEAR as from the mask in the MET layout ($Z_0 \approx 34$ cm), since this would have strongly limited the portion of the convex mirror surface illuminated by the beam. In fact, it was placed at a distance ≈ 124 cm from the last focal point of BEAR. This ensured that the beam cross section was larger than that of M_2, thus implying that the reflectivity measurement results from an average over the whole mirror surface. The consequent variation of incidence angles over the M_2 surface should produce a variation of the centre wavelength of the mirror reflectivity curve, but, as a related analysis has shown, this variation should be very little indeed.

The SO transmission curve $T(\lambda)$ in the EUV range, as measured at ELETTRA with an absolute accuracy of 1%, is displayed in figure 3.3(a) (blue line). The relevant theoretical transmission curves are also shown in the figure; they have been deduced on the basis of the considerations illustrated in the forthcoming subsections. We see that the experimental curve reaches a peak value (almost 7%) at 14.8 nm rather than at 14.4 nm, and displays a mirror reflectivity bandwidth (≈ 0.6 nm FWHM and larger) and the observed wavelength shift. This presumably denotes a wavelength shift of both mirror reflectivities and a relative mismatch as well.

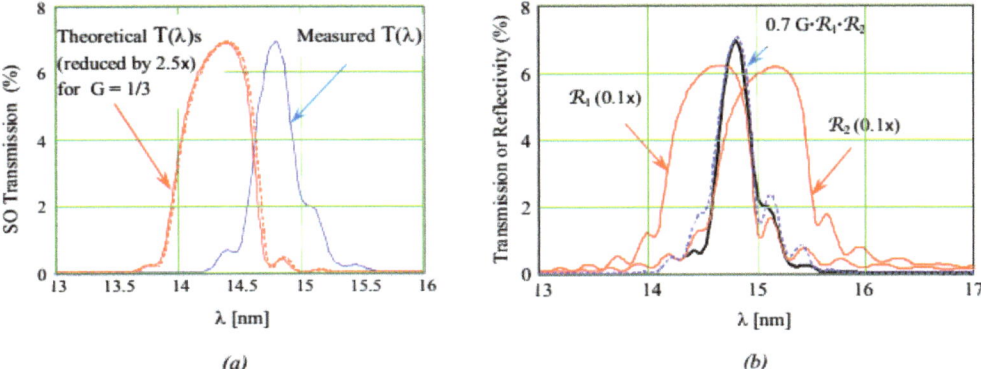

(a)   *(b)*

Figure 3.3 *(a) Experimental SO transmission T(λ) (blue line) and theoretical transmissions of an ideal SO with geometrical transmission factor G = 1/3, corresponding to a source located at the correct mask position (solid red line) and at approximately 90 cm farther (dashed red line). The red lines have been scaled by a factor 2.5 in order to reach the same peak value of the blue one. (b) Wavelength dependent reflectivities of ideal Mo/Si multilayer mirrors centred at 14.66 nm and 15.16 nm having the same peak of 62% (red lines) and transmission of an SO resulting from such reflectivities and a geometrical transmission factor G = 1/3 (blue dotted line), which can be compared with the measured SO transmission (black line), also reported in panel (a).*

Defined by the ratio of the output to the input power, $T(\lambda)$ is related to the product of the single-mirror spatially integrated reflectivities, $\mathcal{R}_1$ and $\mathcal{R}_2$,

$$T(\lambda) = G(\lambda)\mathcal{R}_1(\lambda)\mathcal{R}_2(\lambda) \tag{3.1}$$

through the geometrical transmission factor $G(\lambda)$. According to the measurement conditions and the characteristics of the system, the geometrical transmission factor $G(\lambda)$ arises from several factors. Firstly, since, as mentioned earlier, the beam cross section was larger than that of M_2, part of the incoming rays travelled externally to the mirror surface, thus amounting to an unavoidable loss of the relevant photon flux. Since the BEAR beam divergence depends slightly on wavelength, such a loss yields a smooth dependence of G on the wavelength. Also, some of the rays hitting the convex mirror follow multiple (> 2) reflection paths before converging to the relevant best focus position, whose location on the SO axis is closer to the convex mirror than that of the rays undergoing double reflection paths. Finally, both the rays hitting the three M_2 holding fins and those travelling almost parallel to the optical axis do not contribute to the SO imaging process.

Even though such factors, and hence $G(\lambda)$, could be estimated theoretically, we resolved to perform an experimental estimation of $G(\lambda)$ in order to obtain an estimation of the mirror reflectivity product $\mathcal{R}_1\mathcal{R}_2$ (through equation (3.1)) as error-free as possible. In this connection, two different procedures have been followed. One consists of assuming G be wavelength independent so that its value could be deduced from equation (3.1) on the basis of measurements of the single-mirror reflectivity in the visible (VIS) range, carried out in our laboratory by using a frequency doubled Nd-YAG laser at 532 nm. Accordingly, the BEAR beam has been spatially limited (by using four knives after the monochromator) in order to select its central portion, over which the beam characteristics are fairly wavelength independent. The other procedure consists of recording an image of the cross section of both the input and output beams (at the wavelength of the measured SO transmission peak), which then by comparison allows us to estimate the portion of the incident beam that could not reach the SO exit.

3.2.1 SO mirror reflectivity product as resulting from the estimate of G in the visible range. The SO transmission in the visible range (specifically, at $\lambda_{green} = 550$ nm) was measured by adjusting the BEAR line monochromator at the zero order (no wavelength selection) and by introducing a green interference filter (at λ_{green}) after the last focal point. As a result, we obtained $T(\lambda_{green}) = 4\%$. On the other hand, as already pointed out, the reflectivity of the convex mirror at its central (flat) part was measured (at normal incidence) in our laboratory by using a frequency doubled Nd:YAG laser at 532 nm and estimated to be $\approx 35\%$.

Hence, by assuming $\mathcal{R}_1(\lambda_{green}) \approx \mathcal{R}_2(\lambda_{green}) \approx 35\%$, by equation (3.1) the SO geometrical transmission factor at 550 nm is estimated to be $G_{VIS} = 1/3 = G(\lambda)$, in accord with the value obtained from the aforementioned measurement in ENEA. As a consequence, the SO mirror reflectivity product $\mathcal{R}_1\mathcal{R}_2$ should be three times the corresponding transmission at any wavelength, namely $\mathcal{R}_1\mathcal{R}_2 \approx 21\%$.

The transmission curve for a pair of ideal multilayer mirrors with a reflectivity peak at 14.4 nm, conforming to the estimated $G(\lambda)$, is shown in figure 3.3(a) corresponding to both a source located as in the ENEA MET (i.e. at the mask position) and as in the measurement setup (i.e. at approximately 90 cm further), in the latter case the different incidence angles on M_2 are accounted for. The single-mirror reflectivity curves have been obtained from the Henke website,[49] entering the appropriate incidence angles on the mirrors (as conveyed by specific calculations) and as multilayer periods the values measured by SESO on the samples coated during the tests for the preparation of the mirror coatings (for instance, a period of 7.48 nm on M_2 at 6 mm radially off centre).

Evidently, the SO transmission displays a reduction of both the peak value (by more than a factor 2) and the bandwidth (by a factor ≈ 2) with respect to the expected values as well as an overall shift towards longer wavelengths. Such behaviour is presumably ascribable to different wavelength shifts of the mirror reflectivities with respect to the desired value (14.4 nm), in both cases towards longer wavelengths. In this connection, we may note that, although a small spectral shift toward longer wavelengths might be expected as a consequence of the fact that the source, as said, is not at the prescribed SO object plane, such a shift should be however much smaller than that displayed by the experimental curve, as can be inferred from a comparison between the red lines in the same figure.

For illustrative purposes, figure 3.3(b) shows the transmission of a SO (with $G = 1/3$) composed of two mirrors with reflectivies centred at $\lambda = 14.66$ nm (period of 7.557 nm) and $\lambda = 15.16$ nm (period of 7.85 nm), both having a 62% peak value. The single-mirror reflectivities have been both reduced by a factor of 10 in the plot for a better visualization of the curves. Also, the resulting SO transmission has been reduced by a factor ≈ 1.4 in order to match the peak of the experimental curve (black line).

3.2.2 SO mirror reflectivity product as resulting from the estimate of G in the EUV range. As already said, the geometrical transmission factor G was also estimated in the EUV range, precisely at the wavelength of the SO transmission peak (i.e. $\lambda_{peak} = 14.8$ nm) by comparing the EUV beam spots at the SO entrance and exit, respectively, recorded on Gafchromic HD-810 radiochromic dosimetry film (placed at ≈ 106 cm from the last focal spot of the BEAR line) and an ILFORD Qplate film as shown in figure. 3.4.

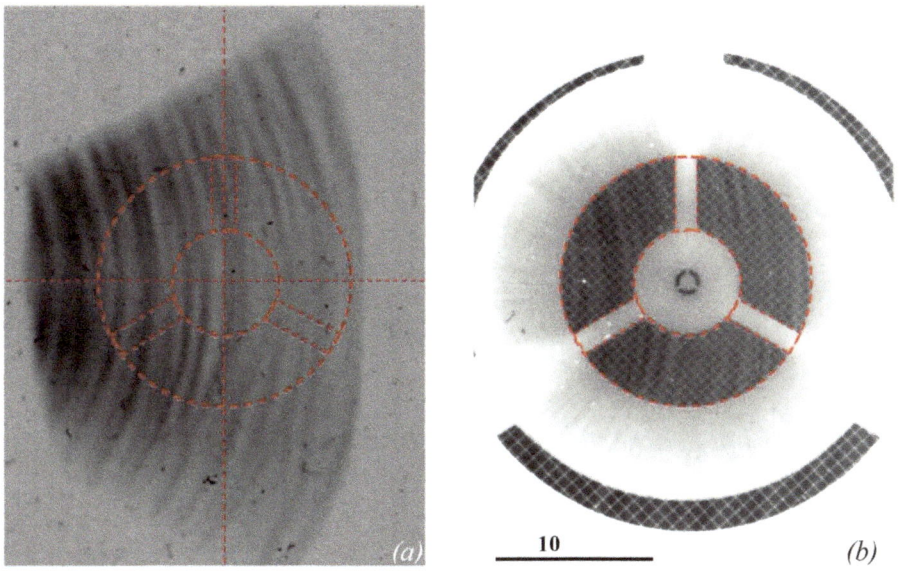

Figure 3.4 *EUV spots at (a) the SO entrance recorded on a Gafchromic HD-810 film and (b) the SO exit recorded on a ILFORD Q-plate photographic plate. Both spots are at $\lambda_{peak} = 14.8$ nm.*

The spot at the entrance is shaped as a portion of an annulus. This is due to the deformation of the squared spot (formed by the four knives at the monochromator exit) produced by the grazing-incidence parabolic mirror of ELETTRA, which introduces also a spatial modulation of the beam intensity, clearly visible in both images. Such a modulation is in this case very useful since it allows to recognize in the input beam (figure 3.4(a)) that portion (highlighted by a red dashed line), which could reach the SO exit and accordingly produce on the Q-plate the same spatial modulation (similarly highlighted in figure 3.4(b)). The mesh shadow viewable in figure 3.4(b) is due to the nickel mesh of a Zr filter (from Luxel Incorporation) used to cut the visible light. No filters were used for the Gafchromic HD-810 film, as it is not sensitive to visible radiation. The white area external to the highlighted region of the Q-plate image corresponds to the shadow of the diverging mirror M_2 (finally surrounded by the direct beam radiation), while the small dark ring in the middle of the highlighted region corresponds to the radiation (closer to the optical axis) which, experiencing multiple reflection paths, has then been focused along the SO axis to a plane at distance from the mirror centre of curvature shorter than the double reflection path ray image position Z_i. The images in the figure were acquired by a 16 bit CANON CanoScan 8400F, for which the relevant relation between the grey level g and the detected light intensity I has been tested to be as $g(I) = I^{1/2}$.

In order to infer the radiation fluence hitting the SO, we need to know the response curve of the Gafchromic HD-810 film at the wavelength of concern. To this end, the film has been exposed at four different EUV doses at a short distance (a few centimetres) from the last focal spot of the BEAR beamline, the dose being incremented by approximately a factor of two at each step (figure 3.5).

For the EUV doses of concern ($\approx$32–320 J), we can reasonably assume the film response to a local fluence $F(x,y)$ (i.e. the energy released on the film per unit area) to be the same as that of a photographic film through the EUV range, namely[50,51] $D(x,y) = D_S \ln(1 + F(x,y)/F_S)$, where $D(x,y)$ is the local film optical density, whilst D_S and F_S denote the EUV saturation values of the film optical density and fluence.

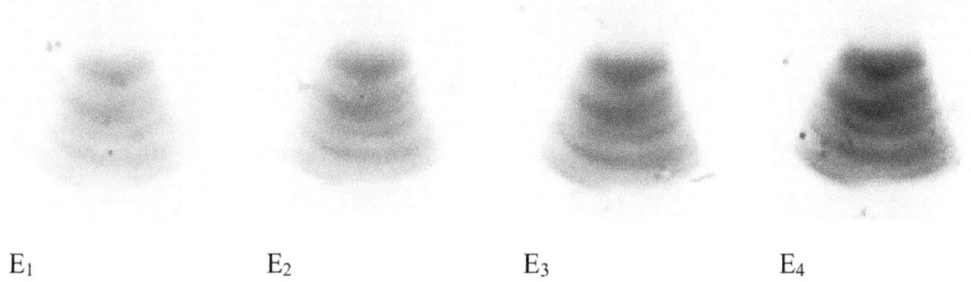

E_1 E_2 E_3 E_4

Figure 3.5 *1×1 cm² images of the Gafchromic HD-810 radiochromic films exposed at the BEAR beam line with different exposure doses: E_1 = 32.37 J, E_2 = 63.30 J, E_3 = 129.8 J, and E_4 = 317.5 J.*

Since, as earlier noted, for our scanner $g(x,y) = I(x,y)^{1/2}$, the relation between the fluence released on the film and the grey levels conveyed by the scanner can be written as

$$\log \frac{g_0}{g(x,y)} = \frac{1}{2} D_S \ln\left(1 + \frac{F(x,y)}{F_S}\right), \tag{3.2}$$

g_0 being the scanner grey level for a non-exposed film ($F = 0$).

Accordingly, the fluence as a function of the image grey level can be expressed in the form

$$F(x,y) = F_S \left(\exp\left[-2\log\left(g(x,y)/g_0\right)/D_S\right] - 1\right). \tag{3.3}$$

Finally, by integrating $F(x,y)$ on the four exposed films, the best fit values for the two parameters F_S and D_S have been deduced as F_S = 4.22 mJ/cm² and D_S = 0.165.

Then, on the basis of the above relation, the grey level map $g(x,y)$, conveyed by the image in figure 3.4(a), yields the input fluence map $F_i(x,y)$. An additional map, say $\overline{F}_i(x,y)$, is then generated from $F_i(x,y)$, by assigning zero values to all the pixels external to the region highlighted by the red dashed line. It is evident that $F_i(x,y)$ represents the geometrical fraction of the input beam effectively contributing to the SO imaging.

The geometric transmission factor at the EUV wavelength, G_{EUV}, is calculated as the ratio of the integrals of $\overline{F}_i(x,y)$ and $F_i(x,y)$, thus eventually obtaining the value

$$G_{EUV} = \frac{\int \bar{F}_i(x, y) \mathrm{d}x\mathrm{d}y}{\int F_i(x, y) \mathrm{d}x\mathrm{d}y} = 0.269. \qquad (3.4)$$

Even though slightly lower than G_{VIS}, it increases the peak value of the SO mirror reflectivity product (in the EUV range) to $\mathcal{R}_1\mathcal{R}_2(EUV) = 26\%$ according to the experimental transmission peak. Also, the accuracy of this estimate is definitely higher than that based on the measurements in the visible range.

Then as G_{EUV} is the value of interest for the ENEA MET, the estimate of G_{VIS} and hence of $\mathcal{R}_1\mathcal{R}_2$ in the visible range represents a useful touchstone.

3.2.3 ENEA MET: SO related limitations. Due to the SO transmission centre wavelength shift with respect to the other three multilayer reflective components of the ENEA MET, the total transmission over the printing line is reduced roughly by a factor of six. The curves in figure 3.3 unambiguously show how poor is the intersection between the SO transmission curve and that of a pair of Mo/Si mirrors working at the operating wavelength. Evidently, this limits the performance of the ENEA MET, which however might be partly recovered by replacing the aforementioned optics with optics peaking at 14.8 nm.

4 CONCLUDING REMARKS

On the basis of previous related works, the features of the ENEA EUVL MET, successfully operated in 2008, have been reviewed. As reported in,[1] exposing PMMA resist to a 14.4 nm radiation fluence of about 20 mJ/cm^2 by means of a modified Schwarzschild objective, 160 nm half-pitch line-space patterns have been reproduced. A simple optical analysis showed that the obtained lithographic resolution is not limited by the optical quality of the projection optics, as the estimated best attainable value is less than 80 nm half-pitch line-space. The MET performance is limited by the SO transmission being below the design value, presumably as a consequence of a mirror reflectivity mismatch, as subsequent analyses seem to indicate. However,[1] the results achieved show that it is possible to attain a nanometre-scale spatial resolution using low-cost Schwarzschild-type projection optics, as long as a proper design, figure error, alignment of the SO, and an excellent damping of both mechanical and thermal oscillations are ensured.

References

1 S. Bollanti, P. Di Lazzaro, F. Flora, L. Mezi, D. Murra and A. Torre, *EPL*, 2008, **84**, 58003.
2 P. Di Lazzaro, S. Bollanti, F. Flora, L. Mezi, D. Murra and A. Torre, *IEEE Trans. Plasma Sci.*, 2009, **37**, 475.
3 G. Baldacchini, A. Baldesi, S. Bollanti, F. Bonfigli, G. Clementi, A. Conti, T. Dikonimos, P. Di Lazzaro, F. Flora, A. Gerardino, R. Giorgi, A. Krasilnikova, T. Letardi, N. Lisi, T. Marolo, L. Mezi, R.M. Montereali , D. Murra, E. Nichelatti, P. Nicolosi, L. Palladino, A. Patelli, M.G. Pelizzo, A. Piegari, S. Prezioso, A. Reale, V. Rigato, A. Ritucci, A. Santoni, F. Sarto, F. Scaramuzzi, E. Tefoue Kana,

G. Tomassetti, A. Torre and C.E. Zheng, The Italian FIRB project on extreme ultraviolet lithography, *XXVII ECLIM Conf.*, 2004, 6-10 September (Rome, Italy).

4 V. Bakshi (Ed.), *EUV Lithography* 2009 (SPIE and John Wiley & Sons)

5 B. Lai, F. Cerrina and J.H. Underwood, *Proc. SPIE* 1985, **563**, 174; K. Hoh and H. Tanino, *Bull. Electrotech. Lab.*, 1985, **49,** 47.

6 H. Kinoshita, K. Kurihara, Y. Ishii and Y. Torii, *J. Vac. Sci. Technol. B*, 1989, **7**, 1648.

7 V. Banine and R. Moors, *International Workshop on EUV and Soft X-ray Sources*, (http://euvlitho.com/2011/2011%20Source%20Workshop%20Proceedings.pdf),   2011, 7-11 November (Dublin, Ireland);
   http://www.asml.com/asml/show.do?ctx=46772&dfp_product_id=842

8 SEMATECH Litho Forum, New York, 2010, 10-12 May. The Forum materials can be found at http://www.sematech.org/meetings/archives.htm.
   See also http://www.sematech.org/corporate/annual/annual10.pdf.

9 V. Bakshi (Ed.), *EUV Sources for Lithography*, 2006, **PM149** (SPIE Press, Bellingham)

10 K.Koshelev, *International Workshop on EUV and Soft X-ray Sources*, 2011, 7-11 November (Dublin, Ireland);
   T. Higashiguchi, *International Workshop on EUV and Soft X-ray Sources*, 2011, 7-11 November (Dublin, Ireland);
   http://euvlitho.com/2011/2011%20Source%20Workshop%20Proceedings.pdf

11 F. Flora, L. Mezi, C.E. Zheng and F. Bonfigli, *Europhys. Lett.*, 2001, **56**, 676.

12 S. Bollanti, F. Bonfigli, E. Burattini, P. Di Lazzaro, F. Flora, A. Grilli, T. Letardi, N. Lisi, A. Marinai, L. Mezi, D. Murra and C. Zheng, *Appl. Phys. B*, 2009, **76**, 277.

13 N. Kandaka and H. Kondo, *Jpn. J. Appl. Phys.*, 1998, **37**, L174.

14 J. Pankert, R. Apetz, K. Bergmann, G. Derra, M. Janssen, J. Jonkers, J. Klein, T. Kruecken, A. List, M. Loeken, C. Metzmacher, W. Ne®, S. Probst, R. Prummer, O. Rosier, S. Seiwert, G. Siemons, D. Vaudrevange, D. Wagemann, A. Weber, P. Zink, O. Zitzen, *Proc. SPIE*, 2005, **5751**, 260.

15 C. Koay, S. George, K. Takenoshita, R. Bernath, E. Fujiwara, M. Richardson and V. Bakshi, *Proc. SPIE*, 2005, **5751**, 279.

16 S.S. Harilal, B. O'Shay and M.S. Tillack, *J. Appl. Phys.*, 2005, **98**, 0361021-3.

17 S.S. Harilal, B. O'Shay, Y. Tao and M.S. Tillack, *Appl. Phys. B*, 2007, **86**, 547-553.

18 R. W. Coons, S. S. Harilal, D. Campos and A. Hassanein, *J. Appl. Phys.*, 2010, **108**, 063306.

19 D. Campos, S. S. Harilal and A. Hassanein, *J. Appl. Phys.*, 2010, **108**, 113305.

20 S. Bollanti, P. Di Lazzaro, F. Flora, T. Letardi, A. Marinai, A. Nottola, K. Vigli-Papadaki, A. Vitali, F. Bonfigli, N. Lisi, L. Palladino, A. Reale and C. E. Zheng, *Proc. SPIE*, 2003, **3767**, 33.

21 http://www.frascati.enea.it/fis/lac/excimer/egeria.html

22 S. Bollanti, D. Amodio, A. Conti, P. Di Lazzaro, F. Flora, L. Mezi, D. Murra, A. Torre, C. E. Zheng, D. Garoli, M.G. Pelizzo, P. Nicolosi, V. Mattarello, V. Rigato, A. Gerardino, *Proc. SPIE*, 2007, **6703**), 670308.

23 P. Di Lazzaro, S. Bollanti, F. Flora, L. Mezi, D. Murra and A. Torre, this volume.

24 S. Bollanti, P. Di Lazzaro, F. Flora, L. Mezi, D. Murra and A. Torre, *Appl. Phys. B*, 2009, **96**, 479.

25 P. Di Lazzaro, S. Bollanti, F. Flora, L. Mezi, D. Murra, A. Torre, *Appl. Surf. Sci.*, 2013, **272**, 13

26 http://www.mathworks.com/products/matlab/

27 http://www.medialario.com

28 http://www.lnl.infn.it

29 http://www.elettra.trieste.it/experiments/beamlines/bear/index.html

30 http://www.inf.cnr.it

31 K. Schwarzschild, *Abh. WIss. Goett. Math. Phys.*, 1905, **K1 NF 4**, 1.

32 http://www.seso.com

33 A. Budano, F. Flora, L. Mezi, *Appl. Opt.*, 2006, **45**, 4254

34 I.G. Artyoukov and K.M. Krymski, *Opt. Eng.*, 2000, **39**, 2163.

35 S. Bollanti, P. Di Lazzaro, F. Flora, L. Mezi, D. Murra and A. Torre, *Appl. Phys. B*, 2006, **85**, 603.

36 S. Bollanti, P. Di Lazzaro, F. Flora, L. Mezi, D. Murra and A. Torre, *Appl. Phys. B*, 2008, **91**, 127.

37 Information on and a demonstration version of the ZEMAX package are in www.optima-research.com

38 S. Bollanti, P. Di Lazzaro, F. Flora, L. Mezi, D. Murra and A. Torre, *Proc. SPIE*, 2005, **5962**, 5962Y-1

39 T. Scheimpflug, *Photogr. Korr.*, 1906, **43**, 516.

40 M. L. Foucault, *C. R. Acad. Sci. Paris*, 1858, **47**, 958.

41 R. B. Korsmeyer, *Sky & Telescope*, 1944, **3**, 18.

42 A.K. Ray-Chaudhuri, W. Ng, S. Liang, S. Singh, J. T. Welnak, J. P. Wallace, C. Capasso, F. Cerrina, G. Margaritondo, J. H. Underwood, J. B. Kortright and R. C. C. Perera, *J. Vac. Sci. Technol A*, 1993, **11**, 2324.

43 A.K. Ray-Chaudhuri, W. Ng, S. Liang and F. Cerrina, *Nucl. Instr. Methods A*, 1944, **347**, 364.

44 F. Barbo, M. Bertolo, A. Bianco, G. Cautero, S. Fontana, T. K. Johal, S. La Rosa, G. Margaritondo and K. Kaznacheyev, *Rev. Sci. Instrum.*, 2000, **71**, 5.

45 S. Bollanti, P. Di Lazzaro, F. Flora, L. Mezi, D. Murra and A. Torre, *Nucl. Instr. Methods A*, 2013, **720**, 168.

46 S.W. Smith, *The Scientist and Engineer's Guide to Digital Signal Processing*, 1997 (California Techn. Pub, 1997), chapter 25.

47 J.D. Gaskill, *Linear systems, Fourier transforms, and optics*, 1978, (John Wiley & Sons, New York)

48 B. J. Lin, *Proc. SPIE*. 1986, **633**, 44.

49 http://henke.lbl.gov/optical constants/

50 B. L. Henke, S. L. Kwok, J. Y. Uejio, H. T. Yamada, and G. C. Young, *JOSA B*, 1984, **1**, 818.

51 G.J. Tallents, J. Krishnan, L. Dwivedi, D. Neely, I.C.E. Turcu, *Proc. SPIE*, 1997, **3157**, 281.

CHARACTERISATION AND MITIGATION OF IONS AND PARTICULATE EMITTED BY SOURCES FOR EXTREME ULTRAVIOLET LITHOGRAPHY

P. Di Lazzaro, S. Bollanti, F. Flora, L. Mezi, D. Murra and A. Torre

ENEA Frascati Research Centre, Technical Unit APRAD, P.O. Box 65 00044, Frascati, Rome, Italy

1 INTRODUCTION

Laser produced plasmas are widely used as extreme ultraviolet (EUV) and soft X-ray radiation sources in many different fields. Lithography is one of the most challenging applications of the EUV spectral region (5-50 nm). The worldwide importance of EUV lithography (EUVL) is basically due to its potential to extend optical projection lithography to higher resolution in integrated circuit manufacturing, thanks to the shorter wavelength and to the availability of high reflectivity normal incidence mirrors.[1]

In addition to the huge technological and financial effort carried out by international consortia to raise EUVL at the industrial level, less expensive laboratory-scale facilities have been built up to perform component testing and metrology. Within a national project on nanotechnologies[2] at the ENEA Research Centre in Frascati a micro exposure tool (MET) has been designed and developed for projection lithography, exploiting the facility EGERIA (extreme ultraviolet radiation generation for experimental research and industrial applications). EGERIA is a EUV/soft X-ray laser produced plasma source equipped with a high efficiency debris mitigation system (DMS).

2 THE EGERIA FACILITY

The EUV/soft X-ray source facility EGERIA, which has been extensively characterized,[3,4,5,6,7] is based on a solid tape target where the plasma is created by a focused XeCl ($\lambda = 308$ nm) excimer laser beam.

An increase of EUV photons can be achieved by decreasing the plasma temperature, thus preventing the emission of harder X-rays. This goal may be achieved by moving the target out of focus. In this way, when the laser irradiance is $I \sim 10^{10}$ W/cm^2, the EUV conversion efficiency is $\approx 0.7\%$/eV over a solid angle of 2π srad and the focused beam diameter is ~ 0.2 mm.

The considerable emission of debris (fast atoms and particles) from the plasma is strongly attenuated by a patented DMS[8] in order to preserve the ellipsoidal mirror located 75 mm from the source. The DMS protects the collector mirror from ions and particulate bombardment, thus increasing its lifetime. It combines a static low-pressure rare gas (typically krypton at ≈ 1 mbar) with a fan rotating at 6000 rpm. The DMS experimental set up is shown in figure 1.

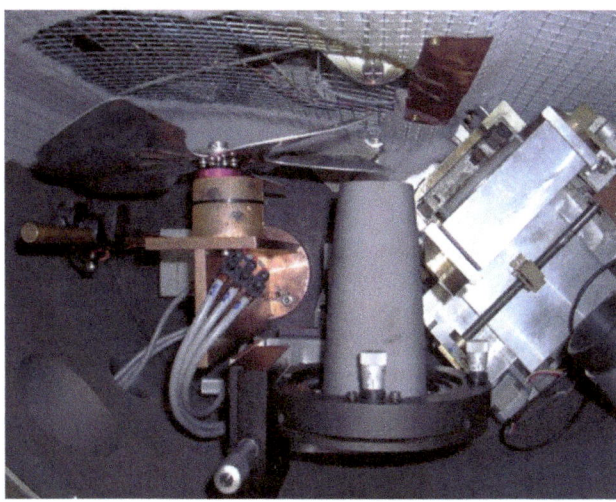

Figure 1 *Top view of the EGERIA source and DMS setup. Middle: ellipsoidal collector mirror. Right: lens focusing the laser beam. Left: DMS fan.*

The buffer gas serves to slow down and possibly stop the debris with diameters up to 1 μm. The mechanical device is in turn aimed at intercepting the arrival on the collector of larger size debris particles (≈ 1–10 μm) which, due to their size and mechanism of formation, are largely unmoderated by the gas. A commercial five-blade aluminium-alloy fan is presently in use, at a minimum distance of 4 cm from the source. A patented fan,[9] specifically designed for optimal performance, will soon be implemented.

The use of krypton is preferred instead of the widely used argon since the mitigating capability of krypton is higher under comparable operating conditions. However, since krypton does not efficiently transmit EUV radiation below a wavelength of 13.8 nm, the use of a krypton based DMS necessitates shifting the operating wavelength of the lithographic setup from the standard 13.5 nm to greater than 13.8 nm. This also has the advantage of naturally inhibiting the transmission of undesired out of band radiation in the EUV and vacuum ultraviolet ranges. For a typical 10 cm source to collector distance, a krypton buffer gas pressure of 1 mbar is expected to ensure an effective mitigation of debris sizes up to 1 μm, transmitting only debris having speeds lower than 100 ms^{-1}. The same transmission coefficient at $\lambda \approx 14$ nm in the same spatial range is achieved by 0.4 mbar argon which can only confine debris with diameters less than 10 nm and transmits debris with speeds ≤ 2 km s^{-1}.[10]

The ENEA DMS has been extensively tested and characterised using metallic tape targets (e.g., copper, tantalum), emitting a huge amount of debris,[10] either ionic, neutral, clusters or particulate. For this reason it can be considered a good test bed for debris characterization and for testing the DMS.

2. DEBRIS CHARACTERIZATION

The optimum design of a DMS requires knowledge of the main debris characteristics. This characterization was performed with different techniques, including a ballistic pendulum, a space-sensitive time-of-flight detector, a plate (glass, paper, plastic) to be analyzed by optical microscopy and by a gated CCD camera, a Faraday cup and an electrostatic analyzer. The following details the experimental results achieved with each of these techniques.

2.1 Faraday Cup and Electrostatic Analyzer

The Faraday-cup detector gives quantitative values of the ion current and hence an estimate of the number of ions. In this experiment the Hercules laser system[11,12] was used, emitting a long laser pulse (120 ns) with a tantalum target at an incidence angle of 45° in order to allow ionic beam detection at the normal to the target surface as shown in figure 2a.[13] Figure 2b shows that the ion kinetic energy reaches many kilo-electron volts with corresponding speeds of tens of kilometres per second.

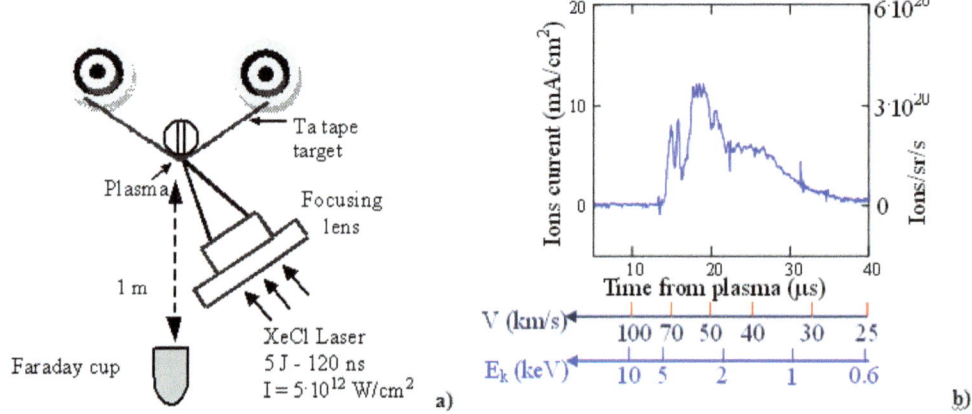

Figure 2 *(a) Experimental setup for ion characterization by a Faraday-cup detector. (b) Oscilloscope trace of the detector signal. Vertical: the absolute values of the ion current and of the number of ions per unit time and per unit solid angle are shown (see text). Horizontal: time elapsed since the onset of the plasma.*

The ion fluxes in figure 2b were obtained assuming an average ion charge state of +2, as measured by the energy analyzer as discussed in the following. Hence, the emitted ion flux was estimated to be of the order of 10^{20} ions/sr/s, so that during the ion signal duration of approximately 15 µs a total of $\approx 2\times10^{15}$ ions/sr is generated.

Unlike the Faraday cup, the energy analyzer (EA) does not give quantitative absolute ion fluxes, but it can distinguish the ionic charge states. Using the EA (see figure 3a) provided the charge state distributions and ion kinetic energies of different target materials, namely aluminium, copper and tantalum, with atomic numbers 13, 29 and 73. The main results are that the charge state distribution was peaked around a charge +2 as shown in figure 3b in the case of tantalum, and that the kinetic energy distribution, for a fixed laser intensity, is almost independent of the target material.

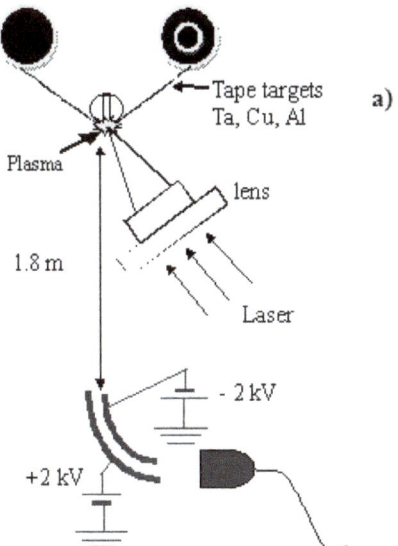 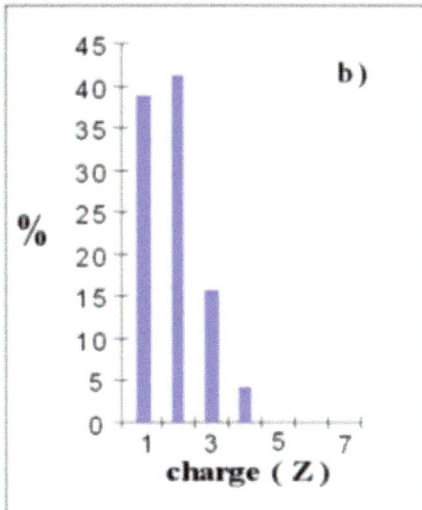

Figure 3 *(a) Experimental set-up of the energy analyzer. (b) Charge state distribution of the tantalum ions, obtained from the EA signals in the experimental configuration of figure 2.[13]*

2.2 Ballistic Pendulum

This novel technique, shown schematically in figure 4a, allows measurement of the total debris momentum. It consists of a thin metallic plate, supported by two thin wires, placed a few centimetres from the plasma source. A large area photodiode that receives a He-Ne laser beam partially obscured by the pendulum acts as a "pendulum position detector", as shown in figure 4a. The pendulum is calibrated by measuring the photodiode signal as a function of the pendulum position. Figure 4b shows the speed of the pendulum pushed by the debris immediately after the laser shot when using a 100 µm thick copper tape target. The pendulum speed is enhanced by a shock wave when the pressure in the chamber is increased above 0.1 mbar. As shown in figure 4b, for pressure values smaller than 0.1 mbar the starting speed does not depend on pressure and is mainly due to the debris. Thus it can be used to deduce the debris momentum.

The measured pendulum starting speed of 10 mm/s (in vacuum, see left part of figure 4b) multiplied by the pendulum weight gives a debris momentum, in the 1 srad solid angle intercepted by the pendulum, of $p_{debris} = 2 \times 10^{-5}$ kg m/s. This value can be compared with that of the copper ions, which can be estimated assuming they are of the same amount $(2 \times 10^{15}/sr)$ and the same average kinetic energy $(E \approx 2$ keV$)$ as the tantalum ions shown in figure 2. This gives $p_{ion} = \sqrt{2ME} = 1.6 \times 10^{-5}$ kg m s^{-1}, where M is the copper atomic weight. The difference between the total debris momentum and the ionic debris momentum is negligible, within the 30% experimental error. Thus it may be concluded that either the neutrals have much smaller energy than the ions, or that their amount is significantly smaller.

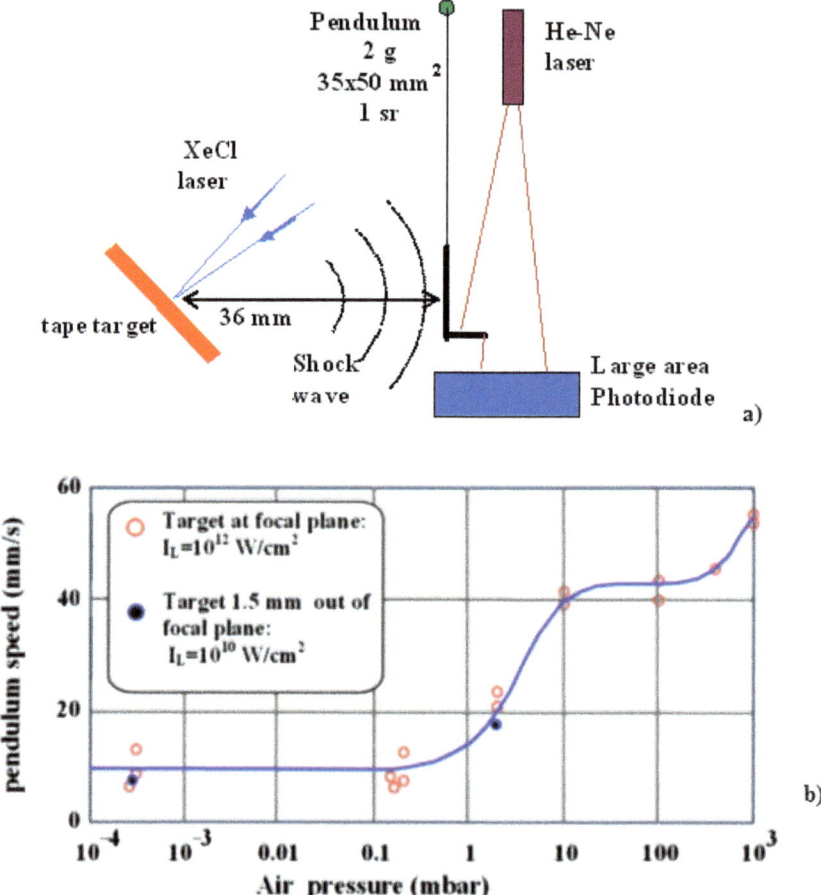

Figure 4 *(a) Experimental setup for debris momentum measurement with the ballistic pendulum. (b) Starting speed of the pendulum after the laser shot, due to the impact of debris (pressures lower than 0.1 mbar) and shock wave plus debris (pressures higher than 0.1 mbar).*

2.3 Space-Sensitive Time of Flight Detector

A novel and low-cost time of flight (TOF) detector has been designed and tested that gives information on both velocity and space distribution of the debris. It consists of a slit in front of a fast spinning disk where debris is collected, as shown in figure 5. The operation principle is simple; by synchronizing the laser shot with one reference point of the spinning disk, a visible track due to the deposited debris layer is generated on the disk surface. This track has a length and a coloured profile which reproduces the time profile evolution of the debris flux.

In general, the TOF device is sensitive to every kind of debris. However, when using a foil of paper for ink jet printers as the collecting disk, the sensitivity to ions and neutrals is much larger than to clusters. In addition, the black colour of the ion and neutral debris tracks is very different to those left by clusters, which are of same colour as the target material, i.e. red in the case of copper. As a consequence, the contributions of the two different type of debris can be easily distinguished. It is likely that the black colour results from a chemical reaction due to the high energies of the ions and neutrals. Interestingly, the black tracks disappear one or two days after the exposure, and it must be analyzed immediately after the exposure.

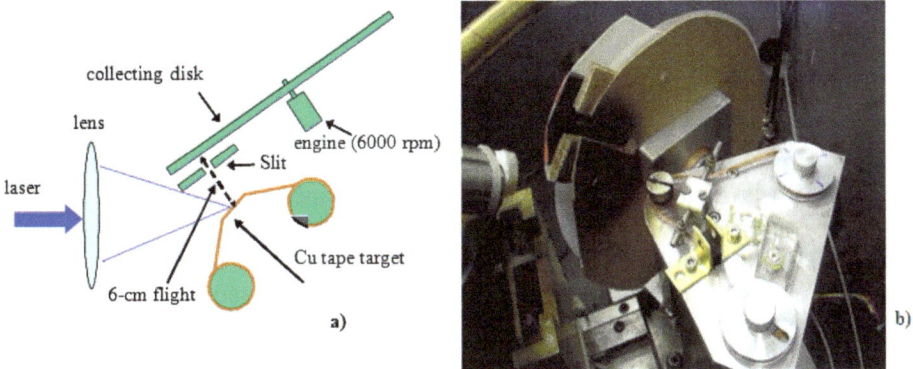

Figure 5 *(a) Schematic of the TOF detector. (b) Photograph of the device.*

Due to the relatively low 6000 rpm turning speed of the collecting disk, the maximum measurable debris speed is about 5 km/s, limited also by the 1 mm wide slit aperture. As a consequence, this detector gives velocity information to some extent complementary with the results of the Faraday cup summarized in figure 2b. The laser shot is synchronized to the slit by an optical trigger (the jitter, < 1 μs, is negligible for the revolution rate used).

The radial orientation of the slit (see figure 5b) allows the time evolution of the debris speed to be obtained from the track profile along the disk circumference, while the track radial profile gives the angular distribution of the debris emission.

Figures 6a and 6b show the results obtained after 10^3 laser shots on a copper tape target, for two different values of the angular direction with respect to the target normal (that is for two different radial distances from the disk centre). In the figures, the track position along the disk circumference has been converted into speed values (assuming the zero timescale is coincident with the arrival of the first ions, in vacuum) while the track darkness is converted to atomic debris flux. The laser incidence angle was 45°.

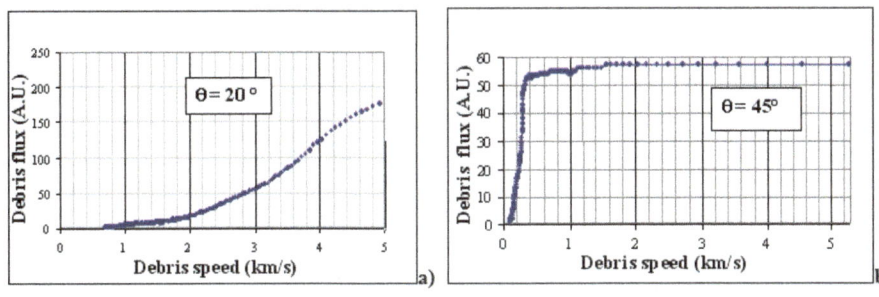

Figure 6 *Debris speed profile obtained from the TOF detector (shown in figure 5) for the case of a copper tape target and integrated over 1000 shots: (a) at 20° from the target normal (i.e., at 65° from the laser axis); (b) at 45° from the target normal (i.e., at 90° from the laser axis). The vertical scales of the two graphs are consistent with each other.*

Comparing figures 6a and 6b it is clear that the flux of the atomic (ions + neutrals) debris faster than 3 km/s is much larger close to the target normal than far away from it, while the amount of the slower debris is larger far away from the normal.

2.4 Spherical Screen

Collecting the debris on a plastic or paper spherical screen is the simplest way to obtain the angular profile of the time-integrated debris flux. Debris were collected on spherically shaped paper strips (as typically used for inkjet printers) fixed on a hemispherical plastic surface, as shown in figure 7a. Figure 7b shows the copper debris emission in vacuum integrated over 1000 shots with a laser intensity $I = 5\times10^{12}$ W/cm^2. The figure shows that the paper region close to the target normal (up to 45°) is dark, because the debris are mainly copper atoms. Moreover, the narrow corona between 45° and 50° is reddish, which means the main debris collected in this angular region are copper clusters.

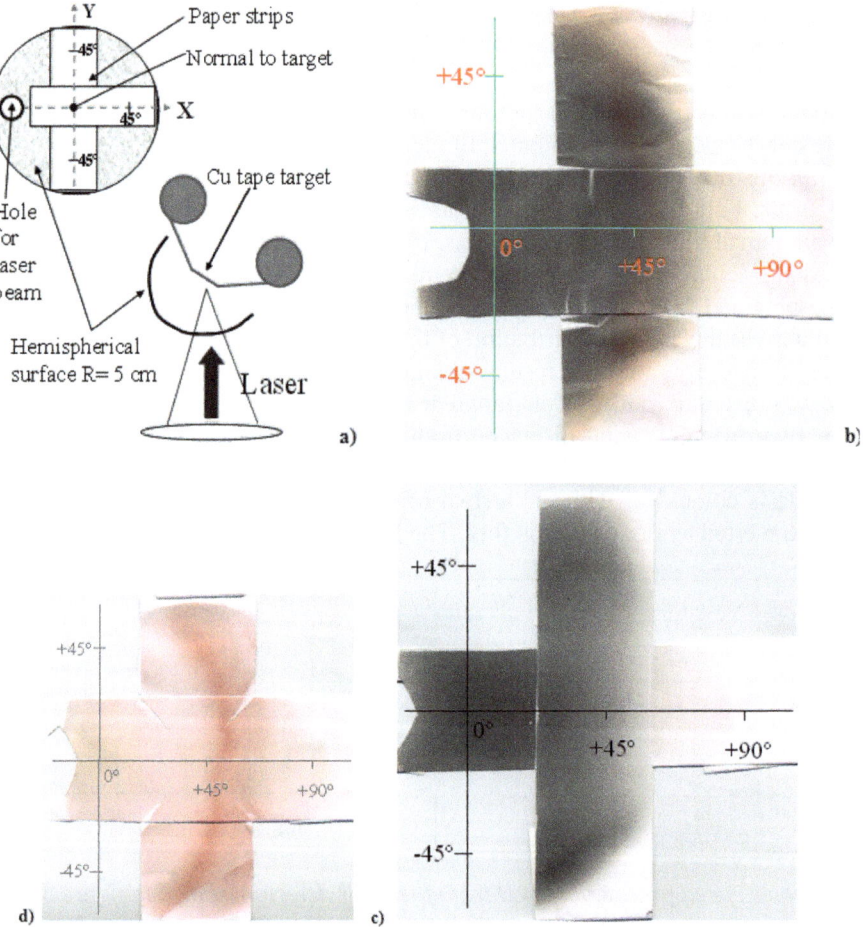

Figure 7 *(a) Experimental setup for the time-integrated detection of the debris angular distribution. (b) Result obtained after 1000 shots in vacuum ($< 10^{-3}$ mbar) at $I = 5\times10^{12}$ W/cm^2. (c) As b) but with 1 mbar of krypton in the vacuum chamber with. (d) As b) but for $I = 5\times10^{10}$ W/cm^2.*

The experiment was repeated with the same number of laser shots, but with 1 mbar of krypton gas in the interaction chamber. The results are shown in figure 7c; the flux of atomic debris is strongly reduced by the gas over the 5 cm flight path from the target, so that mainly cluster debris hits the paper. Finally, figure 7d shows the results of the same experiment of figure 7b (in vacuum) but with the laser intensity reduced to 5×10^{10} W/cm^2. Comparing figures 7b and 7d makes it clear that a lower laser intensity drastically reduces the emission of the cluster debris (in figure 7d the reddish corona almost disappears). The intensity of about 10^{10} W/cm^2 provides the maximum emission of EUV radiation.[7] In addition, this experiment shows that a krypton pressure as low as 1 mbar is sufficient to thermalize the atomic debris.

2.5 Glass Plate

The effect of thermalization of the debris by krypton gas has been confirmed by collecting the debris on a glass plate placed 4 cm from the plasma source. After 500 shots in vacuum the plate had an optical density (OD) in the visible range of 0.6, due to the copper debris deposition, while repeating the experiments with 2.5 mbar static krypton gas, with the same number of laser shots, the final OD was 0.015, 40 times lower than in vacuum, while the amount of particulate debris (measured by analyzing the plate using an optical microscope) was more than 100 times lower.[7]

The higher mitigation factor for atomic than for cluster debris was an unexpected result. In order to gain a deeper insight, theoretical calculations of the thermalization lengths (or range of flight) for both atomic and cluster debris were performed.

2.6 Calculation of Debris Thermalization Length

The very high speed of the atomic debris particles shown in figure 2b prevents the use of mechanical devices (shutters or slotted spinning discs) to stop them. Furthermore, filters cannot stop the debris while providing sufficient EUV transparency. As a consequence, the only way to mitigate the atomic debris is to use a gas at a pressure to be adjusted. To date the most studied gas for debris mitigation has been argon,[14] as it is relatively transparent at 13.5 nm, the wavelength of choice for EUV lithography.

Krypton is much heavier than argon and thus more efficient in debris mitigation. A comparison of the range of flight versus gas pressure of atomic debris produced by different materials (lithium, tin and tantalum) in different gases is shown in figure 8.

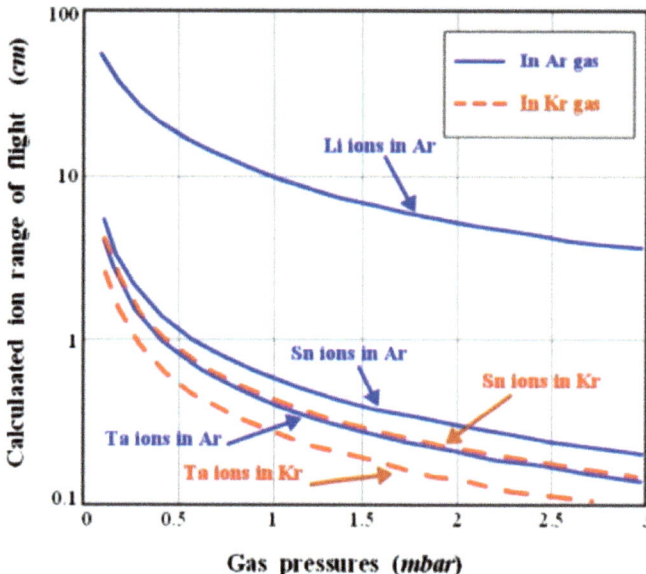

Figure 8 *Calculated thermalization lengths for atomic debris, emitted by plasma sources at a kinetic energy of 4 keV, through argon and krypton.*

Unfortunately, krypton does not have a good transmittance at 13.5 nm, but shows a relatively low absorption coefficient for $\lambda > 13.8$ nm.[14] Figure 9 shows the calculated range of flight of 2 keV tin atoms in argon and krypton versus the required gas transmission at different wavelengths. The transmission was varied by changing the gas pressure. Once the transmission coefficient is fixed, the range of flight in krypton at $\lambda = 14.2$ nm is four times shorter than in argon at $\lambda = 13.5$ nm. It turns out that a krypton pressure 2.5 times larger at 14.2 nm than for argon 13.5 nm can be used to give the same transmission. Consequently, a DMS based on krypton buffer gas will have an ionic debris mitigation factor orders of magnitude larger than for argon.

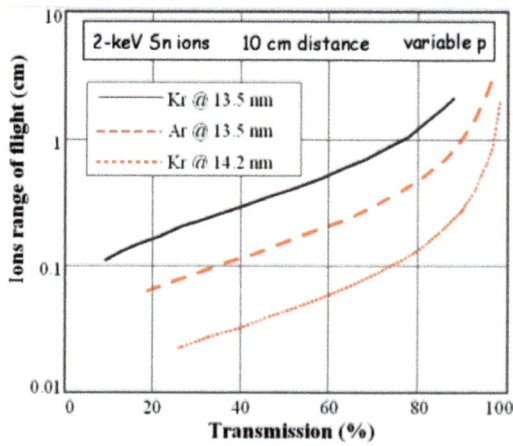

Figure 9 *Calculated thermalization length of tin atoms emitted at 2 keV kinetic energy, as a function of the buffer gas transmission at 13.5 nm and 14.2 nm.*

3 DEBRIS MITIGATION SYSTEM

As a result of the accurate characterization of atomic and particulate debris emitted by the ENEA MET laser-plasma solid tape target source, as described above, a DMS was designed to combine a gas filter for continuous gas cleaning with a buffer gas (argon or krypton) and a fan, as shown in figure 1.

The efficiency of the installed DMS was estimated by observing the debris flux contamination of glass slides after exposure to the plasma generated on a 100 μm thick copper tape by a XeCl laser beam with irradiance $I = 10^{10}$ W cm^{-2} under different conditions, specifically in vacuum for 3×10^3 laser shots and in gas (krypton or argon) plus a fan for 5×10^4 laser shots. Each slide was studied at different positions over the exposed regions by a 100× Leica optical microscope used in reflection mode, with the resultant views being acquired as background corrected gray-level images by a 16 bit Andor CCD camera; the results are shown in figure 10.

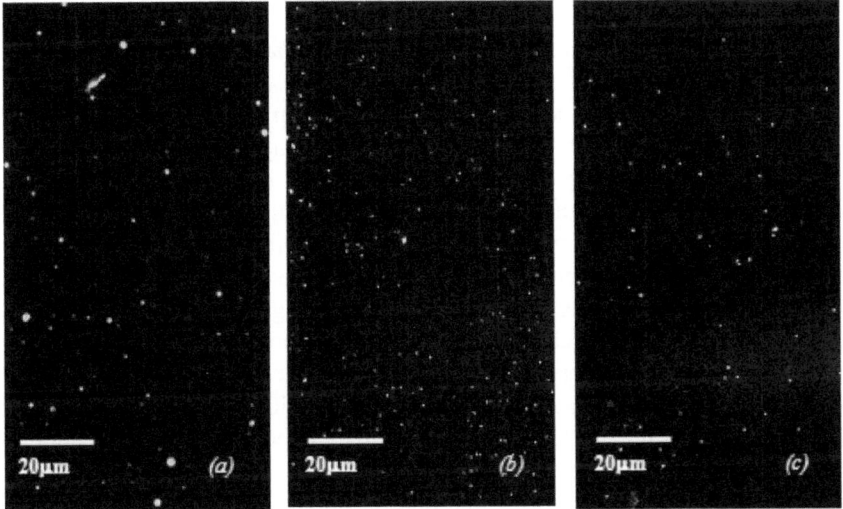

Figure 10 *Examples of glass slide images relevant to the exposure to debris flux in (a) vacuum after 3×10^3 laser shots, (b) argon plus fan after 5×10^4 laser shots and (c) krypton plus fan after 5×10^4 laser shots.*

Overall, the images display dark backgrounds with the deposited debris appearing as light spots, the highest brightness typically pertaining to the spots of the vacuum exposed slides. The analysis of the images was performed by a dedicated code,[15] which uses some basic image processing routines of the MATLAB toolbox (V5.4). It is based on specific procedures, built on the basis of accurate observation of the various images for debris recognition, spot fine-structure identification and debris size evaluation (both above and below the microscope resolution). Evidently, it stands out as a tool for further plasma debris related investigations.

As an example of the results of the code, the debris suppression capability of krypton plus a fan is definitely higher than that obtainable using argon plus a fan.

Concerning the debris size, figure 11 shows a comparison of the debris number versus size (conveyed by the average diameter Φ) in vacuum with those in gas (argon or krypton) plus a fan at a slide position where the relative debris mitigation factors (DMFs) are maximum.

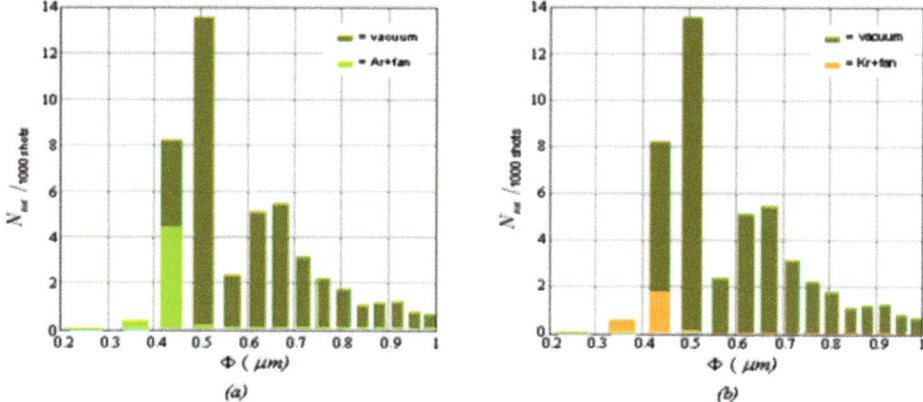

Figure 11 *Histograms of the total debris number per 1000 shots versus the debris diameter at the slide positions where the DMFs are maximum for exposure in (a) argon plus fan and (b) krypton plus fan compared to the vacuum distributions.*

The debris distributions for both gases plus a fan drop very rapidly with debris size, reaching negligible values for $\Phi \geq 1 \mu m$, whilst displaying debris occurrences for $\Phi \leq 0.4$ μm, to an extent slightly greater in krypton than in argon. In contrast, no such small debris was detected in vacuum, as can be seen in figure 11. It has been suggested[15] that the lack of detected debris with $\Phi \leq 0.4$ μm in the images of the vacuum exposed slides might signal the formation of submicrometric clusters as a consequence of the interaction of Cu atoms (neutrals/ions) with the ambient gas (Ar or Kr atoms) due to a homogeneous nucleation-like mechanism favored by the cooling action of the ambient gas. Measured debris mitigation factors using Krypton as buffer gas are DMF ≈ 800 for atoms and nanometre sized clusters, and DMF ≈ 1600 for particles larger than 500 nm, both emitted by a copper plasma at a laser intensity of 10^{10} W/cm^2. These DMF values, which are at the forefront in this field, have been measured both through optical density measurements of the copper layer deposited on glass plates and through the measurement of EUV mirror reflectivity degradation (both placed at the collector entrance). Recently, larger DMF values have been obtained for a tantalum tape target (the target used for the MET experiments) but these results need to be analysed further.

In conclusion, the analysis has confirmed the expectations of theoretical estimates concerning, for instance, the larger debris suppression efficacy of krypton in comparison with argon. It has also highlighted new issues, such as the possible formation of sub-micrometre clusters due to a homogeneous nucleation-like mechanism, favoured by the cooling action of the buffer gas. Further investigations need to be performed to confirm such issues.

4. CONCLUDING REMARKS

A study of the various debris (ions, neutrals, particulate, clusters) emitted by the laser-plasma EGERIA has been presented. A range of diagnostics was used, namely a Faraday cup and an electrostatic analyzer, a ballistic pendulum, a space-sensitive time-of-flight detector, plates (glass, paper, plastic) analyzed using optical microscopy and a CCD camera. The analyses have provided detailed knowledge of the velocity, size, charge, momentum, spectral energy and spatial distributions of the emitted debris. This knowledge was of fundamental importance in the design and operation of a patented debris mitigation system, in order to preserve the expensive elliptical mirror that acts as collector. Suppression factors up to 800 and 1600 have been obtained for atoms and particles, respectively; the factor for the particles is one of the best achieved to date.

The excellent performance of the DMS was a basic prerequisite in achieving a 90 nm resolution pattern on PMMA resist.[16,17]. The obtained resolution is among the best ever achieved by low-cost laboratory-scale EUVL micro exposure tools.

References

1 K. Kemp and S. Wurm, *C. R. Physique*, 2006. **7**, 875.
2 G. Altieri, D. Amodio, G. Baldacchini, A. Baldesi, C. Bellecci, S. Bollanti, F. Bonfigli, G. Clementi, A. Conti, T. Dikonimos, P. Di Lazzaro, P. Dunne, F. Flora, M. Francucci, D. Garoli, P. Gaudio, A. Gerardino, R. Giorgi, T. Letardi, N. Lisi, S. Martellucci, V. Mattarello, L. Mezi, R.M. Montereali, D. Murra, E. Nichelatti, P. Nicolosi, G. Nocerino, L. Palladino, M.G. Pelizzo, A. Piegari, A. Reale, M. Richetta, V. Rigato, A. Ritucci, S. Scaglione, E. Salernitano, A. Santoni, A. Sytchkova, E. Tefouet Kana, G. Tomassetti, A. Torre, P. Tucceri, A. Vincenti, C.E. Zheng and P. Zuppella, *Proc. II Workshop di Ateneo sulle Nanotecnologie*, 2008, 63.
3 S. Bollanti, P.Di Lazzaro, F. Flora, G. Giordano, T. Letardi, G. Schina, C.E. Zheng, L. Filippi, L. Palladino, A. Reale, G. Taglieri, D. Batani, A. Mauri, M. Belli, A. Scafati, L. Reale, P. Albertano, A. Grilli, A. Faenov, T. Pikuz, R. Cotton, *J. X-Ray Sci. Technol.*, 1995, **5**, 261.
4 S. Bollanti, R. Cotton, P. Di Lazzaro, F. Flora, T. Letardi, N. Lisi, D. Batani, A. Conti, A. Mauri, L. Palladino, A. Reale, M. Belli, F. Ivanzini, A. Scafati, L Reale, A.Tabocchini, A.Ya. Faenov, T.A Pikuz and A. Osterheld, *Il Nuovo Cimento D*, 1996, **18**, 1241.
5 G.A. Vergunova, A.I. Magunov, V.M. Dyakin, A.Ya. Faenov, T.A. Pikuz, I.Yu. Skobelev, D. Batani, S. Bossi, A. Bernardinello, F. Flora, P. Di Lazzaro, S. Bollanti, N. Lisi, T. Letardi, A. Reale, L. Palladino, A. Scafati, L. Reale, A.L. Osterheld and W.H. Goldstein, *Physica Scripta*, 1997, **55**, 483.
6 S. Bollanti, P. Albertano, M. Belli, P. Di Lazzaro, A. Ya. Faenov, F. Flora, G. Giordano, A. Grilli, F. Ianzinni, S.V. Kukhlevsky, T. Letardi, A. Nottola, L. Palladino, T. Pikuz, A. Reale, L. Reale, A. Scafati, M.A. Tabocchini, I.C.E. Turcu, K. Vigli-Papadaki and G. Schina, *Il Nuovo Cimento D*, 1998, **20**, 1685.
7 S. Bollanti, F. Bonfigli, E. Burattini, P. Di Lazzaro, F. Flora, A. Grilli, T. Letardi, N. Lisi, A. Marinai, L. Mezi, D. Murra and C.E. Zheng, *Appl. Phys. B*, 2003, **76**, 277.
8 F. Flora, C. Zheng, L. Mezi, *Method of stopping ions and small debris in extremeultraviolet and soft x-rays plasma sources by using krypton*, 2002, European Patent EP1211918
9 D. Murra, S. Bollanti, P. Di Lazzaro, F. Flora, L. Mezi, A. Conti: Italian Patent Office UIBM, No. RM2006A000303.
10 S. Bollanti, D. Amodio, A. Conti, P. Di Lazzaro, F. Flora, L. Mezi, D. Murra, A. Torre, C.E. Zheng, D. Garoli, M.G. Pelizzo, P. Nicolosi, V. Mattarello, V. Rigato and A. Gerardino, *Proc. SPIE*, 2007, **6703**, 670308.
11 S. Bollanti, P. Di Lazzaro, F. Flora, G. Giordano, T. Hermsen, T. Letardi and C.E. Zheng, *Appl. Phys B*, 1990, **50**, 415.
12 P. Di Lazzaro, *Proc. SPIE*, 1998, **3423**, 35.

13 P. Fournier, H. Haseroth, H. Kugler, N. Lisi, R. Scrivens, F. Varela Rodriguez, P. Di Lazzaro, F. Flora, S. Duesterer, R. Sauerbrey, H. Schillinger, W. Theobald, L. Veisz, J.W. Tisch and R.A. Smith, *Rev. Sci. Instrum.*, 2000, **71**, 1405.

14 F. Flora, L. Mezi, S. Bollanti, F. Bonfigli, P. Di Lazzaro, T. Letardi and C.E. Zheng, *Proc. SPIE*, 2001, **4504**, 77.

15 S. Bollanti, P. Di Lazzaro, F. Flora, L. Mezi, D. Murra and A. Torre, *Appl. Phys. B*, 2009, **96**, 479.

16 S. Bollanti, P. Di Lazzaro, F. Flora, L. Mezi, D. Murra and A. Torre, *European Physics Letters*, 2008, **84**, 58003

17 P. Di Lazzaro, S. Bollanti, F. Flora, L. Mezi, D. Murra and A. Torre, *IEEE Trans. Plasma Sci.*, 2009, **37**, 475.

EUV MULTILAYER OPTICS: DESIGN, DEVELOPMENT AND METROLOGY

P. Nicolosi

University of Padova, Dept of Information Engineering, via Gradenigo 6/B   Padova, 35131, Italy and National Research Council-Istituto di Fotonica e Nanotecnologie, IFN-CNR, via Trasea 7, Padova, 35131, Italy

1 INTRODUCTION

Extreme ultraviolet (EUV) multilayer coatings are presently widely used in both science and technology. The most characteristic technological application of multilayers is in lithography (EUVL) for large volume production of electronic chips. According to the well known Moore's law the density of components in integrated circuits doubles every about 18 months; the resolution in the processing of the wafer has to improve correspondingly. Since the lithographic process consists of the projection of a mask on the wafer, according to the optics diffraction limit in order to improve the resolution the wavelength of radiation has to decrease. Accordingly, effort is concentrated on the development of a lithographic tool at 13.5 nm in order to reach a resolution down to about 30 nm.[1] In science the applications of multilayers are widely spread among beam lines of large scale facilities such as synchrotrons and free electron lasers (FELs), and for astronomy. At large scale facilities multilayer optics are used in order to select bandwidth or polarization and to focus the radiation beam. In astronomy multilayers are used in Solar imagers working at selected wavelengths for spectroscopic diagnostic purposes. More recently multilayer coatings for ultrashort pulses in the sub-femtosecond regime have been developed. By designing multilayers capable of reflecting or even compressing ultra-short pulses the "magic" door to atto-physics has been opened. To give a classical perspective a few tens of attoseconds correspond approximately to the time an electron takes to complete one orbit around the nucleus of a hydrogen atom; in the same time the electric field of an optical pulse makes a small fraction of its oscillation. Understanding the physics at such short time scales represents a great challenge both for theory and experiment and can provide amazing results. In this review after a short general discussion of the basic theory of multilayers coatings some recent results in the development of multilayers for ultrashort pulses and astronomical applications will be presented.

In the EUV spectral range transparent materials do not exist; only fluorides have relatively good transmittance at short wavelengths, with LiF presenting the shortest absorption edge at about 105 nm. This means that all materials have complex refractive indices with non- negligible imaginary coefficients, and that it is not possible to make use of refractive optics, such as lenses, prisms, plates or windows, in order to focus or steer the radiation in experimental set-up. Thus, in principle, it is only possible to use optics working in reflection, but in this case a further problem arises as a result of the low normal incidence reflectivity of standard metal coatings such as gold, and platinum. This is because at such short wavelengths the real parts of the refractive indices are very close to unity and, consequently, in order to get efficient reflection the coatings have to be used at very small glancing angles, below the critical angle for total reflection. This means that the optical apertures are very small, affecting the final throughput of the system and giving aberrations that are considerably larger than at near-normal incidence, thus further affecting the final performances. For these reasons, the design of optical coatings with high reflectivity at nearly normal incidence is strongly required. Multilayer coatings (multilayers or ML) consist basically of structures obtained through the repetition of bi-layers made of two materials. The materials are selected in order to get the highest reflectivity at the interfaces and in addition their thickness is optimized in order to provide coherent superposition of the various reflected components. The working principle is similar to that of dielectric coatings widely used in the optical range, for example filters and laser mirrors. However since at such short wavelengths the thickness of the layers scales down to a few nm's new technological problems need to be solved.

2 MULTILAYER WORKING PRINCIPLES

The working principle of a multilayer stack is based on the coherent superposition of the reflected components of the electromagnetic fields at the various interfaces between each pair of layers, as indicated in figure 1.

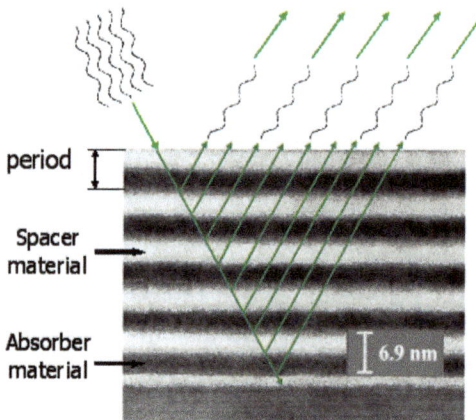

Figure 1 *Transmission electron microscope section image of a Mo/Si multilayer with an illustration of the reflected beam building up.*

At each interface the relation between the incident, reflected and transmitted components of the electromagnetic wave electric field amplitude is given by the Fresnel equations. For p polarization (electric field parallel to the plane of incidence) the reflection coefficient is

$$r_{12}^{p} = \frac{\tilde{n}_1 \cos \theta_2 - \tilde{n}_2 \cos \theta_1}{\tilde{n}_1 \cos \theta_2 + \tilde{n}_2 \cos \theta_1},$$

(1)

and the transmission coefficient is

$$t_{12}^{p} = \frac{2\tilde{n}_1 \cos \theta_1}{\tilde{n}_1 \cos \theta_2 + \tilde{n}_2 \cos \theta_1}.$$

(2)

For s polarization (electric field orthogonal to the incidence plane) the reflection coefficient is

$$r^s_{12} = \frac{\tilde{n}_1 \cos\theta_1 - \tilde{n}_2 \cos\theta_2}{\tilde{n}_1 \cos\theta_1 + \tilde{n}_2 \cos\theta_2} \tag{3}$$

and the transmission coefficient is

$$t^s_{12} = \frac{2\tilde{n}_1 \cos\theta_1}{\tilde{n}_1 \cos\theta_1 + \tilde{n}_2 \cos\theta_2} \tag{4}$$

These values depend on the angles of incidence, θ_1, and transmission (refraction), θ_2, and on the optical constants, i.e. the complex refractive indices $\tilde{n}_i = n_i - i\beta_i$, $i = 1,2$, of the two materials.

The reflection and transmission coefficients of a single layer resulting from multiple reflections at the interfaces are given by

$$r = r_{01} + \frac{r_{12} t_{01} t_{10} \exp(2i\varphi)}{1 + r_{01} r_{12} \exp(2i\varphi)} \tag{5}$$

$$t = \frac{t_{01} t_{12} \exp(2i\varphi)}{1 + r_{01} r_{12} \exp(2i\varphi)} \tag{6}$$

where the subscript (0) refers to the medium from which the radiation is incident on the layer (1), and (2) corresponds to the medium into which the radiation is transmitted. The term in the complex exponent takes into account the phase change φ due to propagation between the two boundaries of the layer, of thickness d

$$\varphi = \frac{2\pi d}{\lambda} \tilde{n}_1 \cos\theta_1 \tag{7}$$

The angle θ_1 is the refraction angle in the film. It can be derived from Snell's law corresponding to the conservation of the photon momentum component parallel to the interface plane,

$$\cos\theta_1 = \sqrt{1 - \left(\frac{\tilde{n}_0 \sin\theta_0}{\tilde{n}_1}\right)^2} \tag{8}$$

For N layers there will be $N + 1$ interfaces, and the ith layer will have thickness d_i.

It is then straightforward to compute the resulting amplitude reflection and transmission coefficients for the film stack by recursive application of equations (5) and (6). By squaring the field amplitudes the intensity can be easily derived.

For a multilayer stack made of two materials with period d the reflectivity peak is given by the well known Bragg relation. This can be derived by making the factor in the exponents of equation (5) equal to $2m\pi$ in order to satisfy coherent superposition of the Fresnel reflected components, leading to

$$2\tilde{n}d \sin \alpha = m\lambda \tag{9}$$

where $\alpha = \pi/2 - \theta$ is the grazing incidence angle. This relation can be corrected for refraction by writing the real part of the refractive index as $n = 1 - \delta$, neglecting absorption, and by taking into account Snell's law which relates the refraction angle and the incidence angle to the refractive index of the material, leading to[2]

$$m\lambda \cong 2d \sin \alpha_0 \left[1 - \frac{2\delta}{\sin^2 \alpha_0} \right]^{1/2}. \tag{10}$$

An approximate relationship between the number of layers contributing to the reflectivity and the spectral bandwidth of the multilayer can be derived by considering that the effective number of layers, N_{eff}, coherently contributing to reflection corresponds to the coherence length,

$$L = dN_{eff} \cong \frac{c}{2\Delta\nu} = \frac{\lambda^2}{2\Delta\lambda}. \tag{11}$$

This shows that there is an inverse proportionality between the bandwidth and the effective number of layers contributing to reflection; the higher the number the narrower the bandwidth. It is clear that there is a limit in the effective number of layers as well as in the bandwidth, so that the reflectivity saturates at a value less than one – adding more layers than N_{eff} will not give any further improvement.

2.1 Interfaces

One further parameter that has to be taken into account, since it can significantly affect the resulting reflection efficiency of the multilayer coating, is the surface quality at the interfaces. Real interfaces between two layers, 1 and 2, of different materials will not show ideal step-like refractive index profiles, corresponding to transition from the index $\tilde{n}_1$ of one layer to the index $\tilde{n}_2$ of the second layer, but will be affected by two major characteristics: roughness and intermixing (physical or chemical). The former results from the combined effects of the original roughness of the substrate surface, the deposition process and the intrinsic characteristics of the materials – some materials grow more or less smoothly according to the deposition technique. Intermixing can result from interdiffusion: molecules, clusters or atoms of one material can penetrate or drift into the adjacent layer[3] or composites can result from chemical interactions between the two materials. Since the thickness of layers in the EUV corresponds to a few atomic layers the interface characteristics depend strongly on the deposition process, i.e. the energy of the ad-atoms sticking on the substrate, and the materials characteristics. Some materials like metals tend to grow in polycrystalline form after the film reach some thickness, for example about 2 nm for Mo, others like Si grow preferentially amorphous. Furthermore materials can react forming compounds with different adjacent layers or can migrate diffusing and penetrating some depth in the adjacent layer. For example it has been verified that different interlayers form at the interfaces of Mo on Si or Si on Mo depending on the thickness and the temperature[4-7]. It is clear then that quality (more or less sharp and smooth) and structure (chemical) of interfaces can affect significantly the expected performances of a multilayer coating. From the optical point of view intermixing can be simulated by a suitable gradient of the refractive index at the interfaces as a function of the depth along the thickness of the multilayer structure.

Besides the fact that mixed interfaces or compounds formation can cause a drop of the amplitude reflection coefficient, Roughness affects reflectivity since radiation is scattered along non-specular directions, referred to the average surface profile of the coating. Non smooth surfaces and buried interfaces can cause a serious drop of reflectivity and of the imaging performances of the optics, in the case of high frequency roughness the radiation is scattered quite far from the specular direction, while low frequency roughness by scattering radiation in a narrow cone within the acceptance aperture of the optics can cause flare affecting resolution and edge sharpness of images. The problem has been widely treated[8,9] and can be approximated by introducing a factor that characterizes the width of a Gaussian broadening of the reflectivity at the boundary. So the overall effect on efficiency of interface imperfections can be approximated by multiplying the Fresnel coefficients by the Debye-Waller factor[10].

$$g = \exp\left(-2\tilde{n}_i \tilde{n}_j \left[\frac{2\pi\sigma_{ij} \cos\theta_i}{\lambda} \right]^2 \right) \tag{12}$$

where σ_{ij} is the root-mean-square roughness of the interface between the ith and jth layers.

2.2 Polarization

As already stated the reflection and transmission coefficients are different for different polarization states of the radiation. For a beam consisting of s- and p-components with polarization factor

$$f = \frac{I_s - I_p}{I_s + I_p} \tag{13}$$

the reflectivities R_s and R_p can be recovered from the reflected signals, R_u and R_d, measured in two mutually perpendicular incidence planes,

$$R_s = \frac{R_u + R_d}{2} - \frac{R_u - R_d}{2f} \tag{14}$$

and

$$R_p = \frac{R_u + R_d}{2} + \frac{R_u - R_d}{2f}. \tag{15}$$

The average reflectivity can then be derived simply using

$$R = \frac{R_s + R_p}{2} = \frac{R_u + R_d}{2}. \tag{16}$$

So far as the phase is concerned those of the reflected and transmitted waves are given by

$$\tan \varphi_R = \frac{\mathrm{Im}[r]}{\mathrm{Re}[r]} \tag{17}$$

and

$$\tan \varphi_T = \frac{\mathrm{Im}[t]}{\mathrm{Re}[t]} \tag{18}$$

where Im[] and Re[] indicate the imaginary and real parts of the reflection or transmission coefficients. This implies that the reflection and transmission coefficients of a multilayer can be expressed by complex numbers, depending on wavelength, of the form

$$r(\lambda) = \sqrt{R_0(\lambda)}e^{i\varphi(\lambda)} \tag{19}$$

where $\sqrt{R_0}$ is a real amplitude term. The meaning of this expression will be clearer in the following. In particular these parameters must be evaluated when the polarizing properties of a multilayer have to be characterized or if it is required to reflect ultra-short pulses.

3 EUV OPTICAL CONSTANTS OF MATERIALS

In order to design a multilayer structure its performance must be simulated. Reliable performance simulation depends on accurate knowledge of the optical constants of the materials as well as the multilayer structure characteristics depending on the film deposition procedures, interface composition and structure, and whether the layers are crystalline or amorphous. As already mentioned the coupling of different thin films can result in interdiffusion or chemical reactions at the interfaces, and some materials can oxidise when exposed to air or be carbon contaminated if exposed to carbon reach environment. In addition, the density of the film can depend on the growing process; the optical properties of a material in the form of a layer a few nanometres thick can differ considerably from those corresponding to its bulk structure. Accordingly, considerable effort has been devoted to deriving reliable measurements of the optical constants of thin films of suitable materials.

    An extensive database can be found at the Centre for X-Ray Optics (CXRO), Lawrence Berkeley Laboratory[11] where the real parts refractive indices and the absorption coefficients of many materials are tabulated in the range 30 eV–30 keV. This compilation is based on the data published by Henke *et al*[12].

    The complex refractive index of a material is

$$\tilde{n} = 1 - \delta - i\beta \tag{20}$$

where $n = 1-\delta$ is the real part of the refractive index and β is related to absorption[*]. The field of a plane electromagnetic wave travelling in a material along z axis with this refractive index is

$$E(z,t) = E_0 e^{i(\omega t - kz)} = E_0 e^{i(\omega t - 2\pi\tilde{n}z/\lambda)} = E_0 e^{-2\pi\beta z/\lambda} e^{i(\omega t - 2\pi n z/\lambda)}, \tag{21}$$

---

[*] other texts report $\tilde{n} = 1 - \delta + i\beta$

where it is clear that the amplitude of the wave is progressively attenuated with increasing thickness of material.

The absorption coefficient, μ, is defined taking into account the intensity of the wave, i.e., the square of the amplitude, leading to

$$\mu = \frac{4\pi\beta}{\lambda} \tag{22}$$

so that the intensity varies with the distance z travelled in a material as

$$I = I_0 e^{-\mu z}. \tag{23}$$

The refractive index of a material can be derived from the free electron scattering model as[1 (chap.2)]

$$\tilde{n} = 1 - \delta - i\beta = 1 - \frac{N_a r_0}{2\pi} \lambda^2 \left(f_1 + i f_2 \right) \tag{24}$$

where $r_0 = e^2 / \left(4\pi\varepsilon_0 m_e c^2 \right)$ is the classical electron radius, N_a is the density of atoms and f_1 and f_2 are the atomic scattering factors corresponding to the effective number of electrons per atom. For high energy radiation the atomic electrons are effectively free, and so in the case of zero loss the scattering factor f_1 approaches the number Z of electrons in the atom; a free electron re-radiates all the energy received. When there are absorption losses the f_2 coefficient is not negligible, since for relatively lower energy photons the electrons cannot be assumed to be totally free. Furthermore, the optical constants near absorption edges and resonances can be strongly affected by the configuration of the atom and its environment.

The values of f_1 can be derived from the f_2 data, which is often easier to obtain, using Kramers-Kronig relations.[13] Accurate knowledge of optical constants of materials is fundamental to design reliable multilayer structures. As already mentioned optical constants of a thin film can differ significantly from the bulk form furthermore can depend also on the deposition process or the following handling and storage and even can change after some time from deposition because of structural or chemical modifications. For this it is fundamental that data are derived from accurate experimental measurements. Different experimental techniques can be applied in order to obtain reliable values of the optical constants of materials in the EUV. Several data reported in the literature have been derived from reflectivity measurements over wide spectral ranges as functions of the angle of incidence,[14] and from transmission measurements of thin films.[15] Since the reflectivity of a ML is the result of an interference process, materials with the lowest absorption coefficient and the highest difference of real refractive index are selected. This means that in a (x,y) plot where the various materials are identified by points with (δ, β) coordinates selection should concentrate on points located at the extreme edges of a narrow strip region close to abscissa. In this way a larger number of layers with the highest reflectivity at each interface can contribute to the reflection process. Often it is useful to choose a material with an absorption edge shifted at slightly higher energies than the operating wavelength. Furthermore the thickness of the layer with the higher absorption (absorber) should be made suitably thinner than that one of the layer with lower absorption (spacer) for optimized performances. However the simultaneous role of absorption and refraction must be carefully evaluated for optimal solution. In fact the uncertainty in the optical constants of materials deposited in very thin layers with different deposition techniques, is influenced also by the coupling between different materials as well as by possible chemical reaction or by environmental effects like oxidation, carbon contamination, furthermore exposure to high EUV radiation fluxes or thermal annealing can affect considerably the actual performances of the coating with respect to the expected simulated ones. The consequence is that a ML has to be characterized by eventually measuring also its reflectivity.

4 DEPOSITION TECHNIQUES

Thin films can be deposited using different techniques, for example thermal, e-beam, ion beam evaporation, laser ablation and magnetron sputtering. The most commonly used among them is magnetron sputtering, which guaranties the growing of very smooth layers with a good control of the deposition rate and of the stability of the process. Some techniques have been also coupled together in some cases in order to assist the deposition process like the case of ion beam assisted deposition. In the deposition chamber the substrate is positioned at some suitable distance from the target of the material to be deposited and often it is mounted on a spinning support in order to get a more uniform deposition rate. In thermal evaporation the material is placed in vacuum in a conductive crucible typically made of graphite or tungsten depending on the material to be evaporated. An electric current flow heats the crucible causing melting and evaporation or sublimation of the material. In e-beam evaporation the material is placed in vacuum in a crucible and is evaporated by heating it with the particle beam focused on it. An electron gun emits electrons by the thermo-ionic emission caused from current flowing through a filament. The electrons are accelerated with high voltage and focused with suitable electric and magnetic fields configuration on the material in the crucible, so releasing their kinetic energy cause the heating and the melting followed by evaporation of the material. By the current flowing through the filament the electrons emission current intensity can be controlled and so the deposition rate. The latter can be monitored via a quartz crystal microbalance. Obviously the microbalance must be suitably positioned and correspondingly calibrated w.r.t. the target and the substrate in order to get thickness data representative of the actually grown thin film. In the figure 2 an example of an e-beam and thermal evaporation deposition set-up is shown. Different materials can be deposited by selecting different crucibles mounted on a carousel. Suitable crucibles materials have to be selected according to the target ones depending on their melting temperatures and reactivity properties.

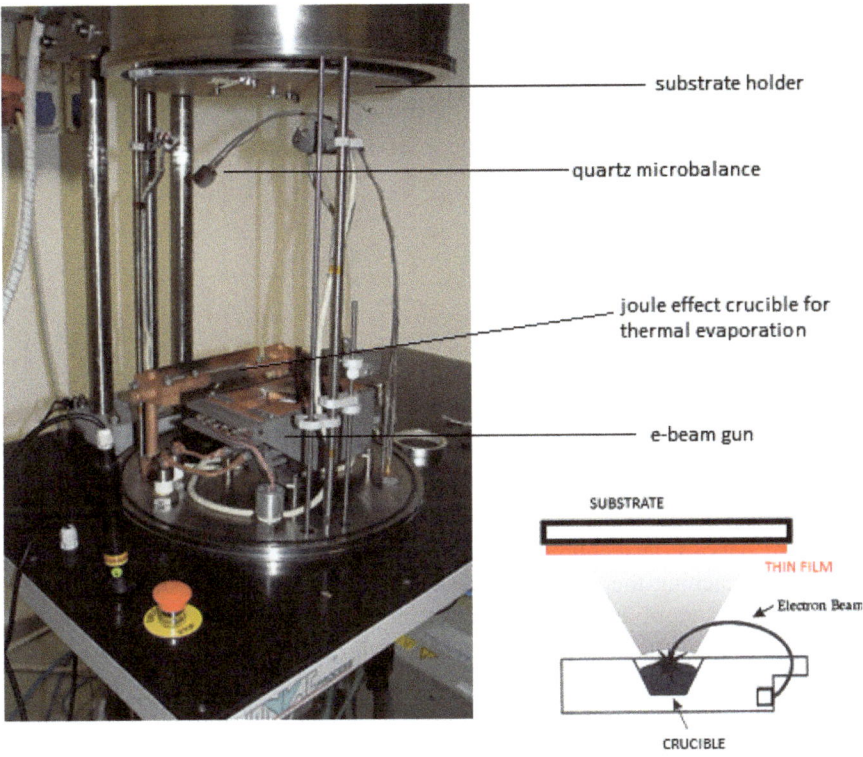

Figure 2 *Thermal evaporation and e-beam deposition set-up.*

In the case of RF magnetron sputtering the material is placed in vacuum on magnetic dipoles generating a magnetic field. The sputtering gas, typically Ar, is injected at low pressure (range of 10-3 mbar), and is excited by RF power (range of 100 Watt) so a plasma is generated. The magnetic field confines the plasma close to the sputtering targets. The ions of the plasma are driven along the field lines impinging onto the target with consequent sputtering of the target material that grows as thin film on the substrate. (Figure 3) More targets can allow the growing of different layers, deposition of the different materials is selected by a shutter rotating on the various targets.

The use of suitably shaped rotating masks can modulate the deposition rate and allows the growing of layers with different thickness on the various zones of the substrate. In this way the period of the ML coating can be optimized like for optics characterized by varying incidence angles of radiation on different positions of their surface, as in the case of large aperture optics.

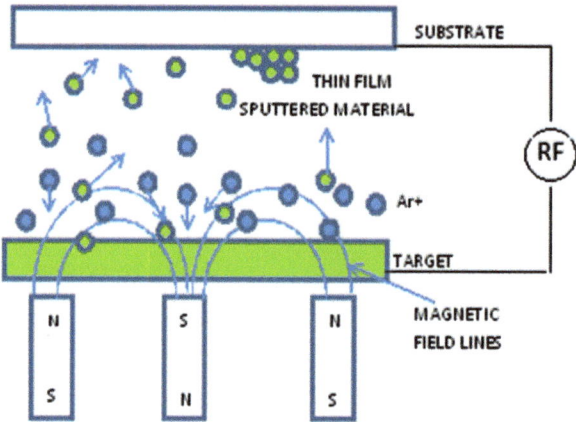

Figure 3. *Scheme of the magnetron sputtering process.*

Different deposition techniques are characterized by different energies of the ad-atoms, the atoms reaching the substrate. These energies range from fraction of eV for e-beam evaporation to about 10 eV for magnetron sputtering. The structure of the single layer as well as the characteristics of interfaces can depend strongly on the deposition parameters. The energy of ad-atoms as well the temperature of substrate can have an important role in the characteristics of the grown film. In fact the atoms sticking to the substrate should have enough energy to spread through the surface area growing in the form of uniform films and avoiding the formation of columnar structures. However to much high energy of ad-atoms can result in interdiffusion of layers of different materials with the consequence of affecting the quality of interfaces. The rise of substrate temperature can increase the mobility of the ad-atoms, however it can cause also diffusion and chemical reactions at interfaces, while it has been verified for Mo/Si ML's that deposition at low (cryo-) temperatures can favour the nucleation process and by reducing the rate of chemical bonding at interfaces improves the smoothness of interfaces. Often the deposition process can be assisted by low energy ion bombardment which can be suitably modulated during the growing of the film. Finally ultrathin barrier layers (e.g. B4C or Si3Ni4 in the case of Mo/Si ML)) can be interposed between different materials in order to avoid diffusion or the formation of compounds at interfaces. All these aspects have been deeply studied particularly in the development of coatings optimized at 13.5 nm for EUV lithography, and correspondingly a new technical field has been developed called "interface engineering" which concerns all technical efforts and solutions for the optimization of the growing process of ML with uniform and smooth layers[16].

For more exhaustive general reviews on this subject the reader is referred to topical bibliography like that in ref 17.

5 MULTILAYER METROLOGY AND CHARACTERIZATION

Any multilayer sample needs to be characterized in order to check its compliance with expected performances according to the project design; many parameters can affect the actual performances. In addition to possible imprecise knowledge of the optical constants of the materials, there are uncertainties about the layer thicknesses and their structural quality as some materials can grow amorphous layers while others will be polycrystalline.

Furthermore the quality of interfaces, due to interdiffusion and the forming of chemical compounds has to be taken into account. Similarly, stresses intrinsic to some materials or due to thermal effects can result in tension or compression strains at interfaces and can affect the stability of the structure. Finally the environment to which the multilayers are exposed can have further effect through chemical or physical processes such as oxidation, carbon contamination, thermal annealing or plasma exposure , like in space with effects due to ion bombardment.

Samples can be monitored in-situ during the deposition process or can be characterized later ex-situ. In the latter case the sample has to be exposed to the atmosphere, with possible contamination and oxidation effects, unless special technical solutions are adopted in order to keep the sample in vacuum while transferring it from the deposition chamber to the test area. In-situ monitoring during multilayer growth using soft X-ray reflectivity measurements can improve thickness and roughness control[18] while ellipsometry can give data about the composition and roughness of the growing film[19] and X-ray photoemission spectroscopy (XPS) and Auger spectroscopy can provide data about composition, i.e. intermixing and contamination[20].

Ex-situ the performance of a multilayer sample can be tested through reflectivity measurements, X-ray reflectance and diffraction, X-ray photoemission spectroscopy (XPS), or with atomic force microscopy (AFM) to characterize the surface quality, or transmission electron microscopy (TEM) for study of the layer structure, although this will imply the cutting of a sliced section of the multilayer. Several other techniques, such as electron diffraction can also be used. EUV reflectometer can be laboratory systems based on laser produced plasma sources, generally of high atomic number material targets (e.g. Cu, Au, W, rear earths). Focusing the beam from a laser system delivering pulses of about 800 mJ peak energy and 10 ns FWHM on such targets can produce a very bright EUV spectrally continuous emission. The EUV emitting plasma is coupled with EUV collector optics, feeding a monochromator, projection optics and test chamber[21].

Alternatively, synchrotron facilities can be used [ref 1 chapter 5]. Bending Magnet Beam-Lines emit continuum radiation in the EUV extending over a wide spectral range. Energy scanning with high resolution of a few thousands is performed with a monochromator system. Through the measurement of the incident and reflected radiation the reflectivity of the multilayers can be derived. A synchrotron source is characterized by a stable and well known emission related to the current in the storage ring. Furthermore the incident beam can be continuously monitored through a metallic mesh, by measuring the current induced by photoelectric emission. Since the bending magnet emission is totally linearly polarized in the orbital plane and elliptically polarized out of plane, the multilayer polarization response can be characterized by suitable selection of the radiation using narrow slits prior to the monochromator.

The multilayer structure, with information on the period and interface quality can be analysed also through grazing incidence X-ray reflectance measurements (XRR); a typical curve is shown in figure 4 . The peaks correspond to the Bragg maxima and so their separation corresponds to the period of the structure, while the intensities of and between the peaks depend on the interface roughness[22].

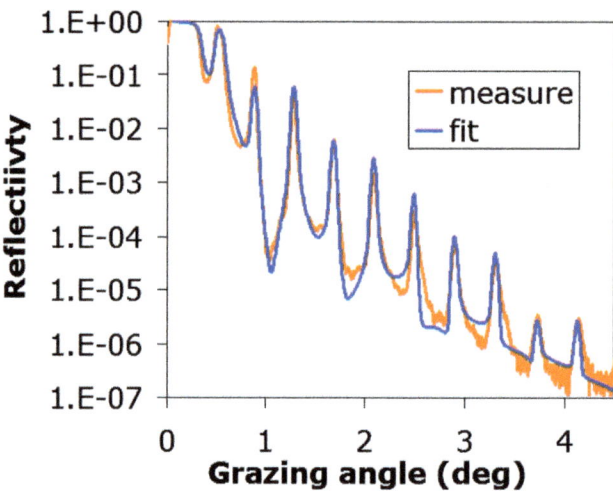

Figure 4 *example of XRR for a ML with d=10.75nm, sigma= 0.45 nm, gamma ratio d1/d= 0.54*

As already mentioned a thin film can develop intrinsic mechanical strain. This is due to the miss-match between the characteristic interatomic distance of the film and the adjacent material. The stress, tensile or compressive, in the coupling between film and substrate can be as high as MPa up to GPa and can affect the performances of the coating, for example by inducing deformation of the optical surface figure or compromising the long term stability of the structure. Stress of thin films can be tested typically via interferometric measurements of the curvature of the substrate. Stress analysis of multilayer performed on several samples in order to study both the relation with the deposition process and possible technical solutions has been reported. For example in ref 23 results of stress analysis on Mo/B4C multilayers subject to thermal annealing have been reported. R.W.E.van de Kruijs et al. have deduced from XRD measurements coupled with interferometric ones the stress in Mo/Si ML as due to the intrinsic strain of crystallites that form in the Mo layers during the growing process[24]. Various other papers reported on the stress and stability analysis as function of the deposition parameters, see for example ref 25 and 26 . Stress relaxation via thermal annealing and stress compensation by growing suitable interlayers have been studied by Kola et al.[27] and by Mirkarimi et al.[28] however the introduction of interlayers in order to compensate for the intrinsic strain of the structure has to be optimized in order to avoid some unacceptable drop of the final reflectivity of the ML[29].

Furthermore as already mentioned the roughness of the ML has to be carefully studied since it can affect the final reflectivity and imaging performance of the structure. Scattering measurements and their relationship with the surface and buried interfaces of the multilayer are quite a challenging task. A great effort has been done connected to the development of projection imaging optics for EUVL systems with measurements performed with the technological equipments of the beam-lines like at Lawrence Berkeley National Lab[30] and at Photon Factory[31]. The roughness of the surface and buried interfaces is characterized by different spatial frequencies. Correspondingly different instruments have to be used and consequently need to be relatively accurately inter-calibrated in order to carry on a complete consistent analysis. AFM, mechanical profilers, laser scanning microscopes, angle resolved scattering and total integrated scattering have been used to derive the power spectral density (PSD) functions of ML structures and correspondingly the associated roughness. The roughness and so the quality of a surface is characterized through its power spectral density (PSD) which is essentially related to the Fourier transform of the roughness. Measurements which relate scattering from a surface with its roughness and the growing process of the coating are very important in order to optimize the deposition processes and the performances of the optics, see for example ref. 32 and 33 and references therein.

Finally last but not least in high reflectivity X-ray multilayer structures, where strong standing wave fields, due to the superposition of incident and reflected fields, are generated near the Bragg angle, XRR is combined with X-ray standing wave (XSW) to exploit the potentialities of the two techniques in order to characterize the multilayer structure. Combined XSW-XRR can determine interface roughness and layer composition. Often Fluorescence from different elements in the multilayer can be simultaneously measured along with the reflectivity, although these measurements can be very difficult for non periodic and low Z elements structures because of the weak standing wave field and the low fluorescence yield[34, 35].

6   MULTILAYERS APPLICATIONS, SOME EXAMPLES

Multilayers made of different materials have been and are currently extensively studied for a wide range of potential applications from photolithography to laboratory and space applications. Most of these structures are based on periodic repetition of different layers but also new designs take into account structures made of a-periodic multilayers. Furthermore new materials and designs have been studied in order to explore the application of ML in different spectral ranges, in more critical operational conditions for instance improving resistance to thermal stresses or high radiation fluences, and to get a tailored reflectivity curve. Here we would like to present a short survey of some worthy examples of ML applications.

6.1 EUV Lithography (EUVL)

We can say that presently the greatest financial investment on ML's technological development is for EUV lithography and related applications. The technique of manufacturing integrated electronic chips is based on the micro-fabrication through a photolithographic process. A mask, reproducing the design of the circuit components and its connections is illuminated and projected on the Si wafer with suitable demagnification. The wafer is coated with thin film of photoresist which is a photosensitive material so that the design of the mask can be reproduced on the wafer after processing through chemical development followed by etching of its latent image. As we have written in the introduction, faster circuits can be realized by increasing the density of components in the chip. However this can be obtained by reducing the size of details of the image projected on the photoresist i.e. increasing the resolution of the optical projection system. Since the diffraction limit for resolution scales with the wavelength, EUV projection systems based on Mo/Si ML optics working at 13.5 nm have been developed, reaching resolution better than 30 nm at 13.5 nm. ASML has already realized the first prototypes of production systems, ALPHA Demo Tool. Several problems have been addressed and solved in order to develop a photolithographic systems suitable for high volume production . These problems concern mostly the source, the collector optics, the illuminator optics, the mask, the projection system and the photoresist. For example, the source, presently laser generated plasmas and discharge plasmas, should have emission concentrated at the operative wavelength with enough power and stability, the requirement being about 100W at intermediate focus[36]. The collector optics, which is very close to the source and so is exposed to high thermal flux and is affected by debris from the source, must have high efficiency and very long life-time, the requirement is that it should stay stable within about 1% for a few 10^4 hours. The mask besides high reflectivity must have very low density of imperfections, this is for obvious reasons, since like all system it works in reflection and imperfections would be reproduced on the wafer affecting the quality of the final production output. The illuminator/projection system consists typically of several mirrors, and it has to have very high efficiency, which results from the scaling with power n of the single coating reflectivity, n being the number of optics. This system must comply, consequently, with very precise matching of the reflectivity peak wavelength of all coatings, in fact the global system efficiency curve results much narrower than the curve for a single coating so that even a slight shift between the reflectivity curves of the single element mirrors can affect seriously the final global efficiency . In addition to this the high aperture and geometry of the system implies that the angle of incidence of radiation changes throughout the optical surface of the various elements and correspondingly the coating must be deposited by varying its period to guarantee highly uniformity of the efficiency throughout the full surface of the optics. Finally the coating must have very low roughness for better efficiency and resolution, as already discussed, and long stability to high thermal and radiation fluxes[16]. All these requiremenst have been met with highly controlled deposition processes of the multilayer coatings. The new NXE:3100 photolithographic system built by ASML and Carl Zeiss represents the next step forward to the standard electronic chips manufacturing via EUV lithography[37,38]. Looking at the next developments in the EUV lithography the effort is going to shift concentrating presently towards shorter wavelengths at about 6.7 nm, i.e. just below the B edge. Consequently, B and B-compounds based multilayer are being more and more intensively studied[39]. For a further updated presentation of recent developments of this technical application the reader is referred to ref. 40,41.

6.2 Multilayers for ultrashort pulses

The generation of ultrashort laser pulses of relatively high energy photons has opened new frontiers to the investigation of fundamental physical phenomena like those typical of the dynamics in atomic and molecular processes and to new technological applications[42]. Pulses in the time regime of attoseconds (1 as= 10^{-18}s) can be generated through the interaction of femtosecond laser pulses with gas jets. When the laser pulse is focused at high power density in the gas the atoms undergo tunnel ionization. The freed electrons are accelerated by the laser field and recombining with the parent atom emit a spectrum of high order laser harmonics extending from the NIR to the extreme ultraviolet and soft x-rays. Since the process takes place in a fraction of the laser period, pulses shorter than 1 fs can be generated[43,44]. The possibility of reflecting and focusing such short pulses in the EUV spectral range has been studied by several groups. The point is that the short time duration of the pulse and the narrow bandwidth of the ML structures do not match. In fact, since time and energy are related by Fourier transform, according to the Heisenberg's Uncertainty Principle the time duration and bandwidth of a pulse of X-rays are governed by the relationship $\Delta E \Delta \tau \geq 1.825$ eV fs, and so very short pulse implies a large bandwidth with the consequence that photons of quite different energy must be reflected by the multilayer coating. If the multilayer is not sufficiently broadband, the reflected pulse will be stretched in time. However, photons of different energy penetrate to different depths in the multilayer depending on the optical constants of the materials, as illustrated in figure 5, and so the reflected pulse broadens in time with different delays depending on the energy of its components. Hence, a standard periodic multilayer cannot reflect an ultrashort pulse while preserving its time profile, since its limited bandwidth and phase behaviour will filter out some spectral components and consequently will alter the shape of the reflected pulse, as demonstrated in figure 6 where a 200 as pulse is reported after reflection by a standard periodic Mo/Si ML.

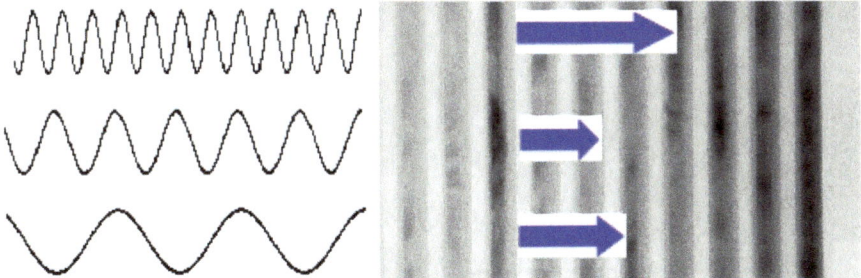

Figure 5 *Components with different energies penetrate to different depths in the multilayer.*

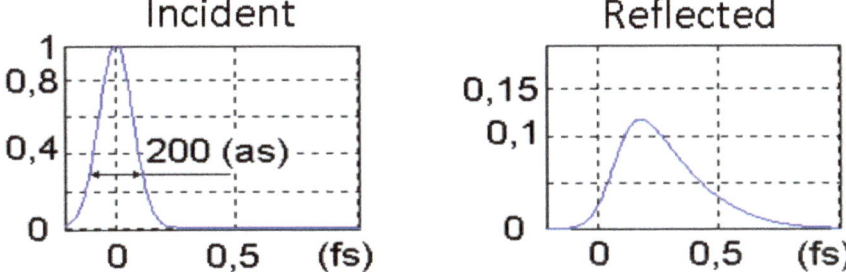

Figure 6 *A 200 as EUV pulse reflected by a standard Mo/Si periodic multilayer.*

This means that the multilayer reflectivity given by equation (19) must be characterized by a suitably large bandwidth and an optimized phase. By considering the Taylor expansion of the phase term (written here in terms of the frequency ν)

$$\varphi(v) = \varphi(v_0) + (v - v_0)\varphi'(v_0) + \frac{1}{2}(v - v_0)^2 \varphi''(v_0) + \frac{1}{3!}(v - v_0)^3 \varphi'''(f_0) + O(v - v_0)^4 \quad (25)$$

and considering an incident Gaussian pulse

$$E(t) = e^{-t^2/2\sigma^2} e^{-i v_0 t}, \quad (26)$$

it can be seen that the reflected pulse shape is not affected by the first, constant, term of the phase expansion. The second, first-order, term also does not affect the shape of the pulse since it only describes the group delay, i.e. the fact that the reflected pulse builds up after some time $\tau_g = \varphi'(v_0)/2\pi$. However as shown in figure 7, the second-order term causes time broadening of the pulse and a reduction of its amplitude, and the third-order term affects the reflected pulse even more, with distortion and temporal dispersion. This shows that in order to reflect a Gaussian pulse the ML should be characterized by a linear phase trend with negligible second and third order terms. Hence this means that the multilayer must be characterized by a suitably large reflectivity bandwidth and a suitable phase trend (figure 8) to compensate for the chirp of the incident pulse in order to obtain a linear phase trend with negligible effect on the reflected pulse shape.

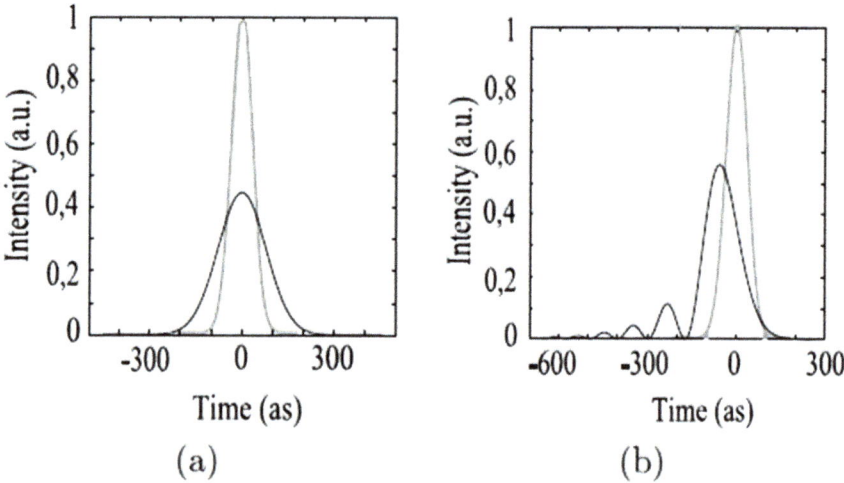

Figure 7 *Effect of (a) the second- and (b) third-order terms on the reflection of a Gaussian incident pulse. Incident pulse, light grey curve; reflected pulse, dark curve.*

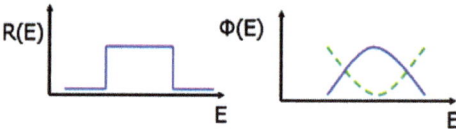

Figure 8 *The optimal multilayer should have large reflectivity bandwidth and phase chirp(continuous line) to compensate for the intrinsic chirp(dashed line) of the incident pulse energy components, the second- and third-order terms of the phase expansion.*

High reflectivity and large bandwidth can be obtained with a-periodic multilayer structures. This problem has been addressed in various works. Beigman et al. have analyzed numerically the conditions to get reflection of ultrashort fs and as pulses with high efficiency and no distortion of the pulse shape and its duration[45], Wonisch et al. have used genetic algorithm to design a-periodic ML structures, then these have been realized, by monitoring with in-situ soft X-ray reflectometry their deposition, and have been characterized by XRR and TEM[46]. Morlens et al. have studied how to compensate with a BB a-periodic ML the second order phase constant chirp of the plateau harmonics, as in the case of those generated from short quantum trajectories[47]. Then a three element a-periodic ML for the 35-50 eV range has been realized[48]. A broadband aML with linear phase trend has been realized and analysed by Wonisch et al.[49]. M.Suman et al.[50] have studied the design of Mo/Si multilayers with a wide bandwidth, 75–105eV centred on 90eV, characterized by high reflectivity, spectral phase compensation and amplitude reshaping. The authors have considered different incident pulse shapes, namely Gaussian with linear and chirped phase, rectangular with linear and chirped phase, and rectangular comb with chirped phase. From this paper as an example of the results, figure 9 shows the phases of a rectangular incident pulse, the computed multilayer and the reflected pulse after compensation. The simulation shows that an incident pulse of about 600as can be compressed to 140as.

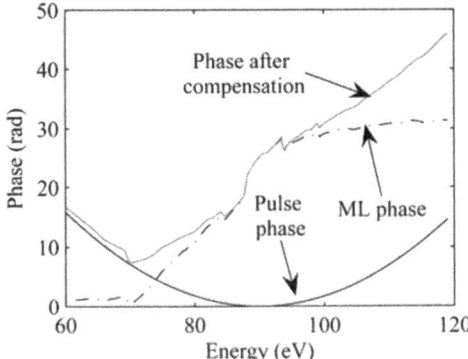

Figure 9 *The phases of a rectangular energy pulse (dark continuous curve), the multilayer (dash-dot curve) and the reflected pulse (light continuous curve) [from ref. 50]*

The a-periodic structure was designed with an optimization procedure based on an evolutive strategy which allows to converge towards structures not critically affected by the actual uncertainty, for example of the layer thickness, due to the deposition process. Through suitable definition of the merit function both efficiency, defined as the ratio between the reflected and the incident intensity, and time duration of the reflected pulse have been optimized[51].

Above 100 eV where the high order harmonics spectrum reaches the cut-off region materials different from Si have to be selected. Recently M Hofstetter et al.[52] have reported on a ML structure with broadband and high reflectivity based on (La/Mo) couples designed for the range 80-130 eV. By applying XUV-pump/IR-probe and frequency-resolved optical gating for complete reconstruction of attosecond bursts (FROG/CRAB) techniques reflection of single 200 as pulses has been demonstrated.

One interesting further technical aspect concerns the characterization of these a-periodic structures, which needs to exploit some specifically targeted metrological techniques. In figure 10 the thicknesses of the various layers in the multilayer structure are shown, confirming the technical feasibility since all layers are thicker than about 2nm but on the other hand it is also clear that the thicknesses do not follow any regular trend.

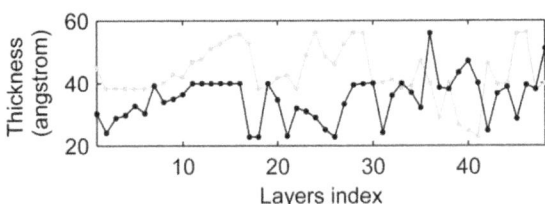

Figure 10 *Layer thicknesses from the top of the structure (low numbers): light silicon, dark molybdenum [from ref 50]*

Since the structures are not periodic X-ray reflection cannot provide straight information without a very complex inversion procedure. In addition the multilayer reflectivity phase, as already discussed eq 25, is critical and needs to be measured. So while the efficiency and bandwidth of the ML can be derived through reflectivity measurements the phase can be recovered from indirect measurement of the amplitude of the signal due to the standing wave resulting from the superposition of the incident and reflected electromagnetic fields in the multilayer structure. Since photoelectron emission at the multilayer surface is related to the intensity of this field, a measurement of the total electron yield (TEY) allows recovery of the multilayer phase through

$$\text{TEY}(E) = C(E)I_0\left[1 + R(E) + 2\sqrt{R(E)}\cos\varphi(E)\right] \tag{27}$$

where C is a constant depending on the material, I_0 is the intensity of the incident wave, R is the reflectivity and φ is the multilayer phase. TEY data can be recovered from drain current measurements on the sample.

Schematics of the TEY measurement and the field intensity distribution in the multilayer are shown in figures11, and figure 12 shows the expected TEY data with marked the points corresponding to nodes and antinodes of the field.

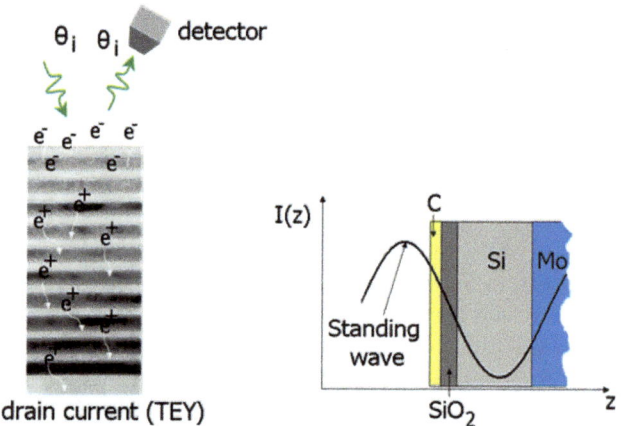

Figure 11 *Measurement of TEY via drain current data and Standing wave distribution in a multilayer. Silicon oxide and carbon layers due to oxidation after exposure to atmosphere and contamination are also shown.*

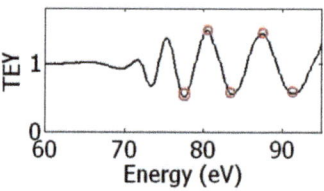

Figure 12 *Expected TEY data distribution, with red circles marking the position corresponding to nodes and antinodes of the field.*

This technique has been applied by A.Aquila et al. [53] and by M. Suman et al. [54]. Suman et al. report on the realization and characterization of an aperiodic multilayer with second order phase chirp designed to reflect a Gaussian pulse of 30eV bandwidth and positive 0.3fs^2 second order phase chirp. The multilayer was designed to have a nearly uniform reflectivity over a wide spectral range. Both reflectivity and phase were measured proving that the ML can efficiently reflect and compress a Gaussian incident pulse from 450as to 130as. Recently the technique of measuring the photocurrent induced by exposing the ML to EUV radiation has been applied also by C Bourassin-Bouchet et al. who deduced the effect of the ML reflection on the carrier-envelope phase of pulses shorter than 50 as [55].

6.3 Multilayers for Space Applications

Recent solar missions have been devoted to the observation of the Sun with high spatial and spectral resolution. In particular, spectral imaging with high temporal resolution can give useful information about the dynamics and energy processes in the solar plasma, such as did past missions like the Solar and Heliospheric Observatory (SOHO), whose payload had the normal incidence EUV telescope EIT[56], and the Transition Region and Coronal Explorer (TRACE)[57]. EIT was designed to provide full disk images of the solar transition region and inner corona. Its design was based on a Ritchey-Chretien telescope whose mirrors are divided in quadrants coated with Mo/Si EUV multilayer structures optimized in different spectral bands. The peak wavelengths of the four sections are 30.4 nm HeII, 17.1 nm Fe IX-X, 19.5 nm Fe XII, 28.4 nm Fe XV. TRACE was devoted to the study of the dynamics and evolution of the solar plasma from the photosphere to the corona. The instrument consists of a Cassegrain telescope with four sections. The optics have efficiency optimized at three EUV wavelengths: 17.1 nm, 19.5 nm and 28.4 nm, and one covering the band from 120 to 700 nm. Further spectral filtering was accomplished by the use of suitable filters.

More recently worth to be mentioned is Solar Dynamic Observatory (SDO) the mission which has been designed in the context of the International Living With a Star program. This mission will provide data about the evolution of the solar atmosphere that will be used to derive information for "Space Weather" events that can affect life on our planet[58]. On board of SDO is the Atmospheric Imaging Assembly (AIA) instrument. It consists of four Cassegrain telescopes made of normal incidence mirrors, in order to perform imaging at seven EUV wavelengths and one UV band. The EUV sections are optimized at 9.4, 13.1, 17.1, 19.5, 21.1, 30.4 and 33.5 nm by coating the optics of the different sections with suitable multilayers Mo/Y, Mo/Si, SiC/Si and then by inserting thin film (Zr, Al) filters. Measurements concerning the characterization of the AIA multilayer coatings are reported in ref 59. Future missions include the Solar Orbiter (SOLO), which will approach the Sun to as close as 0.28AU [60].

The main problems posed by imaging instruments are the efficiency, the spectral resolution and the spectral purity, the latter in particular to avoid contamination of the detected signal due to strong emission from other close spectral bands. If observations in different spectral bands are required, multi-band efficiency is also important, such as the Multi Element Telescope for Imaging and Spectroscopy (METIS) planned for SOLO that was devoted (before descoping) to the observation of the solar corona at the 30.4nm HeII and 121.6nm HI emissions line, and in the visible spectral range. Different spectral lines correspond to different ionization stages of the emitting ions and so to different equilibrium temperatures and plasma parameters. Accordingly, they can give important input data to accomplish detailed spectroscopic diagnostic of the coronal plasma.

Considerable work has been devoted to the development of multilayer coatings with high EUV spectral range efficiency. Material combinations already extensively studied include for example Mo/Y, Mo/Si, Si/B4C/Mo, Si/B4C, SiC/Si, Si/Mo2C, Si/C and SiC/Mg, Ir/Si [56, 57, 59, 61, 62, 63, 64]. Recent results of work performed within the context of the COST Action MP0601 are presented in the following. These will concern the development of multilayer coatings with high efficiency coupled with high spectral purity, the study of multilayer coatings exposed to harsh environments characteristic of those in proximity to the Sun. Some of the most important target lines for solar plasma diagnostics and the corresponding temperatures are shown in table 1. The various lines are characteristic of different plasma regions such as flaring regions, the quiet corona and its transition region, the corona and hot flare plasma, active coronal regions, and the chromosphere. From this short list it is clear that some important lines appear very close to other lines emitted by different ions or elements. For example, observations at 28.4nm or 33.5nm can be overwhelmed by the wings of the strong HeII line at 30.4nm.

Table 1 *Spectral lines for solar plasma diagnostics*

Wavelength [nm]	Ion	Temperature [K]
9.4	Fe XVIII	9×10^6
17.1	Fe IX	10^6
19.5	Fe XII	1.5×10^6
21.9	Fe XIV	2×10^6
28.4	Fe XV	1.6×10^6
30.4	He II	8×10^4
33.5	Fe XVI	2.2×10^6

Selective detection of some lines can be accomplished with ML structures characterized by suitably narrow bandwidth and/or by coupling the use of some suitable thin film filters. An alternative solution can be the design of structures with a customized rejection of some wavelength. In this way the possible efficiency loss due to additional filters can be avoided. This problem has been addressed by Suman et al. with multilayer designs developed to give structures with efficient reflectivity at 28.4nm or 33.5nm coupled with high rejection ratio at 30.4nm. The design was based on an analysis of the standing wave distribution inside the multilayer structure, using capping layer above a periodic multilayer in order to obtain absorption at the HeII line with minimal effect on the working wavelength[65,66]. Without the capping layers, a standard periodic Mo/Si multilayer optimized at 28.4 nm would have such a wide bandwidth that it would reflect a considerable percentage of the HeII neighbour line, seriously affecting the detected signal because of the strongly unfavourable intensity ratio between the two lines. The design sequence is as follows and illustrated in figure 13. First, a multilayer is designed to provide the highest reflectivity at the working wavelength. Next, computation of the standing wave distribution in the region in front of the multilayer structure is carried out both for the working wavelength and for the unwanted wavelength. Third, the capping layer structure is optimized by locating the absorbing layers close to the antinodes of the unwanted wavelength and close to the nodes of the working wavelength. Finally the complete structure is optimized including the capping layers along with the period and thickness ratio of the multilayer in order to obtain a high rejection ratio along with a high reflectivity. Structures have been studied to work at 33.5 and 28.4 nm, as a proof test of the procedure. The final actual optimization should take into account not only the optical properties of the materials but also other parameters that can affect the performances of the structure, like for example the selection of materials according to their stability, quality of interfaces due to interdiffusion and compounds, and characteristic induced stresses.

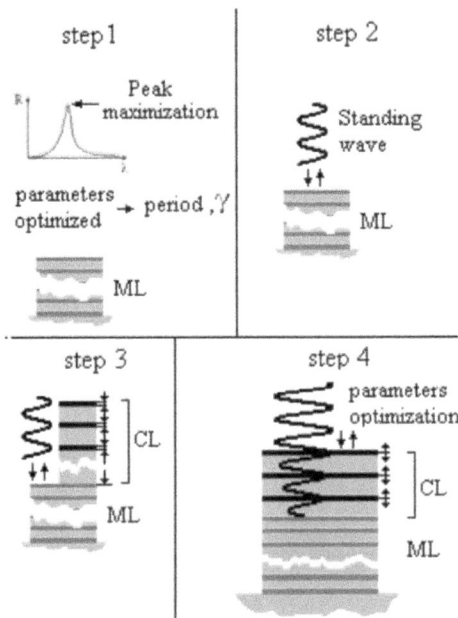

Figure 13 *Optimization procedure of a multilayer (ML) plus capping layer (CL) structure to provide high reflectivity coupled with high rejection ratio for two close wavelengths.*

In ref 66 a Mo/Si periodic structure with a W/Si capping layer designed for the Fe XVI line at 33.5 nm, is reported; this multilayer was designed using data from the Center for X-Ray Optics[11]. The computation shows a reflectivity peak of nearly 17 % at the working wavelength, slightly lower than for a standard periodic multilayer without capping layers, but the rejection ratio is about 10^{-5} compared to about 0.2.

A second periodic Mo/Si structure with a Mo/Si capping layer has been designed for the FeXV 28.4nm line and in this case a test sample has been realized and characterized. This structure was calculated using data from Tarrio et al.[67] and made at Reflective X-Ray Optics (RXO) LLC[68] by magnetron sputtering. The sample was characterized just after deposition at RXO and later at the Advanced Light Source, Berkeley. The latter results of the reflectivity measurement are shown in figure 14.

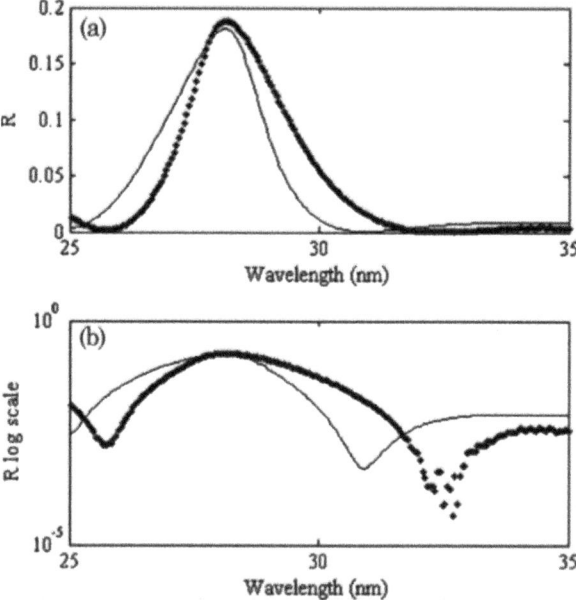

Figure 14 *Reflectivity of the Mo/Si multilayer with Mo/Si capping layer of table 3 as a linear plot (upper figure) and logarithmic plot (lower figure), The points show the experimental data and the lines the theoretical calculation.[from ref 66]*

The measured peak reflectivity is 0.19 with a minimum of about 6×10^{-5}, demonstrating the feasibility of the structure. However the peak reflectivity is slightly higher and at slightly longer wavelength than calculated, and the minimum is not at the correct spectral position. Since analysis shows that the deposition uncertainty negligibly affects the layer thicknesses, in particular that of the thick second a-Si layer in the capping layer structure, which is the most critical one, this suggests that the main concern is with the optical constants of the elements, the two databases used giving quite different values[11,67], and demonstrate that the optical constants can depend critically on the deposition procedure.

One further objective has been the development of multilayer structures capable of operating in harsh environments. The future mission SOLO will fly close as 0.28AU to the Sun and the payload will be exposed to non-negligible fluxes of particles and radiation. One major consequence concerns the lifetime of the coatings because of the temperature gradients and the damage induced by particle bombardment on the upper layers. Studies of the stability of different coatings have been carried out, most of them in the context of EUV lithography[23,29, 69-75]. In fact high temperature can cause structural changes of the materials and mixing or chemical reactions. Besides the highest temperature limit one coating can survive, one further aspect is related to the continuous temperature gradients. For example it has been evaluated that the optics in the METIS experiment, although with active temperature control, will undergo through a 14 days cycling gradient ranging from about 20 to 62 °C. In addition to this, the space environment around the sun is strongly influenced by the solar wind. Its characteristics depend on the solar activity phase for example if the source are quiet regions the plasma is mostly composed by protons, He2+, O6+ and Fe10+ ions with velocities between 300 and 1200 km/s corresponding to keV energy range[76], while, in the case of eruptive events like solar mass ejections or solar flares high energy particles are injected into the heliosphere with energy of MeV and higher. While it has been proved that high energy particles do not affect significantly the performances of EUV ML coatings [77], the effect of low energy particles had not been fully tested before. Tests have been reported by Pelizzo et al. and Corso et al. on ML samples designed to work at 30.4 nm corresponding to the HeII resonance line. These results have been published in ref. 78, 79. They used samples made of Mo/Si ML structures which have been optimized with different capping layers structures made respectively of oxidized Si layer, Ir/Si, Ir/Mo, and Ru/Mo. The corresponding design was optimized taking into account the standing wave distribution of the field in the ML structure as it has been already previously discussed.

The samples have been tested through thermal cycles simulating the gradients experienced during the mission. The samples have been exposed to the heating flux of three quartz halogen lamps, test duration was of five cycles in 570 hours (about 23 days) with temperatures ranging from 20 to 77 °C and leaving the max temperature stable for one hour. To avoid a prolonged time duration of the experiment the test gradients were three times steeper than those expected during the mission. The reflectivity has been tested before and after the thermal annealing. Performances resulted stable within the experimental uncertainty. Plasma effects have been tested through low energy proton bombardment. The laboratory exposure doses have been determined by scaling the solar wind parameters, as deduced from data derived from in situ measurements and reported on ref 76 to the life time exposure dose that is foreseen in a solar mission like SOLO. On this basis the samples have been exposed to protons of 1 keV energy to simulate an in orbit time respectively of three months and one year so corresponding to a total dose respectively of about $9 \cdot 10^{15}$ and $36 \cdot 10^{15}$ particles. Samples have been measured before and after the exposure to proton bombardment. In fig 15 the correspondent reflectivity measurements are reported.

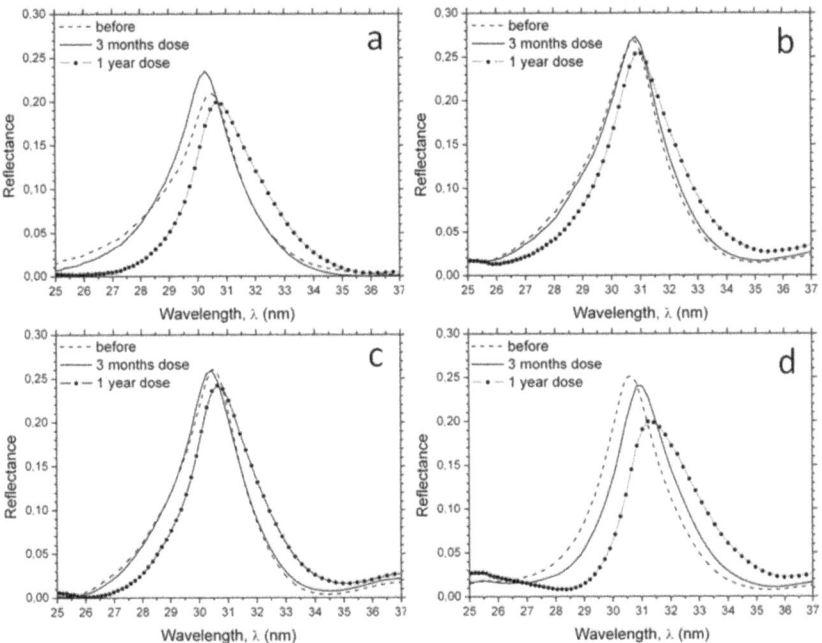

Figure 15 *a: oxidized Si capping layer, b: Ir/Mo capping layer, c: Ir/Si capping layer, Ru/Mo capping layer [from ref 78].*

The reflectivity peaks before the test are respectively 0.21 for the Si/Mo, 0.27 for the Ir/Mo, 0.26 for the Ir/Si, and 0.25 for the Ru/Mo samples. All samples have nearly the same spectral bandwidth, FWHM 2.9 nm. The samples have been tested to be stable after six month time. After exposure the samples have been tested mapping different surface points. After 1 year equivalent dose the Ru/Mo sample shows a considerable decrease of reflectivity while the other samples showed only a slight worsening of performances, 0.20 Si/Mo, 0.25 Ir/Mo, 0.24 Ir/Si coupled with a slight, a few nm, spectral shift. AFM measurement of the surface did not show any increase of roughness while TEM analysis of the most damaged Ru/Mo sample showed a clear delamination of the structure below the second Si layer from the vacuum interface.

One important aspect of the exposure to proton flux of multilayers can be damage resulting in delamination and blisters formation. In fact delamination has been observed in the case of Ru/Mo capped ML sample and blister formation in the case of an Ir/Si ML sample. These effects have been studied recently also by Kuznetsov et al.[80]. Damage can result from the delamination of hydrogen implanted layers triggered by the intrinsic stress, compressive or tensile, typical of thin films and from blistering induced by high energy ions or by hydrogenation from low energy ions. More recently effects induced by exposure to alpha particles bombardment have been reported[81].

7 SUMMARY AND CONCLUSIONS

Design and realization of ML coatings needs the understanding of the optical working principles of these structures as well as to take into consideration the technological problems related to the materials intrinsic properties and to their deposition in thin films form. In this review we have tried to illustrate these various aspects through an introduction on the basic physics of ML coatings, the deposition and characterization techniques and then reporting on some examples of more peculiar applications. After a short review on the EUV lithography for integrated circuits production, it has been discussed and explained the conceptual design of different multilayers: for reflection of ultrashort EUV pulses and for space applications. These structures can be a-periodic, computed with optimization procedures in order to get wide bandwidth and high reflectivity and suitable phase, or making use of capping layers tailored by taking into account the stationary e.m. field configuration in order to get high reflectivity coupled with high rejection ratio in specific spectral bands. Furthermore, the problem of surviving in harsh environment like that one encountered in space close to the sun has been also addressed. Not less important is the problem of realization and testing of ML samples, this has been discussed with particular emphasis also on new metrological techniques like the measurement of the reflectance phase of the multilayer.

ACKNOWLEDGEMENTS

The author gratefully recognizes the contribution of his colleagues, M.G. Pelizzo, P. Zuppella, M. Suman, G. Monaco, A.J. Corso.

References

1. D. Attwood, Soft X-Rays and Extreme Ultraviolet Radiation, Principles and Applications Cambridge Univ . Press 2000, chap 10 and references therein.
2. A.G. Michette, Optical Systems for Soft X-Rays, Plenum Press, N.Y. 1986.

3. Oura, K.; V.G. Lifshits, A.A. Saranin, A.V. Zotov, M. Katayama (2003). Surface Science: An Introduction. Springer-Verlag Berlin Heidelberg. ISBN 3-540-00545-5.

4. S. Yulin, T. Feigl, T. Kuhlmann, N. Kaiser, A.I. Fedorenko, V. V. Kondratenko, O. V. Poltseva, V. A. Sevryukova, A. Yu. Zolotaryov, E. N. Zubarev, "Interlayer transition zones in Mo/Si superlattices", Journal of Applied Physics, 92 (3), pp. 1216-1220 (2002).

5. S. Bajt, D. G. Stearns, P. A. Kearney, "Investigation of the amorphous-to crystalline transition in Mo/Si multilayers", Journal of Applied Physics, 90, 2, pp. 1017-1025 (2001).

6. I. Nedelcu, R.W.E. van de Kruijs, A.E. Yakshin, F. Bijkerk, "Temperature dependent nanocrystal formation in Mo/Si multilayers", Physical Review B 76 (24) (2007) 245404).

7. I. Nedelcu, R. W. E. van de Kruijs, A. E. Yakshin, F. D. Tichelaar, E.Zoethout, E. Louis, H. Enkisch, S. Müllender, F. Bijkerk, "Interface roughness in Mo/Si multilayers", Thin Solid Films 515 (2), pp. 434-438(2006).

8. D.G. Stearns "The scattering of X-rays from non-ideal multilayer structures" J.Appl. Phys. 65, 491-506 (1989).

9. L.Névot and P.Croce, "Caractérization des surfaces par réflexion rasante de rayons X. Application à l'étude du polissage de quelques verres silicates" Revue Phys. Appl. 15, 761-779 (1980); D.G.Stearns, D.P.Gaines, D.W.Sweeney, E.M.Gullikson "Non specular x-ray scattering in a multilayer-coated imaging system" J. Appl. Phys. 84, 1003-1028 (1998); S. Schröder, T. Feigl, A. Duparré, A. Tünnermann "EUV reflectance and scattering of Mo/Si multilayers on differently polished substrates" Optics Express, Vol. 15, 13997-14012 (2007).

10. E.Spiller Soft X-ray Optics (SPIE, Bellingham, WA, 1994).

11. http://www-cxro.lbl.gov

12.   B.L.Henke, E.M.Gullikson, and J.C.Davis "X-ray Interactions: photoabsorption, scattering, transmission, and reflection at E=50-30,000 eV, Z=1-92", Atomic Data and Nuclear Data Tables,vol 54,181-3424, 1993.

13. M. Altarelli, D.L.Dexter, H.M.Nessenzveig "Superconvergence and sum rules for the optical constants", Phys Rev. B 6, 4502-4509, 1972.

14.   D.Garoli, F. Frassetto, G.Monaco, P.Nicolosi, M.G.Pelizzo, F.Rigato, V.Rigato, A.Giglia, S.Nannarone, "Reflectance measurements and optical constants in the extreme ultraviolet-vacuum ultraviolet regions for SiC with a different C/Si ratio", Appl. Opt. 45, 5642-5650 (2006);   G. Monaco, D. Garoli, R. Frison, V. Mattarello, P. Nicolosi, M.G. Pelizzo, V. Rigato,L. Armelao, A. Giglia, S. Nannarone, "Optical constants in the EUV soft x-ray (5÷152 nm) spectral range of B4C thin films deposited by different deposition techniques", in Advances in X-Ray/EUV Optics, Components, and Applications, eds. Ali M. Khounsary and Christian Morawe, SPIE conf. proc. Vol 6317, 631712(1-12) (2006); J.F.Larruquert, R.A.M.Keski-Kuha, Appl. Opt. 39, 2772-2781 (2000); J. I. Larruquert, J. A. Me´ndez, and J. A. Aznarez "Far-ultraviolet reflectance measurements and optical constants of unoxidized aluminum films", Appl. Opt. 34, 4892-4899 (1995).

15.   T.A. Johnson, R. Soufli, E.M. Gullikson, M. Clift, "Zirconium and niobium transmission data at wavelengths from 11-16 nm and 200-1200 nm ", in Optical Constants of Materials for UV to X-Ray Wavelengths, eds. R. Soufly and J.F.Seely, SPIE conf. proc. Vol 5538 119- 124 (2004);   M. Fernández-Perea, J. I. Larruquert, J. A. Aznárez, J. A. Méndez, L. Poletto, D. Garoli, A. M. Malvezzi, A. Giglia, S. Nannarone, "Optical constants of Yb films in the 23-1700 eV range", JOSA A, Vol. 24, 3691-3699 (2007); J. I. Larruquert, J. A. Aznárez, J. A. Méndez, A. M. Malvezzi, L. Poletto, and S. Covini, "Optical Properties of Scandium Films in the Far and the Extreme Ultraviolet", Applied Optics, Vol. 43, 3271-3278 (2004).

16. E.Louis, A.E.Yakshin, T.Tsarfati, F.Bijkerk, "Nanometer interface and materials control for multilayer EUV-optical applications" Prog. Surface Science 86, 255-294 (2011).

17. P.J. Kelly, R.D. Arnell Magnetron sputtering: a review of recent developments and applications, Vacuum 56 (2000) 159-172, and Handbook of Thin Film Deposition, Edited by: Krisna Seshan, William Andrew Publ. (2012).

18. A. Kloidt, H. J. Stock, U. Kleineberg, T. D6hring, M. Pr6pper, B. Schmiedeskamp and U. Heinzmann Thin Solid Films, 228, 154-157 (1993); E. Louis, H.J. Voorma, N.B. Koster, L. Shmaenok, F.Bijkerk, R. Schlatmann, J. Verhoeven, Y.Y. Platonov, G.E. van Dorssen, H.A. Padmore, "Enhancement of reflectivity of multilayer mirrors for soft x-ray projection lithography by temperature optimization and ion bombardment", Microelectronic Engineering23, 215–218 (1994).

19. X.Gao, J.Hale, S.Heckens, J.A.Woollam, "Study of metallic structures, optical properties, and oxidation using in situ spectroscopic ellipsometry," J.Vac Sci. Technol. A16, 429-435 (1998).

20. A. Kanjilal, M. Catalfano, S. S. Harilal, A. Hassanein and B. Rice "Time dependent changes in extreme ultraviolet reflectivity of Ru mirrors from electron-induced surface chemistry ", Jour Appl. Phys. 111, 063518 (2012).

21. D.L.Windt, W.K.Waskiewicz, ""Multilayer Facilities Required for Extreme-Ultraviolet Lithography," Jour. Vac. Sci. Technol. B12, 3826-3832 (1994).

22. E. Spiller, "Characterization of multilayer coatings by x-ray reflection," Rev. Phys. Appl. 23, 1687-1700 (1988).

23. M. Barthelmess and S. Bajt, "Thermal and stress studies of normal incidence Mo/B4C multilayers for a 6.7 nm wavelength", Appl. Opt. 50, 1610-1619 (2011).

24. R.W.E.van de Kruijs, E.Zoethout, A.E.Yakshin, I.Nedelku, E.Louis, H.Enkisch, G.Sipos, S.Muellender, F.Bijkerk, "Nano-size crystallites in Mo/Si multilayer optics, Thin Solid Films 515, 430-433 (2006).

25. D.L. Windt, W.L. Brown, C.A. Volkert, W.K. Waskiewicz, "Variation in stress with background pressure in sputtered Mo/Si multilayer films", Journal of Applied Physics 78, 2423–2430 (1995).

26. P.B. Mirkarimi, "Stress reflectance and temporal stability of sputter-deposited Mo/Si and Mo/Be multilayer films for extreme ultraviolet lithography", Opt. Eng. 38 1246–1259 (1999).

27. R.R. Kola, D.L. Windt, W.K. Waskiewicz, B.E. Weir, R. Hull, G.K. Celler, C.A. Volkert, "Stress relaxation in Mo/Si multilayer structures", Applied Physics Letters 60, 3120–3122 (1992).

28. P.B. Mirkarimi, C. Montcalm, "Advances in the reduction and compensation of film stress in high-reflectance multilayer coatings for extreme-ultraviolet lithography" in Emerging Lithographic Technologies II , Yuli Vladimirsky ed., SPIE conf. proc. Vol. 3331, 133 (1998).

29. E. Zoethout, G. Sipos, R.W.E. van de Kruijs, A.E. Yakshin, E. Louis, S. Muellender, F. Bijkerk, "Stress Mitigation in Mo/Si Multilayers for EUV Lithography", in Emerging Lithographic Technologies VII Roxann L. Engelstad ed., SPIE Conf. Proc. Vol. 5037, 872–877 (2003).

30. E. M. Gullikson, S. Mrowka, B. B. Kaufmann, "Recent Developments in EUV Reflectometry at the Advanced Light Source" in Emerging Lithographic Technologies, V E.A.Dobsiz ed., SPIE conf. proc. Vol 4343, 363-373 (2001).

31. K. Misaki, N. Kandaka, "EUV scattering from Mo/Si multilayer coated Mirrors", EUVL Symposium oct 2010 Japan, http://www.sematech.org/meetings/archives/litho/8939/pres/ML-P03.pdf

32. S. Schröder, T. Herffurth, M. Trost, A. Duparré , "Angle-resolved scattering and reflectance of extreme-ultraviolet multilayer coatings: measurement and analysis", Appl. Opt. 49, 1503-1512 (2010).

33. A. Duparre´ , J. Ferre-Borrull, S. Gliech, G. Notni, J. Steinert, J. M. Bennett, "Surface characterization techniques for determining the root-mean-square roughness and power spectral densities of optical components" Appl. Opt. 41, 154-171 (2002).

34. S. K. Ghose and B. N. Dev, "X-ray standing wave and reflectometric characterization of multilayer structures," Physical Review B, vol. 63, 245409- 11 (2001).

35. S. K. Ghose, D. K. Goswami, B. Rout, B. N. Dev, G. Kuri, and G. Materlik, "Ion-irradiation-induced mixing, interface broadening and period dilation in Pt/C multilayers," Applied Physics Letters, vol. 79, 467–469, (2001).

36. "EUV Sources for Lithography", V.Bakshi SPIE publ. 2006.

37. "Extreme Ultraviolet Lithography" , V. Bakshi and A. Yen eds. SPIE publ. (2012).

38. C. Wagner and N. Harned , "EUV lithography: Lithography gets extreme", Nature Photonics 4, 24 - 26 (2010).

39. Y. Platonov ; J. Rodriguez ; M. Kriese ; E. Gullikson ; T. Harada ; T. Watanabe ; H. Kinoshita "Multilayers for next generation EUVL at 6.X nm", in EUV and X-Ray Optics: Synergy between Laboratory and Space II, SPIE conf.proc. vol. 8076, 80760N(2011).

40. Extreme Ultraviolet (EUV) Lithography III, P. P. Naulleau, O. R. Wood II eds. SPIE conf. proc. Vol 8322 (2012).

41. web site: http://www.euvlitho.com/

42. E. Goulielmakis, V. S. Yakovlev, A. L. Cavalieri, M. Uiberacker, V. Pervak, A. Apolonski, R. Kienberger, U. Kleineberg, F. Krausz, " Attosecond Control and Measurement: Lightwave Electronics", Science 317, 769-775 (2007).

43. M. Lewenstein, P. Balcou, M. Y. Ivanov, A. L'Huillier, P. B. Corkum, "Theory of high-harmonic generation by low-frequency laser fields," Phys. Rev. A 49, 2117–2132 (1994).

44. M. Drescher, M. Hentschel, R. Kienberger, M. Uiberacker, V. Yakovlev, A. Scrinzi, T. Westerwalbesloh, U. Kleineberg, U.Heinzmann, F. Krausz, "Time-resolved atomic inner-shell spectroscopy," Nature 419, 803–807 (2002).

45. I. L. Beigman, A. S. Pirozhkov, E. N. Ragozin, "Reflection of few-cycle x-ray pulses by aperiodic multilayer structures," J. Opt. A 4, 433–439 (2002).

46. A. Wonisch, Th. Westerwalbesloh, W. Hachmann, N. Kabachnik, "Aperiodic nanometer multilayer systems as optical key components for attosecond electron spectroscopy," Thin Solid Films 464–465, 473–477 (2004).

47. A.-S. Morlens, P. Balcou, P. Zeitoun, C. Valentin, V. Laude, S. Kazamias, "Compression of attosecond harmonic pulses by extreme-ultraviolet chirped mirrors," Opt. Lett. 30, 1554–1556 (2005).

48. A.-S. Morlens, R. López-Martens, O. Boyko, P. Zeitoun, P.Balcou, K. Varjú, E. Gustafsson, T. Remetter, A. L'Huillier, S. Kazamias, J. Gautier, F. Delmotte, M.-F. Ravet, "Design and characterization of extreme-ultraviolet broadband mirrors for attosecond science," Opt. Lett. 31, 1558–1560 (2006).

49. A. Wonisch, U. Neuhusler, N. M. Kabachnik, T. Uphues, M. Uiberacker, V. Yakovlev, F. Krausz, M. Drescher, U.Kleineberg, U. Heinzmann, "Design, fabrication, and analysis of chirped multilayer mirrors for reflection of extreme-ultraviolet attosecond pulses," Appl. Opt. 45, 4147–4156 (2006).

50. M. Suman, F. Frassetto, P. Nicolosi, M.-G. Pelizzo, "Design of a-periodic multilayer structures for attosecond pulses in the EUV," Appl. Opt. 46, 8159-8169 (2007).

51. M.G.Pelizzo, M.Suman, D.L.Windt, P. Zuppella, P. Nicolosi, "EUV multilayer coated mirrors for atto-physics, photolithography and space experiments:Software design procedure" Nucl. Instr. and Methods in Phys Res A 623 782-785 (2010).
52. M Hofstetter, A Aquila, M Schultze, A Guggenmos, S Yang, E Gullikso3, M Huth, B Nickel, J Gagnon, V S Yakovlev, E Goulielmakis, F Krausz, U Kleineberg," Lanthanum–molybdenum multilayer mirrors for attosecond pulses between 80 and 130 eV", New Jour of Phys 13, 063038 1-15 (2011).
53. A. Aquila, F. Salmassi, E. Gullikson, "Metrologies for the phase characterization of attosecond extreme ultraviolet optics", Opt Lett 33, 455- 457 (2008).
54. M. Suman, G. Monaco, M. G. Pelizzo, D. L. Windt, P. Nicolosi,"Realization and characterization of an XUV multilayer coating for attosecond pulses", Opt Express 17, 7922 – 7932 (2009).
55. C Bourassin-Bouchet, S de Rossi, J Wang, E Meltchakov, A Giglia, N Mahne, S Nannarone, F Delmotte, "Shaping of single-cycle sub-50-attosecond pulses with multilayer mirrors", New Journ of Phys 14, 023040 1-16 (2012).
56. J. P. Delaboudinière, G. E. Artzner, J. Brunaud, A. H. Gabriel, J. F. Hochdez, F. Millier, X. Y. Song, B. Au, K. P. Dere, R. A. Howard, R. Kreplin, D. J. Michels, J. D. Moses, J. M. Defise, C. Jamar, P. Rochus, J. P. Chauvineau, J. P. Marioge, R. C. Catura, J. R. Lemen, L. Shing, R. A. Stern, J. B. Gurman, W. M. Neupert, A. Maucherat, F. Clette, P.Cugnon and E. L. Van Dessel , 'EIT: Extreme-ultraviolet imaging telescope for the SOHO mission', Solar Physics 162,291 – 312 (1995).
57. B. N. Handy, L. W. Acton, C. C. Kankelborg, C. J. Wolfson, D. J. Akin, M. E. Bruner, R. Caravalho, R. C. Catura, R.Chevalier, D. W. Duncan, C. G. Edwards, C. N. Feinstein, S. L. Freeland, F. M. Friedlander, C. H. Hoffmann, N. E.Hurlburt, B. K. Jurcevich, N. L. Katz, G. A. Kelly, J. R. Lemen, M. Levay, R. W. Lindgren, D. P. Mathur, S. B. Meyer, S. J. Morrison, M. D. Morrison, R. W. Nightingale, T. P. Pope, R. A. Rehse, C. J. Schrijver, R. A. Shine, L. Shing, T. D.Tarbell, A. M. Title, D. D. Torgerson, L. Golub, J. A. Bookbinder, D. Caldwell, P. N. Cheimets, W. N. Davis, E. E.DeLuca, R. A. McMullen, D. Amato, R. Fisher, H. Maldonado, and C. Parkinson, 'The Transition Region and Coronal Explorer,' Solar Physics, 187, 229 – 260 (1999).
58. http://sdo.gsfc.nasa.gov/ see also http://hea-www.harvard.edu/AIA/
59. R. Soufli, D. L. Windt, J. C. Robinson, S. L. Baker, E. Spiller, F. J. Dollar, A. L. Aquila, E. M. Gullikson, B. Kjornrattanawanich, J. F. Seely, L. Golub, "Development and testing of EUV multilayer coatings for the Atmospheric Imaging Assembly instrument aboard the Solar Dynamics Observatory", in Solar Physics and Space Weather Instrumentation SPIE Conf. Proc. Vol 5901, 59010M-1 (2005); R. Soufli, E. Spiller, D. L. Windt, J. C. Robinson, E. M. Gullikson, L. Rodriguez-de Marcos, M. Fernández-Perea, S. L. Baker, A. L. Aquila, F. J. Dollar, J. A. Méndez, J. I. Larruquert, L. Golub, P.l Boerner, "In-band and out-of-band reflectance calibrations of the EUV multilayer mirrors of the Atmospheric Imaging Assembly instrument aboard the Solar Dynamics Observatory", in Space Telescopes and Instrumentation 2012: Ultraviolet to Gamma Ray, SPIE Conf. Proc. Vol 8443 (2012).
60. http://sci.esa.int/science-e/www/area/index.cfm?fareaid=45
61. D. L. Windt, S. Donguy, J. Seely, B. Kjornrattanawanich, E. M. Gullikson, C. C. Walton, L. Golub, and E. DeLuca, "EUV Multilayers for Solar Physics", in Optics for EUV, X-Ray, and Gamma-Ray Astronomy, SPIE conf. proc. vol. 5168, 1-11 (2004).
62. J. Gautier, F. Delmotte, M. Roulliay, F. Bridou, M.-F. Ravet, and A. Jérome, "Study of normal incidence of three-component multilayer mirrors in the range 20-40 nm", Appl. Opt. 44, 384-390 (2005).

63. M. G. Pelizzo, A. J. Corso, P. Zuppella, P. Nicolosi, S. Fineschi, J. Seely, B. Kjornrattanawanich, D. L. Windt, "Long-term stability of Mg/SiC multilayers", Opt. Eng. 51, 023801-9 (2012).

64. P.Zuppella, G.Monaco, A.J.Corso, P.Nicolosi, D.L.Windt, V.Bello, G.Mattei, M.G.Pelizzo, "Iridium/silicon multilayers for extreme ultraviolet applications in the 20–35 nm wavelength range", Opt. Lett. 36, 1203-1205 (2011).

65. M. Suman, M. G. Pelizzo, D. L. Windt, G. Monaco, S. Zuccon, and P. Nicolosi, "Innovative design of EUV multilayer reflective coating for improved spectral filtering in solar imaging," proceedings of the 7th International Conference on Space Optics, 2008.

66. Suman, M. G. Pelizzo, D. L. Windt, and P. Nicolosi, Extreme-ultraviolet multilayer coatings with high spectral purity for solar imaging Appl. Opt. 48, 5432-5437, 2009.

67. C. Tarrio, R. N. Watts, T. B. Lucatorto, J. M. Slaughter, and C. M. Falco, Appl. Opt. 37, 4100–4104 (1998).

68. http://www.rxollc.com/

69. V. V. Kondratenko, Yu. P. Pershin, 0. V. Poltseva, A. I. Fedorenko, E. N. Zubarev, S. A. Yulin, I. V. Kozhevnikov, S. I. Sagitov, V. A. Chirkov, V. E. Levashov, A. V. Vinogradov, "Thermal stability of soft x-ray Mo–Si and MoSi2-Si multilayer mirrors", Appl. Opt. 32, 1811-1816 (1993).

70. T. Feigl, H. Lauth, S. Yulin, N. Kaiser, "Heat resistance of EUV multilayer mirrors for long-time applications", Microelectronic Engineering, 57–58, 3–8 (2001).

71. S. Bajt, D. G. Stearns, "High-temperature stability multilayers for extreme-ultraviolet condenser optics", Appl. Opt. 44, 7735-7743 (2005)

72. S. B. Hill ; I. Ermanoski ; C. Tarrio ; T. B. Lucatorto ; T. E. Madey ; S. Bajt ; M. Fang ; M. Chandhok, "Critical parameters influencing the EUV-induced damage of Ru-capped multilayer mirrors", Emerging Lithographic Technologies XI SPIE Conf. Proc. Vol. 6517, 65170g (2007).

73. M. Ishino, M. Koike, M. Kanehira, F. Satou, M. Terauchi, K. Sano, "Thermal stability of Co/SiO2 multilayers for use in the soft x-ray region", Jour. Appl. Phys. 102, 023513-5 (2007).

74. H. Maurya, P. Jonnard, K. Le Guen, J.-M. André, Z. Wang, J. Zhu, J. Dong, Z. Zhang, F. Bridou, F. Delmotte, C. Hecquet, N. Mahne, A. Giglia, S. Nannarone, "Thermal cycles, interface chemistry and optical performance of Mg/SiC multilayers", Eur. Phys. J. B 64, 193-199 (2008).

75. I. Nedelcu, R. W. E. van de Kruijs, A. E. Yakshin, and F. Bijkerk, "Thermally enhanced interdiffusion in Mo/Si multilayers", J. Appl. Phys. 103, 083549-6 (2008).

76. ESA, "Solar Orbiter environmental specification", Issue 2.0 TEC-EES-03-034/JS (2008).

77. A. D. Rousseau, D. L. Windt, B. Winter, L. Harra, H. Lamoureux, and F. Eriksson, "Stability of EUV multilayers to long-term heating, and to energetic protons and neutrons, for extreme solar missions", in Optics for EUV, X-Ray, and Gamma-Ray Astronomy II SPIE Conf. Proc. Vol. 5900, 590004-9 (2005).

78. M.G. Pelizzo, A.J. Corso, P. Zuppella, D.L. Windt, G. Mattei, P. Nicolosi, "Stability of extreme ultraviolet multilayer coatings to low energy proton bombardment", Opt Express, 19, 14838-14844 (2011).

79. A.J. Corso, P. Zuppella, P. Nicolosi, M.G. Pelizzo, "Long term stability of optical coatings in close solar environment", in Solar Physics and Space Weather Instrumentation IV SPIE Conf. Proc. Vol. 8148, 81480X (2011).

80. A S Kuznetsov, M A Gleeson and F Bijkerk, "Hydrogen-induced blistering mechanisms in thin film coatings", J Phys: Condens. Matter, 24, 052203-6 (2012).

81. M. Nardello, Paola Zuppella, V. Polito, Alain Jody Corso, Sara Zuccon, M.G. Pelizzo, "Stability of EUV multilayer coatings to low energy alpha particles bombardment", Opt Express 2013, 21, 28334.

APPLICATIONS OF KrF LASERS FOR GENERATING COHERENT EUV RADIATION

I.B. Földes[1] and S. Szatmári[2]

[1]Department of Plasma Physics, Association EURATOM, Wigner Research Centre for Physics of the Hungarian Academy of Sciences, H-1121 Budapest, Konkoly-Thege u. 29-33, Hungary
[2]Department of Experimental Physics, University of Szeged, H-6720 Szeged, Dóm tér 9, Hungary

1 INTRODUCTION

There are several difficulties which inhibit the easy extension of lasers toward the vacuum ultraviolet (VUV) and extreme ultraviolet (EUV) spectral ranges. The first is associated with the lifetimes of atomic levels decreasing in proportion to the third power of the transition energy assuming a given transition dipole matrix element. Classically, or in case of a given oscillator strength it decreases with the square of energy. Therefore the transitions relevant for this spectral range have very fast spontaneous decays. Second, the Einstein stimulated emission coefficient B_{ik} is proportional to $\lambda^3 A_{ik}$, which means that for shorter wavelengths the spontaneous emission (coefficient A_{ik}) will be dominant compared to stimulated emission in this spectral range. Consequently, in most EUV and X-ray lasers the radiation is in fact amplified spontaneous emission (ASE) only, hence they are known as ASE lasers. A further difficulty is that in order to be able to excite high-energy levels for laser operation in the EUV range the internal levels of high-Z materials must be excited. Thus highly ionized matter of high temperature is needed and consequently high pumping energy is required. Due to these problems the required pumping energy increases very quickly with decreasing wavelength – faster than the fourth power of the photon energy.

The most prevalent, commercially available short-wavelength lasers are based on the population inversion of excimer molecules in gas discharges, hence the name excimer lasers. That with the most favourable properties is the KrF laser,[1] which has an operation wavelength of 248 nm. Here, the advantages and disadvantages of KrF lasers for the generation of EUV radiation are summarized. Note that incoherent radiation in this spectral range can be obtained from plasmas generated by high-power lasers. Laser heating of solid plasmas results in direct plasma radiation up to the kilo-electronvolt range containing both continuum radiation and line radiation from the excited ionic levels. Also, inner shell radiation can be generated resulting from non-linear processes in the plasma when fast electrons are generated, removing inner-shell, e.g. K-electrons. Whereas direct plasma radiation has at least a picosecond duration, K-shell radiation can be as short as the pulse duration of the interacting laser beam.

There are two ways to generate coherent radiation in the EUV using KrF lasers, one is pumping of X-ray lasers the second is based on the generation of high harmonics in gases or plasmas. Although the properties of KrF systems are not optimal for pumping in conventional collisionally pumped X-ray lasers, an interesting scheme for the realization of lasers in the hard X-ray range has been reported by using ultrashort KrF laser radiation[2]. In addition, short-pulse KrF systems are suitable for generating harmonics, although only low-order harmonics can be generated due to the $I\lambda^2$ scaling. These sources can provide intense coherent radiation in the EUV and can also be used as seed pulses for X-ray laser systems.

The properties of KrF lasers are summarized in section 2, and the sub-picosecond laser system of the HILL laboratory is introduced in detail. Emphasis is given to methods for improving the properties of KrF laser pulses, especially the temporal contrast. In section 3 some properties of KrF generated laser plasmas are introduced, and the possibilities of their use for X-ray laser pumping are discussed. Finally results concerning intense harmonic generation are summarized.

2 PROPERTIES OF KrF LASERS

In order to realize an intense x-ray source a high-power laser system is needed. The saturation energy density of KrF is in the mJ/cm^2 range, i.e. nearly three orders of magnitude lower than that of solid-state systems. The storage time of the medium is in the nanosecond range, less than the 10-20 ns pumping time of discharge-pumped excimer systems, whereas electron-beam systems have pumping times of up to hundreds of nanoseconds. Thus high-energy systems are of ~10 ns pulse duration and they use angular multiplexing. Although the high-energy pulses necessitate large aperture lasers, due to the accessible highly homogeneous beam quality, the high repetition rate (5-10 Hz) and the efficient pumping system make such systems good candidate for a future laser fusion facility based on the success of the Electra Laser of the Naval Research Laboratory.[3] It should be emphasized that the gas host material, with its low density and low non-linearity, results in good propagation of the beam in excimer amplifiers, without the dangers of self-focusing, phase-front distortion and self-phase modulation. The low level of non-linearity is extremely useful as it allows excimer amplifiers to be used for the amplification of ultra-short pulses. For obtaining coherent EUV radiation high power can be more important than high energy, and the following paragraphs summarize the properties that allow KrF systems to be used as short-pulse laser amplifiers.

Due to the low probability of stimulated emission for the UV wavelengths of excimer lasers, it is more advantageous to generate a seed pulse with a laser oscillator in the long wavelength region and then shift it into the short wavelength region through frequency up-conversion. Ultraviolet gain modules are used as optical amplifiers for the amplification of the frequency converted pulses. The laser wavelength is chosen so that the frequency of the up-converted laser beam is matched to the maximum gain of the excimer system. As the seed pulse originates from the long wavelength oscillator and its frequency is converted by an intensity dependent non-linear process, it has good temporal contrast as well as beneficial directional properties.

There are two basic methods to generate seed pulses. For the work described here a hybrid dye/excimer system is used.[4] An excimer-pumped cascade pulsed dye laser generates sub-picosecond pulses, tuneable over the whole visible spectrum, which are amplified to typically 150 μJ. With the wavelength of the dye laser set to be twice that of the excimer amplifier, the frequency-doubled pulses can be used as seed pulses for amplification in the excimer amplifier. Frequency doubling is performed just before the excimer amplifier. The ultraviolet seed pulses, of ≈ 15 μJ energy, are then amplified in the amplifier channel of the twin-tube excimer laser filled with the appropriate gas mix.

Alternatively, the broad bandwidth of solid-state titanium-sapphire lasers can be used to give seed pulses after frequency tripling.

For the present work KrF excimer amplifiers are used for which as high as 75% of the stored energy is extractable for pulses shorter than ≈ 5 ps, i.e. for more than an order of magnitude shorter than the 75 ps relaxation time of the amplifier. This corresponds to a saturation density $\varepsilon_{sat} = 2$ mJ/cm^2. Also, KrF amplifiers allow pulse durations shorter than 100 fs.[1] As shown in figure 1 an off-axis amplification scheme is applied which is capable of resolving the inherent inhomogeneity problem of the amplifiers, caused by the inhomogeneous transverse distribution of the deposited energy in the discharge. In the off-axis mode the more homogeneous longitudinal distribution is what mainly determines the intensity distribution of the output beam. Another advantage of the off-axis amplification scheme is that it improves the temporal contrast of the beam. As the seed pulses are obtained after frequency conversion, they are essentially free from pre-pulses, and therefore pedestals originate only from the ASE of the amplifiers. As the gain contrast is much higher for the off-axis scheme than for on-axis amplification, the latter helps in providing high-contrast sub-picosecond laser pulses.

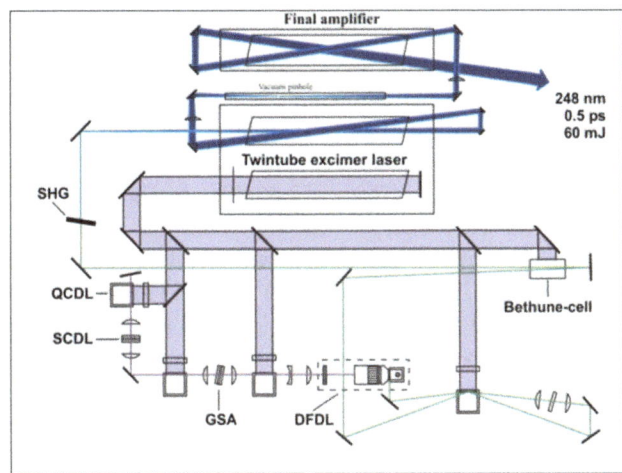

Figure 1 *The KrF laser system of the HILL laboratory*

Figure 1 illustrates the 120 GW KrF laser system of the HILL laboratory, based on a twin-tube excimer laser. The dye-laser chain is pumped by the first, XeCl filled discharge, and after several steps of pulse shortening and amplification the distributed feedback dye-laser (DFDL) provides sub-picosecond laser pulses. These pulses are then directly amplified in the Bethune cell and - after frequency conversion in a BBO crystal (SHG) – they are tuned onto the wavelength for maximum gain of the KrF amplifier. After two off-axis passes in the second tube the beam is passed through a vacuum pinhole spatial filter to the final amplifier. In the output of the laser beam 60-80 mJ is obtained in pulses of 500-600 fs with an energy contrast of ~100, i.e. with a prepulse energy – which originates solely from the ASE of ~15ns duration – being less than 1 mJ.

This system is an upgrade of an earlier one which did not have the the final amplifier but which was based on three-pass amplification in the first amplifier.[1] In the earlier system the 3-pass off-axis amplification resulted in 600 fs pulses of 15 mJ energy. The ASE prepulse had a duration of ≈ 15 ns, therefore the power contrast was $\approx 3 \times 10^{-6}$ even for an energy contrast of ~7%. In the case of tight focusing by an off-axis parabola, an intensity contrast better than 10^{10} could be obtained since the beam could be focused to a 2 μm diameter spot, whereas the ASE spot size was several hundred micrometers, corresponding to an ASE intensity of below 10^7 W/cm^2. In the present arrangement the spatial filter blocks ASE radiation with directional properties different to those of the main beam, and therefore the energy contrast is improved. However, the ASE transmitted through the pinhole is further amplified, resulting in a lower intensity contrast with ~10^8– 10^9 W/cm^2 ASE intensity in the focal plane. In order to further improve the contrast and to provide clean conditions for harmonic experiments and for X-ray spectroscopy of hot and warm dense matter, further methods were employed.

The ultimate method to remove pre-pulses is based on self-induced plasma shuttering or the plasma mirror technique.[6] If the intensity of the laser pulse falling onto a transparent solid material is chosen so that only the leading edge of the main ultra-short pulse is above the threshold for plasma production, pre-pulses or a pedestal of lower intensity will be transmitted. The created plasma does not have sufficient time to expand during the rise time of the main laser pulse, and therefore it interacts with an overdense plasma with a very steep density gradient. In this way, a double plasma mirror can improve the contrast by up to four orders of magnitude. Several successful experiments have been carried out, cleaning the pulses of titanium-sapphire and optical parametric chirped pulse amplifier (OPCPA) laser systems by plasma mirrors[7,8] and recently the applicability of plasma mirrors for KrF systems has also been demonstrated.[9]

In order to investigate the plasma mirror effect an s-polarized laser beam was focused by an F/10 lens onto an antireflection-coated glass plate. Reflection experiments were carried out using angles of incidence 45° and 12.4°. The target was moved from shot-by-shot by stepper motors. The laser energy was kept constant whereas the intensity was varied by shifting the lens relative to the target, which offered more than four orders of magnitude pump intensity range, from less than 10^{12} W/cm^2 to over 10^{16} W/cm^2. The first experiments were carried out at 45° angle of incidence,[9] however the best results were obtained for 12.4° where reflectivities up to 50% could be obtained[10] as illustrated in figure 2. This shows a logarithmic increase of reflectivity above the plasma threshold, and saturation close to a laser intensity of 10^{14} W/cm^2. The saturation behaviour is similar to the observations with infrared (800 nm) radiation.[8] Although the observed reflectivity is lower, the results show that plasma mirrors can be directly applied to increase the contrast of short-pulse high-power KrF systems.

There is another possibility for the use of plasma mirrors in KrF systems due to the direct short-pulse amplification scheme and the saturation of KrF amplifiers. Consequently, if the plasma mirror is applied in front of the final amplifier the output energy will not be significantly reduced and the only remaining pre-pulse source is the ASE from the final amplifier. This ASE does not give a significant contribution at the focus because it has larger divergence and will remain essentially unfocused in the focal plane.

The possible applications of short pulse KrF laser systems extend to a much broader area than the generation of EUV pulses. With the increase in size of discharge pumped excimer systems and by using electron-beam pumped amplifiers KrF systems may become appropriate even for the fast ignition scheme of inertial confinement fusion[11,12] combined with a main driver pulse of nanosecond duration.[3] In the following sections, however, the generation of soft X-ray and EUV radiation is considered.

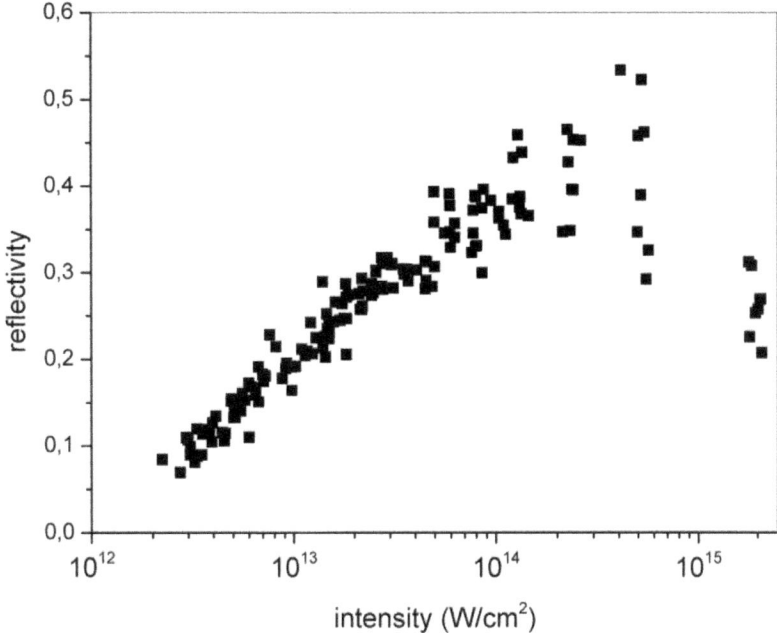

Figure 2 *Intensity dependence of the reflectivity for 12.4° angle of incidence.*

3 X-RAY GENERATION IN KrF LASER PLASMAS

The efficiency of generating X-rays increases as the laser wavelength decreases. This was already well known in the 1980s when experiments were carried out with nanosecond laser pulses.[13] The penetration depth of 248 nm radiation into the plasma is higher than for longer wavelengths as the critical density is proportional to λ^{-2}. For nanosecond pulse durations the dominant absorption mechanism is inverse Bremsstrahlung; this is a three-body collisional process and the photoabsorption coefficient, the opacity scales as

$$Z^2 n_e n_i T_e^{-3/2} \left[1 - \left(n_e / n_c \right) \right]^{-1/2} , \tag{1}$$

where Z is the ionic charge, n_e and n_i are the electron and ion densities, T_e is the electron temperature and n_c is the critical density. Equation (1) shows that absorption occurs most strongly at high electron and ion densities. Line radiation emission processes are most efficient where bound-bound transitions of high-Z matter are strongly excited by plasma electrons, i.e. also at high densities. These are the reasons why high absorption and high conversion efficiency to X-rays were obtained using short wavelength radiation with efficiencies up to 50-70% depending on laser intensity,[13] significantly higher than for long wavelength laser radiation. The X-rays and EUV radiation generated this way are, however, of long pulse duration and incoherent. It is important to note that because non-linear processes scale as $I\lambda^2$ (as can be understood simply from the oscillating motion of electrons in the electric field of the laser), their effect will be low for short wavelength radiation, and consequently the absorption will not be reduced.

As described above, generating X-ray lasing from laser plasmas requires a lot of energy. This is especially true for nanosecond systems, and so the first demonstrations of laser plasma X-ray lasers used high-power systems mainly devoted for fusion studies.[14] The situation changed in the 1990s when X-ray lasing was demonstrated with the so-called transient pumping scheme,[15] which needs significantly lower energy. A first laser, typically of ~1 ns pulse duration, is used to prepare the plasma target which is then heated with a high flux density provided by a picosecond laser to obtain transient population inversion. Both lasers must have energies of several Joules. It is not difficult to achieve such energies with KrF lasers of ~10 ns pulse duration, but for shorter pulses a short seed pulse is necessary along with beam multiplexing in the amplifier, which lead to a large system footprint. For the short pulse system a duration of around 600 fs would be optimal, but with current systems the pulse energy is lower than the required 1 J level.

In the following X-ray and EUV generation using short-pulse KrF systems is summarized. The short wavelength of the KrF radiation allows focusing to a spot of size comparable to the wavelength for a diffraction limited beam, meaning that a spot size smaller than $1\,\mu m$ can be obtained resulting in a focused intensity of more than 10^{19} W/cm^2 with a relatively small laser system[1] and even 10^{20} W/cm^2 can be reached with modest energies. A general problem for high-power laser-plasma interactions is the low absorption of the laser radiation at high intensities. Due to the greater penetration depth of 248 nm radiation this will also be higher for KrF systems compared to longer wavelengths, an advantage of short wavelength systems. Absorption up to 70% has been observed at intensities in excess of 10^{18} W/cm^2 for 380 fs pulses.[16] The same work also found that the efficiency of X-ray generation is higher for p-polarized laser light than for s-polarized radiation even if the absorbed intensity is kept at the same level. The higher efficiency was probably caused by the higher photon energy of X-rays generated by the p-polarized radiation. Consequently this suggests that the absorption is not purely collisional as it is for nanosecond pulses, since for collisionally absorbed radiation the emitted radiation is a function solely of the absorbed intensity. Therefore this observation demonstrates the presence of non-collisional absorption such as resonance absorption[17] or Brunel absorption[18] for sub-picosecond KrF laser pulses.

The durations of the X-ray and EUV pulses generated by femtosecond lasers are always longer than the laser pulses, they are in the picosecond range in case of thermal excitation. The possibility of generating X-ray pulses with durations comparable to the laser pulses can be realized by the so-called K_α generation. Fast electrons generated in laser plasmas as a consequence of non-linear interactions penetrate through the steep gradient of the laser-heated matter and eject inner-shell electrons of the target atoms. This produces fluorescence line radiation as the inner shell vacancy is filled from the outer shells. The inner shell K_α line can be clearly separated spectrally from the K-shell radiation of multiply ionized atoms. Nevertheless, in order to have a strong K_α signal clean from line and continuum radiation from the heated matter,[19] layered targets are generally used in which case the K_α source is heated by hot electrons only, thus separated from the laser-plasma interactions. A simple 1-dimensional model for K_α generation gives the conversion efficiency consistent with the experimental observations.[20] In the model a hot electron component is assumed from the laser-plasma interactions which penetrates into the solid slab. The stopping power for the hot electrons is taken into account, together with the cross-section of K-shell ionization and photoabsorption cross-section for the K_α photons. A simple formula is derived for the number of the generated K_α photons, which can be integrated using the material parameters.

Although the $I\lambda^2$ scaling means that the electron temperature will not be very high for lasers of 248 nm wavelength, multi kilo-electronvolt electrons can still serve as sources of

K_α radiation for materials with K-shells in the soft X-ray region up to several kilo-electronvolts. Although K_α radiation is monochromatic and of of short pulse duration, it is not coherent. Coherent radiation can be provided either by X-ray lasers or high harmonic generation sources.

In the last 10 years much debate has been initiated by the group at the University of Illinois which claims to observed X-ray lasing action at ≈ 0.28 nm from KrF laser irradiated xenon clusters.[21] Instead of using solid targets the ≈ 230 fs pulses of 248 nm laser radiation was focused into a gas-jet target consisting of xenon clusters typically containing 5–20 atoms/cluster. The focus diameter was 4 µm, corresponding to intensities of 1.6×10^{19} W/cm^2. Self-focusing of the laser pulse resulted in a channel of length 1.5-2.5 mm and diameter 1.5-2.5 µm. Strong X-ray line emission was observed along the filament at several wavelengths including 0.271 nm, 0.2804 nm, 0.286 nm and 0.288 nm.[21,22,23,24] The line widths of the emission were significantly narrower than those in the spontaneous emission in the transverse direction to the filament. An interesting feature was the reproducibility of the amplified emission and at the same time the irreproducibility of the wavelength of these emissions, originating from the hollow L-shell radiation from Xe^{32+}, Xe^{34+}, Xe^{35+} and Xe^{37+}. In some cases several lines appeared to be amplified in a single shot, which was considered to be improbable. In the cited works,[21,22,23,24] however, gain coefficients as high as 27-104 cm^{-1} were measured. Other groups working with other types of lasers could not observe similar phenomena. A possible explanation is that using radiation of longer wavelength the filamentary structure will be different, e.g. less homogeneous and probably of different temperature.

Recently a theoretical model was given to explain the observations.[25] In the self-focused filament intensity levels of $\approx 2 \times 10^{20}$ W/cm^2 are predicted, and the steep intensity gradient will create a steep ionization gradient transverse to the channel. Near the channel axis L-shell ionization levels may well be reached but this becomes less likely when further away from the axis. Near the axis the L-shell emission from the highly stripped core will propagate into the less ionized surrounding regions where they can photo-ionize inner-shell 2s and 2p electrons, thus leading to the photo-ionization laser proposed as early as 1967.[26] The calculations confirm earlier conjectures[25] that the amplified emission is a consequence of inner-shell photo-ionization and hole creation within the M-shell ionization states of xenon. The model, as well as giving possible gains up to the observed 50 cm^{-1}, predicts temporal windows for population inversion of order of magnitude 10 fs, which allows the simultaneous observation of more than one amplified line within a single 230 fs laser pulse.

Although these results have to be further confirmed by more experimental and theoretical work, they are encouraging as they suggest the possibility of using short-pulse KrF lasers as unique pumping sources for X-ray lasers with energies of several kilo-electronvolts.

4 GENERATION OF HARMONICS WITH KrF LASERS

4.1 Harmonic Generation in Gases

A commonly used method to obtain coherent radiation in the short wavelength EUV and XUV regions is to use harmonic generation, starting from intense short laser pulses. The interaction of the intense field of a laser pulse with an appropriate medium induces a non-linear polarization leading to the generation of radiation at harmonic frequencies of the inducing pulse. Most such work is done using titanium sapphire lasers, operating at wavelengths in the near infrared. Current commercial systems are available with a wide range of parameters, with pulse durations less than 10 fs, energies from microjoules to hundreds of millijoules and repetition rates, for the lower energies, of up to 250 kHz. With these sources, wavelengths shorter than those of the water window (2.3-4.5 nm) have been achieved through harmonic generation. Short-pulse KrF excimer lasers operating at 248.5 nm, with sub-picosecond pulses and pulse energies up to 100 mJ at repetition rates up to 10 Hz, or 30 mJ at 300 Hz may, however, provide an alternative. Due to the shorter wavelength (higher photon energy) and the good focusability, such systems offer specific advantages over longer wavelength sources.

In the wavelength range below 100 nm the generation of short-pulse coherent radiation by non-linear optical processes, both Ti-sapphire and KrF lasers have their specific regimes and advantages. To penetrate into this spectral range with visible or IR laser systems high-order non-linear processes are required, which become efficient at intensities above 10^{13} W/cm^2. At these intensities the power of the generated harmonic frequencies no longer follow the classical perturbative exponential decrease with harmonic order but a plateau range arises, where the harmonic power remains almost constant. Figure 3 shows a schematic of the dependence of the harmonic power I_n on the harmonic order n for long and a short pump wavelengths at fixed pump intensity.[27]

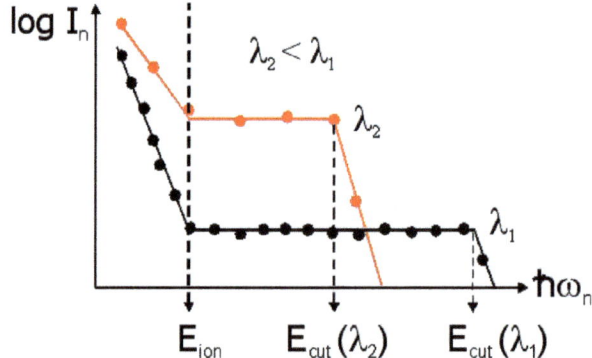

Figure 3 *Schematic behaviour of the harmonic power I_n versus the harmonic photon energy.*[27]

Three regions may be distinguished: the perturbative region at very low orders where the power decreases exponentially, the plateau region with almost constant harmonic power and finally the cut-off region where the power decreases rapidly. The plateau range starts at a harmonic energy corresponding to the ionization energy E_{ion} of the non-linear medium and ends at a cut-off energy E_{cut} given by

$$E_{cut} = E_{ion} + aU_p, \qquad (2)$$

where U_p [eV] is the ponderomotive energy given by

$$U_p = 9.33 \times 10^{-14} I_p \lambda^2, \qquad (3)$$

for a pump intensity I_p [W/cm^2] and pump wavelength λ [μm]. The factor α has a value of 3.17, according to a simple theoretical three-step model.[28]

From the expression for the cut-off energy and from figure 3 it can be seen that for a given medium and intensity the longer wavelength Ti-sapphire laser yields a broader plateau range with shorter wavelengths and higher orders, while the shorter wavelength KrF laser has a less extended plateau but allows higher efficiencies, as the plateau is reached at a lower harmonic order. In the plateau range the process of harmonic generation is far from resonance and typical conversion efficiencies are usually below 10^{-6}. The highest powers are of course obtained at the low-order harmonics in the perturbative regime and thus – for the generation of very high power EUV radiation – the 3rd or 5th harmonic of a KrF laser, corresponding to 82.8 nm or 49.7 nm, seem to be especially attractive.

For harmonic generation at high intensities noble gases are well suited, due to their comparatively high ionization potentials. For the efficient generation of the 3rd harmonic, argon is the best suited non-linear medium as the ionization energy is high enough to allow absorption-free generation in the perturbative regime, and in comparison to neon or helium the static polarizability, i.e. the non-linear susceptibility, is much larger. In addition, argon allows operation in the close vicinity of a three-photon resonance transition, which strongly enhances the non-linear susceptibility. For efficient frequency conversion into the 3rd harmonic, phase matching of the fundamental and harmonic waves is an essential requirement; this means that the phase velocities of the waves are equal. If this condition is fulfilled, the generated harmonic wave travels through the medium with the same velocity as the fundamental wave. Consequently, at every location the generated wavelets add constructively to previously generated wavelets travelling along the fundamental wave, thus leading to an enhancement of the harmonic wave. The wave-vector mismatch can be expressed as

$$\Delta k = k_{3\omega} - 3k_{\omega} = \Delta k_{\text{disp}} + \Delta k_{\text{geom}}, \tag{4}$$

with $k_{3\omega}$ and k_{ω} being the values of the wave vectors of the 3rd harmonic and the fundamental, respectively. The total phase mismatch Δk consists of a dispersive part Δk_{disp} and a geometrical part Δk_{geom}. It was shown by Dölle and co-workers[26] that, due to the resonance in argon, the dispersive part can be tuned so that when no plasma is generated the wave-vector mismatch caused by the atomic part of Δk_{disp} can be zero or even of opposite sign to Δk_{geom}. They carried out experiments with a gas jet target where strong third harmonic generation was observed and claimed a conversion efficiency as high as 1%, providing $\approx$100 μJ at 82.8 nm. The gas jet target was significantly smaller (1 mm) than the geometrical coherence length. In order to improve the conditions, an elongated gas jet target was used for the current experiments, which allowed a longer amplification length and higher harmonic intensity. As shown in figure 4, the strongest signal, obtained for a single stream 9 mm long nozzle, was about three times larger than for the standard cylindrical 1 mm diameter nozzle, and also stronger than for a double stream version of the same length. The optimum position was out of focus because in focus the generated plasma deteriorates the phase matching. This observation is in accordance with the coherence length for harmonic generation obtained from the geometrical mismatch only, which was 4.4 mm for our focusing condition ($l_{coh}=4/b_w$) with b_w confocal parameter. Therefore in this case the contribution of the atomic part and that from the ionization was negligible.[29]

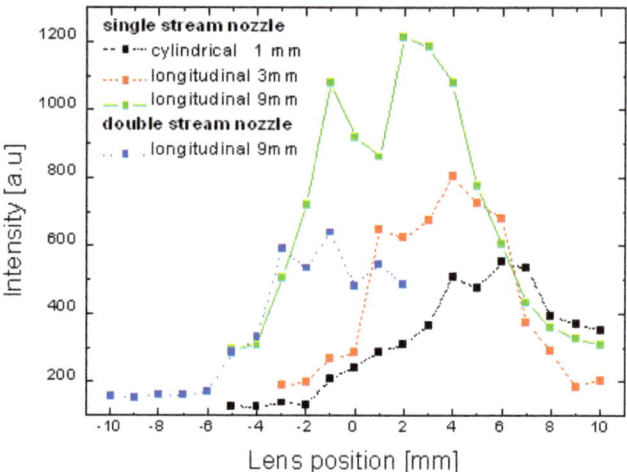

Figure 4 *The intensity of 3ω radiation as a function of focal position for different nozzles.*

Although the amplification length of the harmonic generation was increased compared to previous work,[27] the harmonic energy was lower than the 100 μJ measured therein but higher than the earlier measured[30] 14 μJ. The problem with such measurements is that the detectors used could not distinguish directly between the 248 nm and 82.8 nm signals, thus giving rather indirect evidence. Further experiments have been carried out with diamond detectors which are not sensitive to the 248 nm radiation, and therefore absolute energy measurements can be expected and estimations show harmonics energy of ~30 μJ.[29] Independent to these results it can be confirmed that a very intense coherent source, with tens of microjoules at 82.8 nm, can be realised using a tabletop 10 mJ KrF laser system.

4.2 Harmonic Generation in Plasmas

A method complementary to using gaseous target is the generation of high-order harmonics from the interaction of high-power lasers with solids. Electron oscillations crossing a steep plasma density gradient are anharmonic, and thus serve as a source of harmonics; intense researches started in the 1990s for high-harmonic generation utilising this phenomenon. Numerical simulations[31,32] showed that, in contrast to harmonics generated in gases, there is no fundamental limit due to ionization in this case, especially in the relativistic regime where the oscillation speeds of the electrons in the laser field become comparable with the speed of light. The practical possibilities of this method were clearly demonstrated by experiments in which harmonics with energies of several kilo-electronvolts were generated,[33] with good collimation and focusing properties.[34]

Due to the $I\lambda^2$ scaling of non-linear effects the use of UV laser beams has similar limitations in such methods as with gas harmonics. Although low-order harmonics can be more intense than the same wavelength harmonics from infrared lasers, KrF lasers do not - in general - provide relativistic conditions, which sets the upper limit for harmonic generation at a frequency equal to the plasma frequency in the undisturbed solid. Nevertheless, harmonics have been observed with short-pulse KrF lasers[35] and wavelengths as short as 62 nm have been detected.[36] These are, of course, low orders of the KrF radiation, but they are intense coherent EUV sources and can serve as a basis for understanding non-relativistic short-pulse interactions with matter.

The early experiments[35] with KrF lasers, using intensities of $\approx 5 \times 10^{15}$ W/cm^2 showed harmonics up to the third order. The harmonics propagated in the direction of the specularly reflected light from the target, and both p- and s-polarized radiation generated harmonics and conserved the polarization.[37] As this behaviour at low intensities cannot be explained well by theories based on resonance absorption, a single-electron model was developed. This takes into account the evanescent electric and magnetic fields in the over-dense plasma layer and claims that the $\mathbf{v} \times \mathbf{B}$ force cannot be neglected in the over-dense region even in the non-relativistic regime, and thus it may be a source for high harmonics for either p- or for s-polarized radiation. High harmonics generated by this will have polarization similar to that of the incoming laser radiation.[38]

The question is, what happens at higher intensities? Experiments with 1 ps KrF lasers have only generated harmonics up to the fourth order for intensities 10^{19} W/cm^2, and they also showed that rippling and thus diffuse harmonic propagation starts between 10^{16} and 10^{17} W/cm^2, well above the intensities of previous experiments.[36] Those results were however obtained with laser pulses that were not clean, i.e. with strong prepulses. This meant that it was unclear whether surface rippling is an effect which occurs in the preformed plasma or an intrinsic effect. Also, it is well known that prepulses may strongly affect harmonic generation, and thus it is necessary to determine the effect of an intense beam in the case of an initially clean, steep surface. Hence experiments were carried out with a 248 nm, 700 fs laser beam, with a high contrast and up to an intensity of 1.5×10^{17} W/cm^2 and a nanosecond pedestal lower than 10^7 W/cm^2. Harmonics up to the fourth order were observed for several targets (Al, B or C evaporated on glass plates), and showed some features different from the low intensity case. It was confirmed that above 10^{16} W/cm^2 some diffuse harmonic emission arose, since a significant amount of radiation was detected well outside of the original light cone. The polarization behaviour also changed. At low intensities the harmonics retain the polarization of the laser but as the intensity increases the polarization of the generated harmonics becomes more mixed. For intensities greater than 10^{17} W/cm^2 the generated harmonics contain roughly equal p- and s-polarized components, independent of the polarization of the incoming beam. These observations demonstrate the rippling of the critical surface, which can be understood as a consequence of the unstable balance of light pressure and plasma expansion. One dimensional particle in cell (PIC) simulations show that the light pressure is in equilibrium with the expanding plasma, and therefore the plasma gradient does not change significantly even if the pulse duration changes. Clearly this equilibrium is unstable and Rayleigh-Taylor like instabilities may evolve, especially when the laser pulse duration is at least as long as several hundred femtoseconds.[39]

5 CONCLUSIONS

Due to the short wavelength of KrF laser radiation it penetrates deep into the plasma, therefore laser-plasma interactions with KrF lasers occur in the dense plasma. A consequence of the generated thick layer of hot dense plasma is that efficient generation of x-rays is possible using KrF lasers. These types of lasers have a perspective to be used for pumping x-ray lasers, too.

In the extreme ultraviolet range of spectra they also offer an alternative to solid-state systems by the generating intense low-order harmonics, appropriate for seed pulses of x-ray lasers as well. At the presently available KrF intensities it is worthwhile to study harmonic generation via different absorption mechanisms such as resonance and Brunel absorption. Due to the observed surface rippling an improvement can be expected for KrF intensities in the relativistic range, i.e. well above 10^{19} W/cm^2 for which efficient harmonic generation can be expected[29] in the VUV and soft X-ray spectral ranges and for which the quality of the harmonic beam will probably improve similarly to that of harmonics generated by infrared laser systems.[31,32]

References

1 S. Szatmári, G. Marowsky and P. Simon, *Laser Physics and Applications*, 2007, **215**, (Berlin; Springer).

2 A. McPherson, T.S. Luk, B.D. Thompson, A.B. Borisov, O.B. Shiryaev, X. Chen, K. Boyer, C.K. Rhodes, *Phys. Rev. Lett.* 1994, **72**, 1810

3 S.P. Obenschain, D.G. Colombant, A.J. Schmitt, J.D. Sethian and M.W. McGeoch, *Phys. Plasmas*, 2006, **13**, 056320.

4 S. Szatmári and F.P. Schäfer, *Opt. Commun.*, 1988, **68**, 196.

5 S. Szatmári, *Appl. Phys. B*, 1994, **58**, 211.

6 H. Kapteyn, M. Murnane, A. Szoke and R. Falcone, *Opt. Lett.*, 1991, **16**, 490.

7 G. Doumy, F. Quéré, O. Gobert, M. Pendrix, P. Martin, P. Audebert, J.C. Gauthier, J. Geindre and T. Wittmann, *Phys. Rev. E*, 2004, **69**, 026402.

8 C. Ziener, P.S. Foster, E.J. Divall, C.J. Hooker, M.H.R. Hutchinson, A.J. Langley and D. Neely, *J. Appl. Phys.*, 2003, **93**, 768.

9 I.B. Földes, D. Csáti, F.L. Szűcs and S. Szatmári, *Radiation Effects and Defects in Solids*, 2010, **155**, 429.

10 I.B. Földes, A. Barna, D. Csáti, F.L. Szűcs, S. Szatmári, *J. Phys. Conf. Ser.*, 2010, **244**, 032004

11 V.D. Zvorykin, N.V. Didenko, A.A. Ionin, I.V. Kholin, A.V. Konyashchenko, O.N. Krokhin, A.O. Levchenko, A.O. Mavritskii, G.A. Mesyats, A.G. Molchanov, M.A. Rogulev, L.V. Seleznev, D.V. Sinitsyn, S.Yu. Tenyakov, N.N. Ustinovskii and D.A. Zayarnyi, *Laser and Particle Beams*, 2007, **25**, 435.

12 I.B. Földes and S. Szatmári, *Laser and Particle Beams*, 2008, **26**, 575.

13 W.C. Mead, E.K. Stover, R.L. Kauffman, H.N. Kornblum and B.F. Lasinski, *Phys. Rev. A*, 1988, **38**, 5275.

14 D.L. Matthews, P.L. Hagelstein, M.D. Rosen, M.J. Eckart, N.M. Ceglio, A.U. Hazi, H. Medecki, B.J. MacGowan, J.E. Trebes, B.L. Whitten, E.M. Campbell, C.W. Hatcher, A.M. Hawryluk, R.L. Kauffman, L.D. Pleasance, G. Rambach, J.H. Scofield, G. Stone and T.A.Weaver *Phys. Rev. Lett.*, 1985, **54**, 110.

15 P.V. Nickles, V.N. Shlyaptsev, M. Kalachnikov, M. Schnürer, I. Will and W. Sandner, *Phys. Rev. Lett.*, 1997, **78**, 2748.

16 U. Teubner, J. Bergmann, B. van Wonterghem, F.P. Schäfer and R. Sauerbrey, *Phys. Rev. Lett.*, 1993, **70**, 794.

17 V.L. Ginzburg, *The Propagation of EM Waves in Plasmas*, 1970, (New York: Pergamon)

18 F. Brunel, *Phys. Rev. Lett.*, 1987, **59**, 52.

19 A. Rousse, P. Audebert, J.P. Geindre, F. Fallies, J.C. Gauthier, A. Mysyrowicz, G. Grillon and A. Antonetti, *Phys. Rev. E*, 1994, **50**, 2200.

20 D. Salzmann, C. Reich, I. Uschmann, E. Förster and P. Gibbon, *Phys. Rev. E*, 2002, **65**, 036402.

21 A.B. Borisov, X. Song, F. Frigeni, Y. Koshima, Y. Dai, K. Boyer and C.K. Rhodes, *J. Phys. B*, 2003, **36**, 3433.

22 K. Boyer, A.B. Borisov, X. Song, P. Zhang, J.C. McCorkindale, S.F. Khan, Y. Dai, P.C. Kepple, J. Davis and C.K. Rhodes, *J. Phys. B*, 2005, **38**, 3055.

23 A.B. Borisov, X. Song, P. Zhang, A. Dasgupta, J. Davis, P.C. Kepple, Y.Dai, K. Boyer, C.K. Rhodes, *J. Phys. B*, 2005, **38**, 3935.

24 A.B. Borisov, P. Zhang, E. Rácz, J.C. McCorkindale, S.F. Khan, S. Poopalasingam, J. Zhao and C.K. Rhodes, *J. Phys. B*, 2007, **40**, F307.

25 T.B. Petrova, K.G. Whitney and J. Davis, *J. Phys. B*, 2010, **43**, 025601.

26 M.A. Duguay and P.M. Rentzepis, *Appl. Phys. Lett.*, 1967, **10**, 350.

27 C. Dölle, C. Reinhardt, P. Simon and B. Wellegehausen, *Appl Phys B*, 2002, **75**, 629.

28 P.B. Corkum, *Phys. Rev. Lett.*, 1993, **71**, 1994.

29 R. Rakowski, A. Barna, T. Suta, J. Bohus, I.B. Földes, S. Szatmári, J. Mikołajczyk, A. Bartnik, H. Fiedorowicz, C. Verona, G. Verona Rinati, D. Margarone, T. Nowak, M. Rosiński, L. Ryć; *Rev. Sci. Instrum.*, accepted.

30 B. Wellegehausen, K. Mossavi, A. Egbert, B.N. Chichkov and H. Welling, *Appl. Phys. B*, 1996, **63**, 451.

31 P. Gibbon, *Phys. Rev. Lett.*, 1996, **76**, 50.

32 R. Lichters, J. Meyer-ter-Vehn and A. Pukhov, *Phys. Plasmas*, 1996, **3**, 3425.

33 B. Dromey, S. Kar, C. Bellei, D.C. Carroll, R.J. Clarke, J.S. Green, S. Kneip, K. Markey, S.R. Nagel, P.T. Simpson, L. Willingale, P. McKenna, D. Neely, Z. Najmudin, K. Krushelnick, P.A. Norreys and M. Zepf, *Phys. Rev. Lett.*, 2007, **99**, 085001.

34 B. Dromey, D. Adams, R. Hörlein, Y. Nomura, S.G. Rykovanov, D.C. Carroll, P.S. Foster, S. Kar, K. Markey, P. McKenna, D. Neely, M. Geissler, G.D. Tsakiris and M. Zepf, *Nature Physics*, 2009, **5**, 146.

35 I.B. Földes, J.S. Bakos, G. Veres, Z. Bakonyi, T. Nagy and S. Szatmári, *IEEE J. Sel. Top. Quantum. Electron.*, 1996, **2**, 776.

36 D.M. Chambers, P.A. Norreys, A.E. Dangor, R.S. Marjoribanks, S. Moustaizis, D. Neely, S.G. Preston, J.S. Wark, I. Watts and M. Zepf, *Opt. Commun.*, 1988, **148**, 289.

37 G. Veres, J.S. Bakos, I.B. Földes, K. Gál, Z. Juhász, G. Kocsis and S. Szatmári, *Europhys. Lett.*, 1999, **48**, 390.

38 S. Varró, K. Gál and I.B. Földes, *Laser Phys. Lett.*, 2004, **1**, 111.

39 E. Rácz, I.B. Földes, G. Kocsis, G. Veres, K. Eidmann, S. Szatmári, *Appl. Phys. B*, 2006, **82**, 13-18.

BROADBAND MULTILAYERS: TAILOR MADE MIRRORS FOR LINEARLY POLARIZED X-RAYS FROM A LASER PLASMA SOURCE

M. Krämer[1] and K. Mann[2]

[1]AXO DRESDEN GmbH, Gasanstaltstraße 28, D-01237, Dresden, Germany
[2]Laser-Laboratorium Göttingen e.V., Hans-Adolf-Krebs-Weg 1, D-37077 Göttingen, Germany

1 INTRODUCTION

The wavelength region of the so-called "water window" between 2.34 nm (corresponding to the K absorption edge of oxygen) and 4.4 nm (the carbon K edge) is of high interest for the analysis of organic specimens. However, suitable radiation sources are not as common as in the hard X-ray regime below 1 nm. Laser plasma sources with a solid or gas target are an approach to encompass this region.[1]

Many laboratory sources emit polychromatic, non-polarized radiation; however, linearly polarized radiation with a well-defined energy is advantageous or even mandatory for many experimental techniques such as near edge X-ray absorption fine structure (NEXAFS).[2] Synchrotron radiation can fulfill these requirements, but access to synchrotron beam lines is very limited. Thus, laboratory scale solutions are highly demanded. Using a crystalline or multilayer mirror with a Bragg angle equal to the Brewster angle (which is close to 45° for X-rays) provides the possibility of simultaneously achieving polarization and monochromatization. However, intensity losses due to the reflection are high. Thus dedicated, optimized mirrors need to be developed.

2 PLASMA SOURCE

The experimental setup for the generation and characterization of soft X-rays emitted from laser plasmas in the water window is shown in figure 1. A Nd:YAG laser beam (Innolas, 1064 nm, 1 Hz, 800 mJ, 7 ns) was focused on a pulsed gas puff target centred in a vacuum chamber, as described in detail elsewhere.[1] The laser focus had a diameter of about 60 µm, yielding power densities of up to 4×10^{12} W/cm², sufficient to ignite a hot dense plasma.

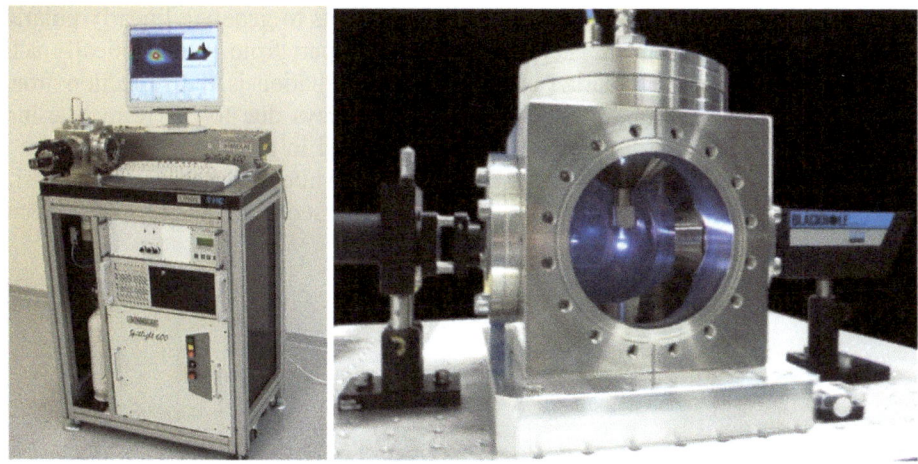

Figure 1 *Experimental setup of the laser-plasma XUV source used for various metrology applications.*

Krypton with a backing pressure of 25 bar was used as the target gas, providing broadband radiation (Kr XXV – Kr XXXVI) in the water window spectral range as shown in figure 2. Due to the small mean free path of soft X-rays at atmospheric pressure the target vacuum chamber was evacuated to approximately 10^{-4} mbar. This table-top soft X-ray source was applied to a variety of metrological applications, in particular NEXAFS spectroscopy as described in the chapter *Near-edge x-ray absorption fine structure measurements using a laboratory-scale XUV source* in the current book.

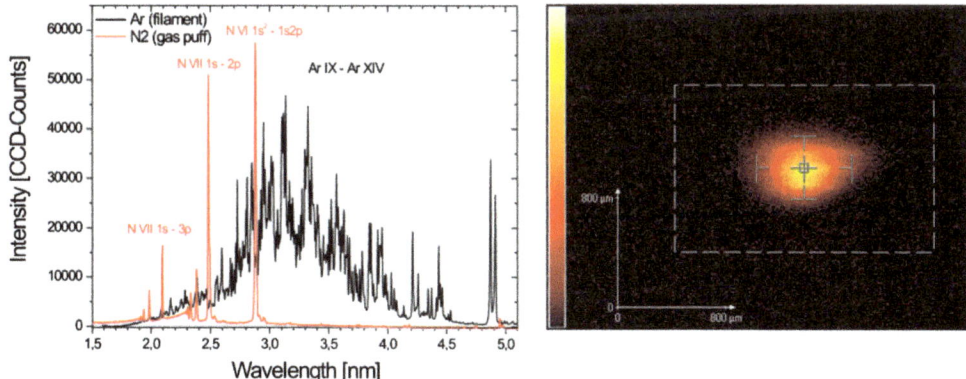

Figure 2 *Water-window emission spectra of laser-induced plasmas from pulsed gas targets of nitrogen (red) and argon (black). The peak brilliance for the isolated nitrogen line at 2.878 nm was 10^{18} photons/s/mrad^2/mm^2. The XUV pinhole camera image (right) of the krypton plasma indicates a source size of 320×160 μm².*

3 POLARIZING MIRRORS

In NEXAFS experiments the use of linearly polarized radiation offers the possibility to analyze orientation and bonding configuration of molecules adsorbed on solid surfaces, as has been demonstrated in various synchrotron experiments.[3] Thus, as plasma-based lab-scale sources are inherently non-polarized, optical elements to generate linearly polarized radiation for table-top NEXAFS systems are required. Apart from using optically active materials, a common way to polarize electromagnetic radiation is the reflection from a surface under the Brewster angle, which is ~45° for X-rays due to the refractive index being close to unity. For this angle, the radiation polarized in the horizontal direction (p-polarization) penetrates completely into the material, while the vertically s-polarized radiation is partly reflected as illustrated in figure 3. In consequence, the reflected beam is linearly polarized.

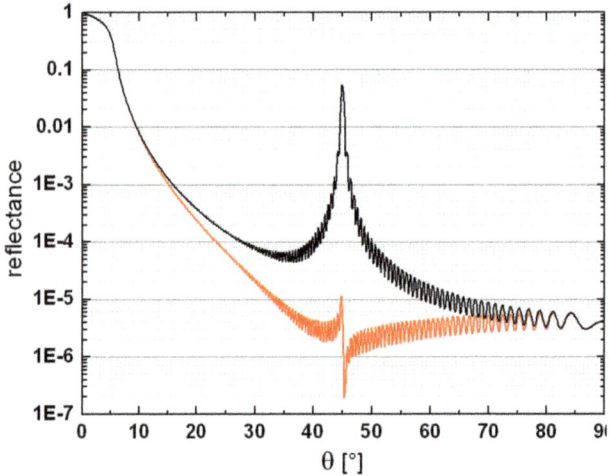

Figure 3 *Reflection coefficients of 300 eV X-rays for a Cr/Sc multilayer, designed for the Brewster angle and with a period of 2.95 nm. It can be seen that the reflection coefficient for waves polarized parallel to the mirror surface (red curve, p-polarization) is reduced by several orders of magnitude at a grazing incidence angle of 45°. Thus, only vertically s-polarized radiation (black curve) is reflected. The curves were simulated using the IMD software.[4]*

However, the peak reflectance even for s-polarized radiation at large grazing angles such as 45° is in the range of 10%, and typical reflectors have a very small energy bandwidth $\Delta E/E$ of ~10^{-3}–10^{-4} or below (crystals) to a few percent (periodic multilayers). Thus, the peak width, and in consequence the integral reflected intensity, is rather small. It is clearly desirable to reflect a larger part of the spectrum in order to obtain a higher integral intensity.

4 OPTICS DESIGN AND FABRICATION

Multilayers (like crystals) reflect radiation of wavelength λ according to the Bragg condition $m\lambda = 2d\sin\theta$, with θ being the grazing incidence angle and d the period (bi-layer) thickness of the two materials the multilayer mirror is composed of; normally the first-order reflection ($m = 1$) is used. Hence, at a given incidence angle only radiation of one specific wavelength (with a resolution of roughly 10^{-2}–10^{-3}) is reflected. If a source emits a larger energy band (as it is the case for Bremsstrahlung from laboratory X-ray tubes, for example) or several characteristic lines of similar wavelength (as in plasma sources), only a part of this emitted radiation can be reflected with much being lost.[5]

Since the period in multilayer stacks can be varied ("graded"), in contrast to crystals which are highly periodic, it is possible to adjust the layer thicknesses to the source spectrum by providing bi-layer thicknesses for each wavelength of interest, as illustrated in figure 4.

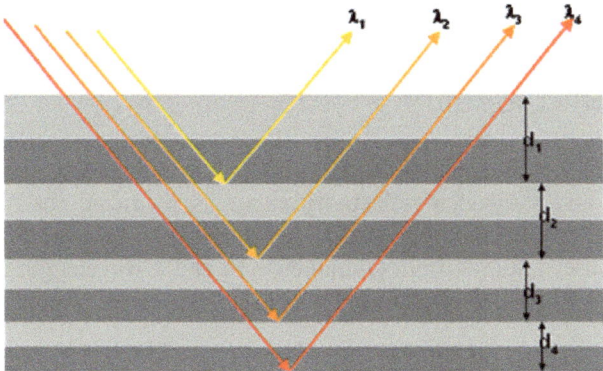

Figure 4 *Principle of a depth-graded broadband multilayer reflector. The bi-layer thickness d varies with depth in the layer stack in order to fulfil the Bragg condition λ = 2d sinθ at a given grazing incidence angle θ for a larger range of wavelengths.*

Typically, the multilayer has thicker layers close to the surface to reflect the longest wavelength in the spectrum, and the layer thickness decreases smoothly with increasing depth in the stack, down to the thinnest layer corresponding to the shortest wavelength emitted from the source. However, this simple approach does not directly provide the best result, as other parameters such as total number of layers and thickness ratio of the two elements in each bi-layer have to be adjusted as well. Thus, a software code was developed to vary the layer parameters iteratively and then calculate the resulting reflectivity.[6] Three steps of this iteration from the narrow peak of a periodic multilayer to a broad band "plateau peak" are shown in figure 5.

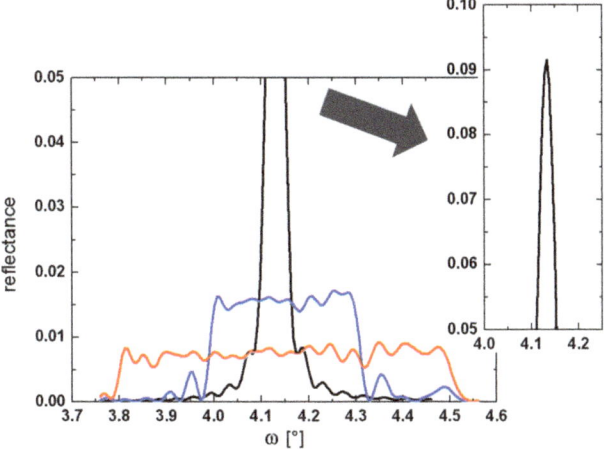

Figure 5 *Development of a broadband mirror for the water-window spectral range. A regular, periodic multilayer (black, narrow peak) has a rather small wavelength bandwidth of ~0.045 nm. Varying the individual bi-layer thicknesses (blue and red) increases the peak width, but at the cost of the peak height. However, the total area under the peak still increases as shown in these simulations.*

After simulating the ideal layer composition it was converted into a format readable by a magnetron sputter deposition machine and used to fabricate a Cr/Sc multilayer with the calculated layer stack. The deposition procedure had to be extremely precise as even small deviations of single layer thicknesses can have a significant impact on the width and especially the plateau shape of the target reflectivity curve.

5 EXPERIMENTAL RESULTS

After deposition the broadband multilayer mirror was characterized with a copper Kα laboratory source, and showed a reflectivity curve as predicted for this energy by the simulation software (figure 6a). Thereafter, measurements were performed at the laser plasma source at Laser-Laboratorium Göttingen e.V, and a wavelength dependent scan of the reflectivity showed very good agreement with the simulated results, as can be seen in figure 6b.

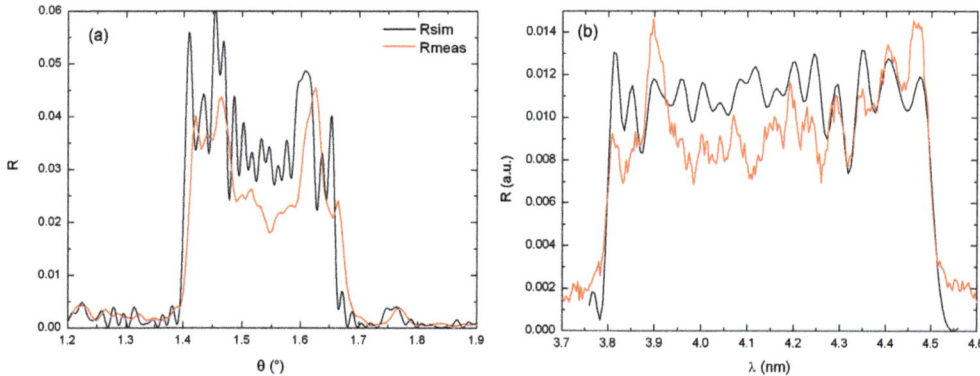

Figure 6 *(a) Test measurement at an X-ray source (copper Kα, 8.04 keV) and (b) wavelength dependent reflectivity at 45° (laser plasma source) of a depth-graded multilayer mirror for wavelengths around 4.1 nm. Calculated curves are shown in black, measurements in red. The plateau widths of 0.26° (a) and 57 eV (b) correspond to an energy bandwidth of 17%. Due to the reflection at 45° the spectrum on the right represents linearly (s-)polarized radiation from a table-top XUV plasma source.*

Even though the peak reflectance of ~1% is not very high, the very large bandwidth permitted the reflection of the entire emission spectrum of interest and sufficient intensity to perform NEXAFS experiments with this linearly polarized radiation from a table-top XUV plasma source. Corresponding work is still in progress.

6 SUMMARY AND OUTLOOK

A prototype tailored broadband multilayer reflector has been developed that accepts the entire emission spectrum of interest of an XUV laser plasma source (≈300 eV) with an energy bandwidth of ≈17%, which is ten times higher than for a typical periodic multilayer. It has been shown that very large energy bandwidths can be achieved with depth-graded multilayer mirrors. These graded layer stacks can be fabricated with very high accuracy and deliver experimental results close to theoretical values.

The application of this technique for higher energies has also been tested, and looks very promising especially in terms of peak reflectivities. Furthermore, improvements to the simulation and fabrication processes seem feasible, permitting precise tailoring of the spectral characteristics of multilayers. NEXAFS experiments with linearly polarized radiation from table-top laser plasma sources are in progress.

ACKNOWLEDGMENTS

The authors would like to thank R. Dietsch, T. Holz and D. Weißbach (AXO DRESDEN GmbH) for invaluable support in the multilayer production.

References

1 S. Kranzusch, C. Peth and K. Mann, *Rev. Sci. Instr.*, 2003, **74**, 969.
2 C. Peth, F. Barkusky and K. Mann, *J. Phys. D: Appl. Phys.*, 2008, **41**, 105202.
3 J. Stoehr, *NEXAFS Spectroscopy*, Springer Series in Surface Science Vol. 25, Springer, Berlin (2003)
4 D.L. Windt, *Computers in Physics*, 1998, **12**, 360-370.
5 R. Dietsch, T. Holz, M. Krämer and D. Weißbach, *Proc. SPIE*, 2010, **7995**, 79951U-1.
6 M. Krämer, R. Dietsch, T. Holz and D. Weißbach, *Proc. SPIE-Optifab*, 2011, **TD07-74**, 1.

Applications

SHORT WAVELENGTH LABORATORY SOURCES FOR SEMICONDUCTOR INSPECTION AND FABRICATION

Davide Bleiner and Mabel Ruiz-Lopez

Institute for Applied Physics, University of Bern, Sidlerstrasse 5, CH-3012, Bern, Switzerland
Present address: EMPA Materials Science & Technology, Überlandstrsse 129, CH-8600 Dübendorf, Switzerland

1 INTRODUCTION

Many breakthroughs in semiconductor science and technology rely on manufacturing or inspection instrumentation being developed and made available in anticipation of forthcoming science cases. Three major challenges are to be mentioned in the development of tools: (i) scaling-down, (ii) improvement of duty cycle, and (iii) reduction of costs. Directly or indirectly, each of such efforts can take full advantage of light sources with shorter and shorter wavelength, especially if these are laboratory-scale facilities. Short wavelength light permits both the manufacturing and inspection at reduced length scales, thus downsizing the integrated circuits. In practice, a new technology becomes *enabling* only if it manages to pass through *alpha* and *beta* testing *in due time*. In fact, tool manufacturers go through two lengthy stages of testing before release. The first stage, or *alpha* testing, is often performed only by users within a developing company. The second stage, or *beta* testing, involves a selected number of external users. These scheduled procedures permit that forthcoming needs are identified and addressed well in advance. In conclusion, it is the technologies under today's research development that are the only option for tomorrow's industry.

Present-day nano-scale manufacturing and inspection industry relies on Deep Ultraviolet (DUV: 190-250nm) laser sources.[1] The research community is thus now anticipating that Extreme Ultraviolet light (EUV: 5-50 nm) may be the required move for enabling next generation photolithography and microscopy. EUV developments are performed in parallel with contending technologies since as long as at least one manufacturing technology option is available when needed, the investment is ultimately successful.

Figure 1 summarizes the principle of (a) photolithography and (b) microscopy, and shows how the two processes can be considered complementary projection approaches. In fact, in photolithography the object ("mask") is demagnified and miniaturized on the image plane, whereas in microscopy the object ("sample") is magnified on the image plane. In terms of the light source adopted, there is a single but remarkable difference between the two, namely the power. For high volume manufacturing the lithography source must provide a few hundred Watt average power. In microscopy instead, especially for high contrast imaging, one is more interested in spectral purity and power stability. Another difference concerns the image "acquisition" which in photolithography is accomplished with a photosensitive compound (the "resist") located at the image plane, whereas in microscopy a CCD or any other light detector is placed at the image plane. From this qualitative picture now we will move on towards a more quantitative understanding.

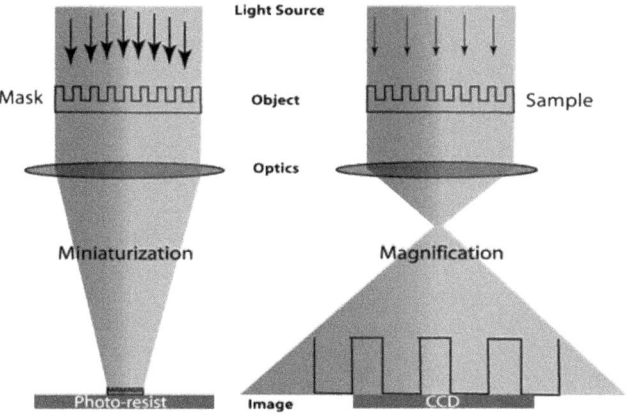

Figure 1 *Principle of (a) Photo-lithography and (b) Microscopy.*

1.1 Resolution: Wavelength and Coherence

Two fundamental laws give the lateral and depth resolution of a focal illumination spot. These are important to define the minimum size that can be visualized/printed. The resolved *critical dimension* (CD) and the *depth of focus* (DOF) are given by the Abbé criterion as a follows

$$CD = k_1\lambda/NA \tag{1}$$

$$DOF = k_2\lambda/NA^2 \tag{2}$$

where k_1 and k_2 are coefficients that depend on the illumination characteristics, λ is the illumination wavelength, and NA is the numerical aperture defined as a function of beam delivery semi-aperture, i.e NA = $n\sin\theta$, with n the refractive index of medium (for air $n = 1.0$). Historically, values for k_2 and k_2 greater than 0.5 have been used in high volume manufacturing,[2] which however degraded the illumination profile. One notices that the DOF scales down more rapidly than CD as a function of NA, thus posing a limit on the aspect ratio (AR = CD / DOF) of the nano features (Figure 2). Resolution enhancement techniques such as phase-shift masks, modified illumination schemes, and optical proximity correction can be used to enhance resolution while increasing the effective DOF, though hardly realizable in the EUV.[3]

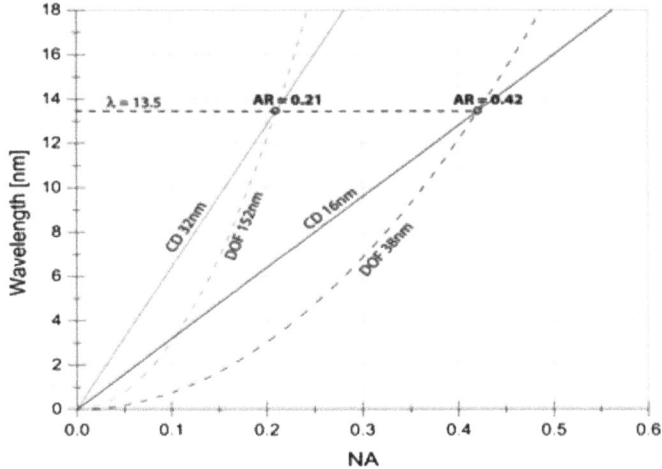

Figure 2 *Illumination wavelength versus numerical aperture for given CD (lateral resolution) and DOF (depth resolution). For a 32 nm CD the largest possible DOF at λ = 13.5 nm is 152 nm. Larger NA are required when the CD is reduced, for identical wavelength. The aspect ratio (AR = CD/DOF) of the features doubles accordingly and this limits the surface modulation that can be resolved. In the plot $k_1 = k_2 = 0.5$.*

Often used in microscopy and lithography, the illumination (in-)*coherence factor* σ is another important parameter. This is defined as $\sigma = NA_c/NA_p$, i.e. the numerical aperture of the condenser optics (NA_c) located before the sample to enhance the photon flux, and the numerical aperture of the imaging optics (NA_p), located after the sample. When $\sigma = 0$, the illumination is considered fully coherent, and on the other end when $\sigma = \infty$ the illumination is said to be totally incoherent. This coherence definition is dependent on the projection setup and is not strictly related to the raw source characteristics, as discussed below.

The importance of the illumination coherence on imaging quality is explained in Figure 3, showing the image intensity profile against a *knife-edge* step feature.[iv] The limiting case $\sigma = \infty$ corresponds to incoherent illumination and gives the profile with the "less edgy" slope. Decreasing σ increases the edge accuracy, known in lithography as NILS (Normalized Image Log-Slope), but "overshoot" – *known as flare* – appears near the edge. If σ is further reduced to as low as 0.2, also to reduce the intensity baseline, the flare becomes excessive. In practical lithography, typical values for σ between 0.4 and 0.7 are chosen as a compromise between NILS and flare.

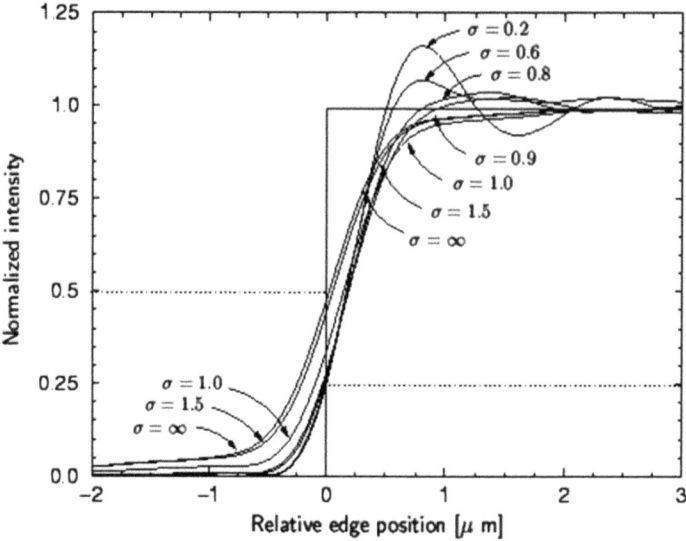

Figure 3 *The degree of partial coherence determines a trade-off between accuracy of a sharp edge and overshoot of the profile.*

1.2 Advantages of Extreme Ultraviolet (EUV)

Diffraction limits the ultimate length scale that one is able to resolve. Resolution improves as the light wavelength decreases. Thus, one needs short wavelength sources in the EUV to push down the target resolution to the low nano-scale, from that currently accessible with DUV, e.g. using 193 nm lasers. In transitioning from DUV to EUV the industry gains as high as 15 times in resolution, even with low numerical aperture (see eq. 1). This large leap ahead is technologically appealing because multiple generations of nanotechnology devices can be put into schedule without additional manufacturing technology changes. Such a transition would also permit to relax pressure on the need to increase the NA, which has been the major exploited parameter in the last few decades of semiconductor breakthroughs. The push has been indeed so excessive that by using immersion projection the barrier of NA > 1 was exceeded. The use of low NA optics provides good depth of focus and linearity for isolated and dense structures simultaneously, eliminating the need for optical proximity or phase shift correction.[v]

Short-wavelength sources are a *"returning interest"* in nano-fabrication. Indeed, already as early as 1977, researchers were able to pattern features as small as 17 nm in silicon wafers, using X-rays.[vi,vii] X-ray lithography did have the same promising benefits of high resolution that motivates the industry nowadays to look at EUV, but also two concurring drawbacks: the use of grazing incidence optics and contact printing. The former was the only viable optics for X-rays, since normal incidence optics (multilayer) is not highly performing with X-rays. Yet, grazing incidence optics is not very practical for lithography applications. Consequently, contact printing was used, that is a 1:1 exposure where the photo-mask is directly pressed against the wafer. The physical contact between mask and wafer was often a source of damage. Besides, the lack of homogenization optics made the X-ray illumination non-uniform, and thus the printing quality was unacceptably poor. These problems resulted in the decision that X-ray lithography was not considered further for high volume manufacturing of integrated circuits,[viii,ix] to the advantage of contending DUV lithography. On the contrary, EUV printing is possible in 4:1 projection using multilayer (normal-incidence) optics, and with the aid of beam homogenization optics the problems mentioned from the past can be overcome.

1.3 Confronting Challenges

EUV is strongly absorbed by all matter, including air, helium and hydrogen above a few mbar ambient pressure. This imposes vacuum operation, which sets some problems in terms of mechanical vibration on the working environment, cleanliness for nano-fab clean rooms such as outgassing of potential contaminants, and oil or exhaust handling.[10,11] Secondly, the high photon energy of the EUV light presents new challenges that are not known in traditional optical lithography. One of these is the generation of photoelectrons ("*proximity effects*") that contribute to the over-exposure of the photoresist and blurring of the aerial image.[12] In this respect, mere optical considerations to evaluate the EUV resolution are incomplete without taking proximity effects into account.

Last but not least, part of the energy that is transferred from the plasma environment to ionize/excite the radiating ions, is also causing ion acceleration. There is a continuum distribution of kinetic energy values documented up to 15-30 keV,[13] which is sufficient to erode, implant or at least veneer the exposed multilayer optics. Such "ion debris" is a cause of increasing technology costs when interacting with EUV collection optics, contributing to aging and need for frequent system servicing. Strategies using deflecting fields have proven successful to reduce the debris load by up to a factor of a few thousands[14], whenever the debris is charged. Neutrals remain a major concern for the protection of EUV optics.

EUV photons are generated in conversion of electrical or optical energy. The energy conversion efficiency ($CE = E_{out} / E_{in}$) is thus a metric of the physical (and economic) yield of the source operation. Obviously, the optimization of the CE is an important research activity. The experimental optimization is typically benchmarked towards computational values that provide theoretical limits. In the case of incoherent EUV sources, the theoretical limit is the Planck distribution (black body). For coherent sources, or EUV lasers, there has been a dramatic improvement in the last 20 years mostly due to the introduction of new and more efficient pumping schemes, such as use of pre-pulses and grazing-incidence pumping (GRIP). This has permitted two other advantages, i.e. the reduction of the source footprint, since less pumping power was required, and the production of shorter and shorter EUV wavelengths.

2 CANDIDATE EXTREME ULTRAVIOLET SOURCES

2.1 Specifications

Three main light source specifications are important for scientific and industrial applications, namely the spectral brightness, the spectral bandwidth and the footprint. The *spectral brightness* is defined as the emitted power per unit solid angle and unit bandwidth (BW). The SI units are Watts per square metre per steradian per nanometre ($W\ m^{-2}\ sr^{-1}\ nm^{-1}$), the last term referring to wavelength, but very often the spectral brightness is given as the number of photons per second per square millimetre per square milliradian in a spectral bandwidth ($\Delta\lambda/\lambda$) of 0.1% (photons $s^{-1}\ mm^{-2}\ mrad^{-2}$ 0.1% BW). The brightness is obtained by the solid angle (sr) of intensity ($W\ cm^2$) emission or, alternatively, the emitted power (W) per unit etendue ($cm^2\ sr$) of the source. Considering the collimated emission within a few milliradians (or even less), "laser sources" have enhanced brightness with respect to "thermal sources", even at modest power. Thermal sources emit isotropically over 4π sr (i.e., 1.3×10^7 mrad^2), which drastically degrades the brightness, even for a few orders of magnitude gain in power. The aperture of light emission is related to the source dimensions by a parameter called spatial coherence, i.e. $I_s = \lambda D/s$, where λ is the wavelength, D is the distance from the source at which the coherence is determined, and s is the source size. The spatial coherence gives the transverse distance over which the light wavefronts maintain their correlation. Ideally, one would prefer to have a laser beam in the EUV for next generation lithography, if this would provide sufficient average power.

The *spectral bandwidth* is the range of wavelengths in the illumination beam. The temporal coherence is related to such "colour purity" (or linewidth) of the radiation by $t_c = \lambda^2/c\Delta\lambda$. The linewidth ($\Delta\lambda$) is important in imaging applications, especially when optics with chromatic aberration are used, such as Fresnel zone plates. Narrow linewidths can in these cases allow sharp contrast and diffraction-limited spatial resolution. In spectroscopy applications it is also important to guarantee high spectral resolution.

Table 1 *EUV sources at the end of 2009[15] comparing the industry requirements and the demonstrated performance of existing R&D sources.*
Legend: IR: Industry Requirement, P: Performance, DPP: Discharge-produced Plasma, HCT: Hollow-Cathode Trigger, LPP: Laser-produced Plasma, ND: No Data, TSR, Table-top Synchrotron Radiation XRL: X-ray Laser, TS: Type of Specification, A: Architecture, λ: Wavelength [nm], ES: EUV Source Power [W], EU: EUV Usable Power [W] B: Brightness [W mm^{-2} sr^{-1}], RR: Repetition Rate [Hz], SE: Energy Stability [%]

	Lasertec	Selete	Zeiss SMT	AMAT	KLA Tencor	AIXUV	Colorado State University	Energetiq	Ritsu-meikan University
TS	IR	IR	IR	IR	IR	P	P	P	P
A	LPP	ND	ND	ND	ND	DPP-HCT	XRL	DPP	TSR
λ	13.5	13.5	13.5	EUV	13.5	11-50	13.2,13.9	13.5	3-40
ES	120	10	ND	ND	40	0.7	10^{-5}	10	0.4×10^{-3}
EU	1	0.1	ND	1	5	ND	10^{-5}	10	0.4×10^{-3}
B	200	15	>30	100	2500	0.7	2000	12	2.3×10^5
RR	2000	>2000	N/D	100	50000	1000	2.5	2000	400
SE	N/D	1	0.1	3	1	2	8	2	1

The spectral characteristics of the light source are crucial, most notably the wavelength (or photon energy), the bandwidth (or temporal coherence), the divergence (or spatial coherence) and the average power. Existing laboratory scale light sources are still at the R&D level (Table 1). For both imaging or fabrication the line shape is very important and if the source does not provide a sufficient degree of spectral purity (in-band versus off-band), a monochromator to limit the transmission bandwidth is used. The latter however degrades the intensity and introduces issues of heat load due to the blocked off-band radiation. Figure 4 shows a selection of sample spectra from a number of short-wavelength sources. The mercury lamp (figure 4a) was the workhorse of the early days of optical lithography, with a number of discrete lines (e.g. "g-line", "i-line"). The laser-produced plasma (figure 4b) is one of the candidate sources for EUV lithography, whose emission is characterized by an unresolved dense array of transitions around 13.5 nm. The EUV laser (figure 4c) is produced in a similar way to the LPP, but using a line focus for coherent photon stimulated emission in the single-pass amplification scheme. This dramatically narrows the bandwidth, which has a significant interest for high quality imaging. A gas DPP spectrum is also shown (figure 4c), which is a high power facility.

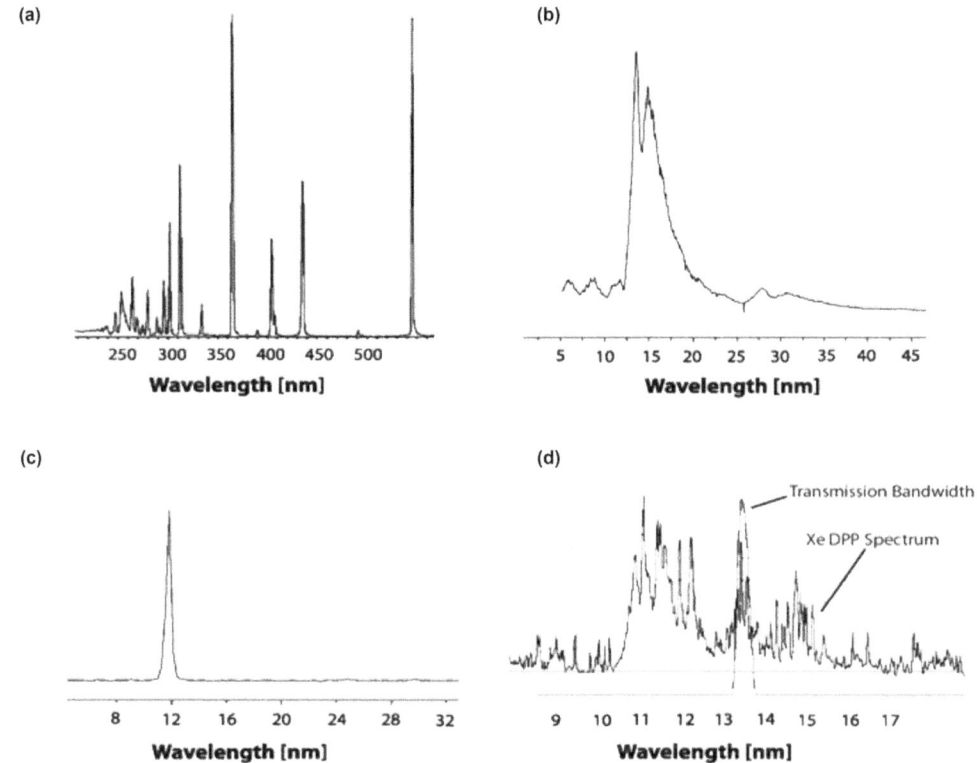

Figure 4 *Spectra from a selection of EUV sources. a) Mercury lamp (lithography used the "g-line" at 436 nm and "i-line" at 365 nm); b) Laser-produced plasma with tin solid target (own data 2009); c) Extreme Ultraviolet Laser (own data 2010); Discharge-produced plasma (courtesy of C. Bruinemann, Scientec).*

Finally, the *source footprint* is important, namely for two reasons. Firstly, the accessibility to the light source and its flexibility to be interfaced to different research or industry installations is enhanced if the source is a compact system. Secondly, the cost-of-ownership typically scales with the size.

2.2 Short-wavelength laboratory-scale source concepts

2.2.1 Laser-produced plasmas. The physics of the laser-target and laser-plasma interactions determines the characteristics and efficiency of laser-produced plasma (LPP) sources. Spontaneous emission from an LPP is typically over 2π sr (planar target) or 4π sr (spherical target). The plasma is pumped by collisions of the hot free-electron cloud with the atoms (to ionize them) and ions (to excite them). The occurrence of debris was initially underestimated but for LPP too limits the source uptime. The reported in-band power is at the moment in the range of a few tens of Watts (figure 4b). The irradiation with high power lasers ($I \gg 1$ GW cm^2) of a target induces a compact (i.e. small etendue) plasma source of a few hundred micrometres diameter that is extremely bright. The drive laser is a pulsed system with pulse durations between 10 ns (e.g. Nd:glass at $\lambda = 1$ µm) and 100 ns (e.g., excimers in the DUV, or CO_2 at $\lambda = 10.6$ µm), thus reaching peak powers of a few megawatts. The target can be any kind of material; a solid,[16] a liquid-jet or droplet,[17] or a gas-puff[18] (Figure 5). The solid target, thanks to its higher radiator density, has proven to give the highest conversion efficiency, i.e. from 0.2 % up to 6%. A conversion efficiency of 3.3% is the black-body maximum for a 2% EUV bandwidth, so that any higher value receives the significant contribution from the characteristic spectrum. The drawback of using a solid target, which is overcome with a fluid target, is that it undergoes rapid consumption during repeated irradiation, which necessitates stopping the source operation for servicing after a certain uptime. A jet or a train of droplets presented at the laser focal spot can extend the source operation to several hours.

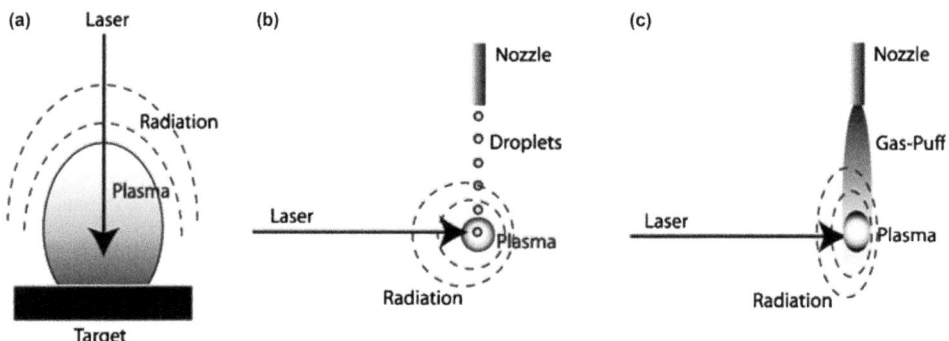

Figure 5 *Laser-Produced Plasma Short Wavelength Source: a) Solid target, b) Droplet-target, c) Gas-puff target.*

2.2.2 Discharge-produced plasmas. An electrical discharge is produced by a sudden current passage across a dielectric material as the impedance transiently drops. In argon or xenon the gas breakdown is accompanied by a radiating plasma (figure 4d). In its simplest form, a discharge-produced plasma (DPP) source consists of two electrodes in a cell held at low pressure (1–10 torr). The DPP is initiated by a seed spark. A number of atomic species, including the radiating "fuel" (e.g. tin), emit but also strike the cathode, which creates debris. Debris is indeed one of the main downsides of the DPP source. For a basic DC glow discharge, three main regions can be distinguished: dark discharge (Townsend-regime) for currents below 1 mA, glow discharge (gas breakdown) for currents up to 1 A, and arc discharge for very high currents up to 10 kA.

Gas DPPs come in different designs, each having specific advantages and disadvantages, which are briefly reviewed in the following (Figure 6). The dense plasma focus (DPF) is characterized by two nested electrodes that are separated by an insulator along their entire length, save at the tip. Two configurations, with either anode or cathode at the centre, are possible. An annular plasmoid sheet forms in the discharge gas and travels down the length. A plasmoid is a coherent structure of plasma and related fields. The Lorentz forces[19] compress the plasmoid to a small dense fireball over the tip. The highly excited species formed during plasmoid-pinching give rise to intense short wavelength emission. CE in the ranges of 0.3—0.4% at 13.5 nm were reported.[20]

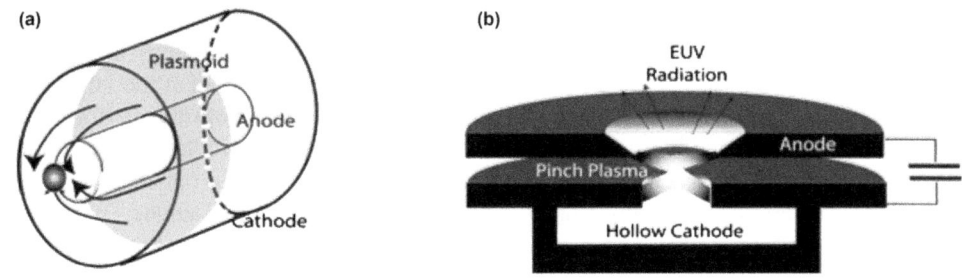

Figure 6 *Gas Discharge Produced Plasma Sources: a) dense plasma focus (DPF),[21] b) hollow cathode triggered (HCT).*

The hollow cathode triggered (HCT) gas discharge is realized with two juxtaposed circular plates with axial boreholes. On the cathode side, the borehole gates to a hollow space filled with low-pressure discharge gas. On the anode side the borehole is the exit for the short wavelength radiation. The two electrodes are connected to a low inductance capacitor to enable fast current buildup during the pinching. The CEs for the HCT design are as high as 0.8% for the EUV lithography wavelength.[22]

Ordinary, Xe, Ar, etc. lamps are too "cold" DPP to emit in the EUV. However, if the DPP is compressed, for instance using magnetic fields, to a hot and high-density plasma, emission in the EUV can be achieved. The compression process is transient and achieved in pulses of a few hundreds of nanoseconds. Considering that for the radial compression (or "pinching") magnetic fields are induced by injecting a current along the z-axis of the discharge (axis of rotational symmetry), the process is given the name of Z-pinch effect. W.H. Bennett provided a mathematical relation between the pressure p_{mag} of the magnetic field and the operating parameters

Typically, to match EUV lithography specifications for $\lambda = 13.5$ nm radiation, currents of 20 kA are required to induce a 20 T magnetic field and thus compress the plasma down to half a millimetre in radius achieving as high as 2 kbar compression pressures. The energies involved are thus around 5–10 J. The EUV power at 13.5 nm (2% BW) that has been demonstrated injecting tin as radiator is well above 400 W.[23] However, such a power is emitted over a 4π sr, and when collected by grazing incidence optics only 10-25% can be extracted. Normal incidence EUV collector optics is with DPP not an option because the source has a large *etendue*, whereas normal incidence optics require point sources. The requirement to use collection optics introduces two main drawbacks for DPP sources. First, the growth in plasma size as a function of input power deteriorates the fraction that can be collected, since the acceptance angle is fixed by the optics size and, second, the occurrence of hardware or radiator debris damages the optics and degrades the reflectivity and lifetime.

Studies have shown[24] that, depending on the plasma to optics distance, the source module lifetime can vary between 10^7 pulses (ca. 3 hours) and 10^{10} pulses (ca. 115 days) at a few thousand pulses per second (1–4 kHz). Another important problem is that of cooling. In fact, both the pulsation cycling and the high temperatures of a few thousands degrees Kelvin achieved cause mechanical deformations, which in optical components is particularly problematic.

2.2.3 Compact accelerator short-wavelength sources. X-ray (soft or hard) tubes[25] (Figure 7a) are traditional short wavelength laboratory sources. The technology involved in the production of X-rays has come a long way since the days of Röntgen at the end of the nineteenth century, with various techniques being used to increase both the performance and duty cycle. Still X-ray tubes are very limited in CE (<1%) by severe heat load and also have large bandwidths. X-rays are obtained by converting electrical into optical power, i.e. by means of accelerating electrons from a cathode towards a "target" anode. When the electrons strike the target a rapid loss of energy occurs, which is converted into heat and Bremsstrahlung. For kilo-electronvolt electron bunches the radiation that is given off is in the hard X-ray spectral range. The electrons bombarding the anode produce radiation in all directions. This becomes critical when using the X-ray source for imaging or lithography, as the intensity varies depending on the line of sight. Beam collimators can be implemented for aperturing the radiation. These are lead restrictions placed near the anode to control the width of the X-ray beam allowed to propagate across. Nevertheless, this collimation method drastically reduces the output source power.

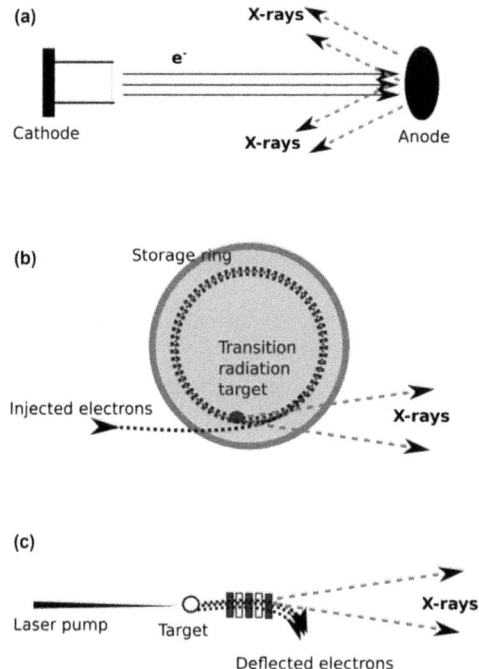

Figure 7 *Schematics of electron-based X-ray sources. a) Röntgen tube, b) table-top synchrotron,[26] c) betatron.[27]*

Also based on relativistic electron acceleration, but exploiting the "*transition radiation effect*", Yamada[26] developed a table-top short-wavelength source (figure 7b) with a brightness of the order of 10^8–10^{11} photons s^{-1} mm^{-2} mrad^{-2} nm^{-1}, comparable to first generation large scale light sources. In general, the generation of 100 keV hard X-rays in a "building-size" facility requires at least 200 M € and a synchrotron with an energy of a few Giga-electronvolts.[26] Transition radiation is induced by relativistic charged particles when they travel across media with different dielectric constants. Since the medium alters the particle's field, the excess energy is given off as radiation. The radiation is emitted in a solid angle of $1/\gamma$, where γ is the Lorentz factor. Typically, a thin foil target is exposed to a 6 MeV electron beam to induce the transition radiation. Due to collisions, Bremsstrahlung radiation partly contributes to the signal background.

A similar concept building upon the principle of synchrotron radiation in a laser-induced undulator quivering is that of betatron radiation (figure 7c).[27] Focusing a laser with intensity $>10^{18}$ W cm^{-2} onto a gas jet target, the pulse excites a wakefield in the plasma in which electrons begin to quiver and perform so-called betatron oscillations at a frequency $\omega_0 = \omega_p/(2\,\gamma)^{0.5}$ where ω_p is the plasma frequency. The betatron oscillation leads to a broadband X-ray emission, collimated within tens of milliradians and pulses of a few femtoseconds duration. The *"Compact Synchrotron Light Source"* (Lynchean Inc.) has scaled down the architecture of a third generation synchrotron, where inverse Compton scattering radiation is induced using a wiggler. If a laser beam with a wavelength of 1 μm is used as "a wiggler" (or rather an undulator considering the ultrashort oscillation wavelength), the electron beam energy necessary for 1 keV hard X-ray radiation is about 25 MeV. This shrinks footprint of the source by a factor of about 200 and results in a storage ring compatible with laboratory dimensions. Unfortunately the price is still in the few millions order of magnitude.

3. EUV METROLOGY: IMAGING AT THE NANO-SCALE

3.1 Actinic Inspection

In nano-fabrication using optical lithography, the mask is extremely important since any defect on it will be replicated systematically. Thus, mask defect inspection is a critical quality control procedure to be done prior to implementation in high volume manufacturing. There are two typologies of defects in an EUV lithography mask,[28] amplitude defects and phase defects. The former are related to particles or any kind of contaminations on the mask surface, while the latter are bumps or pits in the depth. Mask defect inspection can in principle be performed with either well-established DUV equipment or with actinic inspection instruments. The latter has the advantage of having full-scale resolution, and can thus highlight even the tiniest defects.

The Greek origin of the word *"actinic"* refers to "ray" (of light). The term was first introduced in photography for monochrome films that did not require darkroom exposure, since the paper was selectively sensitive to UV light. In EUV lithography it is common to speak of *"actinic inspection"* when referring to imaging performed with the very same illumination wavelength that is used for fabrication. Thus, mask defect inspection performed at 13.5 nm is actinic inspection for EUV lithography. This is important because the assessment of mask defectivity in so-called actinic mode permits a robust high volume manufacturing throughput.[29] In fact an ambitious yield of 80–100 wafers per hour may be significantly degraded by unseen defects when screening with non-actinic inspection methods. For this purpose, EUV metrology sources do not have to comply with the excessive average power requirements as in the case of EUV printing sources (table 1). Yet, although mask-defectivity surveys are run on an imaging quality main basis, *throughput* is however important for rapid inspection.

3.2 EUV Setups for Nano-Imaging

Defects are typically identified by means of amplitude modulation of the bright field (specular light). Tiny defects with dimensions comparable to a few EUV wavelengths are however visualized by means of dark field (scattered or diffracted light), because ultra-small features diffract more effectively into large angles. For dark-field imaging, the intense bright-field signal should be blocked off, in order to prevent detector saturation. Bright and dark field are thus complementary with respect to the signal-to-noise ratio, resolution, speed and the sensitivity for small and ultra-small defects.

A systematic survey of the reported actinic imaging systems is given below. The following classification is motivated only by the idea of highlighting some specific imaging setup concepts. As a matter of fact multiple concepts are often combined in a single setup, and thus the presented grouping is somewhat arbitrary. The first EUV imaging setup discussed was based on direct mask illumination and detection of the bright/dark field without any significant optics load.[30] The objectives of this EUV *Limited Liability Company* (EUV LLC) *programme* were to focus on the basic engineering of an alpha tool denoted as the Engineering Test Stand (ETS). Although limited in scanning speed, the instrument was the first to demonstrate actinic inspection of phase defects, native defects 30 nm wide and 3 nm tall, and other important features.[29] The system was illuminated with synchrotron radiation and used Kirkpatrick-Baez optics to have a fixed beam focused on a 2D translating mask. The bright-field signal was recorded with a channeltron, while the dark-field was measured with a microchannel plate.

3.2.1 Reflective Optics Systems. Based on the use of multilayer optics the systems discussed in this group are characterized by a high throughput (50-70% reflectivity per optical element) and freedom from chromatic aberration. The main limitation is spherical aberration that limits the magnification power to a few ×10x. Very often a Schwarzschild setup is implemented, where using a concave (primary) and a concentric convex (secondary) reflective optics pair, the spherical aberration can be corrected for. In a *imaging* Schwarzschild objective radiation enters from behind the small secondary, illuminates the large primary, is directed onto the secondary and exits on-axis through a hole in the primary mirror. Due to the on-axis geometry, source radiation hitting the central portion is stopped on the back-side of the secondary ("central obscuration"), causing a reduction of throughput (and image illumination). Sometimes off-axis projection is used. In this case the resolution is dictated by the NA of the illuminated mirror segment, which is typically below 0.1. On the other hand, the spherical aberration is still dictated by the full NA of the entire Schwarzschild system, even if used in off-axis illumination.

The following EUV imaging setups have been described exploiting reflective optics.

3.2.1.1 MIRAI-ASET.[31,32] With an EUV source the light is condensed and delivered to a mask blank using an ellipsoidal and a planar mirror. After illumination, the reflected dark field is imaged with a 20× Schwarzschild objective onto an EUV sensitive CCD camera (figure 8). The system demonstrated a resolution of 70 nm over a field of view of 0.5×0.5 mm². The MIRAI-II[33] and the SELETE tool[34] are upgrades of this baseline concept. The latter in particular was designed to improve the throughput.

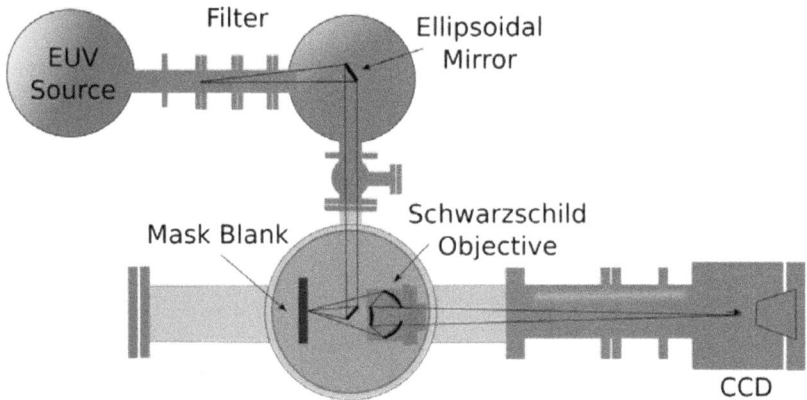

Figure 8 *MIRAI-ASET actinic inspection setup.*[31]

3.2.1.2 SEMATECH/LBNL.[35] Synchrotron light is focused onto the mask with a Schwarzschild objective and the reflected bright field is detected with a photodiode (figure 9). The system demonstrated a resolution below 100 nm. The instrument was able to measure the dark field as well. This is an important option since it is complemented with bright-field capability: in fact, certain defects (surface contamination, multilayer coating damage) may not generate a measurable dark-field signal.[29]

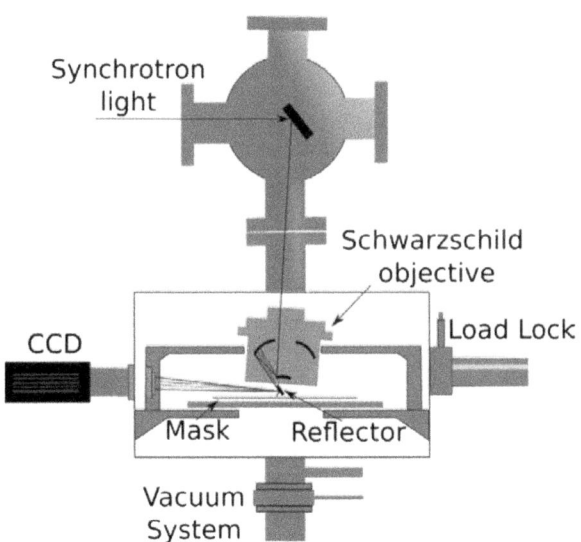

Figure 9 *SEMATECH/LBNL actinic inspection setup.*[35]

3.2.1.3 EXITECH Co.[36] The system is similar to the MIRAI one with one major difference: the detection of visible light, after EUV conversion by means of a scintillator. The 10× off-axis Schwarzschild objective deliver the dark-field signal to the scintillator, which is observed through an optical 50× lens (figure 10). The overall resolution was reported as low as 80 nm.

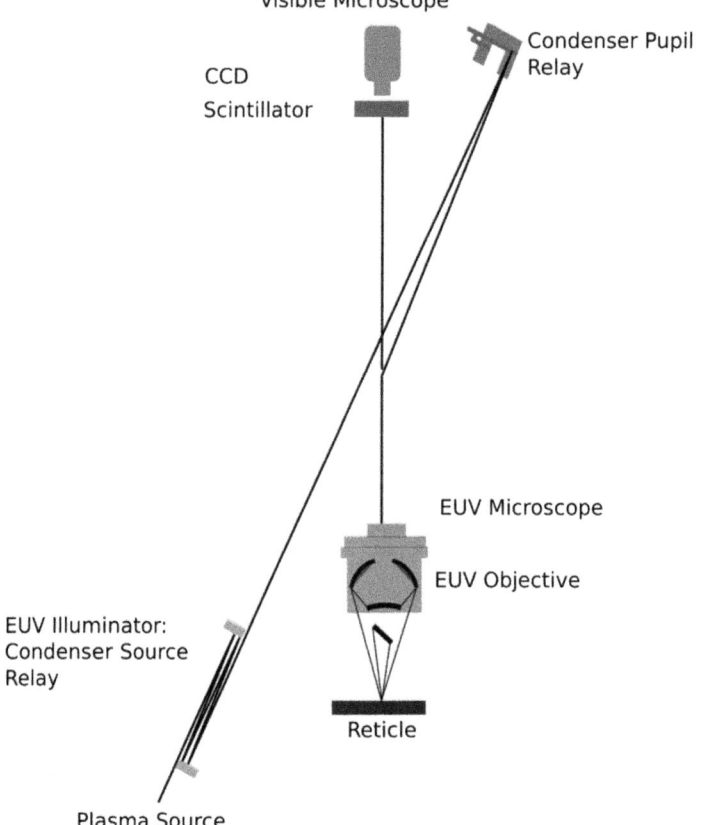

Figure 10 *EXITEC's reticle aerial image microscope: a) optical layout, b) system integration.*[36]

3.2.1.4 Aerial Image Microscope. ZEISS has announced that in 2013 the AIMS ("Aerial Image Microscope") will be commercially available for actinic metrology. Feldmann *et al.*[37] have discussed the specifications of the proposed platform and indicated that an all-reflective system with 10×10 mm^2 field of view and 750× magnification is technically accessible.

3.2.2 Diffractive Optics Systems. Fresnel zone plates (FZPs) are optical elements that exploit diffraction and interference. They act like lenses, but the latter are different as they exploit refraction. The FZP is realized with an alternation of opaque and transmissive annular grooves on a membrane (e.g. silicon nitride) with a very specific width progression. The grooves do not have to be centred on the axis of the membrane disc, and in some cases (i.e. for reflective imaging) off-axis elements are necessary. At the margin of the progression the grooves are the narrowest and the outermost groove width that is illuminated determines the optical capabilities of the FZP, such as the resolution. FZPs can achieve high magnification since the focal lengths can be as short as a few mm. However they suffer from chromatic aberration and modest throughputs of a few percent. To enhance the throughput, free-standing FZPs have been realized. To avoid chromatic aberration these optics require narrow-band sources. This has so far hampered their popularity with EUV plasma sources that are broad band. Other difficulties relate to alignment. The following EUV imaging setups have been described exploiting diffractive optics.

3.2.2.1 AIT/SEMATECH.[29] This was the first Fresnel zone plate based system to offer flexibility in the selection of the objective, namely an array of five FZPs with NAs between 0.25 and 0.35 and magnifications up to 1000× (figure 11). On the downside, 45 s per image are required for the EUV imaging to accumulate sufficient signal-to-noise contrast.

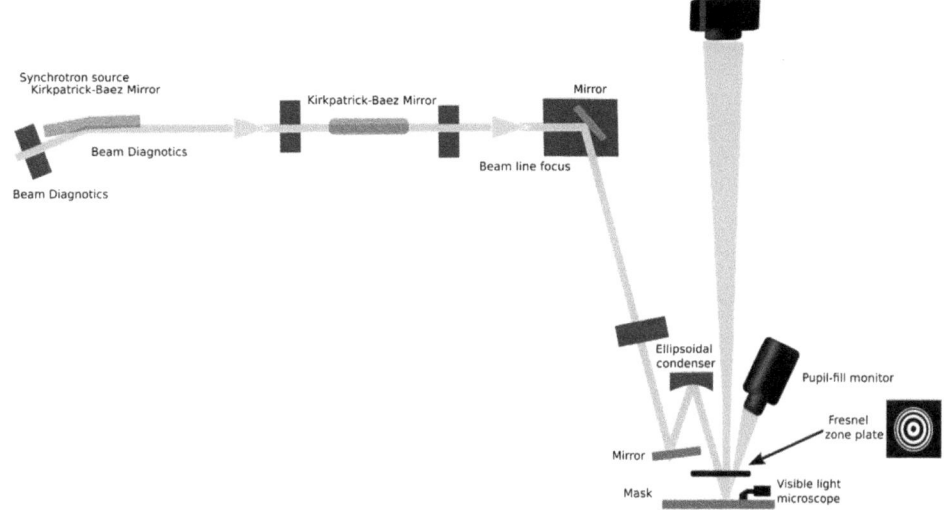

Figure 11 *Actinic inspection tool (AIT) from the SEMATECH/LBNL consortium.*[29]

3.2.2.2 Colorado State University.[38] Using Fresnel Zone Plate (FZP) optics in combination with an EUV laser has the advantage that for the EUV laser the linewidth is extremely narrow ($\Delta\lambda/\lambda < 10^{-4}$), which makes the FZP chromatic aberration of no concern. The EUV laser has also sufficient power to get high-quality images in a few minutes, when operated at 1 Hz. Two FZPs are used, one being an on axis and used as a condenser, the second acts as the objective carrying an off-axis pattern that is imaged onto a CCD (figure 12). The overall magnification is 660× and the demonstrated resolution was as small as 80 nm.

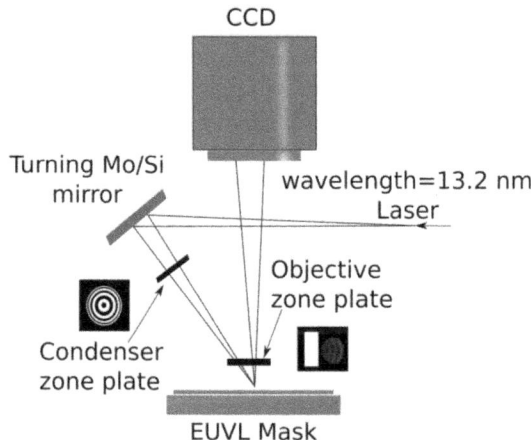

Figure 12 Actinic *inspection setup using the Colorado State University EUV laser.*[38]

3.2.2.3 RTW/Aachen.[39] A system powered with a xenon gas DPP with two-step magnification using Schwarzschild optics has been developed to pre-magnify (21.34×) and a zone plate to image (10–20×) on a CCD (figure 13). With the addition (removal) of a central beam stop directly after the Schwarzschild, the single-stage magnification can be used as a dark-field (bright-field) microscope. The system could be operated in transmission mode. The dark-field resolution obtained was as small as 100 nm, and absorbing defects as small as 40 nm could be imaged.

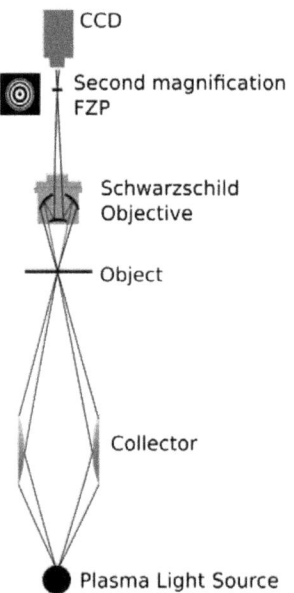

Figure 13 *Actinic microscope of the Fraunhofer ILT/Aachen.[39]*

3.2.3 Interference methods. Other inspection techniques are based on interferometry. For example, MIRAU/NTT[40] was a proposed synchrotron-based system for the first actinic imaging microscope for EUV lithography. A 15x Schwarzschild objective was used for a two-step magnification, where the high magnification was achieved with electron optics (10-200×), with an overall magnification up to 3000×. The incident radiation was beam split, bya freestanding multilayer, into reference and scanning beams. The interference of the two beams down the optical path permits to probe phase properties of the mask (figure 14). Phase and amplitude defects can be separated by means of focus tuning. A resolution of 800 nm for absorptive patterns was reported, and as low as 5 nm step feature resolution.

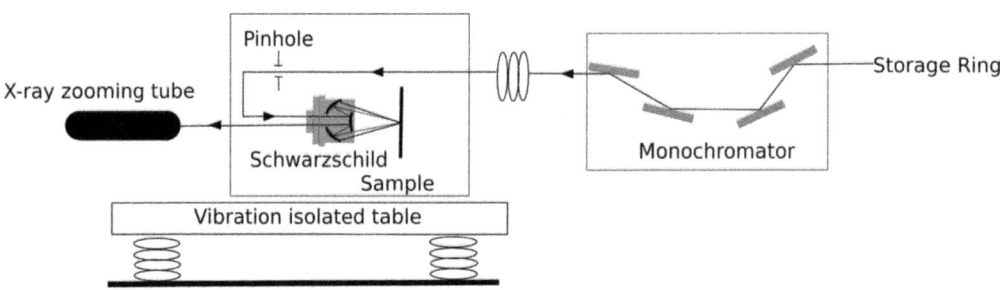

Figure 14 *Actinic microscope of the MIRAU/NTT.[40]*

3.2.4 Lensless imaging. For overcoming drawbacks of physical optics (e.g. aberration, resolution, throughput, cost, etc.) lensless imaging is growing in importance also thanks to its simplicity in the experimental setup. The underlying idea is that of retrieving the mask image by processing the acquired diffraction pattern. On the flip side the source must provide very high intensity and spatial coherence that at the moment are not common in laboratory-scale sources. The following EUV imaging setups have been described exploiting lensless imaging.

3.2.4.1 Brookhaven.[41] The system inspected masks for defects by means of intensity changes on the EUV-exposed photoresist, with defect indicative variations as low as 1.8%. Control of the EUV doses and optimization of the resist for high-contrast imaging were important. Defects as small as 200 nm could be identified.[42]

3.2.4.2 HYOGO.[43] A synchrotron-based system exploited an EUV Schwarzschild objective (30×) to then convert the image by 10–200× using a scintillator of an X-ray zooming tube. The overall magnification was thus in the range 300-6000×. Line defects as small as 90 nm and 4 nm tall, or at depth 100 nm and 2 nm tall, were observed.

3.2.5 Photoelectron emission. The use of high photon energy sources (>> 3 eV), can efficiently trigger photoelectrons. In this respect, photoemission can be exploited for mapping the surface of a sample by detecting the electron signal distribution. An EUV imaging setup *Bielefeld* has been implemented, which uses a photoemission electron microscope (PEEM)[44] to inspect the EUV amplitude across a lithography mask. The EUV-induced photoelectrons are amplified using a microchannel plate and collected on a fluorescent screen coupled to a CCD camera over < 1 s exposure time. EUV synchrotron radiation is focused to a 100 μm field of view (range 2.3 to 1000 μm), with 4° angle of incidence and a spatial resolution of 29 nm was achieved.

4. EUV LITHOGRAPHY: FABRICATION AT THE NANO-SCALE

4.1 Baseline for Extending Moore's Law

Despite 40-year successes, the semiconductor industry is looking with concern beyond current DUV sources, since the latter are not able to push the CD limit beyond the 32 nm node. Alternative approaches based on charged particle beams are limited in processing throughput by the low beam fluxes required to prevent space charge and thus beam spread (resolution throughput trade off).[45] Thus, though in principle electron beams can be used to manufacture masks of high resolution, the production yield of approximately 10 wafers/hour is not interesting for high volume manufacturing. The baseline need for maximized high volume manufacturing and minimized cost of ownership also rules out costly facilities like synchrotrons or free electron lasers (FELs).[46] The candidate enabling sources for next generation lithography are therefore optical laboratory-scale sources.

Quite surprisingly, improvement in optics and projection have been able to extend the capabilities of DUV down to 45 nm node size. However straightforward Abbé's criterion is in relating the source wavelength and CD resolution, one has to note that the "real world" progress has been on a different track (figure 15). In fact, since the same light source may be used for several technology generations and thanks to improvements in optics and projection methods that allowed exploiting larger NA, there has been a rather mild down scaling of the source specifications. With the invention of the immersion technique, i.e. with NA > 1, the NA has been pushed to unexpectedly high values. Figure 15 shows the growing process/design gap as a function of years. Immersion DUV (iDUV) lithography is a resolution enhancement technique in which the gap between the imaging lens and the wafer is filled with a liquid medium. The latter has a refractive index greater than 1, such that the final aperture is further expanded as a consequence of $NA_{immersion} = n\,NA_{air}$. ASML, Nikon and Canon are currently the only manufacturers of immersion lithography systems. Nevertheless, without a change in operating wavelength, sub 32 nm node lithography will face a drastic increase of costs, if not a stall.

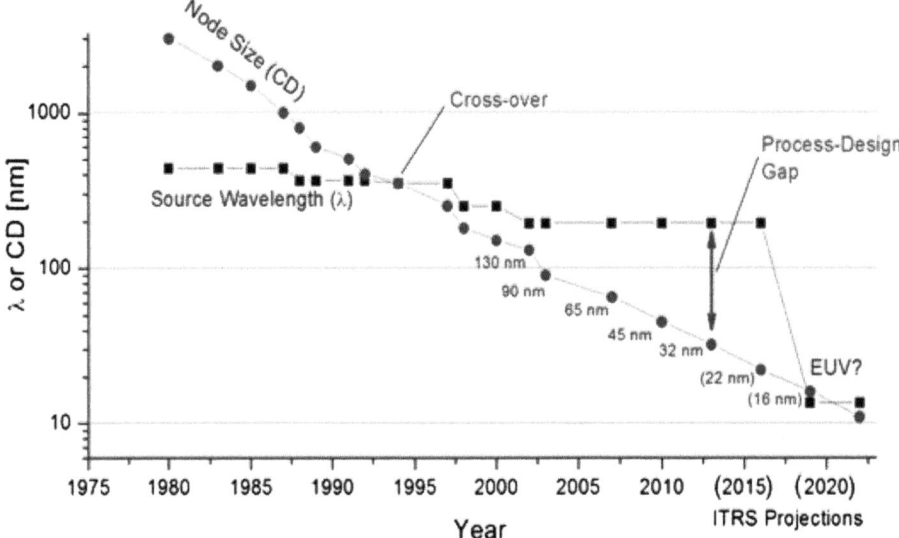

Figure 15 *Down scaling comparison between the light source wavelength for lithography systems and the node resolution. Most of the improvements in feature size have been obtained by increasing the NA. The process/design gap for sub 32 nm, however, is intolerable without a short wavelength light source implementation (EUV).*

Double patterning at iDUV for instance has demonstrated the feasibility of 32 nm node size. The idea is to enhance the resolution by repeating the exposure process after several runs of resist coating/processing have been repeated. During wafer transfer and re-exposure it is critical that the overlay between the first and following exposure cycles is extremely accurate. Clearly, by replicating the exposure process one is able to scale down the feature size yet only by concomitantly scaling up the cost of fabrication. In fact, Moore's law highlights the size miniaturization and the corresponding performance increase of devices as a function of years. In the perspective of the 40 years of Moore's law, another important consideration is to be made: the cost for patterning equipment has grown factor of 4.5 every 10 years.[47] The possibility of one step high-resolution photolithography, i.e using EUV, is therefore seen with enormous commercial interest by the industry.

EUV lithography was proposed already in the late 1980s[48] as the enabling technology to keep pace with Moore's law. Ever since the 1960s the semiconductor industry has striven for more and more integrated circuit miniaturization. The driving force has been the enhancement of clock speed (table 2) and small-scale integration of ICT devices[49] fostering portability of task oriented devices. As a matter of fact, the expectation was matched by a remarkable progress, with the number of transistors per die doubled every 12-18 months (Moore's law). Moore's law held its original trend for the whole semiconductor era thanks to the anticipated development for more advanced lithography tools (for both inspection and fabrication) able to cope with the next feature size reduction step.

The International Technology Roadmap for Semiconductors (ITRS) is the organization that should anticipate such trends. It presents 15 year predictions on key technology innovations and the related challenges. Based on the history of integrated circuit innovations, table 2 shows a representative selection of the state of the art processors in the last few decades. For future integrated circuits, it is possible to keep up the Moore's law trend if next generation lithography manufacturing equipment, especially powerful EUV light sources, are developed in due time, which at the status of the latest ITRS plans is expected not earlier than 2017.

Nanoscale lithography must therefore be based on laboratory EUV sources with the required average power to guarantee high volume manufacturing. In fact, equipment for mass production must combine duty lifetime, i.e. >30,000 hours of uptime operation, throughput, i.e. 80-100 wafers per hour, and accessible cost of ownership. The cost of ownership of lithography systems is related to three main items: equipment cost per throughput, mask cost per mask usage and resist cost.

Table 2 *Summary of the state of the art processors, with corresponding performance and manufacturing radiation, in the last 30 years.*
Legend: iDUV: immersion deep UV, DP: double patterning.

Decade	Model	Supplier	Clock	Node	Transistors	Litho source
1970s	8088	Intel	5 MHz	3 µm	29 k	Hg UV lamp
	68000	Motorola	8 MHz	4 µm	68 k	Hg UV lamp
1980s	SPARC	Sun	40 MHz	0.8 µm	800 k	Hg UV lamp
	80486	Intel	25 MHz	1 µm	1,180 k	Hg i-line lamp
1990s	PowerPC 7400	Motorola	0.35–0.5 GHz	0.2 µm	10.5 M	Excimer KrF
	Athlon	AMD	0.5–1 GHz	0.25 µm	22 M	Excimer KrF
2000s	Opteron Athens	AMD	1.6–3 GHz	90 nm	114 M	Excimer ArF
	Pentium D	Intel	2.8–3.2 GHz	90 nm	115 M	Excimer ArF
	Core i7 Lynnfield	Intel	2.66–3.2 GHz	45 nm	774 M	Excimer iDUV
2010s	Itanium	Intel	2 GHz	65 nm	2 G	Excimer iDUV
	Power 7	IBM	2–4.14 GHz	45 nm	1.2 G	Excimer iDUV
	Opteron MC	AMD	1.7–2.4 GHz	45 nm	1.81 G	Excimer iDUV
	Core i7 Sandy Bridge	Intel	1.6–3.4 GHz	32 nm	995 M	Excimer iDUV+DP
	Core i7 Westmere	Intel	1.86–3.33 GHz	32 nm	1.17 G	Excimer iDUV+DP
Future			>3 GHz	<32 nm	>1.2 G	EUV (?)

4.2 Setups for Nano-Fabrication

In photolithography, three main printing techniques can be distinguished. In *contact printing* the wafer is directly pressed against the mask. The technique was introduced in the 1960s for 0.2 mm node integrated circuits. Contact printing carries optimum resolution, for a given wavelength, by preventing far-field diffraction in the mask wafer interspacing. The resolution is thus only limited by scattering effects across the photoresist thickness. Contact printing is prone however to defect generation or damaging the mask/wafer elements. Because of this drawback this technique is used in research but not in industry. Contact printing is the basis for nano-imprint lithography. In order to mitigate the risk of defects or damage, *proximity printing* was introduced. As the name suggests, a small gap of $d_{gap} = 10–50$ µm is kept between the mask and the wafer. Transition distance from near to far field is given by the Fresnel criterion as the ratio between the squared radius of the aperture and the wavelength (i.e., the near field is for $D_{Fr} < a^2/\lambda$, a is the aperture width). A nitrogen gas flow keeps the mask away from the wafer surface. Unfortunately, this gap degrades the resolution due to diffraction. The achievable resolution scales as $CD = k(l d_{gap})^{0.5}$, where k is a resist dependent coefficient. The square root behaviour is a consequence of the Fresnel diffraction theory valid in the near field region just below the mask.[50]

Both contact and proximity printing methods need homogenization optics before the mask to get a collimated illumination, and imply a full-field printing on the wafer. Projection printing instead is an alternative technique that operates at a certain magnification factor (e.g. 4:1 in EUV lithography), which increases the mask to wafer gap. Optics is also implemented between the mask and the wafer to condense onto the wafer the diverging beam emerging from the mask. Here the illuminated spot is scanned over the wafer. Using DUV, or even EUV radiation, will not suffice for optical proximity printing to compete with projection printing.

4.2.1 System design considerations. EUV sources for lithography are influenced by a number of factors, i.e. collector acceptance, multilayer reflectivity, ambient gas transmission, spectral purity filter transmission, etendue, etc. EUV source development for nanofabrication has so far focused on incoherent plasma radiation, generated either by gas discharge produced plasmas (DPP) or laser produced plasmas (LPP). Although the physics and technologies associated with DPP and LPP are rather different, the industry specifications for the candidate EUV source are identical and are dictated by, first, the EUV collection optics (Mo/Si multilayer reflectors), optimized for operation at $\lambda = 13.5$ nm, and the photoresist sensitivity, that requires fluences of 5–10 mJ/cm^2. Optimization of the EUV source emission to match the industry requirements (115–180 W in band based on 2% bandwidth centred on 13.5nm) is of course the main challenge. Besides, the mechanical stability of the optics (stigmatic projection) and long term operation at constant reflectivity (Mo/Si multilayers offer a maximum reflectivity of approximately 70%)[51] are also critical for the delivery of a commercial EUV lithography source.

The pre-production EUV systems being tested to date implement a source collector that projects the radiation via the intermediate focus (IF) to the wafer projection unit. The latter contains two condenser multilayer mirrors, 6-10 projection multilayer mirrors, and a multilayer chip template (mask).[52] Since the optics already absorb 40-50% of the available EUV radiation, the candidate EUV source needs to be sufficiently bright to compensate for that. The mirror responsible for collecting the radiation is directly exposed to the plasma and is therefore vulnerable to damage from the kilo-electronvolt energy debris.[53,54,55] The wafer throughput of an EUV exposure tool is a critical metric for manufacturing capacity. Given that EUV is a technology requiring high vacuum, the throughput is limited mainly by the transfer of wafers into and out of the equipment chamber[56] and the source runtime/brilliance. Other aspects of the pre-production EUV tools is the need for off-axis illumination on a multilayer or stray light transport[57], since the strong absorption requires tools to work in reflection mode. Off-axis illumination is needed to ensurethat the illumination optics does not block off the return light. The resulting asymmetry in the illuminated pattern causes shadowing effects which degrade the pattern fidelity.[58] The use of EUV in lithography is unfortunately more prone to flare, for flare dependency on wavelength scales as $1/\lambda^2$ and is mostly caused by surface roughness. Calculations have shown that 1% absolute change in flare results in increased line width on the projected integrated circuits of 0.9 nm.[59] Nevertheless, the predicted optical resolution of EUV capability has been demonstrated.[60] In 1996, a joint project between Sandia National Laboratories, University of California at Berkeley, and Lucent Technologies, produced

NMOS transistors with gate lengths from 75 nm to 180 nm, by means of EUV lithography.Subsequently, a collaboration including IBM and AMD, based at the College of Nanoscale Science and Engineering (CNSE) used EUV lithography to pattern the first metal layer of a 45 nm node test chip.[61] Nevertheless, as of 2013 no commodity EUV sources for high volume manufacturing exist.

References

1 K. Jain, *Excimer Laser Lithography*, 1990, (SPIE Press).
2 J.E. Bjorkholm, *Intel Technology Journal*, 1998, **Q3**, 1.
3 K. Suzuki, B. Smith (Eds), *Microlithography: Science and Technology, Second Edition*, 2007, (CRC Press).
4 M.M. O'Toole and A.R. Neureuther, *Proc. SPIE*, 1979, **174**, 22.
5 C.W. Gwyn, R. Stulen, D. Sweeney and D. Attwood, *J. Vac. Sci. Technol. B*, 1998, **16**, 3142.
6 P.M. Eisenberger, *US Patent 4028547*, 1977.
7 E. Spiller and R. Feder, *Topics in Applied Physics*, 1977, **22**, 35.
8 G. Zorpette, *IEEE Spectrum*, 1992, **29**, 33.
9 K. Self, *Proc. IEEE*, 1993, **81**, 1248.
10 J. Waterman, C. Mbananaso and G. Denbeaux, *Proc. SPIE*, 2008, **6921**, 69213K.1.
11 B.M. Mertens, B. van der Zwan, P.W.H. de Jager, M. Leenders, H.G.C. Werij, J.P.H. Benschop and A.J.J. van Dijsseldon, *Microel. Eng.*, 2000, **53**, 659.
12 M. Kotera, K. Yagura, H. Tanaka, D. Kawano and T. Maekawa, *Jpn. J. Appl. Phys.*, 2008, **47**, 4944.
13 J. Sporre, C.H. Castaño, R. Raju and D. N. Ruzic, *J. Appl. Phys.*, 2009, **106**, 4.
14 D.C. Brandt, I.V. Fomenkov, A.I. Ershov, W.N. Partlo, D.W. Myers, N.R. Böwering, G.O. Vaschenko, O.V. Khodykin, A.N. Bykanov, J.R. Hoffman, Ch.P. Chrobak, S.N. Srivastava, D.A. Vidusek, S. De Dea and R.R. Hou, *Proc. SPIE*, **7140**, 71401E.
15 A. Wueest, *SEMATECH EUV Lithography Symposium*, 2009 (Prague).
16 R. Rakowski, J. Mikołajczyk, A. Bartnik, H. Fiedorowicz, F. de Gaufridy de Dortan, R. Jarocki, J. Kostecki, M. Szczurek and P. Wachulak, *Appl. Phys. B: Lasers & Optics*, 2011, **102**, 559.
17 M. Richardson, D. Torres, Ch. DePriest, F. Jin and G. Shimkaveg, *Opt. Comm.*, 1998, **145**, 109.
18 H. Fiedorowicz, A. Bartnik, R. Jarocki, J. Kostecki, J. Mikolajczyk, R. Rakowski and M. Szczurek, *Microproc. & Nanotechn.*, 2003, **7954075**, 314.
19 W. Kruer, *The Physics Of Laser Plasma Interactions*, 2003, (Westview Press).
20 W. Partlo, I. Fomenkov and D. Birx, *Proc. SPIE*, 1999, **3676**, 846.
21 www.holoscience.com.
22 J. Pankert, K. Bergmann and J. Klein, *Proc. SPIE*, 2002, **4688**, 87.
23 V. Bakshi (Ed.), *EUV Sources for Lithography*, 2006, (SPIE Press).
24 U. Stamm, G. Schriever and J. Kleinschmidt, in V. Bakshi (Ed), *EUV Sources for Lithography*, 2006, 413, (SPIE Press).
25 F. Zink, *Radiographics*, 1997, **17**, 1259.
26 H. Yamada, *Nucl. Instrum. .Methods Phys. Res. B*, 2003, **199**, 509.
27 S. Fourmaux, S. Corde, K. Ta Phuoc, P.M. Leguay, S. Payeur, P. Lassonde, S. Gnedyuk, G. Lebrun, C. Fourment, V. Malka, S. Sebban, A. Rousse and J.C. Kieffer, *New J. Phys.*, 2011, **13**, 033017.
28 H. Kinoshita, K. Hamamoto, N.Sakaya, M. Hosoya and T. Watanabe, *Jpn. J. Appl. Phys.*, 2007, **46**, 6113.

29 K. Goldberg and I. Mochi, *J. Vac. Sci. Technol. B*, 2010, **28**, C6E1.

30 M. Yi, T. Haga, C. walton, C. Larson and J. Bokor, *Jpn. J. Appl. Phys.*, 2002, **41**, 4101.

31 T. Terasawa, Y. Tezuka, M.Ito and T. Tomie, *Proc. SPIE*, 2004, **5446**, 804.

32 Y. Tezuka, M.Ito, T. Terasawa andT. Tomie, *Proc. SPIE*, 2004, **5446**, 870.

33 T. Yamane, T. Iwasaki, T. Tanaka, T. Terasawa, O. Suga and T. Tomie, *Proc. SPIE*, 2008, **7122**, 7122D.

34 T. Terasawa, T. Yamane, T. Tanaka, T. Iwasaki, O. Suga and T. Tomie, *Jpn. J. Appl. Phys.*, 2009, **48**, 06FA04.

35 A. Barty, Y.Liu, E.Gullikson, J.S. Taylor and O. Wood, *Proc. SPIE*, 2005, **5751**, 651.

36 M. Booth, O. Brioso, A. Brunton, J. Cashmore, P. Elbourn, G. Elliner, M. Gower, J. Greuters, P. Grunenwald, R. Gutierrez, T. Hill, J. Hirsch, L. Kling, N. McEntee, S. Mundair, P. Richards, V. Truffert, I. Wallhead, M. Whitfield and R. Hudyma, *Proc. SPIE*, 2005, **5751**, 78.

37 H. Feldmann, J. Ruoff, W. Harnisch and W. Kaiser, *Proc. SPIE*, 2010, **7636**, 76361C.

38 F. Brizuela, Y. Wang, C.A. Brewer, F. Pedaci, W. Chao, E.H. Anderson, Y. Liu, K.A. Goldberg, P. Naulleau, P. Wachulak, M.C. Marconi, D.T. Attwood, J.J. Rocca and C.S. Menoni, *Proc. SPIE*, 2009, **7271**, 72713F.

39 L. Juschkin, R. Freiberger and K. Bergmann, *J. Phys. Conf. Ser.*, 2009, **186**, 012030.

40 T. Haga, H. Takenaka and M. Fukuda, *J. Vac. Sci. Technol. B*, 2000, **18**, 2916.

41 S.J. Spector, D.L. White, D.M. Tennant, P. Luo and O.R. Wood II, *Proc. SPIE*, 1998, **3546**, 548.

42 S.J. Spector, D.L. White, D.M. Tennant, L.E. Ocola, A.E. Novembre, M.L. Peabody and O.R. Wood II, *J. Vac. Sci. Technol. B*, 1999, **17**, 3003.

43 K. Hamamoto, Y. Tanaka, S.Y. Lee, H. Hosokawa, N. Sakaya, M. Hosoya, T. Shoki, T. Watanabe and H. Kinoshita, *J. Vac. Sci. Technol. B*, 2005, **23**, 2852.

44 U. Kleineberg, J. Lin, U. Neuhaeusler, J. Slieh, U. Heinzmann, N. Weber, M. Escher, M. Merkel, A. Oelsner, D. Valsaitsev, G. Schoenhense, *Proc SPIE*, 2006, **6151**, 615120.

45 V. Sidorkin, A. van Run, A. van Langen-Suurling, A. Grigorescu and E. van der Drift, *Microelectr. Eng.*, 2008, **85**, 805.

46 G. Dattoli, A. Doria, G.P. Gallerano, L. Giannessi, K. Hesch, H.O. Moser, P.L. Ottaviani, E. Pellegrin, R. Rossmanith, R. Steininger, V. Saile and J. Wuest, *Nucl. Instr. Methods Phys. Res. A*, 2001, **474**, 259.

47 http://ismi.sematech.org/modeling/meetings/20001109/docs/05Litho.pdf

48 A.M. Hawryluk, L.G. Seppala, *J. Vac. Sci. Tech. B*, 1988, **6**, 2126.

49 P. Harsha, S. Mosier, D. Bruggeman, N. Yousefi, L. Woellert, E. Fisher, J.K. Jesse, *Chasing Moore's Law*, 2004, (SciTech Publishing).

50 H. Kirchauer, *Photolithography Simulation*, 1998 (PhD Thesis, TU Wien).

51 T. Feigl, S. Yulin, N. Benoit and N. Kaiser, *Microel. Eng.*, 2006, **83**, 703.

52 F.T. Chen, *Proc. SPIE*, 2003, **5037**, 347.

53 H. Komori, G. Soumagne, H. Hoshino, T. Abe, T. Suganuma, Y. Imai, A. Endo and K. Toyoda, *Proc. SPIE*, 2004, **5374**, 839-846.

54 B.A.M. Hansson, R. Lars, M. Berglund, O.E. Hemberg, E. Janin, J. Thoresen, S. Mosesson, J. Wallin and H. M. Hertz, *Proc. SPIE*, 2002, **4688**, 102.

55 S.N. Srivastava, K.C. Thompson, E.L. Antonsen, H. Qiu, J.B. Spencer, D. Papke and D.N. Ruzic, *J. Appl. Phys.*, 2007, **102**, 023301.

56 A. Brunton, J. Cashmore, P. Elbourn, G. Elliner, M. Gower, P. Grunewald, M. Harman, S. Hough, N. McEntee, S. Mundair, D. Rees, P. Richards, V. Truffert, I. Wallhead and M. Whitfield, *Proc. SPIE*, 2004, **5448**, 681.

57 L. Peters, *Semiconductor International*, 2007 (October 18).

58 M. Sugawara, A. Chiba and I. Nishiyama, *J. Vac. Sci. Technol. B*, 2003, **21**, 2701.

59 M. Chandhok, S. H. Lee and T. Bacuita, *J. Vac. Sci. Technol. B*, 2004, **22**, 2966.

60 K.B. Nguyen, G.F. Cardinale, D.A. Tichenor, G.D. Kubiak, K. Berger, A.K. Ray-Chaudhuri, Y. Perras, S. J. Haney, R. Nissen, K. Krenz, R.H. Stulen, H. Fujioka, C. Hu, J. Bokor, D.M. Tennant and L.A. Fetter, *J. Vac. Sci. Tech. B*, 1996, **14**, 4188.

61 R. Haavind and J. Montgomery, *Solid State Technology*, 2008 (February 27).

CARBON-NANOTUBES FIELD EMITTER TO BE USED IN ADVANCED X-RAY SOURCE

M. Fratini,[1]* S. Iacobucci,[1,2] A. Rizzo,[2] F. Scarinci,[1] Y. Zhang,[3] W.I. Milne,[3] A. Cedola,[1,4] G. Stefani,[3] S. Lagomarsino[4]

[1]CNR Istituto Fotonica e Nanotecnologie, V. Cineto Romano 42, 00156 Rome, Italy
[2]Dipartimento di Fisica, Università RomaTRE ,Via della Vasca Navale 84, 00146 Rome , Italy
[3]Engineering Department, University of Cambridge, Cambridge CB2 1PZ, UK
[4]CNR Istituto Processi Chimico-Fisici, UOS Roma, c/o Dip. Fisica, Università Sapienza P.le A. Moro 2, 00185 Rome, Italy
*present address: "Enrico Fermi" CentreMARBILabc/o Fondazione Santa LuciaVia Ardeatina, 30600179 Rome, Italy

1 INTRODUCTION

X-rays are one of the most traditional tools for the characterization of materials, able to provide structural information by means of diffraction, morphological information by means of radiography and tomography, compositional information by means of fluorescence analysis and chemical information by means of spectroscopy. Medical radiography, security in sensitive sites, quality control of industrial products and the discovery new drugs using protein crystallography are only few examples of the pervasive use of X-rays.

In recent years a strong impulse to laboratory based X-ray techniques has arisen due to the development of X-ray micro-sources with spot sizes of the order of few micrometres and improved brilliance with respect to conventional X- ray sources. Traditional X-ray tube micro-sources are limited in power and photon flux because of the thermal load problem on the anode. The limited brilliance of the electron guns, based on traditional hot cathodes, makes the task of focusing electrons in a micrometre or sub-micrometre spot with good spatial stability difficult. In addition, standard X-ray sources based on hot cathodes work only in continuous mode and with no easy route to a pulsed operation mode. A new generation of cathodes is now emerging, namely field emitting cold cathodes based on carbon nanotubes (CNTs). They offer a solution to these issues. Indeed, carbon nanotubes have many advantages for cold field emission compared to metal and diamond tips: their inertness and stability to long periods of operation, low threshold voltage for cold field emission, low operation temperature, fast response times, low power consumption and small size.[1] Prototype devices using the superior field emission properties of nanotubes have already been demonstrated. These devices include X-ray tubes,[2] scanning X-ray sources,[3] flat panel displays[4] and lamps.[5] CNT field emitting cathodes have nearly one order of magnitude higher brilliance than any other electron source, and they are natural candidates for improved X-ray sources.

Already two companies in the USA commercialize X-ray tubes based on CNTs, however with characteristics not really dissimilar from those of standard hot cathodes, and with problems with current stability. In these cathodes, the CNTs are randomly deposited by electrophoresis, with no specific control of size and dimensions. An aim of the European FP7 AXIS project was to obtain substantial improvements to the state of the art by fabricating X-ray sources based on regular arrays of CNTs, making it possible to obtain precise control of the emitting properties of the cathode. This would result in unprecedented performances concerning current density, stability, modularity, time structure (i.e., pulse width and frequency) and brilliance.

Electron gun simulations were carried out using Runge-Kutta method for computing electron trajectories in three-dimensions, with the aid of the software package Simion 3D V.7.0[a]. Simulations of electrostatic lenses were performed in two steps. The first step simulated the CNT cathode area where high spatial resolution is required to compute the local field in the emitting nano-tip region, while the second step dealt with the rest of the electron beam at the macroscopic level. From the simulations it appears that the angle of emission from a CNT is of paramount importance for obtaining a good re-focusing of the electron beam. Figure 1 shows the spot size at the anode for a 100×1000 μm cathode as a function of the divergence angle θ_G at the gate (the extraction grid) which in turn depends on the extraction angle θ_{CNT} of the emitting tips. As can be seen from the plot in figure 1, increasing θ_{CNT} and θ_G increases the spot dimensions.

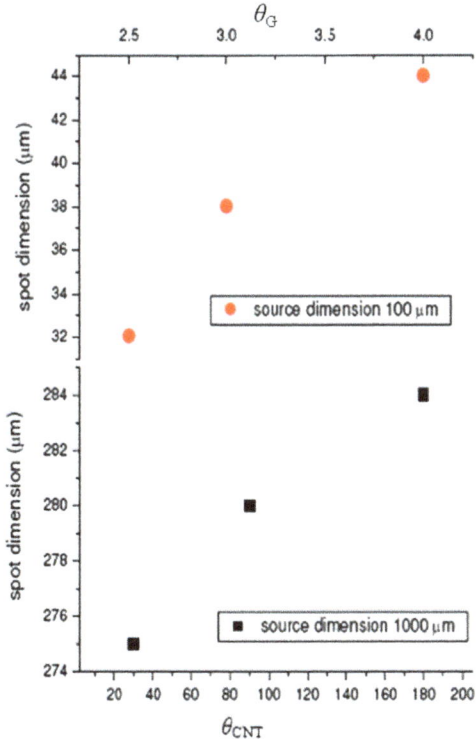

Figure 1 *The spot size at the anode as a function of the angles at the gate, θ_G, and of the emitting tips, θ_{CNT}, for a 100 μm×1000μm cathode.*

[a] Idaho National and Environmental Laboratory - 95/04037.1

1.1 Field Emission

Field emission is the extraction of electrons from a conducting solid by an electric field. A very high electric field is required for electrons tunnelling through the surface potential barrier,[6] but when the solid is shaped as a tip the equipotential lines are concentrated around the tip itself and the local field is enhanced. Compared to thermionic emission this is the preferred mechanism for certain applications because no heating is required and the emission current is primarily controlled by the external field.

The basic physics of field emission is well understood. The emission current I from a metal surface is determined by the Fowler–Nordheim equation[7]

$$I = aV^2 \exp\left(-\frac{b\varphi^{3/2}}{\beta V}\right) \tag{1}$$

where V is the applied voltage, φ is the work function of the material and β is the field enhancement factor which depends on the height and radius of the tip, a and b are 1.541434X10-6 A eV V^{-2} and 6.830890X109 eV$^{-3/2}$V m^{-1} respectively.

With the advent of micro- and nano-fabrication techniques, it has been possible to make tips of nanometre dimensions with the extraction electrode very close to the tip. Both these factors lead to higher field concentration at the tip and hence the extraction voltage can be reduced to a few hundred volts.[8]

For a metal, with a work function of a few electronvolts, and a flat surface the threshold field is typically around 10^4 V/µm, which is impractically high. The work function is a basic material property that cannot be varied significantly.[1] All field emission sources rely on field enhancement due to sharp tips/protrusions, and so that they tend to have smaller virtual source sizes because of the primary role of the field enhancement factor β. The larger the value of β, the higher the field concentration, and therefore the lower the effective threshold voltage for emission.

Since the 1990s emission from nanostructured diamond has also been studied systematically.[9] However, diamond structures are unstable at high current densities (>30 mA/cm^2)[1] These conventional cold field emission sources usually have comparatively poor current stability due to the variation of β and φ in the presence of strong adsorption at reduced temperatures.

CNTs, discovered in 1991,[10] consist of graphite sheets rolled into cylinders a few micrometres long and a few nanometres in diameter. Electron field emission from CNTs was first demonstrated in 1995,[11,12] and has since been studied extensively. Table 1 lists the field emission thresholds of several materials, showing that the threshold for CNTs is significantly lower.[8]

Several properties of CNTs make them favourable for field emission, as listed below.

1. Their shape is optimum for field emission. CNTs have larger field enhancement factors and, thus, lower threshold fields for emission than conventional emitters. This is the result of the atomically sharp tips and of the large aspect ratios.

2. Depending on the exact arrangement of the carbon atoms, single wall CNTs can be either semiconductors or (metallic) conductors. Multiwall CNTs are semi-metallic in nature.

Their resistance decreases with increasing temperature since they are semi-metallic.

Table 1 *Threshold fields of cathode materials.*

Cathode material	Threshold field (V/μm) for 10 mA/cm^2
Molybdenum tips	50-100
Silicon tips	50-100
Amorphous diamond	30-120
Graphite powders	10-20
Nano-diamond	3-5
Carbon nanotubes	1-2

1. They could be heated up, by its field emitted current, to 2000 K and remain stable, allowing high field emission currents.
2. They are robust materials in terms of their mechanical, thermal and chemical properties.

These properties lead to the conclusion that the best possible choice for a cathode with maximal current output for a given applied voltage is a cathode based on CNTs.

Moreover for the best efficiency the CNTs should have uniform orientation and spatial distribution and high spatial density,[13] which makes them excellent electron emitters. However it is important to underline that in order to design and fabricate cathodes with optimized performance, it must be taken into account that when several emitters are assembled to form an array there is electrostatic screening between them. Thus a compromise must be found between the CNT-CNT distance and the CNT density.

2 EXPERIMENTAL DETAILS

2.1 Growth Process

CNT dot arrays were grown on highly doped n-type silicon in a plasma-enhanced chemical vapour deposition (PECVD) system.[b] A 7 nm thick layer of nickel was deposited by direct current (DC) magnetron sputtering. Electron beam lithography and lift off were then employed to produce nickel nanodots 100 nm in diameter to act as catalyst arrays. The substrate, placed on a resistively heated graphite stage, was thermally ramped, at a pressure of 10^{-2} mbar, to 700 °C at 100 °C per minute. During heating, gaseous ammonia was introduced to etch the surface of the nickel catalyst islands, raising the reactor pressure to 2.4 mbar. Acetylene was chosen as the carbon source, and introduced in the deposition chamber once the temperature reached 750 °C. A dc voltage of 640 V was applied between the gas shower head and the heating stage to induce a 35 W plasma to align the CNTs.

The array took the form of a 300×300 μm^2 square containing about 14400 CNTs. The resultant pitch requirement was 2.5 μm and the CNTs were grown to a height of 1.25 μm. The pattern was centred on a 10×10 mm degenerately n-doped silicon substrate. Figure 2 shows a high magnification scanning electron microscope (SEM) image of the CNTs

---

[b] AIXTRON Nanoinstruments Ltd. Black Magic

grown directly on the silicon substrate. The samples exhibited excellent uniformity over large areas, a key requirement for the successful completion of the project.

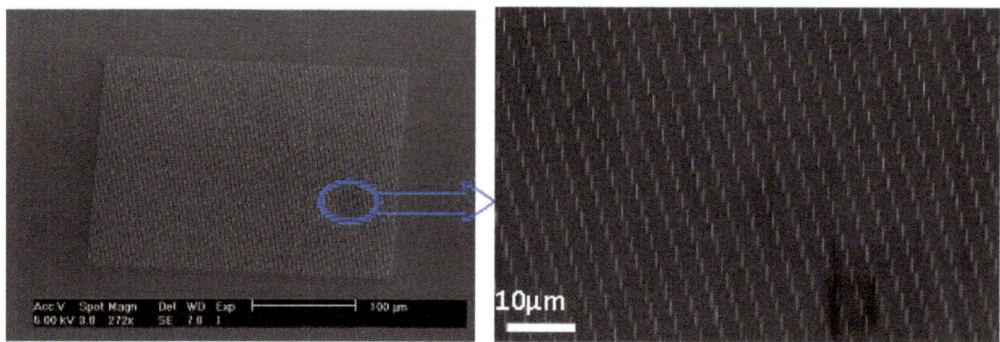

Figure 2 *CNTs grown directly on a Si substrate. A high degree of uniformity was consistently observed;*

2.2 Experimental set up for cathode characterization

Figure 3 shows a schematic diagram of the measurement setup for the CNT dot array samples. The extractor element, optimized to obtain a suitable field at the tip of the CNTs was made of two stainless steel plates, separated by ruby spheres. The high field region was to 1.5 mm diameter by adopting a hemispherical profile for the extraction plate. The use of ruby spheres reduced misalignments, thus guaranteeing easy assembly and reliable mounting. Accurate fabrication ensured good parallelism of the two plates and firm assembly.

It was very important that the two plates were accurately polished surfaces. A clean silicon chip was used to characterize the extraction device; the dielectric strength of the mounting was determined and thermal tests could be carried out without damaging the CNT array. A negative voltage of up to –3.5 kV was applied to the empty cathode, verifying that no parasitic currents were present.

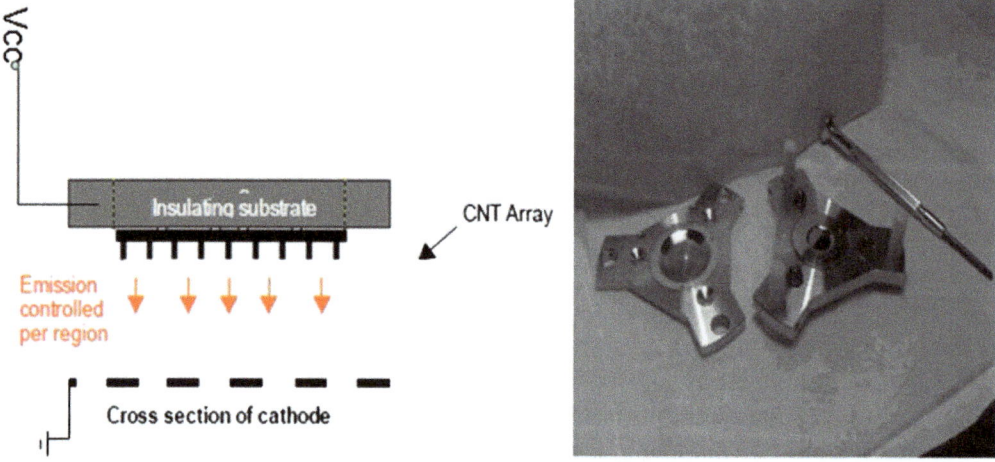

Figure 3 *Details of the emitter. A schematic of the cathode electrode and of the extractor electrode is shown (left), together with a photograph of the two disassembled plates (right).*

3 RESULTS AND DISCUSSION

When the CNT cathode was measured for the first time, a conditioning procedure was carried out through the following steps:

1. A voltage ramp from 0–3000 V was carried out in steps of about 5 V/min.
2. The maximum voltage was maintained for about ten hours.
3. The voltage was rapidly ramped up and down (from 0–3000 V and back).

Figure 4 shows the *I–V* characteristics of a CNT array of 10000 tubes deposited on a silicon substrate in a 300 μm square pattern. The maximum emission current, measured on the extractor, was ≈1.2 mA at 3000 V, corresponding to an applied electric field of 15 V/μm.

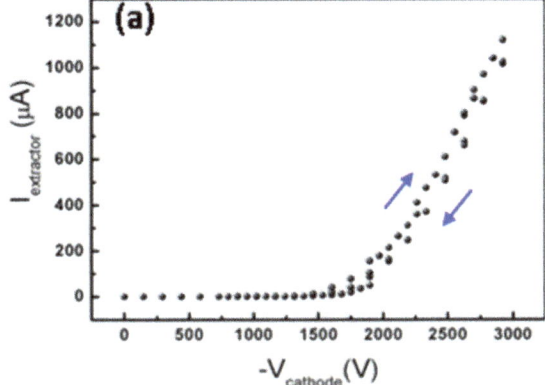

Figure 4 *I-V characteristic (ramp up and down) of a 300 μm square CNT array deposited on a silicon substrate.*

The extracted current I_E was measured, in a range where the power on the grid was not high enough to destroy it (the typical maximum being 300 mW), by using a commercial grid connected to ground through a pico-ammeter, The total current I_{FC} transmitted through the extractor equipped with a grid with 20 μm square holes separated by 20 μm bars was measured by an in-vacuum movable Faraday cup with a 2 mm aperture. The fraction of the emitted current transmitted through the grid is then $I_{FC}/(I_E + I_{FC})$. Figure 5 shows the stability measured three days after the system was switched on, at a fixed voltage and in good vacuum conditions (a pressure of 7×10^{-8} mbar). There are two main observations. First, there is an instability of about ± 10% on the timescale of a few minutes and, second, there is significant drift on a longer timescale.

To improve the stability of the CNT cathode emission current, it is important to graphitize the cathode high temperature annealing, Measurements performed on a square of 300 μm 10 K CNT array deposited on a graphitized silicon substrate yielded a maximum current of 600 μA at 10 V/μm, an instability of about ± 5% and good reproducibility (about 10%) after four months.

After studying the emission current characteristics, we measured the angular distribution of the emitted electrons, which is an important parameter in order to obtain a high-brilliance source. The angular distribution was measured by scanning the Faraday cup in the direction perpendicular to the beam in a field-free region after the extraction grid. The measurements were repeated at different distances from the extractor grid in order to measure the divergence. With this experiment, angle divergences of the electron beam of the order of few degrees were measured, thus indicating that, by minimizing aberrations, the ultimate dimensions of the anode spot reported in figure 1 can be achieved.

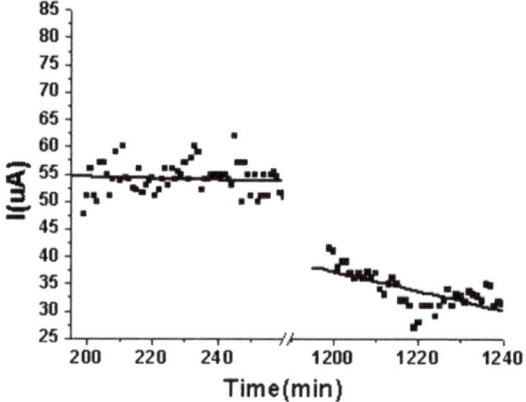

Figure 5 *Plot of the transmitted current as a function of time,measured for a square 300 μm CNT array deposited on a silicon substrate.*

4 X-RAY SOURCE

The cathode described above has been then assembled on an open-type x-ray source, composed of a vacuum chamber, vacuum pumps, e-gun with HV connector and water cooled anode, shown in fig. 6. Tests have demonstrated that the electrons emitted by the carbon nanotube cathode are efficiently focused on the anode, giving rise to an x-ray spot 30x300 micrometers wide. The detailed description of the x-ray source features goes beyond the scope of this paper, and will be the object of further publications.

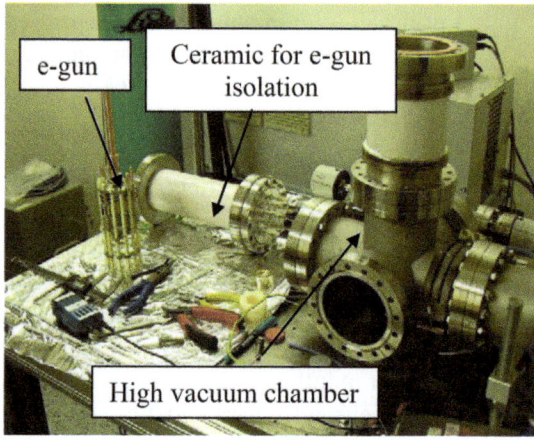

Figure 6 *Open type x-ray source holding the e-gun with the carbon-nano tube cathode.*

5 CONCLUSIONS

The European project AXIS aims at fabricating an innovative X-ray source based on a field emitting CNT cathode. The basic components of such a source which determine the final properties have been presented here, i.e., the CNT cathode, the extractor and the electron optics. The results show that good control of the emission properties can be achieved, in agreement with electron trajectory simulations. These encouraging results form the basis for the realization of a source with improved brilliance, better stability, and pulsed operation

ACKNOWLEDGMENTS

The project AXIS (EU project 222260) is gratefully acknowledged for financial support. All the partners of the project are acknowledged for their support.

References
1 Y. Cheng *et al*, Electron field emission from carbon nanotubes, *C. R. Phys.*, 2003, **4**, 1021–1033.
2 H. Sugie *et al*, Carbon nanotubes as electron source in an X-ray tube, *Appl. Phys. Lett.*, 2001, **78**, 2578.
3 J. Zhang *et al*, Stationary scanning x-ray source based on carbon nanotube field emitters, *Appl. Phys. Lett.*, 2005, **86**, 184104.
4 W.B. Choi *et al.*, Fully sealed, high-brightness carbon-nanotube field emission display, *Appl. Phys. Lett.*, 1999, **75**, 3129.
5 M.Croci *et al.*, A fully sealed luminescent tube based on carbon nanotube field emission, *Microelectronics J.*, 2004, **35**, 329.
6 R. Gomer, Field Emission and Field Ionization, Harvard University Press, Cambridge, MA, 1961.
7 N. De Jorge *et al.*, Carbon nanotube electron sources and applications, *Phil. Trans. R. Soc. Lond. A*, 2004, **362**, 2239–2266.
8 W.I. Milne *et al.*, Carbon nanotubes as field emission sources, *J. Mater. Chem.*, 2004, **14**, 933–943.
9 *Vacuum Micro-Electronics*, ed. Zhu, Wiley, 2001.
10 S. Iijima *et al.*, Helical microtubules of graphitic carbon, *Nature*, 1991, **354**, 56–58.
11 A.G. Rinzler, J.H. Hafner, P. Nikolaev, L. Lou, S.G. Kim, D. Tomanek, D. Colbert, R.E. Smalley, Unraveling nanotubes: field emission from an atomic wire, *Science*, 1995, **269**, 1550–1553.
12 W.A.D. Heer, A. Chatelain, D. Ugarte, A carbon nanotube field-emission electron source, *Science*, 1995, **270**, 1179–1180.
13 Carbon Nanotubes: Synthesis, Structure, Properties, and Applications, in *Topics in Applied Physics*, ed. M.S. Dresselhaus, G. Dresselhaus and P. Avouris, Springer-Verlag, Heidelberg, 2000, vol. 80.

LASER-PLASMA EUV SOURCE FOR MODIFICATION OF POLYMER SURFACES

A. Bartnik[1], H. Fiedorowicz[1], R. Jarocki[1], J. Kostecki[1], L. Pina[2], M. Szczurek[1], and P. Wachulak[1]

[1]Institute of Optoelectronics, Military University of Technology, 00-908 Warsaw, Poland
[2]Faculty of Nuclear Sciences and Physical Engineering, Czech Technical University, 180 00 Prague 8, Czech Republic

1 INTRODUCTION

Organic polymers such as polyethylene (PE), polytetrafluoroethylene (PTFE), polypropylene (PP), polyethylene terephthalate (PET) and polyvinyl fluoride (PVF) are considered to be important materials in various biomedical applications ranging from conventional cell growth to construction of hybrid tissues and artificial organs. These polymers can be used to form the extracellular matrix (ECM) that provides structural support to the animal or human cells which colonize the polymer surface. Because these polymers are too inert to ensure appropriate cell adhesion it is necessary to modify the polymer surface to render it attractive for cell colonization.

Various chemical and physical methods for functionalization of the polymer surfaces have been developed, including chemical and plasma treatment, ion implantation, UV-irradiation. However, some of these are often associated with undesirable side effects such as the degradation of the internal parts of the material by penetrating UV radiation. This problem can be solved using short wavelength radiation in the extreme ultraviolet (EUV) range for surface modification. Such radiation is absorbed in very thin (less than 100 nm) layers of the polymers. The interaction mechanism is similar to ultraviolet laser ablation in which energetic UV photons cause chemical bonds of the polymer chain to be broken,[1] however, the EUV interaction region is limited to very thin surface layer.

A new laser-plasma EUV source for modification of polymer surfaces has been developed. The source is based on the double-stream gas-puff target approach.[2] The target is formed by pulsed injection of high-Z gas (xenon or krypton) into a hollow stream from a low-Z gas (helium) using an electromagnetic valve system equipped with a double-nozzle setup. The outer stream of gas confines the inner stream of gas improving the gas puff target characteristics (higher gas density at larger distance from the nozzle output). Using this new approach strong enhancement of EUV production has been demonstrated.[3] The use of a gas-puff target instead of a solid target makes it possible to generate EUV light with high efficiency, however, without debris production associated with the laser ablation. The source dedicated for modification of polymer surfaces was designed and built within the EUREKA project E!3892 ModPolEUV. Investigations on the modification of surfaces of various polymers have been performed as well as preliminary studies on the cultivation of biological cells on EUV modified polymer surfaces.

2 LASER PLASMA EUV SOURCE

A schematic of the laser-plasma EUV source for modification of polymer surfaces is shown in figure 1. EUV radiation in the wavelength range of about 5–50 nm is produced from a laser-irradiated double-stream gas-puff target. The target is produced with the electromagnetic valve system mounted in a vacuum chamber which is composed of three sections, each pumped separately by oil-free vacuum pumps (differential pumping). The valve system is mounted in the first section using *xyz* translation stages which allow the target to be placed in the required position with an accuracy of about 10 µm. The gas-puff targets are irradiated with laser pulses from a compact commercial Nd:YAG laser system (EKSPLA). The laser pulses are about 3 ns long with energy up to 800 mJ. The gas-puff targets can be formed with a repetition rate up to 100 Hz, but in practice the operation rate of the system was determined by that of the laser, which was 10 Hz. The target was irradiated with the laser beam focused perpendicularly to the gas flow and the distance between the laser focus and the nozzle output was about 1 mm.

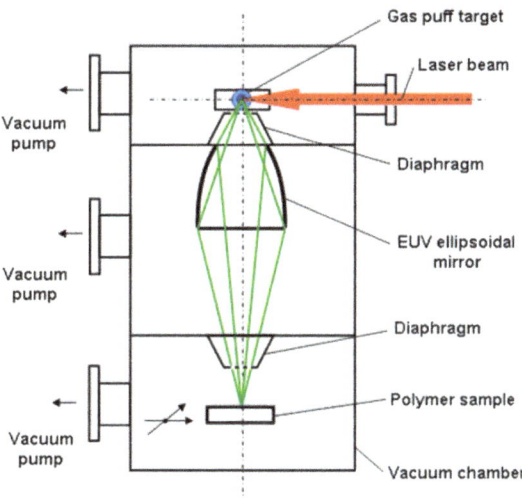

Figure 1 *A schematic diagram of the laser-plasma EUV source for the modification of polymer surfaces.*

The source is equipped with a grazing incidence axisymmetrical ellipsoidal mirror to focus EUV radiation onto the polymer samples. The mirror mounted in the second section of the chamber was developed in co-operation with the Czech Technical University, Prague and was made by the Reflex company, Prague. The second section is separated from the first and the third sections with small diaphragms of about 2mm in diameter. The mirror focuses EUV radiation onto a sample mounted in the third section of the chamber using *xyz* translation stages. Diameter of the EUV spot at the focus was about of 1 mm with a maximum fluence approaching 100 mJ/cm^2. The characterization measurements and the source parameters have been described in detail elsewhere.[4] The details of the EUV source design and its basic elements are shown in figure 2.

The laser plasma EUV source for modification of polymer surfaces was designed and built in close co-operation with PREVAC (www.prevac.eu). A view of the EUV source chamber with the mounting table design is presented in figure 3, and figure 4 is a photograph of the source.

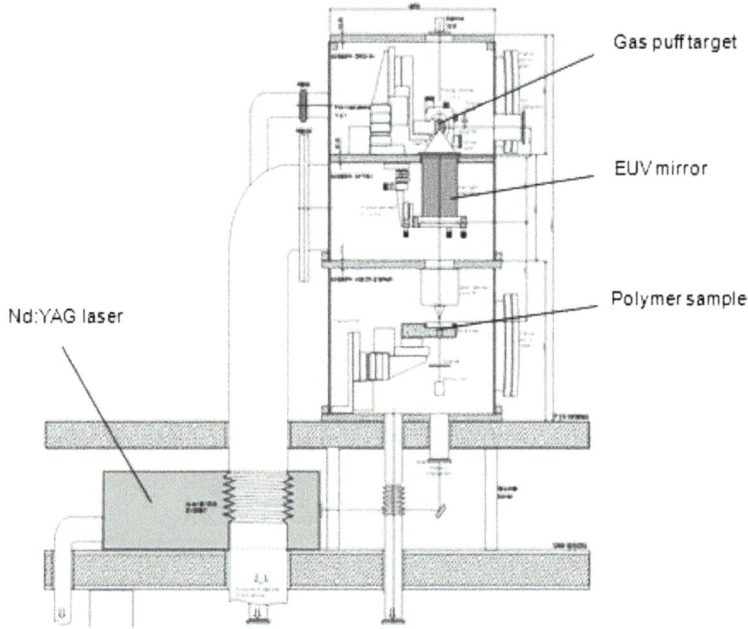

Figure 2 *Details of the EUV source and its basic elements.*

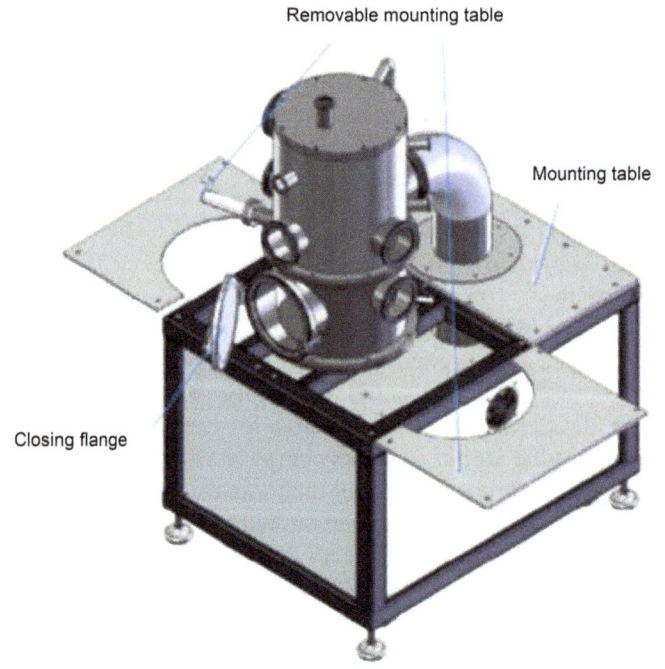

Figure 3 *Designs of the EUV source chamber and mounting table.*

Figure 4 *Laser-plasma EUV source for modification of polymer surfaces.*

3 SURFACE MODIFICATION OF POLYMERS

Different kinds of polymers were irradiated in different irradiation conditions. In case of some polymers, like for example PMMA, PTFE or FEP, the polymer surface after ablation remains smooth while for other polymers the ablation is accompanied by pronounced modification of their surface morphologies. In some cases the surface modifications can be clearly visible in SEM or AFM images even after a single EUV pulse exposure. In other cases irradiation with multiple pulses is necessary. Formation of typical microstructures usually requires 10 - 50 EUV pulses. Examples of such microstructures obtained for polyvinylidene chloride (PVDC) and polyethylene naphthalate (PEN) are shown in Fig. 5a,b respectively. In some cases conical structures are formed. Their heights increase with number of pulses and are connected with material ablation around them. Examples of such cones obtained for polyether imide (PI - Kapton HN) and polyethylene terephthalate, (PET) are shown in Fig. 5c,d respectively. These microstructures were obtained for the EUV fluence close to maximum.

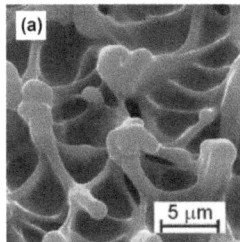 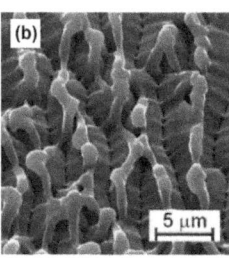  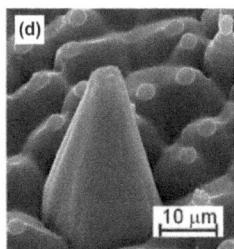

Figure 5 *SEM images of microstructures obtained in different polymers: a) PVDC 20 EUV pulses, b) PEN 50 EUV pulses, c) PI 600 EUV pulses, d) PET 600 EUV pulses.*

Irradiation of polymers with a large number of EUV pulses with a fluence below an ablation threshold usually results in formation of different kinds of nanostructures. Examples of such structures obtained for polymethyl methacrylate (PMMA), polytetrafluoroethylene (PTFE), PET and fluorinated ethylene propylene (FEP) are shown in Fig. 6. In case of PMMA some nodular structures are visible on the surface while the PTFE surface is strongly eroded forming characteristic beads-on-a-string. In both cases a characteristic size of the structures is about 100 nm. Origin of the nanostructures is connected with non-uniform releasing of some amount of volatile fractions, forming numerous nanobubbles beneath the surface in case of PMMA, or leaving some porous structure in case of PTFE. Origin of the structures formed in PET and FEP is similar to PTFE, in these cases however, the near-surface layers are not so porous. More detailed description of the research can be found in the separate papers. [5,6,7,8]

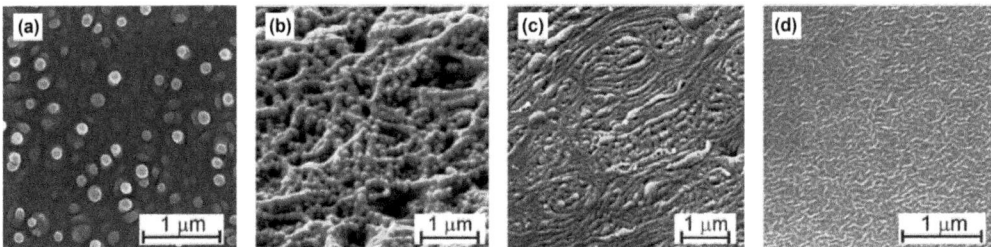

Figure 6 *SEM images of nanostructures obtained in different polymers: a) PMMA - 600 EUV pulses, b) PTFE - 1800 EUV pulses, c) PET - 1200 EUV pulses, d) FEP - 600 EUV pulses.*

Chemical changes in EUV irradiated polymers were investigated using X-ray photoelectron spectroscopy (XPS). It was shown that in a case of high irradiation fluence resulting in smooth ablation there were almost no chemical changes in the remaining material. In case of low fluence or additional injection of a gas into the interaction region the chemical changes are significant. The most interesting results were obtained for polyvinylidene fluoride (PVDF), especially in a presence of nitrogen gas in the interaction region. The XPS corresponding spectra are shown in Fig. 7. First of them was obtained for the pristine PVDF sample where only peaks corresponding to fluorine (F1s) and carbon (C1s) are visible. Presence of two C1s peaks is connected with C-F and C-C bonds. A spectrum shown in Fig. 7b was obtained for a PVDF sample irradiated with low intensity EUV pulses. In this case relative intensity of the F1s peak decreased and one of the C1s peaks corresponding to C-F bonds almost vanished. It is an evidence of defluorination of PVDF being a result of the polymer decomposition. Apart from that an O1s peak corresponding to oxygen atoms appeared. These atoms are incorporated into the polymer structure after exposure to the air. The third spectrum (Fig. 7c) was obtained for PVDF sample irradiated in presence of a nitrogen gas injected into the interaction region. It contains an additional peak (N1s) corresponding to nitrogen atoms incorporated into the polymer molecular structure during irradiation. Its contribution in the material is relatively high reaching a value of 13%. The results of the research on chemical and physical changes in various polymers caused by EUV radiation have been presented and discussed in several separate papers. [8,9,10,11,12,13,14]

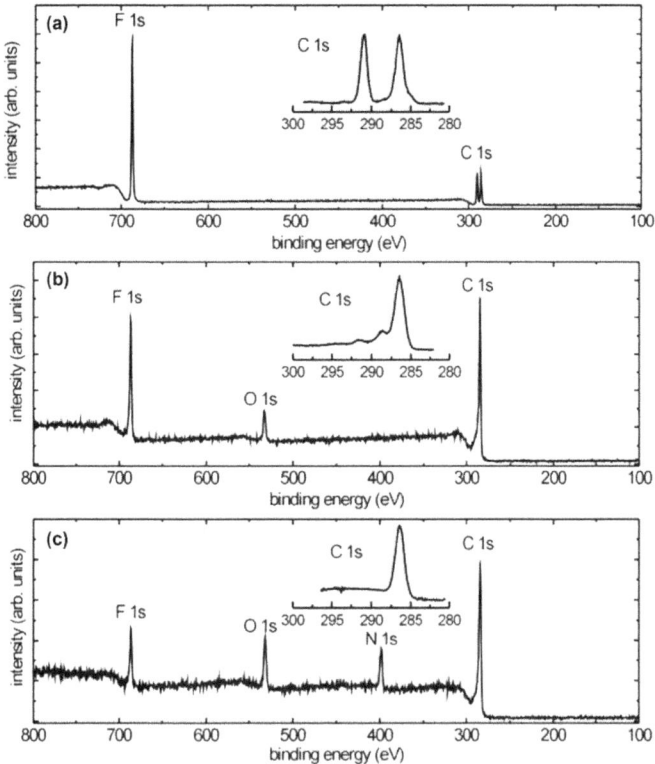

Figure 7 *XPS survey spectra of PVDF surface: a) pristine sample, b) sample irradiated with 150 EUV pulses, c) sample irradiated with 150 EUV pulses in presence of nitrogen. Sub-images show the high-resolution XPS spectra of C 1s.*

The source has been also used successfully in preliminary studies on the cultivation of biological cells on EUV modified polymer surfaces in which good adhesion and aligns of the cells along the oriented wall- and ripple-type microstructures on PET surfaces produced by the EUV irradiation were demonstrated.[15] Comparison of surface modification of polymers for biocompatibility via exposure to EUV with other techniques is given in the review article.[16]

4. CONCLUSIONS

A new laser-plasma EUV source dedicated for modification of polymer surfaces developed during the COST Action MP0601 *Short Wavelength Laboratory Sources* has been presented. The source is based on the double-stream gas-puff target approach and equipped with an axisymmetrical grazing incidence ellipsoidal mirror for EUV beam focusing. The source was designed and built under the EUREKA project E!3892 ModPolEUV in close cooperation with industrial partners (EKSPLA, Reflex, PREVAC). The source has been used successfully in investigations on modification of surfaces from various polymers and preliminary studies on the cultivation of biological cells on EUV modified polymer surfaces. The source should be useful in the research on biocompatibility control of polymer materials.

ACKNOWLEDGEMENTS

The research was partially performed through the support of the EUREKA project E! 2892 ModPolEUV supported by the Ministry of Science and Higher Education of Poland (decision Nr 120/EUR/2007/02) and under the COST Action MP0601 (decision Nr 816/N-COST/2010).

References

1 D. Bauerle, *Laser Processing and Chemistry*, 2000, (Berlin Springer Verlag,).

2 H. Fiedorowicz, A. Bartnik, R. Jarocki, R. Rakowski and M. Szczurek, *Appl. Phys. B*, 2000, **70**, 305.

3 H. Fiedorowicz, A. Bartnik, H. Daido, I.W. Choi and M. Suzuki, *Opt. Commun.*, 2000, **184**, 161.

4 A. Bartnik, H. Fiedorowicz, R. Jarocki, J. Kostecki, M. Szczurek and P.W. Wachulak, *Nucl. Instrum. Methods Phys. Res. A*, 2011, **647**, 125.

5 A. Bartnik, H. Fiedorowicz, R.Jarocki, J. Kostecki, R. Rakowski, A. Szczurek and M. Szczurek, *Acta Physica Polonica A*, 2009, **116**, S108.

6 A. Bartnik, H. Fiedorowicz, R. Jarocki, J. Kostecki and M. Szczurek, *Appl. Phys. B*, 2009, **96**, 727.

7 A. Bartnik, H. Fiedorowicz, R.Jarocki, J. Kostecki, R. Rakowski, A. Szczurek and M. Szczurek, *Acta Physica Polonica A*, 2010, **117**, 384.

8 A. Bartnik, H. Fiedorowicz, R. Jarocki, J. Kostecki and M. Szczurek, *Appl. Phys. A*, 2010, **98**, 61.

9. A. Bartnik, H. Fiedorowicz, R. Jarocki, J. Kostecki, M. Szczurek, A. Biliński, O. Chernyayeva and J.W. Sobczak, *Appl. Phys. A*, 2010, **99**, 831.

10. A. Bartnik, H. Fiedorowicz, S. Burdyńska, R. Jarocki, J. Kostecki and M. Szczurek, *Appl. Phys. A*, 2011, **103**, 173.

11. A. Bartnik, H. Fiedorowicz, R. Jarocki, J. Kostecki, M. Szczurek, O. Chernyayeva, J.W. Sobczak, *J. Electr. Spectr. Rel. Phenom.* 2011, **184**, 270.

12. A. Bartnik, H. Fiedorowicz, S. Burdyńska, R. Jarocki, J. Kostecki, M. Szczurek, *Appl. Phys. A*, 2011, **103**, 173.

13. A. Bartnik, H. Fiedorowicz, R. Jarocki, J. Kostecki, A. Szczurek, M. Szczurek, P. Wachulak, *Acta Physica Polonica A,* 2012, **121**, 445.

14. A. Bartnik, H. Fiedorowicz, R. Jarocki, J. Kostecki, M. Szczurek, P.W.Wachulak, *Appl. Phys. A*, 2012, **106**, 551-555.

15. B. Reisinger, M. Fahrner, I. Frischauf, S. Yakunin, V. Svorcik, H. Fiedorowicz, A. Bartnik, C. Romanin, J. Heitz, *Appl. Phys. A*, 2010, **100**, 511

16. I.U. Ahad, A. Bartnik, H. Fiedorowicz, J. Kostecki, B. Korczyc, T. Ciach, D. Brabazon, *J. Biomed. Mater. Res. Part A*, 2013, Article first published online: 16 OCT 2013, DOI: 10.1002/jbm.a.34958

A SUB-PICOSECOND PLASMA SOURCE FOR TIME-RESOLVED X-RAY MEASUREMENTS

T. Kämpfer[1,2], S. Höfer[1,2], R. Loetzsch[1,2], I. Uschmann[1,2] and E. Förster[1,2]

[1]Institute for Optics and Quantum Electronics, Friedrich Schiller University Jena, Max-Wien-Platz 1, 07743 Jena, Germany
[2]Helmholtz-Institute Jena, Fröbelstieg 3, 07743 Jena, Germany

1 INTRODUCTION

X-rays have been very important and successful in determining atomic and structural properties of matter. The main advantage, over visible radiation, is that X-rays interact with free, valence and core electrons. The latter, located very close to the nucleus, define the atomic structure of the material. Measuring the interference of X-rays with the periodic structure of a crystal allows direct determination of the elementary structure and even very subtle changes of the lattice caused by external parameters such as temperature and pressure.

Fundamental processes in nature occur on the timescales of the natural periods of oscillation of atoms and molecules, i.e. femtoseconds to picoseconds. The new and exciting potential of femtosecond X-ray pulses, that have become accessible due to the recent rapid development of laser technology, provides the feasibility of real-time measurements of the structural properties of matter.

The direct and real-time study of structural processes has been very successful in revealing timescales of phase transitions[1-3], non-thermal melting,[4-8] excitation of acoustic waves,[1,8-12] optical phonons[13] and charge transfer in solids.[14] Time-resolved X-ray diffraction (TRXD) is a rapidly growing research area of current interest. The most important requirement for such a challenging task is to provide a source of hard X-rays with sub-picosecond pulse duration. In this chapter the development and application of a table-top laser system to generate hard X-rays with sub-picosecond pulse durations for time-resolved X-ray measurements are described.

2 REQUIREMENTS

The laser plasma X-ray source at Friedrich Schiller University Jena uses a high-power titanium-sapphire laser system (Quantronix-ODIN-MPA) consisting of two multi-pass amplifiers that provide femtosecond pulses by using chirped pulse amplification (CPA). The system delivers pulse energy of 3.5 mJ with a repetition rate of 1 kHz and a pulse duration of 60 fs at a mean wavelength of 815 nm. The amplified beam has a diameter of about 15 mm and can be focused, using an off-axis parabola with a focal length of 16 cm, to a minimum spot size of 12 μm. However, the intensity on target is then not high enough to obtain a significant conversion rate of laser energy into X-rays. Therefore the beam diameter was expanded by a factor of three using a reflective telescope; with this improvement a focal spot diameter of 4 μm was achieved, resulting in intensity on target greater than 10^{17} W/cm^2 and a conversion efficiency to X-rays a factor of ten higher than without beam expansion.

The laser pulses were focused, using a parabolic mirror in a vacuum chamber, onto a tape target similar to[22], which was constructed at the University of Jena. To provide a new position for each pulse the tape has to move at a few centimetres per second, and to achieve high conversion efficiency the position of the tape surface must be stable within the mirror's Rayleigh length of about 20 μm in order to obtain the highest intensities on target. The source emits over 4π sr; for the work described here the radiation from the rear side of the target was used. Copper, titanium and iron were used as target materials, with thicknesses between 10 μm and 25 μm as a compromise between X-ray absorption and the tape strength for spooling. For titanium an additional vacuum chamber was used to reduce the absorption in air before the X-rays were incident on the sample.

The X-ray yield was studied as a function of the pulse duration, intensity and angle of incidence. As shown in figure 1 the copper K_α yield depends on the tape target speed. The yield was measured using the Laue reflection of a c-cut quartz crystal and a deep depletion back-illuminated CCD. Figure 1 shows that the K_α yield, as well as the spectrally integrated signal above an energy of about 15 keV, increases significantly as the tape speed decreases. Even for tape speeds lower than 4.5 cm/s, when the tape is cut by the laser pulses, the yield still increases. A reduction of the tape thickness by laser ablation resulting in reduced absorption cannot explain this observation, as the dose rate observed from the front side should not increase. A possible explanation is the absorption of the short laser pulse causing conversion of laser radiation to hot electrons. It is well known that the density gradient of the plasma is important in optimizing laser absorption, hot electron production and x-ray yield.[15]

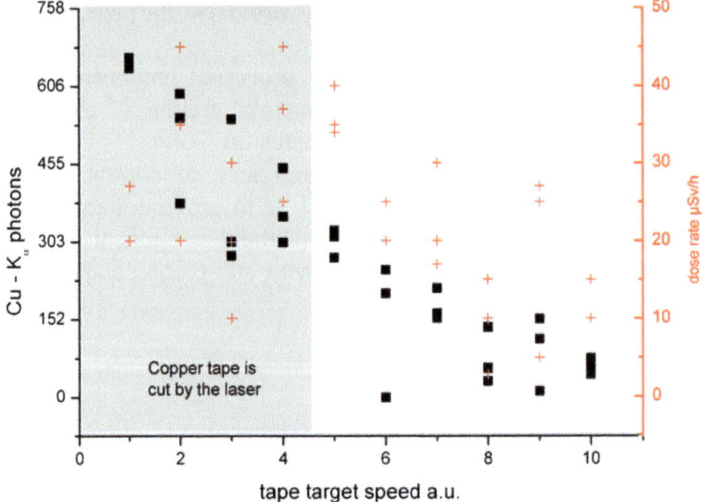

Figure 1 *Dependency of the X-ray flux on tape speed. The copper K_α yield was measured using a CCD camera and 040 Laue reflection from a quartz crystal. The dose rate was measured inside the X-ray shielding close to the plasma source.*

A major problem with laser-plasma sources is debris; in the current work the off-axis parabolic mirror surface was coated within minutes by ablated target material without shielding. Thus the yield of K_α photons was decreased dramatically. For this reason a debris shield was installed between the plasma and the mirror to protect the optic. In the first version this consisted of a 600 μm thick quartz plate which had to be changed and cleaned after 15 minutes. In a revised construction the quartz plate was replaced by a moving 60 μm thick polymethylmethacrylate (PMMA) foil. This allowed stable running of the source without a break for up to four hours.

X-ray CCD cameras were used to record the results of experiments with the laser plasma source, either a deep depletion back-illuminated CCD from Andor or a front-illuminated CCD from Roper. The cameras were used in the single-photon regime as non-dispersive spectrometers. The energy resolution of 300 eV allowed separation of the characteristic radiation from the Bremstrahlung background. Thus, with a post-processing computer algorithm the images were corrected for split events (one photon illuminates multiple pixels). To improve the signal to noise ratio, unwanted hard x-ray photons (i.e. Bremsstrahlung or fluorescence from the sample) were removed.[16]

3 CHARACTERIZATION

To characterize the source the X-ray CCD data were compared with those from an absolutely calibrated X-ray diode. A spectrum is shown in figure 2. The total numbers of K_α and K_β photons per second were obtained from the spectra; the maximum conversion rate from laser energy to K_α radiation was estimated to be $E_{K\alpha}/E_{laser} = 5.4 \times 10^{-6}$. The source size was determined by scanning a knife edge across the X-ray CCD. From the sharpness of the edge shadow a source size below 10 μm was estimated.

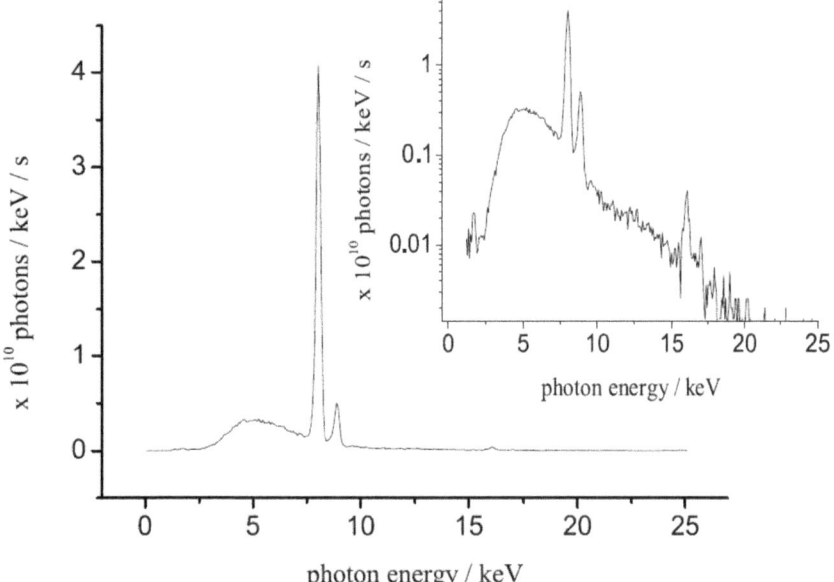

Figure 2 *An X-ray spectrum of the 1 kHz X-ray source with a copper tape target. Spectra were recorded by a back-illuminated deep depletion CCD in front of the tape target. The inset shows the spectrum on a logarithmic scale.*

Figure 3 shows the dose rate close to the plasma source plotted over a time of 8 hours; the dosimeter only detected X-rays with energies above about 15 keV. The pulse to pulse stability of the laser energy was better than 5% and the pointing of the laser monitored by a far-field camera was stable over the whole period. However the yield of the plasma source fluctuated considerably. Possible causes are variations of the tape surface, movement of the tape through the focus and instabilities in the plasma.

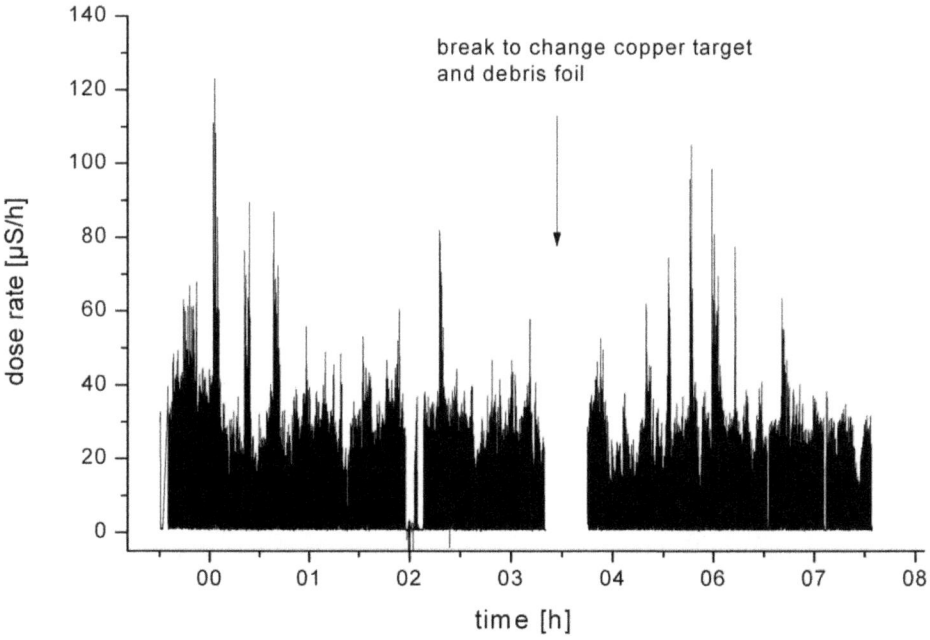

Figure 3 *Dose rate measured inside the X-ray shielding close to the plasma source.*

To be able to measure small changes below a few percent of the diffracted intensity from a sample in pump–probe experiments, normalization is necessary, requiring knowledge of the incident X-ray flux which was measured using a photodiode and the X-ray CCD camera. With the CCD two arrangements were used; the first utilized a second (germanium) crystal in Bragg reflection geometry to detect the diffracted X-rays. In the second, when only part of the diffracting region on the probe crystal was perturbed by the pump beam, the unperturbed part of the reflected signal was used for normalization.

Results from the two CCD arrangements are compared in Figure 4; the probe crystal was quartz in 001 orientation. The symmetric Laue diffraction of the 040 reflection was used for transient measurements. The diffracted intensity of the pumped crystal area was normalized both by the diffracted intensity from a germanium crystal and by diffraction of the unpumped area of the quartz crystal. In pump-probe experiments alternation of pumped and unpumped exposures is necessary in order to detect non-reversible changes of the sample. The unpumped exposures were normalized by both methods which should be consistent if the normalization is working properly. It turned out that the use of the probe crystal for normalization was the more accurate method (see figure 4). The ±2% fluctuation of the signal was comparable to the photon statistics, compared to ±6% when the germanium crystal is used.

4 ACOUSTIC PHONONS IN GERMANIUM

The application of laser produced, sub-picosecond, kilo-electronvolt X-ray sources to study time-resolved processes in solids has been successfully demonstrated in the last few years.[5–7,11–14] The initial experiments revealed fundamental phenomena arising from transient atomic motion, i.e. ultrafast excitation of the crystal lattice, non-thermal melting of semiconductors and acoustical and optical phonons. In these experiments a monochromatic X-ray beam, tuned to a K_α line, was used as a probe to study the structural rearrangement of crystals on an atomic scale by time-resolved X-ray diffraction.

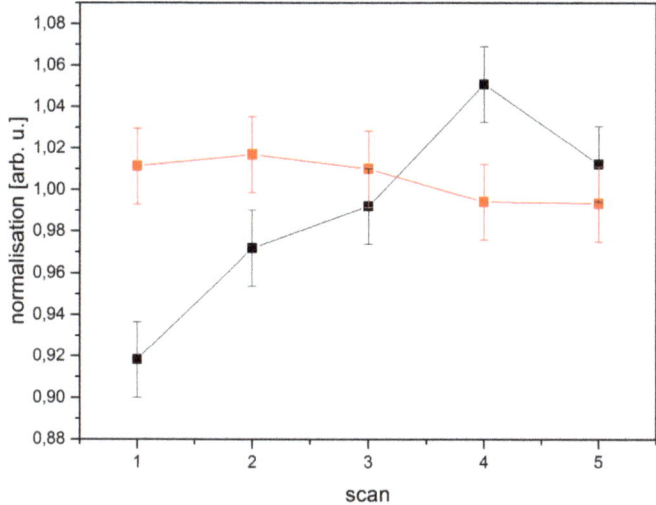

Figure 4 *Comparison between the two methods of normalization of the x-ray diffracted signal (the sample was not excited). Red: normalization by using diffraction of the unperturbed area of the same sample crystal. Black: normalization by the diffracted intensity of an additional germanium crystal.*

The experimental setup is shown in figure 5. The femtosecond laser pulse was split with an energy ratio of 90:10 into two beams as needed for pump-probe measurements. The time delay could be well defined and varied by a translation stage in the excitation arm. The first beam was sent to a vacuum chamber and focused by an off-axis parabolic mirror to provide intensities over 10^{17} W/cm^2 on a titanium tape target. The absorption of the laser pulse on the surface leads to the formation of a hot plasma and accelerated high-energy electrons are slowed down by inelastic collisions in the solid. The isotropically emitted K_α radiation was focused by a toroidally bent gallium arsenide (004) crystal[17] onto the germanium sample. The time resolution of the optical pump/X-ray probe experiment was limited by the duration of the X-ray pulse of < 500 fs.[4,18] The X-ray signal diffracted by the crystal was accumulated by a deep depletion X-ray CCD camera.[16] The second, weaker, femtosecond laser pulse was focused to the same area of the sample. The energy density of the exciting pump beam on the sample could be varied using a tuneable attenuator.

Figure 6 (left) shows the undisturbed rocking curve of the germanium 400 reflection, recorded over 200 s, and the rocking curve of the same crystal 100 ps after laser excitation at a flux of 3 mJ/cm^2.

A large lattice deformation due to an expansive strain wave is indicated by additional contributions to the rocking curve at small diffraction angles. With increasing temporal delay between pump and probe pulses, the expansion of the probed region relaxes towards the initial lattice state. The temporal response of the rocking curve at different delays after laser excitation is shown in figure 6 (right).

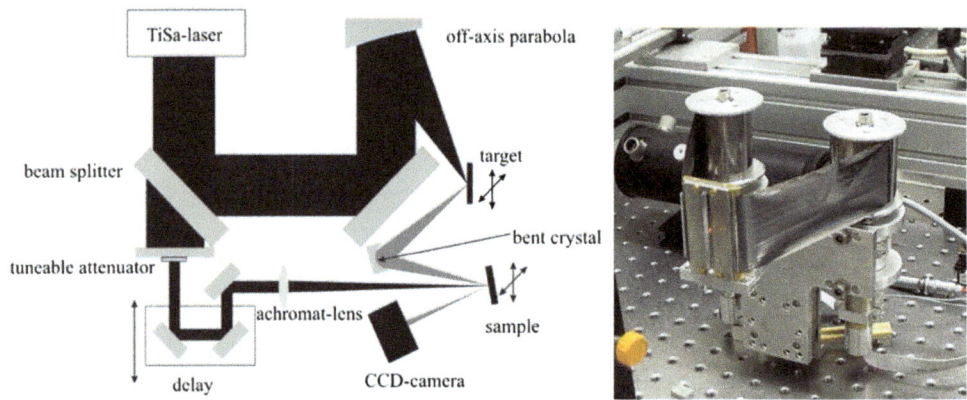

Figure 5 *Left: Experimental setup for laser pump and X-ray probe measurements. Right: Titanium tape target for the plasma source used in time-resolved X-ray diffraction measurements.*

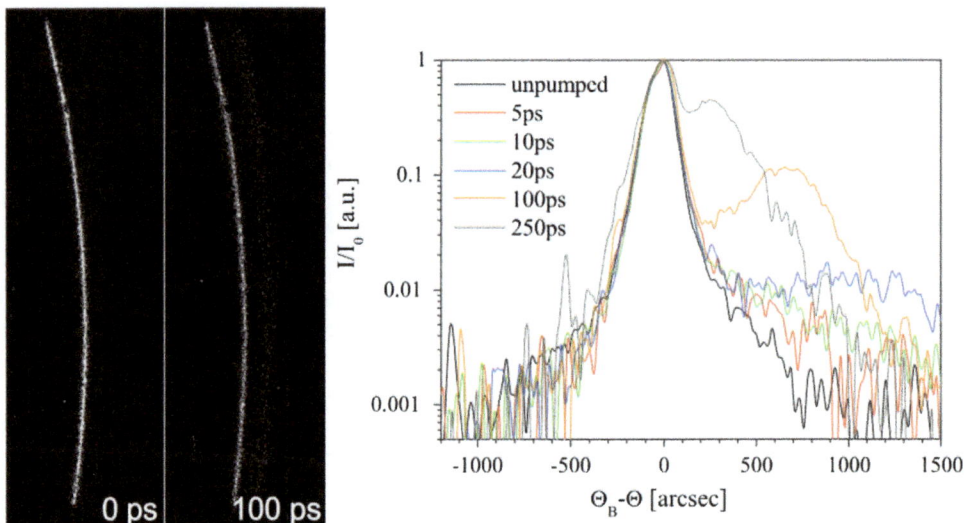

Figure 6 *Time resolved X-ray diffraction of laser excited germanium: fluence 3 mJ/cm², 400 reflection, titanium K_α (4.5 keV), with the 1 kHz x-ray source. Left: CCD images of the recorded Kossel cone (integration time 200 s) at fixed delay between pump and probe pulse (0 ps and 100ps). Right: Measured temporal response of the rocking curve at different delays after ultrashort laser excitation.*

Absorption of the infrared laser pulse drives an acoustic strain wave that propagates from the excited region into the crystal. A range of models is available to describe the emerging strain pulse but there are several steps needed to calculate the spatio-temporal properties of the strain wave and to finally understand the coupling between excited electrons and the energy relaxation into the lattice.[9]

To compare the time-resolved X-ray diffraction data with theoretical results, the physical system was simulated. A detailed description of the formalism, which includes carrier excitation by absorption (single or multiple photons), thermalization of the carrier distribution by carrier–carrier scattering and by radiative and non-radiative recombination, carrier diffusion and finally energy transfer by carrier–phonon scattering, has been presented previously.[9]

The spatio-temporal strain profile was calculated from the thermoelastic equations, and the final diffraction signal was calculated using the Takagi–Taupin theory.[19,20,21] The results of the simulations are shown in figure 7. Comparison of the measured and simulated rocking curves shows good agreement between the experimental findings and the simulation of the lattice strain.

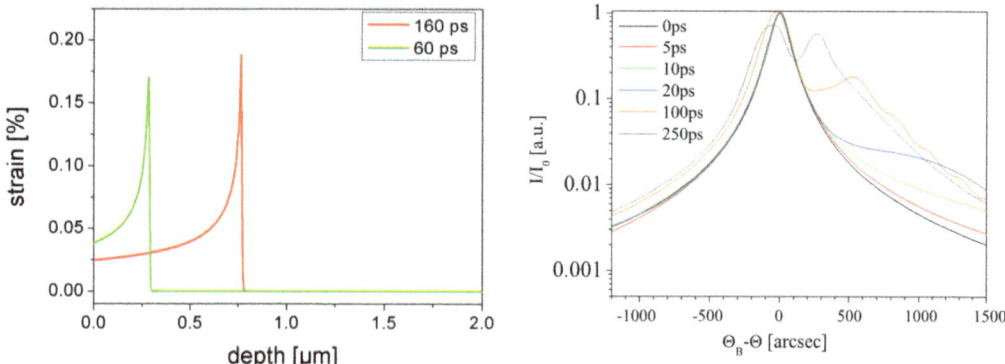

Figure 7 *Simulation of the acoustic strain wave in germanium that propagates from the surface into the bulk after femtosecond laser excitation. Left: Strain profiles of the germanium lattice for different delays after laser excitation. Right: Calculated temporal response of the rocking curve based on a simulation with a microphysical model[9] (400 reflection, titanium K_α radiation, fluence 10 mJ/cm^2).*

5 CONCLUSIONS

Progress in various scientific fields, i.e. laser development, X-ray optics and laser-plasma studies has been utilised to establish ultrafast X-ray techniques and to extend the use of the method to new applications. Diffraction techniques, well established with X-ray tubes and synchrotrons, can be combined with new X-ray sources based on laboratory high-temperature plasmas driven by high-power table top lasers. By using suitable X-ray optics and high efficiency semiconductor detectors, transient diffraction patterns can now be recorded which have exposure times comparable to those of conventional sources. However, fundamental parameters of this technique such as average X-ray output, X-ray photon emissivity from the point-like source and photon density on the sample could still be improved. Even so, it has been shown that the technique of laser pump X-ray probe experiments can yield new insights into fast electronic processes in semiconductors.

6 ACKNOWLEDGEMENT

The authors acknowledge the support of the technical staff in the X-Ray Optics group at Institute for Optics and Quantum Electronics. The work was supported by Deutsche Forschungsgemeinschaft and Bundesministerium für Bildung und Forschung.

References

1 I. Uschmann, T. Kämpfer, F. Zamponi, A. Lübcke, U. Zastrau, R. Loetzsch, S. Höfer, and E. Förster, *App. Phy. A*, 2009, **96**, 91.

2 A. Cavalleri, C. Tóth, C.W. Siders and J.A. Squier, *Phys. Rev. Lett.*, 2001, **87**, 237401,

3 E. Collet, M.H. Lemée-Cailleau, M. Burlon-Le Cointe, H. Cailleau, M. Wulff, T. Luty, S.Y. Koshihara, M. Meyer, L. Toupet, P. Rabiller and S. Techert, *Science*, **300**.

4 T. Feurer, A. Morak, I. Uschmann, Ch. Ziener, H. Schwoerer, C. Reich, P. Gibbon, E. Förster and R. Sauerbrey, *Phys. Rev. E*, 2001, **65**, 016412.

5 C. Rischel, A. Rousse, I. Uschmann, P.A. Albouy, J.P. Geindre, P. Audebert, J.C. Gauthier, E. Förster, J.L. Martin and A. Antonetti, *Nature*, **390**, 490.

6 A. Rousse, C. Rischel, S. Fourmaux, I. Uschmann, S. Sebban, G. Grillon, P. Balcou, E. Förster, J-P. Geindre, P. Audebert, J.C. Gauthier and D. Hulin, *Nature*, 2001, **410**, 65.

7 C.W. Siders, A. Cavalleri, K. Sokolowski-Tinten, C. Tóth, T. Guo, M. Kammler, M. Horn von Hoegen, K.R. Wilson, D. von der Linde, C.P.J. Barty, *Science*, 1999, **286**, 1340.

8 K. Sokolowski-Tinten, C. Blome, C. Dietrich, A. Tarasevitch, M. Horn von Hoegen and D. von der Linde, *Phys. Rev. Lett.*, 2001, **87**, 225701.

9 A. Morak, T. Kämpfer, I. Uschmann, A. Lübcke, E. Förster, R. Sauerbrey, *Physica Status Solidi B*, 2006, **243**, 2728.

10 E. Förster, A. Morak, T. Kämpfer, I. Uschmann, O. Wehrhan and F. Zamponi, *Proc. SPIE*, 2005, **5958**, 59581Z.

11 M. Bargheer, N. Zhavoronkov, Y. Gritsai, J.C. Woo, D.S. Kim, M. Woerner, T. Elsaesser, *Science*, 2004, **306**, 1771.

12 C. von Korff Schmising, M. Bargheer, M. Kiel, N. Zhavoronkov, M. Woerner, T. Elsaesser, I. Vrejoiu, D. Hesse and M. Alexe, *Phys. Rev. B*, 2006, **73**, 212202.

13 K. Sokolowski-Tinten, C. Blome, J. Blums, A. Cavalleri, C. Dietrich, A. Tarasevitch, I. Uschmann, E. Förster, M. Kammler, M. Horn von Hoegen and D. von der Linde, *Nature*, 2003, **422**, 287.

14 M. Braun, C. von Korff Schmising, M. Kiel, N. Zhavoronkov, M. Bargheer, T. Elsaesser, C. Root, T.E. Schrader, P. Gilch, W. Zinth and M. Woerner, *Phys. Rev. Lett,* 2007, **98**, 248301.

15 S. Bastiani, S. Bastiani, A. Rousse, J.P. Geindre, P. Audebert, C. Quoix, G. Hamoniaux, A. Antonetti and J-C. Gauthier, *Phys. Rev. E.*, 1997, **56**, 7179.

16 F. Zamponi, T. Kämpfer, A. Morak, I. Uschmann and E. Förster, *Rev. Sci. Instrum.*, 2005, **76**, 116101..

17 T. Missalla, I. Uschmann and E. Förster, *Rev. Sci. Instrum.*, 1999, **70**, 1288.

18 C. Reich, P. Gibbon, I. Uschmann and E. Förster, *Phys. Rev. Lett.*, 2000, **48**, 4846.

19 S. Takagi, *Acta Crystallographica*, 1962, **15**, 1311.

20 S. Takagi, *J. Phys. Soc. Japan*, 1969, **26**: 1239.

21 D. Taupin, *Bull. Soc. Fr. Minér. Crist.*, 1964, **87**, 69.

22 E. Fill, J. Bayerl, and R. Tommasini, *Rev. Sci. Instrum.*, 2002, **73**, 2190.

APPLICATION OF FOCUSED X-RAY BEAMS IN RADIATION BIOLOGY

J. Lekki[a], J. Bielecki[a], S. Bożek[a,b], Z. Stachura[a]

[a]Institute of Nuclear Physics, Polish Academy of Sciences,, Kraków, Poland
[b]Jagiellonian University Medical College, Faculty of Pharmacokinetics and Physical Pharmacy, Kraków, Poland

1 INTRODUCTION

Radiation biology investigates the interaction of ionizing radiation with living organisms. Ionizing radiation is commonly used in cancer therapy, and radiation biology studies contribute to this technique of therapy in several ways: by explaining the nature of cancer transformation of cells, by a search for the most effective way of killing the cancerous cells and by developing methods of dosimetry in a scale starting from a single cell to the whole irradiated body organ. Dosimetry methods are needed to evaluate the necessary therapeutic doses of radiation prior to irradiation and for controlling the actual dose absorbed during the irradiation process.

Ionizing radiation can damage DNA and thence trigger transformation from healthy to cancerous cells. Radiation biology methods can be used to estimate the hazards of such transformations caused by therapy, the application of radiation sources in medical diagnostics, the everyday exposure to background radiation from natural radioactivity or cosmic rays and the exposure to higher doses during intercontinental flights, space flights, or occupational exposure of workers in the nuclear industry and medical and scientific laboratories.

The molecular basis of transformation from a normal to a malignant cell has been explained, for example, in a review by Bertram.[1] The human body is composed of about 10^{15} well organized, different cells. The cells are required to divide and differentiate in a controlled way in order to repopulate organs and tissues. Cells in some organs, such as the liver, are exchanged only a few times during the lifetime of a body. Cells in other tissues, for example the epithelial layer of intestines, must be replaced every few days or even hours. Damaged cells in these tissues must be removed in a controlled way by a mechanism of programmed death, apoptosis. A balance between the death and birth of cells is controlled by a network of molecular mechanisms which govern cell proliferation and apoptosis. Any cell which escapes this control and is still proliferating can change this balance, which is manifested clinically as a neoplasm, the most malignant form of which is cancer.

Cell functions are regulated by proteins which are produced in cells according to prescriptions coded in the DNA chain. A change in the coding part of the DNA structure within a cell that is still able to replicate is the starting point for the creation of a colony of mutant cells with reduced or over expression of some proteins. This changes the biological properties of the cells. However, a single mutation in a cell is usually not sufficient to cause a cancerous transformation. Several consecutive mutations in the same cell are needed before the organism will lose control over the mutant colony leading to uncontrolled growth of the colony and the formation of a neoplasm. According to Bertram[1], in the genesis of a cancer the following changes in the cell properties should occur:

- development of independence in growth stimulatory signals;
- development of a refractory state to growth inhibitory signals;
- development of resistance to apoptosis;
- development of an infinite proliferative capacity, i.e., overcoming cellular senescence; and
- development of angiogenic potential, i.e., the capacity to form new blood vessels and capillaries.

Even in a normal, healthy, environment DNA damage is a very frequent event. DNA is a very large molecule and its chemical bonds suffer damage due to spontaneous thermal fluctuations or as a consequence of chemical attack by reactive molecules, such as free radicals or singlet oxygen and other reactive oxygen species (ROS). Free radicals and ROS are produced during normal metabolism in a body or are introduced by exposure to external agents, from which the major contributions are tobacco products.

Fortunately, the human body is well prepared for elimination of potentially dangerous cells with damaged DNA. In the development of the cycle of a cell, before the division phase (mitosis) is reached, it must pass multiple preliminary phases during which appropriate enzymes control its readiness to step into the next phase. Each phase is finished with a checkpoint when both the cell development and the existence of external signals, stimulating or prohibiting further development, are checked. If DNA damage is found, the enzymes start to repair it using the undamaged part of the second strand of the DNA helix as a template. In the case of multiple damage, when the corresponding fragments in both strands of the helix are damaged, the enzymes usually try to find as a template a similar sequence further in the DNA chain. Such processes often introduce erroneous end joining and may lead to genomic instability. Cells with changed parts of their DNA structure are usually recognized in the body as intruders and are eliminated by macrophages or other phagocytes.

If DNA repair failure is sensed in the cell then the secondary response pathways are activated. Either the cell cycle is stopped at the checkpoint and its further growth is prohibited or a programmed cell death (apoptosis) is initiated. The apoptotic cell starts to produce enzymes destroying the cell from inside and remnants of the cell are scavenged by phagocytes. Eventually, from a very large number of cells with damaged DNA, only a few cells per day remain mutated and are still able to proliferate.

One of the external factors leading to DNA damage in cells is ionizing radiation from occupational hazards, medical treatment and background sources. While biological hazards, and especially risk of cancer from high radiation doses as delivered during nuclear accidents or radiotherapy is fairly well understood, the risks from low doses due to background radiation such as cosmic rays and natural radioactivity is still not well established. Linear extrapolation of the risk versus dose dependence can be wrong at low doses because it can be affected considerably by biological pathways triggered by signal exchange between cells in the organisms. DNA damage in cells caused by background radiation is a relatively rare occurrence. Only a few hundreds cells per day in a body are influenced by this radiation, which has to be compared with $\sim10^8$ of cells per day with chemical damage to DNA due to normal body metabolism. Two questions arise. First, is background ionizing radiation still an important factor for estimating the risk of cancer disease? Second, how can cells with DNA damaged by low doses of ionizing radiation be distinguished from the many cells damaged by other, much more frequent, causes?

The answer to the first question is yes, it is. Chemical damage due to metabolism usually results in the breaking of a single bond in one of the two strands of the DNA chain (single strand break, SSB), damage that can be easily repaired by the cell as there is a second, undamaged, strand as a template. In contrast, ionizing radiation usually creates many closely spaced breaks of the chemical bonds in both DNA strands (double strand break, DSB) which are often erroneously repaired or are recognized by the cell as un-repairable.

There are many ways by which ionizing radiation damages DNA strands:[2]
- ionization and excitation by primary ions or electromagnetic quanta;
- for X-rays and gamma rays further ionization/excitation by photoelectrons and, at higher energies, by Compton scattered electrons;
- excitation by delta electrons (that is, electrons ionized form atoms via interactions with charged particles); and
- dissociation of water molecules, leading to the generation of free radicals (reactive oxygen species, ROS) that chemically damage the DNA chain.

Multiple breaks of bonds in the DNA strands are created along the ionizing path in a narrow channel with diameter comparable to the DNA helix width. Because of the large probability of multiple DSB creation along the ionizing path, biological consequences of even very low doses, such as caused by background radiation, cannot be neglected.

In addition to the large background of chemical damage created by metabolism in the body, in investigations of radiation damage at the cellular level there are also other difficulties, such as the problem of correctly determining the radiation dose (defined as the ratio of deposited energy to the object mass). The interaction of ionizing radiation with biological tissue is well localized along the path of the ion, electron or X-ray. During irradiation of the tissue by a broad beam, most cells remain intact, while those hit are given a radiation dose many times larger than the average for the whole irradiated sample. A solution to this problem, especially in experiments performed *in-vitro* with cell cultures, is the use of radiation microprobes.

Electron and ion microprobes for radiation biology investigations have been used for many years in numerous laboratories. Two groups are pioneers in this field: the Gray Cancer Institute in London[3] (now relocated to Oxford) and the Center for Radiological Research at Columbia University.[4] Many European groups with ion microprobes dedicated to radiation biology investigations have united their efforts to develop methods of single particle irradiation of individual cells by creating a Marie Curie Research and Training Network, CELLION.[5]

Mechanisms of DNA damage by ionizing radiation of different types are similar, but the biological consequences of irradiation, even at comparable doses, may differ significantly for light or heavy ions and for X-ray or gamma-ray beams. This is because double strand breaks of DNA are the most difficult to repair, and the probability of such damage depends on the density of ionization events along the beam and this is very different for ions and X-rays. This is the rationale for extending radiation biology investigations with ion microprobes to X-ray microprobes. The latter are not so common because of the difficulty of focusing of X-ray beams. In the following section some methods of X-ray focusing will be discussed and examples of existing microprobes will be described.

2 METHODS OF REDUCING THE X-RAY SPOT SIZE

Irradiation of single cells with X-rays requires beam shaping to a spot comparable to the dimension of a cell. To enable good interpretation of results and easy evaluation of the irradiation dose either a monochromatic beam or one with small energy range ("pink" beam) should be used. X-ray beams from modern synchrotrons, very intense even after passing monochromating, can be shaped to small sizes without focusing, using only collimation by ultra-precise slits. Such collimation was used, for example, at the Tsukuba Photon Factory,[6] where a monochromatic 5.35 keV beam was collimated down to a size of less than 10 μm (FWHM) and provided an irradiation dose rate of 0.3 Gy/s, corresponding to a collimated beam intensity of about 10^4 photons per second. The Tsukuba beam line used for radiobiology experiments is able to collimate X-ray beams of energy 4–20 keV to sizes of about 5 μm.

Collimation instead of focusing has also been used in a laboratory source X-ray experiments.[7] An electron beam bombarded a carbon target, inducing emission of C K_α X-rays (277 eV). The emitted X-rays were screened by a metal mesh filter with openings of 2.2–3.5 μm. The shaped X-ray beam striking the sample was thus composed of multiple microbeams of size equal to that of the mesh filter windows.

A third collimated X-ray microprobe, designed for hard X-rays, is working at the Department of Nuclear Engineering, Seoul National University.[8] An X-ray tube with acceleration voltage up to 450 kV emits radiation from a spot of 2.5×5.5 mm^2. The Bremsstrahlung with energy below 20 keV is attenuated by a 3 mm thick aluminium plate. High energy X-rays are collimated to a beam of 150 μm diameter and the dose rate delivered to the sample may be adjusted in the range 0.001–3 Gy/s.

Other single cell irradiation systems use focused X-ray beams. The focusing method is usually chosen depending on the energy range of the radiation. The most suitable method to focus soft X-rays ($E < 5$ keV) is to use Fresnel zone plates (FZPs)[9], which, in their simplest form, have alternating opaque and transparent concentric rings with radially decreasing widths Δr according to

$$\Delta r = r_n - r_{n-1} \approx \sqrt{nf\lambda} - \sqrt{(n-1)f\lambda}, \tag{1}$$

where r_n is the radius of the nth zone, λ is the irradiation wavelength. and f is the first order focal length given by

$$f = \frac{Dd}{\lambda}; \tag{2}$$

D is the zone plate diameter and d is the width of the outermost zone ($r_N - r_{N-1}$), where N is the total number of zones). The focal length is determined by constructive interference of rays diffracted by the zones, and, as can be seen from equation (2), increases with decreasing wavelength (increasing energy). If a polychromatic beam is used, the FZP must be used with a system of apertures to remove radiation of undesirable energies; the slits apertures also eliminate higher order diffraction orders of the FZP. A central opaque circular stop eliminates undiffracted radiation close to the optical axis of the focusing system. An alternative type of zone plate shifts the phase of the transmitted radiation in alternate zones to increase the diffraction efficiency; in the absence of absorption the phase shift would be π rad but since all materials absorb X-rays the.optimum value is material dependent.[10] Both types of zone plate are produced by lithographic methods.[11]

Several groups have reported the use of FZP focused microbeams for radiation biology experiments, including

- The Gray Cancer Institute (GCI), UK.[12] This was the first X-ray microprobe used to irradiate individual cells. The microprobe consisted of an electron gun with electomagnetic focusing. The electron beam was focused onto a carbon target and the emitted X-rays were reflected from a flat silica mirror to reduce the Bremsstrahlung component; the resultant beam, mainly characteristic carbon K X-rays (277 eV), was focused with a zone plate, via a silicon nitride vacuum window, onto cells supported on thin Mylar foils on a moveable stage. The cell positions were determined prior to irradiation using an ultraviolet assay microscope. The measured focal spot size was below 1 µm and the achieved dose rate was about 1 Gy/s, corresponding to about 10^4 C K X-ray photons/s.

- Queen's University Belfast (QUB), UK.[13,14] Following the work at GCI, a similar microprobe was constructed, delivering a focused beam of either carbon K, aluminium K_α (1.49 keV) or titanium K_α (4.51 keV) X-rays. A similar size focal spot size was achieved.

- Nagasaki University, Japan.[15] An essentially identical source to that at QUB was used with carbon K radiation, with similar results.

- Radiological Research Accelerator Facility (RARAF) at Columbia University, USA.[16,17] Proton beams generated using 5.5 MV Singletron accelerator were focused on a titanium target. Titanium K_α X-rays induced by the proton impact, were focused using a FZP with a diameter of 120 µm and outer zone width of 50 nm, corresponding to a focal length of 21.7 mm. An X-ray focal spot of diameter 5 µm was achieved and the delivered dose rate was up to 10 mGy/s. The use of protons instead of electrons to induce X-rays reduced the continuum Bremsstrahlung X-rays to a negligible level.

- Central Research Institute of Electric Power Industry (CRIEPI), Tokyo.[18] A microfocus X-ray tube providing aluminium K_α radiation was used.; the Bremsstrahlung component was removed by reflection from a flat mirror. The aluminium X-rays were focused using a FZP with outer diameter 150 µm, giving a focal spot of diameter about 1.8 µm and a dose rate to cells of about 1 Gy/s.

Zone plates for hard X-rays must have either very long focal lengths or very small diameters — which limits the aperture and hence the focused intensity, or very small outer zone widths. The use of zone plates produced by lithography to provide small focal spots is thus limited to soft X-rays as it is not possible to obtain the high aspect ratios (zone thickness divided by zone width) needed for optimum diffraction efficiency at high energies. Zone plates with narrower outer rings and very high aspect ratios can be produced by sputtering alternating layers on a thin rotating wire and then cutting a disk of the required thickness.[19] Such zone plates can be used to focus X-rays with energies as high as 100 keV [Kamijo-2001].[20] However, single cells are nearly transparent to radiation of such high energy so that the absorbed dose in a single cell is negligible. Therefore, this and other methods of focusing suitable for very hard X-rays are used rather to irradiate tissue samples and not single cells.An X-ray energy of about 5 keV is the limit at which zone plates are the best choice for focusing. At higher energies two other focusing methods are more appropriate, tapered capillaries or elliptical mirrors.

The phase velocities of X-rays in matter are normally slightly larger than those in vacuum; the refractive index is

$$n = 1 - \delta + i\beta, \tag{3}$$

where δ is small and positive. X-rays passing through vacuum or air can be reflected from a mirror surface with small loss of intensity if the incidence angle is lower than the critical angle $\theta_C = \sqrt{2\delta}$. This means that hollow capillaries can be used as light guides with high transmission if the grazing incidence angle is smaller than θ_C; tapered capillaries can thus be used for focusing. Capillaries are made usually from glass, or recently from higher Z materials, and can be used as light guides,[21] beam splitters,[22] beam concentrators[23] and focusing elements.[24] Focal spots as small as 1 μm can be achieved.[25] The critical angle depends on the mirror material and on the X-ray energy; for boron silicate glass $\theta_C \approx 0.032/E$ with θ_C in radians and E, the X-ray energy, in kiloelectronvolts. The transmission and focusing properties of mono- and poly-capillaries (ensembles of capillaries with a common focal point) have been discussed previously.[26,27] Capillaries have no energy selection except the condition that the grazing angle must be smaller (or, at least, not much larger) than the energy-dependent critical angle. Therefore, if a monochromatic beam is required, the beam path must include a monochromator.

Three microprobes with capillary focusing have been reported in radiobiology experiments:

- Japan Atomic Energy Agency, Advanced Photon Research Centre, Kyoto.[28] A titanium sapphire laser beam with 790 nm,70 fs, 150 mJ pulses at a repetition rate of 10 Hz was focused onto a thin copper foil, producing a plasma. Electrons generated from the interaction of the laser with the plasma induced Bremsstrahlung and characteristic Cu K_α (8.05 keV) X-rays. The emitted radiation was focused with a polycapillary lens to a spot of 400 μm (FWHM) with a dose rate delivered to cells of 0.1 mGy/pulse (1 mGy/s).

- Osaka University.[29] The beam from a tabletop microfocus X-ray tube with a rhodium target was focused with a glass capillary to a spot of about 10 μm diameter, 11 mm from the exit of the capillary. The strong Rh L peak (2.7 keV) was absorbed in the intervening medium. The maximum ionization in the targeted cells was due to Bremsstrahlung radiation in the energy range 4–12 keV. The delivered dose rate was estimated to be about 0.05 Gy/s.

- A new high energy X-ray microprobe is under construction at Queen's University Belfast.[30] The instrument consists of an ultra-bright X-ray tube (80 W, 60 kV) with a copper target and an emission spot diameter of 40 μm, and either a polycapillary giving a focal spot of 15 μm, or a single glass tapered capillary with an expected spot size of 0.1 μm. Other targets, such as molybdenum, silver, tungsten and alloys can also be used.

Fine focusing of X-ray beams can also be performed with curved mirrors, with focal spots of some tens of nanometres reported.[31] The mirrors can be arranged in different geometries, most popular being the Kirkpatrick-Baez (KB) system with two crossed mirrors. Each mirror is flat in one direction and elliptical in the perpendicular direction. If the X-ray source is placed at one focus of the ellipse, the emitted rays converge at the other focus. This produces a line focus, but when reflected again by the second, perpendicular, mirror, a point-like focus is formed (figure 1). For high reflection efficiency, grazing incidence is required, with angles of incidence at most close to the critical angle. The mirrors are not energy selective and usually are preceded with a monochromator.

In some implementations the mirror surfaces of mirrors are coated with alternating thin layers, with a period d, differing in their refractive indices – multilayer mirrors. Due to constructive interference a high reflection efficiency occurs at the angle θ satisfies the Bragg law, $n\lambda = 2d\sin\theta$, where n is an integer. The highest reflection efficiency is for $n = 1$ (first order) and can reach 90% for a multilayer consisting of 100 or more layer pairs. The layer spacing is chosen so that the angle of incidence of the beam at the elliptical surface is also the Bragg angle for the desired wavelength. The angle of incidence changes along the surface and so the layer spacing modified be correspondingly.

Mirrors are usually made from small parts the total elliptical surface. If the central part, equidistant from both focal points of the ellipsoide, is used, the focal spot is a 1:1

image of the source. If other parts are used, magnified or de-magnified images of the source are formed at the focal position.

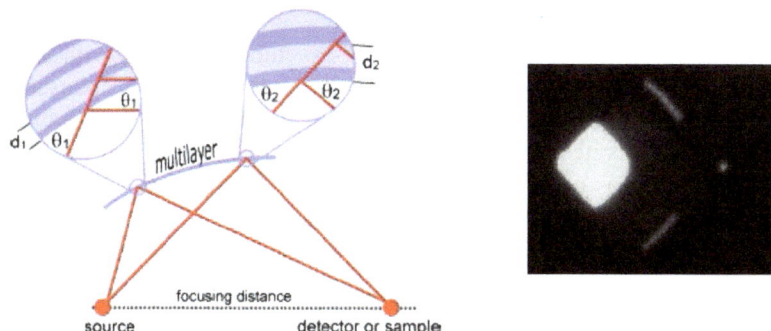

Figure 1 *The operation principle of a multilayer elliptical mirror (left) and the resulting X-ray image (right). The bright rectangular spot is the direct beam, the two lines are formed by reflection from one of the mirrors only and the rightmost spot is due to double reflection from perpendicular mirrors.*

There are three X-ray microprobes using mirrors in the KB geometry and devoted to radiobiology.

- Institute of Nuclear Physics, Polish Academy of Sciences, Kraków.[32] The beam from a microfocus X-ray tube was focused with a pair of multilayer elliptical mirrors, in KB geometry with layer spacings was designed for the Ti K_α energy of 4.51 keV. Due to the very short focal distance of 15 mm a large collection solid angle was achieved. The best resolution of the microprobe was 8 μm (FWHM) and the dose rate delivered to irradiated cell nuclei was about 0.5 Gy/s. However, the microprobe is usually used at the highest focal spot intensity, which is achieved for a beam size of about 20 μm, still comparable to the size of a single cell. The dose rate delivered to whole cells at this resolution is about 0.7 Gy/s.
- SPring-8 synchrotron beam line BL29XUL, Hyogo.[31] The synchrotron radiation beam was focused with achromatic mirrors in the KB geometry. X-rays in the energy range 4.4–19 keV were focused to a spot with size in the range 29×48–2000×2000 nm^2 with corresponding photon fluxes at focus of 6×10^8–8×10^{11}/s. A beam of 15 keV was used for elemental mapping of single cells.
- XFM beam line of the Australian Synchrotron, Melbourne.[33] The beam, with energy in the range of 4.1–25 keV and a photon flux of 1.5×10^{10}/s/μm^2 is focused with KB mirrors to a spot of 5 μm^2. An X-ray beam designed for radiobiology experiments is currently under construction.

3 THE KRAKÓW X-RAY MICROBEAM FACILITY

X-ray microprobes used for radiobiological studies for the investigation of cellular damage must satisfy two main criteria: precise control of the dose delivered to irradiated cells and the spatial location of the focused X-ray beam within the cell. In order to fulfill both requirements technologically advanced components must be used in radiation generation, the X-ray focusing process and sample positioning. In this section the X-ray microbeam facility at the Institute of Nuclear Physics, Polish Academy of Sciences. in Kraków is presented.

3.1 The X-ray Source

The facility is based on a Hamamatsu L9191 X-ray tube, a source with exchangeable transmission targets. In a microfocus X-ray tube the electron beam is focused to a small spot at the anode (target), emitting radiation composed of continuum radiation (Bremsstrahlung) and discrete characteristic lines. For radiobiology experiments an energy of 4.5 keV (Ti K_α line) was used, chosen because experiments to irradiate tissue samples, as well as individual cells, are foreseen – the 1/e range of 4.5 keV X-rays in tissue is 145 ˙μm. The construction of the source enables experiments at different energies since the accelerating voltage can be in the range 20–160 kV, with the maximum tube current of 200 μA. The intensity ratio of Ti K_α line radiation to Bremsstrahlung is maximum at an accelerating voltage of 40 kV, and so. this voltage was normally used in routine experiments.

For single cell irradiation experiments, the crucial parameter is the focal spot size, which was measured using a test object (JIMA RT RC–02 micro-chart, consisting of several tungsten slit patterns of different width, manufactured using ultra-fine semiconductor lithography). A minimum spot size of about 2 μm was achieved for a voltage of 100 kV; at lower voltages the spot size increases,[32] as shown in figure 2.

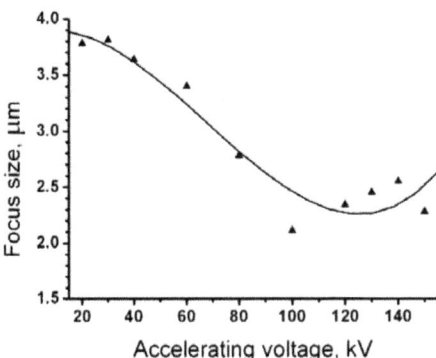

Figure 2 *Emission spot size of the Hamamatsu L9191 X-ray source as a function of acceleration voltage.*

Radiation was emitted into a cone with an angular spread $\Delta\varphi = 120°$. The focal spot was located 0.5 mm from a beryllium output window, enabling the use of optical elements with short focal distances. The vacuum system included a roughing pump to evacuate the X-ray tube and a turbo-molecular pump to provide the low pressure for optimum X-ray tube operation. For radiation protection, the whole microprobe facility is enclosed inside a container ($1 \times 1 \times 2$ m³) with walls covered with 2 mm thick Pb/Cu layer and a large front entrance window made of 1 cm thick lead glass.

3.2 X-ray Focusing

Since X-ray tubes emit divergent beams of radiation with intensities many orders of magnitude lower than synchrotron radiation, it is necessary to use efficient focusing optics. From the available methods of focusing, described in section 2, elliptical multilayer mirrors (Rigaku, USA), operating in the KB geometry, were chosen, due to their durability and higher focused intensity compared to zone plates. The source to focus distance was 30 mm and the magnification was unity. The reflecting surfaces were coated with 80 Cr/C bilayers, and the thickness of a single bilayer in the central part of the mirror was 3.1 nm. The layer thicknesses varied across the mirrors so that the incidence angles were equal to the Bragg angle for an energy of 4.5 keV, hence also acting as a monochromator. Figure 3 shows the emission spectrum and that after reflection from the multilayer mirror system.

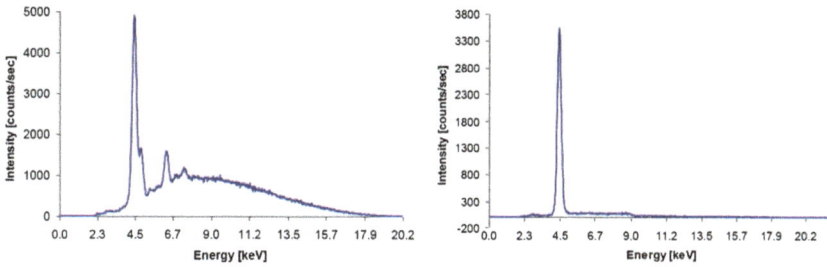

Figure 3 *The spectrum emitted by the X-ray tube (left) b) and that after reflection from the KB multilayer system (right).*[32]

To protect the mirrors from radicals created by irradiation of water vapour in air they were enclosed in a helium chamber sealed by thin beryllium windows. The positions and angles of the mirrors inside the chamber could be adjusted with piezo actuators, and the overall position of the optics was controlled remotely using stepper motors.

The size of focal spot was measured using a knife edge and measuring the intensity with an X-ray sensitive CCD camera, then fitting it to a Gaussian distribution. The FWHM size of the focal spot should be comparable to the size of the source since the magnification was unity, but the smallest focus size achieved so far was about 8 μm. This higher than expected value was probably caused by shape imperfections and deterioration of the mirror surfaces. Typically, focused beams with FWHM 20 μm and fluxes of about 1.75×10^5 photons/second were used during experiments.

3.3 Sample Treatment and the Irradiation Process

The experiments were performed on cultured cell lines of human prostate cancer (PC3). Cells were seeded on 35 mm Petri dishes with 10 mm round holes covered by 1.5 μm thick Mylar foil. A population of 10^5 cells in 4 μl of culture medium was seeded onto the central part of the Mylar foil 16-18 hours prior to the experiments. Before irradiation the medium was removed and the top of the Petri dish was covered with a layer of Mylar foil to isolate the cells from environmental contamination and the positions of the cells were recorded using an optical microscope controlled with precise stepper motors. The control software used image recognition algorithms to irradiate individual cells with precise doses of X-rays. After irradiation, necrotic and apoptotic cells were observed and analysed using fluorescence microscope over a period of time ranging from minutes to tens of hours (figure 4).

Because the energy deposited in a single cell was delivered by a monochromatic focused X-ray beam and the irradiation time was strictly controlled during the experiments it is possible to calculate dose absorbed in single cell.

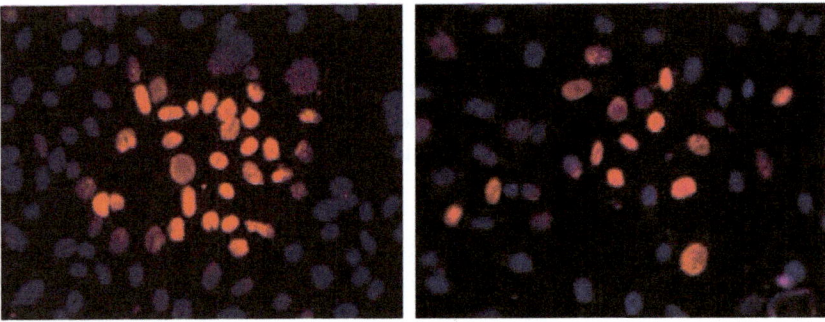

Figure 4 *Image of irradiated PC3 cells (60 Gy per cell, average cell size ~30 μm) obtained by fluorescence microscopy using γH2AX histone (DSB damage visualisation): (left) 30 minutes after irradiation and (right) 7 hours after irradiation. The righthand image shows that some irradiated cells were able to repair the damage (done in collaboration with A.Wiecheć, Institute of Nuclear Physics, IFJ, Kraków).*

3.4 Dose calculations

Currently there are no direct experimental methods to determine the dose deposited in a small volume of a single cell. An approach that can be applied is tissue equivalent proportional counters.[34] The other possibility is to use Monte Carlo (MC) calculations of radiation interactions within the cell volume and its surroundings. There are many MC codes dedicated to (micro)dosimetry, including MOCA-14,[35] NOREC,[36] Phole-2[37] and TRION.[38] GEANT4, the code developed for high energy physics,[39] has been extended with libraries for low energy physics (PENELOPE)[40] and a toolkit for simulation of interactions of radiation with biological systems at the cellular and DNA level (GEANT-4DNA).[41] This code was chosen for the dose calculations. The simulated system was a cell with shape, dimensions and composition as shown in table 1.[42] The cell dimensions were determined from atomic force microscope profiles (figure 5), and the cell was assumed to be covered by 1.5 μm thick Mylar foil.

Table 1 *Parameters of the model cell used in the dose calculations.*

	Shape	Semi-major axes [μm]	Percentage Composition				
			H	O	C	N	P
Cytoplasm	Half ellipsoid	20,15,3.7	59.6	24.24	11.11	4.04	1.01
Nucleus	Ellipsoid	5,5,1.5	10.64	74.5	9.04	3.21	2.61

The interaction with a Gaussian radiation beam of FWHM 20 μm and flux of 1.75×10^5 Ti K_α photons/s was simulated. The average energy deposition within the cell was calculated to be 5.3 ± 0.2 MeV (nucleus) and 4.2 ± 0.1 MeV (cytoplasm), a total for the whole cell of 9.5 MeV or 1.52×10^{-12} J, while that deposited in the Mylar foil was 6.2 ± 0.2 MeV. The mass of a PC3 cell was calculated to be 2.3×10^{-12} kg, assuming a cell density equal to that of water. Thus the deposited dose rate in a single cell is about 0.65 Gy/s.

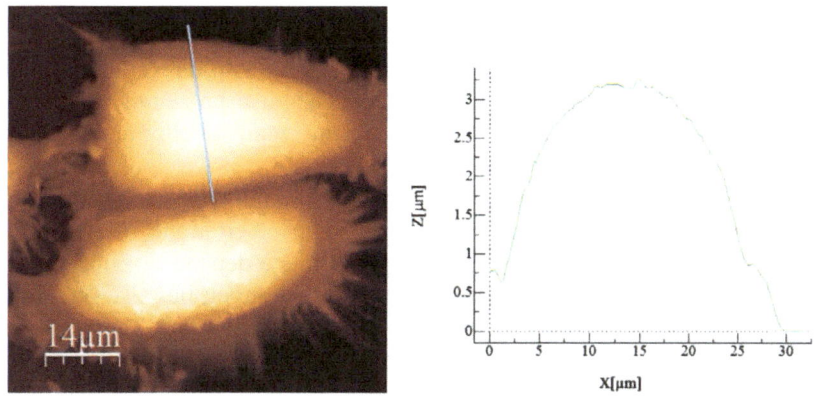

Figure 5 *Left: AFM image of a PC3 cell (courtesy of K.Pogoda, IFJ PAN). Right: cell height profile along the marked line.*

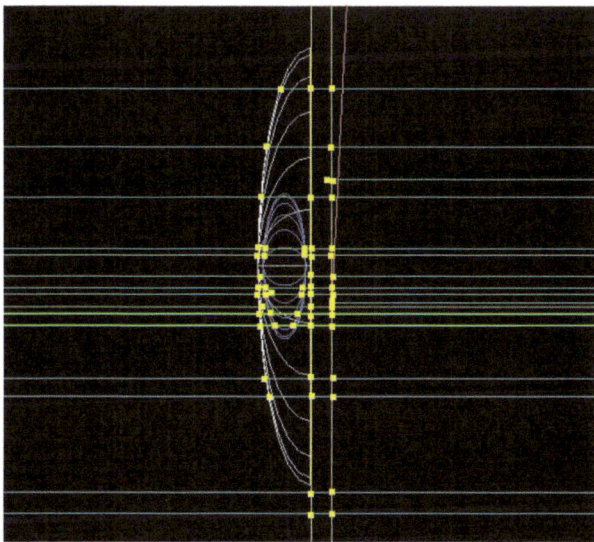

Figure 6 *Simulation of 20 events. The cell is shown in white, with the nucleus in blue, and the Mylar foil in yellow. The green lines are photon tracks and the yellow points show where ionization has occurred. The red line is a delta electron track.*

The ratio of energy deposited in the protecting Mylar foil to that deposited in the cell was determined to be 64%. No significant scattering within the foil region was observed (see figure 6), and thus deterioration of the focal spot size due to the foil can be neglected. The average path length of delta electrons produced during X-ray interactions with the cell material is 260 nm with a maximum range slightly above 0.5 µm.

4. CONCLUSIONS

Microprobes using ionizing radiation have contributed greatly to a recent rapid increase in knowledge of biological pathways and the physical/chemical processes occurring during radiation induced damage of cells and the subsequent repair or elimination of damaged cells. Initially, ion microprobes were used to investigate cell survival as a function of accumulated radiation dose, counting the numbers of apoptotic cells, irradiated cells forming micronuclei and surviving cells with chromosome aberrations.[4,43] It was very soon observed that the biological endpoints of radiation damage depend not only on radiation dose, i.e., the deposited energy divided by the mass of the target, but also on the radiation quality – especially on the density of ionization events and hence on the value of Linear Energy Transfer (LET). Experiments performed with light ions have subsequently been extended to heavier ions of different energy, electron beams, γ- and X-rays and UV radiation. The first X-ray microprobe used routinely to radiation biology operated at the Gray Cancer Institute in Northwood[44] and a review of results obtained in X-ray beam irradiation experiments was given by Hawkins in 2006.[45]

Experiments have shown that the most sensitive part of the cell to radiation damage is DNA coiled in chromatin, and for a majority of cell killings by irradiation the apoptosis process, triggered after unsuccessful attempts to repair the damage, is responsible. New methods of simulating such physical damage to cells induced by ionizing radiation were developed and used in microdosimetry calculations.[2,46] Microprobe experiments have thus contributed to molecular biology studies of the DNA damage repair processes.[47] When one or both strands of DNA are broken, a MRE11-RAD50-NBS1 (MRN) protein complex present in the cell recognizes the loose ends of the strands and within seconds relocates to them, forming nuclear foci and starting a long chain of repair processes. Within this chain, at a much slower rate of minutes to hours the Ataxia Telangiectasia-Mutated (ATM) protein attracts to the foci phosphorylates a large number of H2AX histone molecules into the γ-H2AX that binds the damaged loose ends of DNA. This successively recruits and binds numerous other catalytic proteins, among them cell cycle controlling proteins such as p53 or p21. In a series of reactions the recruited proteins within the complex help to rejoin broken ends of DNA chain and rebuild the damaged parts of it. For a DSB this process may last for hours or even days. If fluorescent antibodies are added to proteins recruited around the damaged regions, the locations of the foci can be observed using a microscope.[48] This method is commonly used in the microprobe experiments to visualize ion tracks or photoelectron tracks in the irradiated cell nuclei. Most frequently staining of the γ-H2AX complex is used,[49] but often radiation induced foci are observed by immunostaining of p21,[50,51] p53, ATM[52] and other proteins. By counting individual tracks or by measuring total fluorescence intensity emitted from the irradiated cell nucleus the biological efficiency of the irradiation can be evaluated. Time dependent development of the track numbers gives information on the kinetics of repair processes.[53,52] Observation of the movement of individual foci or their broadening in time provides information on the diffusion of chromatin in a nucleus that occurs during the repair process.[54].

Arguably the largest impact of microprobe based experiments on the development of radiobiology has been the investigation of non-targeted effects; radiation induced damage was observed in some unirradiated cells in the vicinity of irradiated ones.[55]. Chinese hamster ovary cells were irradiated with a broad beam of α-particles from a radioactive source. Although fewer than 1% of the cell nuclei were traversed by an α-particle, about 30% of the cells showed genetic damage, namely a sister chromatid exchange. Later it was reported that the culture medium from a dish where cells were exposed to radiation (but afterwards removed) has a toxic effect, killing cells when added to unirradiated cell culture.[56]

Microprobes using ionizing radiation offer a unique tool for investigation of non-targeted effects, as pre-defined cells can be irradiated with controlled doses, with other cells left unirradiated. Many such experiments have been performed but the first results were not conclusive. Under similar experimental conditions, in some experiments the effects were observed but in others results were ambiguous. Usually the effect was seen when only a small fraction of the cells was irradiated, reaching saturation when more cells were exposed to the beam. Sometimes the damage was induced in cells in physical contact with an irradiated cell; sometimes very distant cells were also affected. However, the existence of non-targeted effects is now well established, but these phenomena encompass many different biological processes and the biological pathways are mostly either not known or not well documented. A review of non-targeted effects and a discussion of experiments performed in their investigation proposed the following classification.[57]

- Bystander effects: Radiation-induced, signal-mediated effects in unirradiated cells within an irradiated volume.
- Abscopal effects: Radiation-induced effects in unirradiated tissue outside an irradiated volume.
- Cohort effects: Radiation-induced, signal-mediated effects between irradiated cells within an irradiated volume.

The nature of signals mediating the bystander effect and other non-targeted effects is not clear. As intercellular signal candidates several molecules have been proposed, such as reactive oxygen species, interleukins, tumor growth factors, membrane signalling and nitric oxide molecules. Possible biological pathways of creation, transmittance and reception of these signals have been put forward.[58] Many microprobe-based experiments on the bystander effect[57] suggest that it may be caused by different biological pathways with a common end-point, and can be mediated by different signals. The best documented are signals transmitted by the nitric oxide molecule.[59,60,61,62]

Clinical medicine urgently needs experiments delivering reliable knowledge about radiation induced non targeted effects. As explained elsewhere[58] these effects can introduce significant changes to expected radiotherapy results, either beneficial or dangerous to the patient. Additionally, if a better understanding of the bystander effect from the future investigations permits the elucidation of methods to increase or reduce the sensitivity of cells to the effect, this could radically change radiotherapy practise.

References

[1] J.S Bertram, The molecular biology of cancer, Molecular Aspects of Medicine 21 (2001) 167-223

[2] H. Nikjoo, S. Uehara, W.M. Wilson, M. Hoshi, D.T. Goodhead, Track structure in radiation biology: theory and applications, Int. J. Radiat. Biol. 73 (1998) 355-364

[3] M. Folkard, B. Vojnovic, K.M. Prise, A.G. Bowey, R.J. Locke, G. Schettino and B.D. Michael, A charged-particle microbeam: I. Development of an experimental system for targeting cells individually with counted particles, Int. J. Radiat. Biol. 72 (1997) 375

[4] C.R. Geard, D.J. Brenner, G. Randers-Pehrson and S.A. Marino, Single-particle irradiation of mammalian cells at the Radiological Research Accelerator Facility: induction of chromosomal changes, Nucl. Instr. Meth. in Phys. Res. B 54 (1991) 411

[5] Marie Curie Research and Training Network —CELLION‖, Studies of cellular response to targeted single ionsusing nanotechnology‖, EU FP7, Contract No. MRTN-CT-2003-503923

[6] Y. Tanno, K. Kobayashi, M. Tatsuka, E. Gotoh and K. Takakura, Mitotic arrest caused by an X-ray micro-beam in a single cell expressing EGFP-Aurora kinase B, Radiation Protection Dosimetry 122 (2006) 301

[7] C. van Oven, P.M. Krawczyk, J. Stap, A.M. Melo, M.H.O. Piazetta, A.L. Gobbi, H.A. van Veen, J. Verhoeven and J.A. Aten, An ultrasoft X-ray multi-microbeam irradiation system for studies of DNA damage responses by fixed- and live-cell fluorescense microscopy, Eur. Biophys. J. 38 (2009) 721

[8] E.H. Kim, K.M. Lee and S.R. Kim, Utilization of hard X-ray microbeam for low-dose study: Micron-scale dose profiling and bystander response, Program and Abstracts of the 10th International Workshop "Microbeam Probes of Cellular Radiation Response", ed. A. Bigelow, RARAF, Columbia University, March 15-17, 2012, page 27

[9] Michette AG 2010 Zone plates, chapter 40 (pp.40.1–40.11) in: McGraw Hill Handbook of Optics Vol V 3rd ed. (New York, McGraw Hill)

[10] J. Kirz, Phase zone plates for X rays and the extreme UV, J. Opt. Soc. Am., 64, 301, 1974

[10a] A.G. Michette, C.J. Buckley and S.J. Pfauntsch, Phase modulating zone plates for X rays of energy
1–8 keV, Opt. Comm. 141 (1997) 118-122

[11] C. Khan Malek, A review of microfabrication technologies: Application to X ray optics, J. X-Ray Sci. Technol. 3 (1991) 45

[12] M. Folkard, G. Schettino, B. Vojnovic, S. Gilchrist, A.G. Michette, S.J. Pfauntsch, K.M. Prise and B.D. Michael, A focused ultrasoft X-ray microbeam for targeting cells individually with submicrometer accuracy, Radiation Research 156 (2001) 796

[13] G. Schettino, M. Folkard, B. Vojnovic and K.M. Prise, The Queen's University variable-energy X-ray microbeam and its use in studying the dynamics of DNA damage and repair, J. Radiat. Res. 50 (2009) Suppl. A96

[14] G. Schettino, M. Folkard, B. Vojnovic, A. Michette and K.M. Prise, X-ray microbeams for radiobiological studies: Current status and future challenges, PIERS Online 6 (2010) 207

[15] Y. Kobayashi, T. Funayama, N. Hamada, T. Sakashita, T. Konishi, H. Imaseki, K. Yasuda, M. Hatashita, K. Takagi, S. Hatori, K. Suzuki, M. Yamauchi, S. Yamashita, M. Tomita, M. Maeda, K. Kobayashi, N. Usami and L. Wu, Microbeam irradiation facilities for radiobiology in Japan and China, J. Radiat. Res. 50 (2009) Suppl. A29 –A47

[16] A.D. Harken, G. Randers-Pehrson, G.W. Johnson and D. Brenner, The Columbia University proton-induced soft X-ray microbeam, Nucl. Instr. Meth. in Phys. Res. B 269 (2011) 1992

[17] A. Harken, B. Ponnaiya, G. Randers-Pehrson and D. Brenner, Proton-induced soft X-ray microbeam at Columbia University, Program and Abstracts of the 10th International Workshop "Microbeam Probes of Cellular Radiation Response", ed. A. Bigelow, RARAF, Columbia University, March 15-17, 2012, page 17.

[18] Y. Kobayashi, T. Funayama, N. Hamada, T. Sakashita, T. Konishi, H. Imaseki, K. Yasuda, M. Hatashita, K. Takagi, S. Hatori, K. Suzuki, M. Yamauchi, S. Yamashita, M. Tomita, M. Maeda, K. Kobayashi, N. Usami and L. Wu, Microbeam irradiation facilities for radiobiology in Japan and China, J. Radiat. Res. 50 (2009) Suppl. A29 –A47

[19] D. Rudolph, B. Niemann and G. Schmahl, Status of the sputtered sliced zone plates for X-ray microscopy, SPIE Proc. 316 (High Resolution Soft X-Ray Optics) (1981) 103-105

[20] N. Kamijo, Y. Suzuki, M. Awaji, A. Takeuchi, K. Uesugi, M. Yasumoto, S. Tamura, Y. Kohmura, A. Duevel, D. Rudolph and G. Schmahl, Characterization of the sputtered-sliced zone plate for high energy X-rays, Nucl. Instr. Meth. in Physics Research A 868 (2001) 467-468

[21] M.A. Kumakhov, Multiple reflection from surface X-ray optics, Phys. Rep. 191 (1990) 289

[22] V. Arkadiev, A. Bzhaumikhov, A. Erko, F. Schafers, P. Chevallier, P. Populus, Synchrotron radiation beam splitting and filtering by a polycapillary array, Nucl. Instr. Meth. in Physics Research A 384 (1997) 547

[23] W. Jark, A. Cedola, S. Di Fonzo, M. Fiordelisi, S. Lagomarsino, N.V. Kovalenko and V.A. Chernov, High gain beam compression in new-generation thin-film X-ray waveguides, Appl. Phys. Lett. 78 (2001) 1192

[24] A. Bjeoumikhov, M. Erko, S. Bjeoumikhova, A. Erko, I. Snigireva, A. Snigirev, T. Wolff, I. Mantouvalou, W. Malzer, B. Kanngiesser, Capillary µfocus X-ray lenses with parabolic and elliptic profile, Nucl. Instr. Meth. in Physics Research A 587 (2008) 458

[25] J.L. Carrascosa , F.J. Chichón, E. Pereiro, M.J. Rodríguez, J.J. Fernández, M. Esteban, S. Heim, P. Guttmann, G. Schneider, Cryo-X-ray tomography of vaccinia virus membranes and inner compartments, J. Struct. Biol. 168 (2009) 234

[26] E.I. Denisov, V.I. Glebov and N.K. Zhevago, Focusing of X-rays using tapered waveguides, Nucl. Instr. Meth. in Phys. Res. A 308 (1991) 400

[27] S.A. Hoffman, D.J. Thiel and D.H. Bilderback, Developments in tapered monocapillary and polycapillary glass X-ray concentrators, Nucl. Instr. Meth. in Phys. Res. A 347 (1994) 384

[28] M. Nishikino, K. Sato, N. Hasegawa, M. Ishino, S. Ohshima, Y. Okano, T. Kawachi, H. Numasaki, T. Teshima and H. Nishimura, Note: Application of laser produced plasma Kα X-ray probe in radiation biology, Rev. Sci. Instruments 81 (2010) 026107

[29] T. Kuchimaru, F. Sato, Y. Higashino, K. Shimizu, Y. Kato and T. Iida, Microdosimetric characteristics of micro X-ray beam for single cell irradiation, IEEE Transactions on Nuclear Science 53 (2006) 1363

[30] H. McQuaid, K.M. Prise, F. Currell and G. Schettino, Development of a new high energy X-ray microbeam based on reflective X-ray optics, Program and Abstracts of the 10th International Workshop "Microbeam Probes of Cellular Radiation Response", ed. A. Bigelow, RARAF, Columbia University, March 15-17, 2012, page 23

[31] Matsuyama, M. Shimura, H. Mimura, M. Fujii, H. Yumoto, Y. Sano, M. Yabashi, Y. Nishino, K. Tamasaku, T. Ishikawa and K. Yamauchi, Trace element mapping of a single cell using a hard X-ray nanobeam focused by a Kirkpatrick-Baez mirror system, X-ray Spectrom. 38 (2009) 89

[32] S. Bożek, J. Bielecki, J. Baszak, H. Doruch, R. Hajduk, J. Lekki, Z. Stachura and W.M. Kwiatek, X-ray microprobe – A new facility for cell irradiations in Kraków, Nucl. Instr. Meth. in Phys. Res. B 267 (2009) 2273

[33] D. Paterson, M.D. de Jonge, D.L. Howard, W. Lewis, J. McKinlay, A. Starritt, M. Kusel, C.G. Ryan, R. Kirkham, G. Moorhead, D.P. Siddons, The X-ray Fluorescence Microscopy Beamline at the Australian Synchrotron, AIP Conf. Proc. 2011, 1365, 219–222.

[34] Rossi H.H., Specification of Radiation Quality. Radiat.Res. 10 (1959), 522

[35] W.E. Wilson, J.H. Miller, H.G. Paretzke, Microdosimetric aspects of 0.3 to 20 MeV proton tracks—I. Crossers, Radiat. Res., 115 (1988), 339–352

[36] V.A. Semenenko, J.E.Turner, and T.B.Borak, NOREC, a Monte Carlo code for simulating electrons tracks in liquid water Radiat. Environ. BioPhys. 42 (2003) 213

[37] J.E. Turner, R.N. Hamm, H.A. Wright, J.T. Modolo, G.M.A.A. Sordi, Monte Carlo calculation of initial energies of Compton electrons and photoelectrons in water irradiated by photons with energies up to 2 MeV, Health Phys., 39 (1980), 49–55

[38] Lappa AV, Bigildeev EA, Burmistrov DS, Vasilyev ON. "Trion" code for radiation action calculations and its application in microdosimetry and radiobiology, Radiat Environ Biophys., 32(1), (1993), 1-19

[39] S. Agostinelli et al., GEANT4 – a simulation toolkit, Nucl. Instr. Meth. in Phys. Res. A 506 (2003) 250

[40] J. Sempau, J.M. Fernández-Varea, E. Acosta and F. Salvat Nucl. Instrum. Meth. B 207 (2003) 107

[41] Z. Francis, S.Incerti, R.Capra , B.Mascialino, G.Montarou , V.Stepan , C.Villagrasa APPL RADIAT ISOTOPES , vol. 69, no. 1, pp. 220-226, 2011

[42] J.P. Alard, V. Bodez, A. Tchirkov, M.L. Nenot, J. Arnold, S. Crespin, M. Rapp, P. Verrelle and C. Dionet Simulation of neutron interactions at the single-cell level. Radiat Res 158 (2002) 650–656

[43] M. Folkard, K.M. Prise, B. Vojnovic, H.C. Newman, M.J. Roper, K.J. Hollis and B.D. Michael, Conventional and microbeam studies using low energy charged particles relevant to risk assessment and the mechanisms of radiation action, Radiation Protection Dosimetry 61 (1995) 215

[44] G. Schettino, M. Folkard, B. Vojnovic, A.G. Michette, D. Stekel, S.J. Pfauntsch, K.M. Prise and B.D. Michael, The ultrasoft X-ray microbeam: A sub-cellular probe of radiation response, Rad. Res. 153 (2000) 223

[45] R.B. Hawkins, Mammalian cell killing by ultrasoft X rays and high-energy radiation: An extention of the MK model, Radiation Research 166 (2006) 431

[46] S.D. Clarke and T. Jevremovic, MCN5 evaluation of dose dissipation in tissue-like media exposed to low-energy monoenergetic X-ray microbeam, Radiat. Environ. Biophys. 44 (2005) 225

[47] H. van Attikum and S.M. Gasser, Crosstalk between histone modifications during the DNA damage response, Trends in Cell Biology 19 (2009) 207

[48] S. Bekker-Jensen, C. Lukas, R. Kitagawa, F. Melander, M.B. Kastan, J. Bartek and J. Lukas, Spatial organization of the mammalian genome surveillance machinery in response to DNA strand breaks, J. Cell Biol. 173 (2006) 195

[49] G. Dollinger, V. Hable, A. Hauptner, R. Krücken, P. Reichart, A.A. Friedl, G. Drexler, T. Cremer and S. Diezel, Microirradiation of cells with energetic heavy ions, Nucl. Instr. Meth. in Phys. Res. B 231 (2005) 195

[50] M. Scholz, B. Jakob, G. Taucher-Scholz, Direct evidence for the spatial correlation between individual particle traversals and localized CDKN1A (p21) response induced by high-LET radiation, Rad. Res. 156 (2001) 558

[51] B. Jakob, J.H. Rudolph, N. Gueven, M.F. Lavin and G. Taucher-Scholz, Live cell imaging of heavy-ion-induced radiation responses by beamline microscopy, Radiation Research 163 (2005) 681

[52] R. Ugenskiene, K. Prise, M. Folkard, J. Lekki, Z. Stachura, M. Zazula and J. Stachura, Dose response and kinetics of foci disappearance following exposure to high- and low-LET ionizing radiation, Int. J. Radiat. Biol. 85 (2009) 872

[53] R. Ugenskiene, J. Lekki, W. Polak, K.M. Prise, M. Folkard, O. Veselov, Z. Stachura, W.M. Kwiatek, M. Zazula and J. Stachura, Double strand break formation as a response to X-ray and targeted proton-irradiation, Nucl. Instr. Meth. in Phys. Res. B 260 (2007) 159

[54] N. Hamada, G. Schettino, G. Kashino, M. Vaid, K. Suzuki, S. Kodama, B. Vojnovic, M. Folkard, M. Watanabe, B.D. Michael and K.M. Prise, Histone H2AX phosphorylation in normal human cells irradiated with focused ultrasoft X rays: Evidence for chromatin movement during repair, Radiation Research 166 (2006) 31

[55] H. Nagasawa and J.B. Little, Induction of sister chromatid exchanges by extremely low doses of α-particles, Cancer Res. 52 (1992) 6394

[56] C. Mothersill and C. Seymour, Medium from irradiated human epithelial cells but not human fibroblasts reduces the clonogenic survival of unirradiated cells, Int. J. Radiat. Biol. 71 (1997) 421

[57] B.J. Blyth and P.J. Sykes, Radiation-induced bystander effects: what are they, and how relevant are they to human radiation exposures?, Radiation Research 176 (2011) 139

[58] K.M. Prise and J.M. O'Sullivan, Radiation-induced bystander signalling in cancer therapy, Nature Reviews : Cancer 9 (2009) 351

[59] C. Shao, V. Steward, M. Folkard, B.D. Michael and K.M. Prise, Nitric oxide-mediated signaling in the bystander response in individually targeted glioma cells, Cancer Res. 63 (2003) 8437

[60] C. Shao, K.M. Prise and M. Folkard, Signaling factors for irradiated glioma cells induced bystander responses in fibroblasts, Mutation Research 638 (2008) 139

[61] S. Chen, Y. Zhao, W. Han, G. Zhao, L. Zhu, J. Wang, L. Bao, E. Jiang, A. Xu, T.K. Hei, Z Yu and L. Wu, Mitochondria-dependent signalling pathway are involved in the early process of radiation-induced bystander effects, Br. J. Cancer 98 (2008) 1839

[62] M. Maeda, M. Tomita, N. Usami and K. Kobayashi, Bystander cell death is modified by sites of energy deposition within cells irradiated with a synchrotron X-ray microbeam, Radiat. Res. 174 (2010) 37

TIME-RESOLVED X-RAY DIFFRACTION OF CRYOGENIC SAMPLES USING A LASER BASED PLASMA SOURCE

R. Loetzsch[1,2], A. Lübcke[1,3], F. Zamponi[1,4], T. Kämpfer[1,2], I. Uschmann[1,2] and E. Förster[1,2]

[1]Institute of Optic and Quantum Electronics, Friedrich Schiller University Jena, Max-Wien-Platz 1, 07743 Jena, Germany
[2]Helmholz-Institute Jena, Fröbelstieg 3, 07743 Jena, Germany
[3]Max-Born Institute for Nonlinear Optics and Short Pulse Spectroscopy, Max-Born-Straße 2A, 12489 Berlin, Germany
[4]Institute of Physics and Astronomy, University of Potsdam, Karl-Liebknecht-Strasse 24-25, 14476 Postdam

1 INTRODUCTION

The application of sub-picosecond laser produced X-ray sources, providing radiation in the kilo-electronvolt range, has been successfully demonstrated in time-resolved studies of solid matter in the past. The first experiments showed transient atomic movement during fundamental processes such as ultrafast excitation of crystal lattices, non-thermal melting of semiconductors and acoustical and optical phonons.[1–7] In all studies characteristic line radiation was used in laser pump / X-ray probe experiments to investigate the collective atomic displacements in crystals on timescales of a few hundred femtoseconds. Thus, even the rearrangement of atoms triggered by fast electronic processes or chemical reactions becomes observable.

The interaction of high intensity laser pulses with matter at intensities in the range 10^{16}–10^{18} W/cm^2 is now widely used for excitation of short bursts of kilo-electronvolt X-rays. Previously, titanium:sapphire laser systems have been used to focus ~100 fs laser pulses with energies of several hundreds of millijoules and repetition rates of 10 Hz onto solid or liquid targets, thus creating a plasma. By collisionless processes, electrons are thereby accelerated to high energies (kilo- to mega-electronvolts) and emit Bremsstrahlung as well as characteristic line radiation, typically K_α.[8] An alternative approach is to use high repetition rate lasers (kilohertz), but with lower energies of a few millijoules.[9–11]

To date all these experiments have been done at room temperature. To extend such ultra-fast time resolved X-ray diffraction measurements to a wider class of phenomena, this chapter presents investigations of solids at cryogenic temperatures. The setup is described, and results of static measurements of the temperature dependency of $SrTiO_3$ lattice constants, as well as results on the transient behaviour of a system consisting of a $SrTiO_3$ single crystal covered by a thin film of the high temperature superconductor $YBa_2Cu_3O_{7-\delta}$ are presented.

2 EXPERIMENTAL SET UP

The experimental setup is shown in Figure 1. Two femtosecond laser beams were used with variable time delay. The first beam was sent to a vacuum chamber and focused, by an off-axis parabolic mirror to the metallic tape target, to intensities up to and above 10^{17} W/cm^2, producing a high temperature plasma. To avoid damage of the parabolic mirror debris protection, frequently refreshed, had to be used. This is either a glass slide or a plastic tape. The energy density of the pump beam on the sample could be changed by a tuneable attenuator.

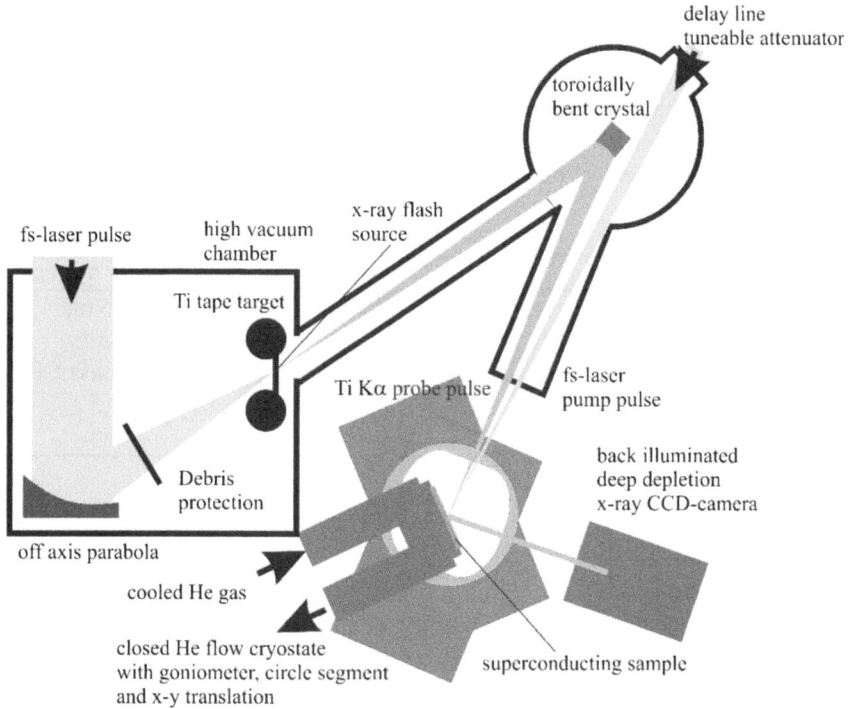

Figure 1 *Experimental setup for the time resolved X-ray diffraction of superconductors.*

The X-rays produced were focused with a toroidally bent crystal onto the sample. They were guided in vacuum to avoid absorption. The X-ray signal diffracted by the sample was detected by a deep depletion X-ray CCD camera placed 100 mm behind the sample. The second, much weaker, femtosecond laser beam was focused to the same area on the sample, which was positioned in a cryostat combined with a goniometer, a circle element and two stages to align the sample. More detailed descriptions of the key elements, namely the X-ray source, the X-ray optic and the cryostat are given in the following sections.

2.1 Laser Plasma X-ray Source

Two kilo-electronvolt X-ray sources have been developed in Jena using the titanium:sapphire laser systems listed in table 1. The laser systems provide high power femtosecond pulses by using chirped pulse amplification (CPA). The initial experiments were done using the Jena 10 Hz titanium:sapphire laser system JETI, which was delivering 180 mJ on the target at that time. Later experiments could be improved by higher laser pulse energy of 360 mJ. Several experiments were performed to optimize the source using the 10 Hz system.[12,13,14,15] A setup was developed for time resolved X-ray diffraction experiments such as a study of nonthermal melting,[6] acoustic phonons[16] and anomalous transmission of X-rays.[17]

The second, 1 kHz, source system (Quantronix-ODIN-MPA) consisted of two multi-pass amplifiers delivering a pulse energy on target of 3.5 mJ and a pulse duration of 55 fs at a central wavelength of 803 nm. This kilohertz X-ray source is described in detail in the chapter by Kämpfer et al. in the current volume.

With both laser systems characteristic line radiation was produced from titanium at 4.5 keV and copper at 8 keV. Measured characteristic X-ray fluxes for the 1 kHz system were 5×10^{10} phonons s^{-1} for titanium and 1.2×10^{10} phonons s^{-1} for copper. For the 10 Hz source the fluxes were 3×10^{12} phonons s^{-1} for titanium and 5×10^{11} phonons s^{-1} for copper.[10] In all cases the X-rays were emitted isotropically.

Table 1 *Important parameters of the Ti:Sapphire CPA laser systems used for the K_α X-ray source.*

	10 Hz system	1 kHz system
Energy on target	180 mJ / 360 mJ	3.5 mJ
Pulse duration	70 fs	55 fs
Amplifier type	Regenerative and multi-pass CPA	Multi-pass CPA
Laser energy	1.6 J* / 3.2 J**	3.5 J
Pulse to pulse stability	5%	5%

*After the second amplifier.
**After the third amplifier.

2.2 X-Ray Optics

To make efficient use of the emitted X-rays for diffraction experiments the K_α radiation from the target was focused by an optic such as a bent crystal or a multilayer mirror onto the sample. The X-ray optic must have a large integrated reflectivity and a large vertical aperture to collect as much radiation as possible. The full width at half maximum (FWHM) of the diffraction curve of the optic should match to the angular resolution required in the diffraction experiment, and it should focus the X-rays to a small spot on the sample. For the experiments described here, titanium K_α radiation was focused by a toroidally bent crystal to a focal spot size of 80 μm FWHM.

Two different crystals were used in the static and the time resolved experiments. The first was a silicon crystal used in 311 reflection and the second was gallium arsenide crystal used in 400 reflection. Their properties are listed in table 2.

Table 2 *Comparison of the two crystal optics used in the experiments.*

	Silicon 311	Gallium arsenide 400
Bragg angle, θ_B	57.1°	76.6°
Horizontal bending radius, R_h	149.8 mm	400 mm
Vertical bending radius, R_v	105.6 mm	378.5 mm
Crystal width, x_h	11 mm	15 mm
Crystal height, x_v	11 mm	40 mm
Source-crystal distance, l_a	126 mm	389 mm
Horizontal opening angle	5.0°	2.1°
Vertical opening angle	5.0°	5.9°

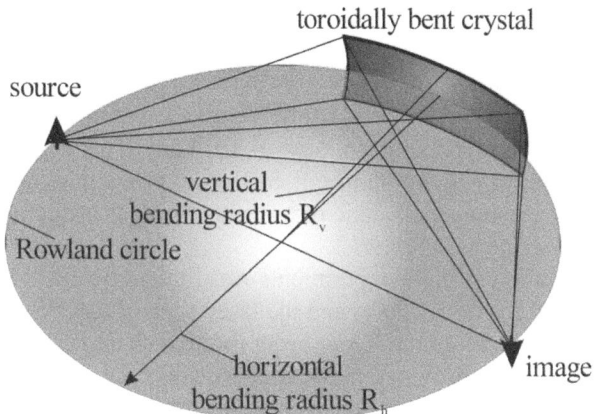

Figure 2 *X-ray imaging with a toroidally bent crystal.*

The method of focusing X-rays with toroidally bent crystals[18] is demonstrated in figure 2. To provide both point to point focusing as well as quasi-monochromatic imaging of the source, the bending radii in horizontal and vertical direction (R_h and R_v) as well as the source-crystal distance l_a have to be chosen appropriately: The bending radii must fulfil[18]

$$R_v/R_h = \sin^2 \theta_B \tag{1}$$

where θ_B is the Bragg angle. l_a is given by[18]

$$l_a = R_h \sin \theta_B \tag{2}$$

and is equal to the crystal-focus distance l_b. The horizontal divergence of the reflected radiation determines the range of incidence angles on the sample and is given by $\Delta\theta = x_h(\sin\theta_B)/l_a$. The vertical divergence should be as great as possible to collect maximal photon numbers, but is limited by the elastic properties of the crystal, as larger crystals will break.

The spectral selection of the crystal is limited by two effects, first by the rocking curve width $\Delta\theta_{RC}$ according to

$$\frac{\Delta\lambda}{\lambda} = \frac{\Delta\theta_{RC}}{\tan\theta_B} \tag{3}$$

and second through spectral broadening due to the horizontal crystal width,

$$\frac{\Delta\lambda}{\lambda} = \frac{x_h^2}{8R_h^2 \tan^2\theta_B} \tag{4}$$

For the gallium arsenide crystal, the rocking curve effect contributes 6.2×10^{-5} while spectral broadening contributes 1.0×10^{-5}. For silicon, the values are 2.8×10^{-5} and 2.5×10^{-5}, respectively. These values must be compared with the natural line width of the titanium $K_{\alpha1}$ emission, which is $\Delta\lambda/\lambda = 3\times10^{-4}$. Thus only a small part of the $K_{\alpha1}$ line is reflected, making the setup sensitive to very small lattice parameter changes or to subtle changes of diffraction profiles.

The capability for measurements of small static or dynamic lattice parameter variations is given by the monochromaticity of the incoming radiation and the angular resolution $\Delta\theta$ achievable in the experiments, which results from a finite source size. For static measurements a standard fine-focus tube with a source size of 400 μm was used, corresponding to $\Delta\theta = 4\times10^{-3}$ rad. For the time resolved measurements the source had a FWHM size of 50 μm, corresponding to $\Delta\theta = 1\times10^{-4}$ rad. This high angular resolution, together with the monochromaticity and high collection efficiency, means that the optics are very well suited to measure small static or transient changes of rocking curves, i.e. peak shifts or rocking curve shape variations.

The temporal resolution of these experiments was limited by the following effects:
1. the time durations of the X-ray pulse and the laser pump pulse;
2. the finite penetration depth of the X-ray beam in the focusing crystal; and
3. the time smearing induced by geometrical broadening, i.e. by the X-ray source size, beam tilt, and the pump-probe delay smearing caused by non-collinearity of the two beams.

For the crystals and the x-ray lines used, the effects mentioned in points 2 and 3 are in the range of around 10–50 fs. For shorter X-ray wavelengths the second effect can cause pulse broadening of several hundred femtoseconds due to less absorption and consequently a larger penetration depth. For the experiments described here the time resolution was limited by the laser pulse duration of 90–120 fs and by the duration of the X-ray pulse. For the latter, the upper limit was 300–600 fs.[6]

2.3 X-ray Diffractometer Cryostat Combination

In order to cover a wide temperature range, two cooling devices were combined with the X-ray diffractometer. The first was a Stirling machine to cool the sample down to a temperature of about 70 K. With independent temperature control it was possible to keep the temperature constant to within 1 K. To allow the X-rays and the laser pump pulse to hit the sample in the vacuum chamber, there were two flanges each having a beryllium window for the X-rays and a quartz window for the optical pump and for optically aligning the sample. These two flanges enclosed an angle of 90°. This arrangement allowed for recording X-ray diffraction patterns at Bragg angles between 42 and 48°. For the high temperature superconductor $YBa_2Cu_3O_{7-\delta}$ that was used as a sample, the (006) reflection was suitable for use with titanium K_α radiation.

A second cryostat based on closed helium flow was implemented in order to increase the cooling power and to reach lower temperatures, close to that of liquid helium. The advantages of this cryostat were higher mechanical stability and more stable electronic diagnostics of the superconducting samples. In addition, larger windows and more flexible window positions and orientations allowed Bragg angles over a much wider range to be used; 0–66° and 76–90° were possible. For this cryostat the sample holder was also freely rotatable with respect to the X-ray windows, allowing many asymmetric reflections to be used. The sample vacuum chamber was placed onto a circle segment, to align the sample tilt, combined with a Huber goniometer type 410 to align the Bragg angle. This goniometer was fixed on two mechanical translation stages, one used for positioning the sample into the focus of the bent crystal and the other to align the correct lateral source position. The angle between the two translation stages could also be varied.

Two diodes are placed close to the sample to measure the temperature, which could be varied via a resistive heater and controlled by a proportional integrated derivative (PID) controller. The accuracy of the temperature measurement was better than 1 K and the stability was in the milli-Kelvin range.

3 STATIC MEASUREMENTS

As a first example we describe static investigations of the structural phase transition of $SrTiO_3$.[19] At temperatures below 105 K, the high temperature cubic lattice becomes tetragonally distorted. Due to twin formation in the low temperature phase, the cubic Bragg reflections split into two reflections in the tetragonal phase. Figure 3(a) shows diffractograms of the 002 reflection at 44.5° for temperatures between 10 K and 290 K; for clarity the curves at different temperatures are shifted along the ordinate. The integration time was 90 s for each diffractogram, yielding 10^5 photons in the main peaks of the reflection curves. The two peaks at high temperatures are due to diffraction of $K_{\alpha 1}$ and $K_{\alpha 2}$ radiation. Below 105 K both peaks split into two, corresponding to reflections from the different twin domains. It should be noted that the setup did not allow absolute Bragg angle measurements, but relative shifts of Bragg angles were easily recorded. The extracted lattice parameters are shown in figure 3(b). Here, the Bragg angle was normalized using the known room temperature lattice constant of 0.3905 nm. For this particular Bragg reflection, the accuracy for detecting lattice parameter changes $\Delta d/d$ was better than 1×10^{-4}. This demonstrates the capability of the setup to measure small lattice parameter changes over a wide temperature range.

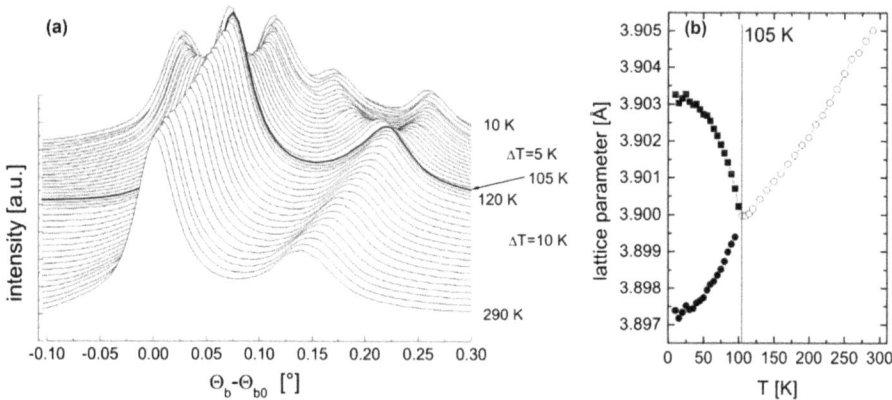

Figure 3 *Temperature dependence of the $SrTiO_3$ lattice parameters. (a) The diffraction profiles of the 002 reflection with Ti K_α radiation at the Bragg angle of ~44.5° between room temperature and 10 K. (b) The extracted lattice parameters; the open symbols are for the cubic lattice and the filled symbols are for the two tetragonal lattice parameters.*

4 TRANSIENT ROCKING CURVES UPON OPTICAL PUMPING

The above described system was used to investigate the structural response of a high temperature superconductor and, in particular, the $SrTiO_3$ substrate upon optical switching the superconducting state. A near-infrared pump pulse breaks cooper pairs in the superconducting $YBa_2Cu_3O_7$ thin film, and the transient lattice response of the thin film and the substrate was investigated by time-resolved x-ray diffraction. The system studied was a

SrTiO$_3$ single crystal covered by an epitaxially grown, 250 nm thick top layer of superconducting YBa$_2$Cu$_3$O$_7$.[20] The superconductor was grown by pulsed laser deposition. The extinction depth for X-ray pulses was only 0.5 μm in each material, and thus only the near interface region was probed. The superconducting transition of the sample was monitored by a resistivity measurement. For this purpose a current of 100 μA was sent through the sample and the voltage was measured. To ensure, that only the investigated sample volume is monitored, the sample was laterally structured. Within the experimental precision no difference was found between resistivity measurement during and without exposure. The sample was found to be superconducting at temperatures below 89 K and the measurements were performed at 70 K. Figure 4 shows the temporal evolution of both the rocking curve width and the peak position. Both show similar temporal behaviour, a sub-picosecond decrease in line width and shift of the Bragg angle, followed by re-establishment of the initial values after 2 ps. These changes in the diffraction profiles were modelled using the dynamical theory of X-ray diffraction.[21,22] In agreement with static measurements on this sample[22], the superconducting state of the YBCO top-layer induces a strain field in a thin interfacial layer of the substrate. From our model, we can best describe the results of optically breaking the superconducting state by a reduction of the strained layer thickness. However, a convincing microscopic picture of the underlying physics explaining our observations is still lacking.

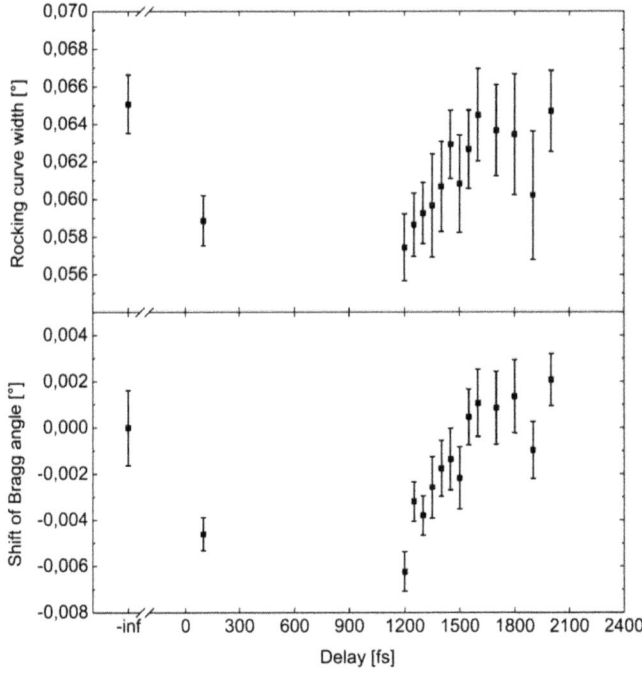

Figure 4 *Temporal dependence of the SrTiO$_3$ 002 rocking curve width (upper panel) and peak position (lower panel) following optical breaking of Cooper pairs in an YBa$_2$Cu$_3$O$_7$ top layer (from ref. 20).*

ACKNOWLEDGEMENTS

The authors gratefully acknowledge the JETI laser staff in Jena; B. Beleites, F. Ronneberger and W. Ziegler for experimental support, F. Schmiedl, T. Köttig and M.

Thuerk for assistance with the cryostats, and V. Große for the preparation of the superconducting sample.

The work was supported by the Deutsche Forschungsgemeinschaft within priority programm 1134.

References

1 C. Rischel, A. Rousse, I. Uschmann, P.A. Albouy, J.P. Geindre, P. Audebert, J.C. Gauthier, E. Förster, J.L. Martin and A. Antonetti, *Nature*, 1997, **390**, 490.

2 C. Rose-Petruck, R. Jimenez, T. Guo, A. Cavalleri, C.W. Siders, F. Raksi, J.A. Squier, B.C. Walker, K.R. Wilson and C.P.J. Barty, *Nature*, 1998, **398**, 310.

3 C.W. Siders, K. Sokolowski-Tinten, C. Toth, T. Guo, M. Kammler, M. Horn van Hoegen, K.R. Wilson, D. von der Linde and C.P.J. Barty, *Science*, 1999, **286**, 1340.

4 A. Rousse, C. Rischel, S. Fourmaux, I. Uschmann, S. Sebban, G. Grillon, P. Balcou, E. Förster, J.P. Geindre, P. Audebert, J.C. Gauthier and D. Hulin, *Nature*, 2001, **410**, 65.

5 K. Sokolowski-Tinten, C. Blome, C. Dietrich, A. Tarasevitch, M. Horn von Hoegen, D. von der Linde, A. Cavalleri, J. Squier and M. Kammler, *Phys. Rev. Lett.*, 2001, **87**, 225701.

6 T. Feurer, A. Morak, I. Uschmann, C. Ziener, C. Reich, P. Gibbon, E. Förster, and R. Sauerbrey, *Phys. Rev. E*, 2002, **65**, 016412/1-4.

7 K. Sokolowski-Tinten, C. Blum, M. Kammler, M. Horn van Hoegen, D. von der Linde, I. Uschmann and E. Förster, *Nature*, 2003, **420**, 1340.

8 P. Gibbon, *Short Pulse Laser Interactions with Matter*, 2005, (London: Imperial College Press)

9 M. Silies, H. Witte, S. Linden, J. Kutzner, I. Uschmann and E. Förster, *Appl. Phys. A*, 2009, **96**, 59.

10 I. Uschmann, T. Kämpfer, F. Zamponi, A.Lübcke, U. Zastrau, R. Loetzsch, S. Höfer, A. Morak and E. Förster, *Appl. Phys. A*, 2009, **96**, 91.

11 F. Zamponi, Z. Ansari, C. v. Korff Schmising, P. Rothhardt, N. Zhavoronkov, M. Woerner, T. Elsaesser, M. Bargheer, T. Trobitzsch-Ryll and M. Haschke, *Appl. Phys. A*, 2009, **96**, 51.

12 C. Reich, P. Gibbon, I. Uschmann and E. Förster, *Phys. Rev. Lett.*, 2000, **84**, 4846.

13 C. Ziener, I. Uschmann, G. Strobrawa, C. Reich, P. Gibbon, T. Feurer, A. Morak, S. Düsterer, H. Schwoerer, E. Förster and R. Sauerbrey, *Phys. Rev. E*, 2002, **65**, 066411.

14 A. Morak, Thesis, Jena-university (2003). "X-ray diffraction on sub-picosecond timescales"

15 F. Ewald, H. Schwoerer and R. Sauerbrey, *Europhys. Lett.*, 2002, **60**, 710.

16 A. Morak, T. Kämpfer, I. Uschmann, A. Lübcke, T. Feurer, H. Schwoerer, E. Förster and R. Sauerbrey, *Phys. Stat. Sol.*, 2006, pssb.200642387.

17 A. Lübcke, I. Uschmann, A. Morak, H. Schwörer, E. Förster and R. Sauerbrey, *Appl. Phys. B*, 2005, **80**, 801.

18 T. Missalla, I. Uschmann, E. Förster, G. Jenke and D. von der Linde, *Rev. Sci. Instrum.*, 1999, **70**, 1288.

19 R. Loetzsch, A. Lübcke, I. Uschmann, E. Förster, V. Große, M. Thuerk, T. Koettig, F. Schmidl and P. Seidel, *Appl. Phys. Lett.*, 2010, **96**, 071901.

20 A. Lübcke, F. Zamponi, R. Loetzsch, T. Kämpfer, I. Uschmann, V. Große, F. Schmidl, T. Koettig, M. Thuerk, H. Schwoerer, E. Förster, P. Seidel and R. Sauerbrey, *New Journal of Physics*, 2010, **12**, 083043.

21 A. Authier, *Dynamical Theory of X-Ray Diffraction*, 2001, (Oxford: Oxford University Press).
22 J. Wark, R.R. Whitlock, J.E. Swain and P.J. Solone, *Phys. Rev. B*, 1989, **40**, 5705.

NEAR-EDGE X-RAY ABSORPTION FINE STRUCTURE MEASUREMENTS USING A LABORATORY-SCALE XUV SOURCE

K. Mann

Laser-Laboratorium Göttingen e.V., Hans-Adolf-Krebs-Weg 1, D-37077 Göttingen, Germany

1 INTRODUCTION

Progress in the development of laboratory-scale soft X-ray sources in recent years has enabled experimental techniques that could earlier only be performed almost exclusively at synchrotron sources. Table-top soft X-ray sources of high brilliance, such as laser-produced plasmas,[1–3] high harmonic radiation[4,5] and X-ray lasers,[6] are now used for various applications, e.g., x-ray microscopy,[7] lens-less diffractive imaging,[8] photoelectron spectroscopy[9] and absorption spectroscopy.[10,11] The latter includes the investigation of near-edge X-ray absorption fine structure (NEXAFS), which is a well established method for elemental and compositional analysis of a sample, also yielding surface sensitive information.[12] In particular, NEXAFS is used to study the structure of intermolecular bonds of polymers by probing the electronic transitions from the core level to unoccupied states. Since each element has a characteristic core binding energy, NEXAFS spectra contain element specific information. Furthermore, the energy levels of both initial and final states are strongly dependent on the molecular bonds involved, resulting in strong spectral features of the near-edge fine structure. The analysis of these unique spectroscopic fingerprints allows the identification and distinction of different polymers.[13]

This chapter presents NEXAFS measurements that were obtained using a laboratory-scale setup based on a laser driven plasma source. It has already been shown that the generation of broad band emission in the spectral range of the "water window" (wavelengths of 2.2–4.4 nm) can be achieved by using solid state targets such as gold or copper.[14] Here, a flexible setup to generate broad band soft X-rays based on a gas puff target, where new target material is supplied continuously with the advantage of low debris, is presented. The table-top setup, consisting of the laser plasma source and a flat field spectrometer can be used for NEXAFS experiments in transmission as well as reflection under grazing incidence conditions.[11] For transmission measurements thin films have to be used due to the low penetration depth of soft X-rays into matter.

2 EXPERIMENTAL SETUP

The experimental setup of the NEXAFS spectrometer for the water window spectral range is shown schematically in figure 1 and in the photograph of figure 2. For the generation of a soft X-ray plasma a Nd:YAG laser (Innolas, 1064 nm, 1 Hz, 800 mJ, 7 ns) is focused into a pulsed gas puff target centred in a vacuum chamber, as described in detail elsewhere.[15] The laser focus has a diameter of about 60 µm, yielding power densities of up to 4×10^{12} W cm^{-2}, sufficient to ignite a hot dense plasma. Krypton is used as the target gas (with a backing pressure of 25 bar), providing broad band radiation (Kr^{24+}–Kr^{33+}) in the water window spectral range (figure 3). Because of the small mean free path of soft X-rays at atmospheric pressure, the target vacuum chamber is evacuated to approximately 10^{-4} mbar. The plasma is monitored with a pinhole camera, consisting of a CCD chip with an X-ray to visible converter and a 30 µm diameter pinhole coated with a titanium foil of thickness ≈ 200 nm to block off out of band radiation. The full width half maximum (FWHM) size of the krypton plasma is about 250 µm in the horizontal and 150 µm in the vertical direction (see the inset to figure 1).

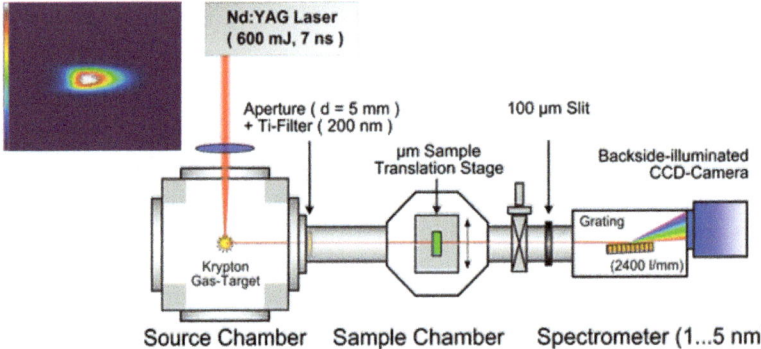

Figure 1 *Experimental arrangement of the laser plasma XUV source used for NEXAFS experiments on thin samples. The inset on the left shows the krypton plasma recorded by the XUV pinhole camera.*

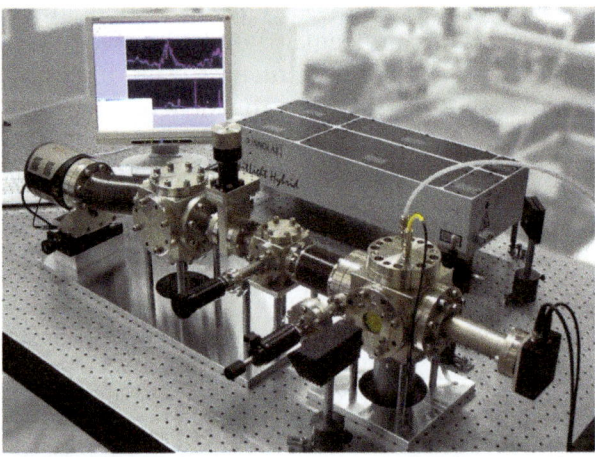

Figure 2 *Photograph of the experimental setup of the laser plasma XUV source and the spectrometer used for NEXAFS experiments.*

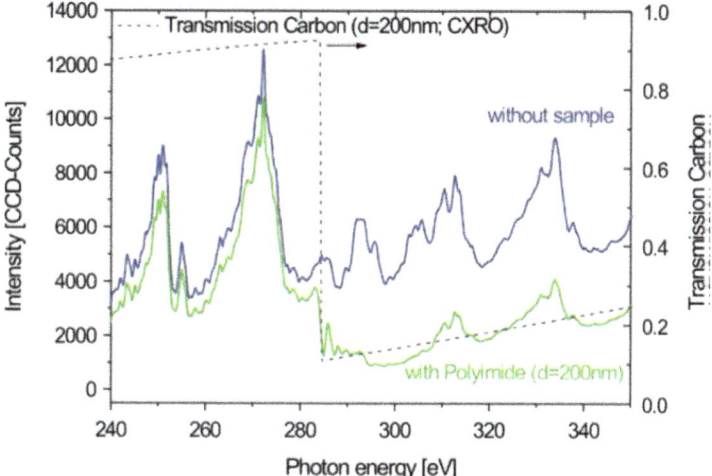

Figure 3 *Emission spectra, averaged over 60 pulses, of the krypton plasma used as a broad band emitter in the water window with and without a 200 nm thick polyimide sample. For comparison the calculated transmission of carbon[16] is shown (dashed line).*

An XUV spectrometer ($\lambda = 1$–5 nm) is used both for the spectral investigation of the plasma source and for the NEXAFS experiments. The spectrometer has a 100 μm entrance slit, an aberration corrected flat field grating (Hitachi, 2400 lines/mm) and a back side illuminated CCD camera (Roper Scientific, pixel size 13 μm). The resolution of this spectrometer is experimentally determined to be $\lambda/\Delta\lambda \approx 200$ at $\lambda = 2.87$ nm. To block visible radiation from the plasma and scattered laser light a 200 nm thick titanium foil is positioned between the plasma source and the sample. To calibrate the spectrometer nitrogen was used as the target gas. The spectrum of the nitrogen plasma, ignited under the same experimental conditions as krypton, consists of several lines, for example the $1s^2$–$1s2p$ transition of N^{5+} at 2.8787 nm,[17] that were used for spectral calibration. For adjustment in the XUV beam the samples are mounted on a linear translation stage. The setup is used for NEXAFS experiments in both transmission and reflection mode. For the latter grazing incidence conditions are chosen ($\theta = 2°$).

3 RESULTS AND DISCUSSION

3.1 NEXAFS on Polyimide

For thin samples it is possible to determine the X-ray absorption fine structure by measuring the transmitted flux. As an example, 200 nm thick polyimide films (PI 2545, HD Microsystems) were used. Figure 3 shows the emission spectra of the krypton plasma both with and without the polyimide sample, each obtained by integrating over 60 pulses. The transmitted spectrum clearly indicates an overall decrease in intensity above the carbon K absorption edge. From the data in figure 3 the optical density can be evaluated using the Beer-Lambert law

$$\mu(E)d = -\ln(I/I_0),\qquad(1)$$

where $\mu(E)$ is the linear energy dependent absorption coefficient, d is the sample thickness, and I_0 and I are the incident and transmitted intensities. The optical density shown in figure 4 evaluated using equation (1) features several sharp peaks below the carbon K absorption edge; these can be attributed to C1s–π* transitions of the benzene rings that are clearly resolved. The broad features above the edge belong to C1s–σ* transitions. To evaluate the data a multi-Gaussian fit was performed. The identified features are summarized in table 1 and compared with reference data obtained from synchrotron experiments. As can be seen, the obtained energy values of the peaks deviate by less than 0.4 eV from the corresponding synchrotron data.

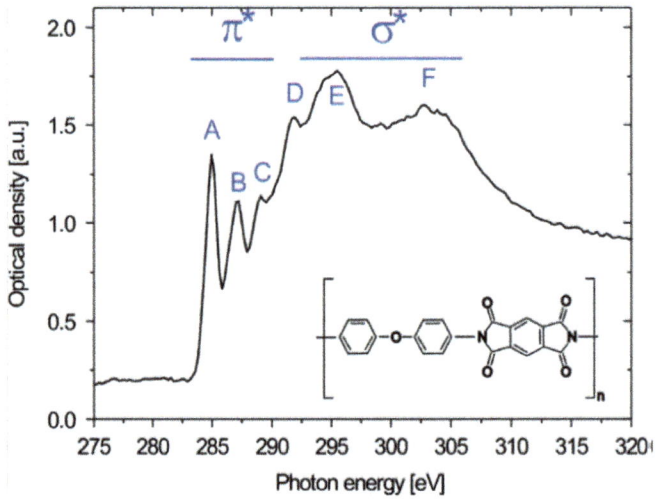

Figure 4 *NEXAFS spectrum and chemical structure of a 200 nm thick polyimide film. The peak positions deviate by less than 0.4 eV from corresponding synchrotron data.*

Table 1 *Energies and assignments of features in the NEXAFS spectrum of polyimide at the carbon K-edge.*

Feature*	Energy [eV]	Assignment	Reference data [eV][18]
A	285.2	1s → π* (C=C)	285.2
B	287.3	1s → π* (C=O)	287.4
C	289.3	1s → π*	289.2
D	291.6	1s → σ* (C–O, C–N)	291.9
E	295.0	1s → σ* (C=C)	295.4
F	303.0	1s → σ* (C=O)	303.1

*See figure 4 for the definition of the spectral features.

These results demonstrate that the table-top laboratory system can be employed to acquire XUV absorption spectra for surface sensitive chemical analysis of different organic materials, yielding information on molecular orbitals, oxidation states and the local environments of carbon atoms. Some examples are given in the following sections.

3.2 NEXAFS on Polymethylmethacrylate

In order to study the growth process and the chemical environment of the involved C-atoms, NEXAFS spectra at the carbon K absorption edge were measured on thin polymethylmethacrylate (PMMA) films.[19] PMMA was deposited on thin silicon nitride substrates by pulsed laser deposition (PLD), using a KrF excimer laser (248 nm, 30 ns) at a fluence of about 125 mJ/cm². Smooth coatings without any droplets could be grown with thicknesses between 100 nm and several micrometres. A typical NEXAFS spectrum is shown in Figure 5 (blue curve), identifying all the absorption maxima and features typical for PMMA bulk material.[19] However, two main differences are apparent. First, the intensity of the sharp peak of the 1s→π*(C = O) transition at 288.5 eV, which is highest in bulk PMMA, is reduced compared to the literature data. Second, a peak of the 1s→π*(C = C) transition at 285 eV exists, indicating C = C bonds which are not present in bulk material (cf. chemical structure formula). Both differences can be explained by incubation processes at the PLD target surface, leading to a loss of methyl ester side groups and the formation of C = C double bonds.

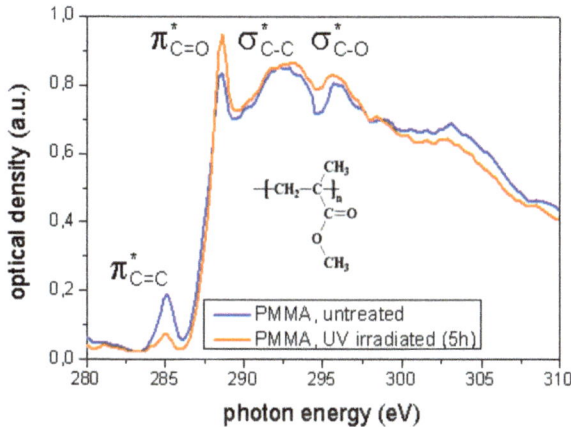

Figure 5 *NEXAFS measurements at the carbon K edge on PMMA films (200 nm thickness) grown by pulsed laser deposition on silicon nitride substrates.*

After UV irradiation of the sample an increase of the 1s→π*(C = O) peak and a reduction of the 1s→π*(C = C) peak were observed (figure 5, red curve), both indicating a re-polymerization of the PMMA film.

3.3 NEXAFS on Lipid Multilayers

A quantitative understanding of the interactions between nanoparticles and biological interfaces, in particular, the cell membrane, is a prerequisite for the design of drug delivery systems. For this reason phospholipid layers were investigated, applying a variety of analytical techniques.[20] NEXAFS spectra of selected dried phospholipid multilayers 1,2-Dioleoyl-sn-Glycero-3-Phosphatidilserine (DOPS), 1,2-Dioleoyl-sn-Glycero-3-Phos-phatidilcholine (DOPC) and 1,2-Dimyristoyl-3-Phosphatidylcholin (DMPC) were recorded, differing in the head group structure and in the presence of C = C double bonds in the hydrocarbon tails. The data displayed in Figure 6 indicate that π*(C = C), σ*(C = C), σ*(C – C) and Rydberg resonances as well as C1s→σ*(C – N) transitions can be resolved. In agreement with their chemical structures, the unsaturated DOPS and DOPC phospholipids exhibit strong π*(C = C) peaks, not present for the saturated DMPC.[20] Similarly consistent results were also obtained from NEXAFS measurements on humic acids, which are main bonding partners of toxic substances in soil. All of the samples were clearly distinguishable in relation to three major carbon-binding groups, namely aromatic, phenolic and carboxylic.[21]

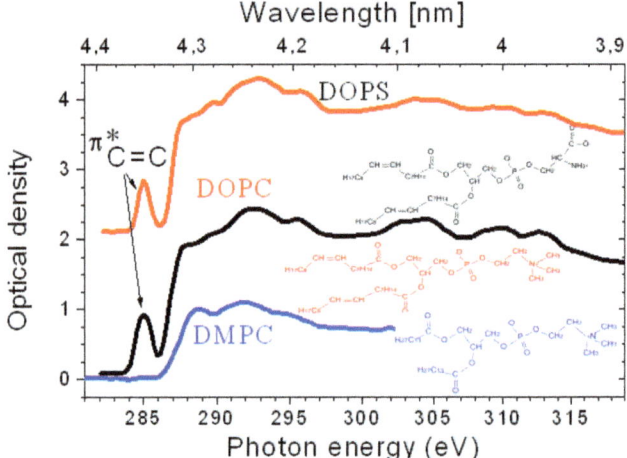

Figure 6 *NEXAFS spectra of unsaturated DOPS and DOPC and of a saturated DMPC phospholipid layer. The spectra are shifted vertically for clarity.*

4 CONCLUSIONS

The results presented in this chapter demonstrate that it is possible to investigate the near edge absorption fine structure of thin solid samples by utilizing broadband radiation from a laboratory-scale laser driven plasma source. NEXAFS spectra were recorded for surface sensitive chemical analysis of different organic materials (polymers, lipid films, humic acids, etc.), yielding information on molecular orbitals, oxidation states and the local environment of the carbon atoms involved. The spectra obtained from thin polymer films are in excellent agreement with synchrotron data. In comparison with the latter, the presented laboratory scale system offers the advantage of acquiring the entire NEXAFS spectrum in single pulses. This makes it ideally suited for time resolved experiments in pump probe schemes. Work in this direction is in progress. Moreover, recent improvements of the XUV source accomplish also the acquisition of NEXAFS spectra at athmospheric pressure, being of particular importance for organic samples in aqueous solution.

References
1 P. Jansson, U. Vogt and H. Hertz, *Rev. Sci. Instrum.*, 2005, **76**, 043503.
2 H. Fiedorowicz, *Laser and Particle Beams*, 2005, **23**, 365.
3 C. Peth, A. Kalinin, F. Barkusky, K. Mann, J.P. Toennies and L.Yu Rusin, *Rev. Sci. Instrum.*, 2007, **78** 103509.
4 J. Seres, V.S. Yakovlev, E. Seres, C. Streli, P. Wobrauschek, C. Spielmann and
F. Krausz, *Nature Phys.*, 2007, **3** 878.
5 M. Zepf, B. Dromey, M. Landreman, P. Foster and S.M. Hooker, *Phys. Rev. Lett.*, 2007, **99**, 143901.
6 Y. Wang, E. Granados and J.J. Rocca, *Phys. Rev. Lett.*, 2006, **97**, 123901.

7 P. Takman, H. Stollberg and H. Hertz, *J. Microsc.*, 2007, **226**, 175.

8 R.L. Sandberg, A. Paul, D.A. Raymondson, S. Hädrich, D.M. Gaudiosi, J. Holtsnider, R.I. Tobey, O. Cohen, M.M. Murnane and H.C. Kapteyn, *Phys. Rev. Lett.*, 2007, **99**, 098103

9 H. Kondo, T. Tomie and H. Shimizu, *Appl. Phys. Lett.*, 1996, **69**, 182.

10 U. Vogt, T. Wilhein, H. Stiel and H. Legall, *Rev. Sci. Instrum.*, 2004, **75**, 4606.

11 C. Peth, F. Barkusky and K. Mann, *J. Phys. D: Appl. Phys.*, 2008, **41**, 105202.

12 J. Stoehr, *NEXAFS Spectroscopy, Springer Series in Surface Science*, 2003, 25 (Berlin: Springer).

13 O. Dhez, H. Ade and G.S. Urquhart, *J. Electron Spectrosc. Relat. Phenom.*, 2003, **128**, 85.

14 M. Beck, U. Vogt, I. Will, A. Liero, H. Stiel, W. Sandner and T. Wilhein, *Opt. Commun.*, 2001, **190**, 317.

15 S. Kranzusch, C. Peth and K. Mann, *Rev. Sci. Instrum.*, 2003, **74**, 969.

16 Center for X-Ray Optics, http://www-cxro.lbl.gov/

17 NIST Atomic Spectra Database, http://physics.nist.gov/ PhysRefData/ASD/index.html

18 J.L. Jordan-Sweet, C.A. Kovac, M.J. Goldberg and J.F. Morar, *J. Chem. Phys.*, 1988, **89**, 2482.

19 B. Fuchs, F. Schlenkrich, S. Seyffarth, A. Meschede, R. Rotzoll, P. Vana, P. Großmann, K. Mann and H.-U. Krebs, *Appl. Phys. A*, 2010, **98**, 711.

20 E. Novakova, G. Mitrea, C. Peth, J. Thieme, K. Mann and T. Salditt, *Biointerphases*, 2008, **3**, FB44.

21 J. Sedlmair, S.-C. Gleber, C. Peth, K. Mann and J. Thieme, *J. Phys.: Conf. Series*, 2009, **186**, 012034.

NANOMETER SCALE IMAGING USING A DESK-TOP LASER PLASMA EUV SOURCE

P.W. Wachulak, A. Bartnik and H. Fiedorowicz

Institute of Optoelectronics, Military University of Technology, 00-908 Warsaw, Poland

1 INTRODUCTION

Future developments in nanoscience demand tools capable of capturing images at a nanometre scale spatial resolution. A direct path to improve the spatial resolution in photon based microscopy is to use short wavelength radiation. Various imaging methods and techniques are currently under active pursuit worldwide, one of them being extreme ultraviolet (EUV) and soft X-ray (SXR) microscopy, based on Fresnel zone plates;[1] a spatial resolution of 12 nm has been demonstrated in synchrotron based SXR microscopy.[2]

Demonstrations of actinic aerial EUV microscopes for mask inspection have been conducted at synchrotron facilities to provide the required short wavelength illumination from bending magnets.[3,4] These microscopes are capable of imaging small objects with spatial resolutions better than 100 nm,[5] but the size and cost of their operation is very high. The demonstration of bright EUV and SXR laboratory scale sources started the development of compact microscopes which can obtain images of objects with exposures of a few seconds and spatial resolutions approaching those of synchrotron based microscopes.[6,7] The development of bright, compact, short wavelength sources is a significant step forward in the commercialization of high resolution imaging tools in the near future. The work done so far in EUV and SXR imaging is very extensive; here, a few examples using coherent and incoherent EUV/SXR sources are presented as pointers for further research.

Images with 700 nm half-pitch resolution using an EUV recombination laser at $\lambda = 18.2$ nm have been reported in early imaging work,[8] and 75 nm resolution was reported employing a $\lambda = 4.48$ nm SXR laser pumped by the fusion class NOVA laser, limiting image acquisition due to the repetition rate of several shots per day.[9] Recently different approaches to sub-micrometre resolution imaging have emerged due to the development of smaller scale short wavelength sources such as high-order harmonics,[10] SXR lasers[11] and incoherent laser-plasma based sources.[12] Using radiation from a capillary discharge laser at a wavelength of 46.9 nm, EUV images were obtained with a spatial resolution of 120–150 nm.[13] Using a 1J, 8ps pump laser, radiation at a wavelength of 13.2 nm from a Ni-like cadmium EUV laser allowed for 55 nm imaging in reflection mode[14] and sub-38 nm resolution in transmission mode.[6] With a capillary discharge laser, microscopy in transmission mode has been demonstrated with a single EUV laser pulse, leading to 54 nm half-pitch spatial resolution and, additionally, temporal resolution of ~1 ns.[15] Quasi-monochromatic emission from an incoherent SXR source based on liquid nitrogen, at $\lambda = 2.88$ nm in the "water-window" range, allowed the demonstration of SXR microscopy with sub-50 nm spatial resolution ($\approx 17\lambda$).[16] Using a xenon based gas discharge EUV source, a Schwarzschild objective and Fresnel zone plate objective, EUV imaging has been demonstrated reaching a spatial resolution of about 100 nm.[17]

2 EXPERIMENTAL DETAILS AND RESULTS

This section describes a desk-top microscope using a laser plasma EUV source based on a double stream gas-puff target, providing 50 nm spatial resolution with a very compact setup. The source used in these experiments was developed for EUV metrology applications within the MEDEA+ project,[18] later modified and optimized for efficient, quasi-monochromatic emission from argon plasma at 13.8 nm wavelength,[19] and quasi-continuum emission in the 13–14 nm wavelength range from xenon plasma.[20] The use of the gas-puff target eliminates the debris production problem associated with solid targets.

Unpolarized EUV radiation from the plasma was collected, focused and spectrally filtered by an ellipsoidal, off-axis Mo/Si multilayer condenser mirror, designed to image the plasma with unity lateral magnification in order to illuminate the entire object. The condenser was developed in cooperation with REFLEX s.r.o. (mirror substrate) and the Fraunhofer Institute for Applied Optics and Precision Engineering, Jena (multilayer coating). A Fresnel zone plate objective was used in transmission mode to form a magnified image onto an EUV sensitive CCD camera. The thickness of the object and the bandwidth of the illuminating radiation were studied in order to estimate their influence on the spatial resolution of the microscope. Objects with thicknesses much larger than the depth of focus (DOF) of the microscope objective will degrade the resolution, while if the objective is highly dispersive, as zone plates are, the illumination bandwidth also plays a key role, due to the wavelength dependence of the focal length. EUV images obtained by illumination of the object with variable bandwidth radiation were compared to study the bandwidth influence on the spatial resolution.

In the experiment two distinct objects were imaged – a copper mesh about 4 μm thick and a holey carbon foil coated with a thin layer of gold to give a total thickness of 70 nm. The mesh thickness was about 11 times the microscope DOF, while that of the foil was only 0.2 times the DOF. A typical EUV image of the holey foil object taken using quasi-monochromatic radiation from argon plasma is shown in figure 1a, with the boxed region magnified in figure 1d. Knife-edge resolution measurements carried out using the magnified region indicate a half-pitch resolution 51±11 nm (about 3.7λ), based on six independent measurements.[21] Figures 1b and 1e show images of the same object when illuminated by quasi-continuous (13–14 nm) radiation from xenon plasma. In this case the measured resolution was 140±15 nm, worse than the quasi-monochromatic case because of the ten times larger bandwidth radiation that illuminated the sample.

Although less prominent the influence of the object thickness on the spatial resolution was also measured using quasi-monochromatic radiation and the copper mesh object, as shown in figures 1c and 1f; a spatial resolution of 73±5 nm was obtained, corresponding to the previously reported spatial resolution of the microscope of 69±4 nm.[22] All the images shown in figure 1 were acquired with a few tens of EUV pulses.

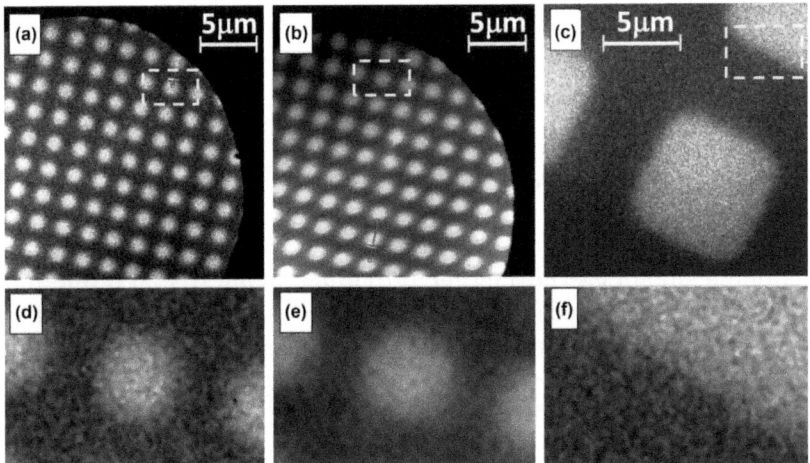

Figure 1 *(a) An EUV image of a "thin" foil object illuminated by a quasi-monochromatic radiation from argon plasma. (d) The boxed area of (a). (b,e) An image of the same object obtained using quasi-continuous illumination from xenon plasma. (c,f) EUV image of a "thick" copper mesh object illuminated by quasi-monochromatic radiation from argon plasma.*

Another experiment was related to the possibility of single-shot operation with temporal resolution equal to the duration of the EUV pulse, in this case about 3 ns. Using the copper mesh object a sequence of images was obtained with decreasing numbers of EUV pulses, as shown in figure 2. The signal to noise ratio clearly decreases as the number of pulses decreases, but the main features of the mesh can be seen even with a single EUV pulse, although the image is very noisy and grainy, indicating the necessity to increase the number of photons illuminating the CCD detector. The signal to noise ratio can be improved by integrating the signal over small areas in the CCD, binning 4×4 pixel areas of the image, for example, as can be seen in the rightmost image in the sequence depicted in figure 2. This procedure, however, results in worse spatial resolution, since the equivalent pixel size after binning becomes larger.

3 CONCLUSIONS

A near-50 nm spatial resolution desk-top EUV transmission microscope based on diffractive optics and a laser plasma source has been demonstrated. This microscope, with incoherent illumination, allowed the capture of images at a wavelength of 13.8 nm with a spatial resolution of 51 nm and exposure times of 50s, comparable to larger table-top systems and synchrotron based installations.

The influence of object thickness and illumination bandwidth on spatial resolution, as well as single-shot operation capabilities were also addressed.

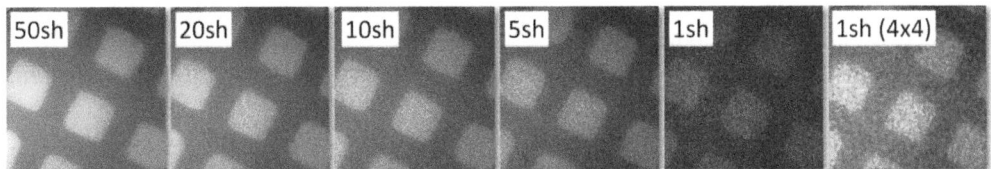

Figure 2 *EUV images of a copper mesh with decreasing numbers of EUV pulses. Last rightmost image in the sequence was obtained after integration of the signal from 16 CCD pixels (4×4 binning) to improve the signal to noise ratio.*

Practical, compact, high-resolution EUV full-field microscopes will allow the study of ultrafast processes with nanometre-scale resolutions and have the potential to become enabling tools for a wide range of nanoscale scientific and technological applications.

ACKNOWLEDGEMENTS

The research was partially performed in the frame of the EUREKA project Σ!3892 ModPolEUV supported by the Ministry of Science and Higher Education of Poland (decision Nr 120/EUR/2007/02), by the Foundation for Polish Science under the HOMING 2009 Programme (grant number HOM2009/14B), the EC's 7. Framework Program (LASERLAB-EUROPE - grant agreement n° 228334 and COST Action MP0601, also co-funded by Polish Ministry of Science and Education, decision number 816/N-COST/2010/0)

References

1 D. Attwood, *Soft X-Rays and Extreme Ultraviolet Radiation*, 1999, Cambridge, Cambridge University Press.
2 W. Chao, J. Kim, S. Rekawa, P. Fischer and E.H. Anderson, *Opt. Exp.*, 2009, **17**, 17699.
3 K.A. Goldberg, *Proc. SPIE*, 2007, **6730**, 67305E-1.
4 H. Kinoshita, K. Hamamoto, N. Sakaya, M. Hosoya and T. Watanabe, *Jpn. J. Appl. Phys. Part 1*, 2007, **46**, 6113.
5 K.A. Goldberg, P. Naulleau, I. Mochi, E.H. Anderson, S.B. Rekawa, C.D. Kemp, R.F. Gunion, H.-S. Han and S. Huh, *J. Vac. Sci. Technol.*, 2008, **B26**, 2220.
6 G. Vaschenko, C. Brewer, F. Brizuela, Y. Wang, M.A. Larotonda, B.M. Luther, M.C. Marconi, J.J. Rocca and C.S. Menoni, *Opt.Lett.*, 2006, **31**, 1214.
7 P.A.C. Takman, H. Stollberg, G.A. Johansson, A. Holmberg, M. Lindblom and H.M. Hertz, *J. Microsc.*, 2007, **226**, 175.
8 D.S. DiCicco, D. Kim, R. Rosser and S. Suckewer, *Opt. Lett.*, 1992, **17**, 157.
9 L.B. Da Silva, J.E. Trebes, S. Mrowka, T.W. Barbee Jr., J. Brase, J.A. Koch, R.A. London, B.J. MacGowan, D.L. Matthews, D. Minyard, G. Stone, T. Yorkey, E. Anderson, D.T. Attwood and D. Kern, *Opt. Lett.*, 1992, **17**, 754.
10 M. Wieland, C. Spielmann, U. Kleineberg, T. Westerwalbesloh, U. Heinzmann and T. Wilhein, *Ultramicroscopy*, 2005, **102**, 93.

11 M. Kishimoto, M. Tanaka, R. Tai, K. Sukegawa, M. Kado, N. Hasegawa, H. Tang, T.Kawachi, P.Lu, K. Nagashima, H.Daido, Y.Kato, K.Nagai and H. Takenaka, *J. Phys. IV*, 2003, **104**, 141.

12 I.A. Artioukov, A.V. Vinogradov, V.E. Asadchikov, Y.S. Kasyanov, R.V. Serov, A.I. Fedorenko, V.V. Kondratenko and S.A. Yulin, *Opt. Lett.*, 1995, **20**, 2451.

13 G. Vaschenko, F. Brizuela, C. Brewer, M. Grisham, H. Mancini, C.S. Menoni, M.C. Marconi, J.J. Rocca, W. Chao, J.A. Liddle, E.H. Anderson, D.T. Attwood, A.V. Vinogradov, I.A. Artioukov, Y.P. Pershyn and V.V. Kondratenko, *Opt. Lett.*, 2005, **30**, 2095.

14 F. Brizuela, Y. Wang, C.A. Brewer, F. Pedaci, W. Chao, E.H. Anderson, Y. Liu, K.A. Goldberg, P. Naulleau, P. Wachulak, M.C. Marconi, D.T. Attwood, J.J. Rocca and C. S. Menoni, *Opt. Lett.*, 2009, **34**, 27.

15 C.A. Brewer, F. Brizuela, P. Wachulak, D.H. Martz, W. Chao, E.H. Anderson, D.T. Attwood, A.V. Vinogradov, I. A. Artyukov, A.G. Ponomareko, V.V. Kondratenko, M.C. Marconi, J.J. Rocca and C.S. Menoni, *Opt. Lett.*, 2008, **33**, 518; this article was also published in *Virtual Journal for Biomedical Optics*, 2008, **3**, 4, and in *Virtual Journal of Nanoscale Science & Technology*, 2008, **17**, 15.

16 K.W. Kim, Y. Kwon, K.Y. Nam, J.H. Lim, K.G. Kim, K.S. Chon, B.H. Kim, D.E. Kim, J.G. Kim, B.N. Ahn, H.J. Shin, S. Rah, K.H. Kim, J.S. Chae, D.G. Gweon, D.W. Kang, S.H. Kang, J.Y. Min, K.S. Choi, S.E. Yoon, E.A. Kim, Y. Namba and K.H. Yoon, *Phys. Med. Biol.*, 2006, **51**, N99.

17 L. Juschkin, R. Freiberger and K. Bergman, *J. Phys.: Conf. Ser.*, 2009, **186**, 012030.

18 H. Fiedorowicz, A. Bartnik, R. Jarocki, J. Kostecki, J. Krzywinski, J. Mikołajczyk, R. Rakowski, A. Szczurek and M. Szczurek, *J. Alloys Compd.*, 2005, **401**, 99.

19 P.W. Wachulak, A. Bartnik, H. Fiedorowicz, T. Feigl, R. Jarocki, J. Kostecki, R. Rakowski, P. Rudawski, M. Sawicka, M. Szczurek, A. Szczurek and Z. Zawadzki, *Applied Physics B* **100**, 3, 461-469, (2010)

20 R. Rakowski, A. Bartnik, H. Fiedorowicz, F. de Gaufridy de Dortan, R. Jarocki, J. Kostecki, J. Mikołajczyk, L. Ryc, M. Szczurek, P. Wachulak, *Appl. Phys.*, 2010, **B101**, 773.

21 P.W. Wachulak, A. Bartnik, H. Fiedorowicz and J. Kostecki, *Opt. Exp.*, 2011, **19**, 9541.

22 P.W. Wachulak, A. Bartnik and H. Fiedorowicz, *Opt. Lett.*, 2010, **35**, 2337.

LASER-PLASMA EUV AND SOFT X-RAY SOURCES FOR MICROSCOPY APPLICATIONS

P.W. Wachulak,[1] A. Bartnik,[1] H. Fiedorowicz,[1] T. Feigl,[2] R. Jarocki,[1] J. Kostecki,[1] L. Pina,[3] M. Szczurek,[1] A. Szczurek[1] and Z. Zawadzki[1]

[1]Institute of Optoelectronics, Military University of Technology, 00-908 Warsaw, Poland
[2]Fraunhofer Institut für Angewandte Optik und Feinmechanik (IOF), 07745 Jena, Germany
[3]Faculty of Nuclear Sciences and Physical Engineering, Czech Technical University, 180 00 Praha 8, Czech Republic

1 INTRODUCTION

The recent development of compact short wavelength laboratory sources has promoted the study of matter properties in the extreme ultraviolet (EUV) region. The strong demand for such quasi-monochromatic sources results from the semiconductor industry striving for ever smaller, faster and less power consuming computer chips, and has also motivated the development of compact, high resolution imaging tools. These sources have a huge potential impact on the speed of nanotechnological development since experiments previously restricted to large-scale facilities can now be performed in the laboratory environment. The importance of the development and commercial availability of EUV sources, in particular the lack of commercial tools for actinic inspection for the semiconductor industry, has already been noted by Intel.[1]

The usability of existing EUV sources has been proven by many examples of experiments involving compact table-top sources including capillary discharge lasers for interference lithography and wavelength resolved holography,[2,3] high-harmonic generation[4,5] for lensless imaging[6] and surface deformation studies,[7] optically pumped EUV lasers for microscopy using zone plates,[8] and xenon discharge produced plasmas (Energetiq) for development of actinic full-field EUV mask blank inspection tool at MIRAI-Selete.[9]

One such compact source is based on a laser plasma gas-puff target (LPGPT source). The suitability of gas-puff targets for efficient soft X-ray generation in the "water window" range is well established, providing strong line emission from transitions in H-like and He-like nitrogen ions at wavelengths of 2.478 nm and 2.879 nm, for example using a Nd:glass laser with 5 J per pulse[10] or more recently in a very compact, desk-top Nd:YAG laser setup employing a much smaller pulse energy of 0.74 J.[11] A significant enhancement of X-ray production in the 1 keV energy range was achieved by modification of the original single nozzle gas-puff target to a double-stream gas puff.[12] Using radiation in the wavelength range 6–20 nm, produced via irradiation of the double-stream gas-puff target with a Nd:YAG laser, direct and very efficient photo-etching of PTFE was observed.[13] The gas-puff source was also used for studying EUV fluorescence radiation from aluminium andb silicon[14] and microstructuring of PMMA.[15] It was also shown that by employing a double stream gas-puff target the energy conversion efficiency from the infrared pumping laser to EUV radiation emitted in the wavelength region of 13.5 nm band can be significantly improved. The calculated EUV production is up to 8.7×10^{13} photons per steradian corresponding to 8 mJ of emitted energy over a solid angle of 2π, a conversion efficiency of 1.6%.[16]

In the following sections laser plasma soft X-ray and EUV sources using a double gas puff target are described. The sources have been characterized and optimized for applications in nanoscale imaging using Fresnel zone plate optics.

2 LASER PLASMA SOURCE FOR NANOSCALE IMAGING AT 13.8 NM

In imaging employing Fresnel zone plate objectives monochromaticity of the radiation is a key parameter. For this purpose a laser-plasma argon gas-puff target source was optimized for efficient EUV photon production. In the experiments,[17] the source was characterized and optimized for quasi-monochromatic emission near 13.5 nm. The source was based on a double nozzle gas-puff target later employed in applications requiring narrow -bandwidth EUV radiation, such as microscopy with zone plate objectives.[18,19] Using argon/helium gas-puff target and a collinear double stream electromagnetic valve, efficient EUV production was achieved in the wavelength range 6–20 nm. Using an elliptical mirror with Mo/Si multilayer coating as a spectral filter (for 13–14 nm wavelengths) and EUV radiation collector, the bandwidth was drastically narrowed. As a result, an efficient EUV source was developed emitting quasi-monochromatic radiation at 13.8 nm with the spectrum shown in figure 1a.

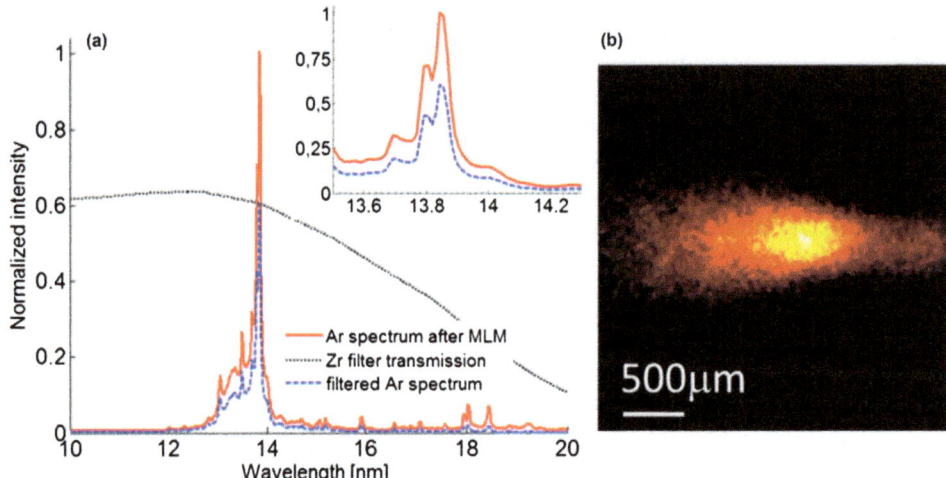

Figure 1 (a) *Emission from the laser plasma gas-puff EUV source equipped with a multilayer mirror showing quasi-monochromatic emission at wavelengths around 13.8 nm. The result of further filtration with a 141 nm thick zirconium filter is shown by the dashed line. (b) A pinhole camera image of the emitting region imaged using a condenser optic.*

Argon plasmas emit radiation in the 13–14 nm band mostly in the two strongest lines of Ar^{7+}, corresponding to the transitions $2p^63p$-$2p^65d$ at $\lambda = 13.793$ nm and

$2p^63s-2p^65p$ at $\lambda = 13.844$ nm, with $\lambda/\Delta\lambda \approx 140$. The number of photons in the 13–14 nm band was measured to be $(8.8\pm0.5)\times10^{10}$ in a single EUV pulse. The EUV source shape and size were measured using a pinhole camera. The source was elliptical with major and minor axes of 1090 μm and 390 μm, as can be seen in figure 1b.

3 "WATER-WINDOW" LASER PLASMA SOFT X-RAY SOURCE

X-ray sources emitting in the "water-window" region, $\lambda = 2.3$–4.4 nm,[20] are important for imaging of biological samples in near-natural environments. High contrast in this spectral range is obtained due to different absorption of different constituents; water has a relatively small absorption coefficient while carbon has much higher absorption, resulting in a very good image contrast.

Soft X-ray (SXR) microscopy has been successfully employed, mainly in transmission mode using diffractive optics, such as zone-plates,[21,22,23] and raster scanning the sample across the focused SXR beam.[24,25,26] Another technique is contact microscopy, where the sample is placed on a recording medium, such as a photoresist, and illuminated by the SXR beam to make an image in the surface of the recording medium.[27,28,29] However in most of these important achievements synchrotron sources were used for sample illumination, due to their obvious advantages. The main motivation for development of compact, table-top SXR sources is to open the possibility to perform experiments without the necessity to employ large "photon factories" such as synchrotrons or free electron lasers. These sources are the state-of-the-art devices with tuneable, wide wavelength output ranges, high photon fluxes and brightnesses. Although desk-top sources cannot match these capabilities they are still important alternatives that can perform some of the experiments in a university laboratory, for example. While synchrotrons are always attractive SXR sources for cutting-edge experiments, compact EUV and SXR sources are demonstrably growing more important.

A previous experiment demonstrating strong X-ray emission in the water-window from a nitrogen gas puff target has been reported.[30] Strong emission was observed at $\lambda = 2.478$ nm and $\lambda = 2.879$ nm from 2–1 transitions in H- and He-like nitrogen ions. This experiment used a relatively large Nd:glass system with a pulse energy of 1–5 J, a pulse duration of ~1 ns and a single stream gas-puff target. More recently the characterisation and optimisation of a desk-top water-window source, suitable for SXR microscopy of biological specimens, were reported.[11] This used a commercial Nd:YAG laser NL 303HT (Ekspla, Lithuania) with a smaller laser pulse energy of 737 mJ at $\lambda = 1064$ nm and a longer, 4 ns, pulse.

The argon and nitrogen sources are complementary, in that the broadband emission from an argon plasma with $\lambda = 2$–4 nm can be used for contact microscopy and the quasi-monochromatic emission from a nitrogen plasma, at $\lambda = 2.88$ nm, is suitable for imaging using diffractive optics. To spectrally narrow the broad emission from gaseous targets two filters were used, one 750 nm thick aluminium and the other 1 μm thick Mylar ($C_{10}H_8O_4$). The spectra for argon and nitrogen plasmas are shown in figures 2a and 2b.

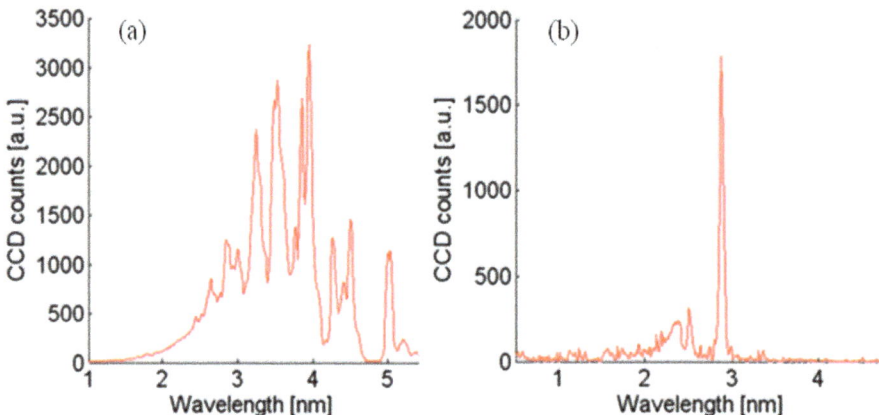

Figure 2 *(a) The quasi-continuous spectrum of argon in the 1-5.5 nm spectral range, including the water-window, and (b) the quasi-monochromatic nitrogen spectrum in the 0.5-5 nm spectral range.*

Argon plasmas emit in a broad spectral range in the water-window range. The dominant groups of spectral lines are Ar^{11+} $2s^22p^23d$–$2s^22p^3$ with $\lambda = 3.065$–3.274 nm, Ar^{10+} $2s^22p^33d$–$2s^22p^4$ with $\lambda = 3.353$–3.631 nm and Ar^{9+} $2s^22p^43d$–$2s^22p^5$ with $\lambda = 3.656$–3.888 nm.[31] Nitrogen plasmas, on the other hand, allow for quasi-monochromatic emission with one dominant spectral line, N^{5+} $1s2p$-$1s^2$ ($1P^0$–$g1S$) at $\lambda = 2.8787$ nm. Other nitrogen lines are N^{6+} $1s$–$2p$ (J–J 1/2–3/2 and 1/2–1/2), at $\lambda = 2.4779$ nm and $\lambda = 2.4785$ nm, N^{6+} $3p$–$1s$, $4p$–$1s$ and $5p$–$1s$ at wavelengths of 2.0910 nm, 1.9826 nm and 1.9361 nm respectively. The plasma spatial distributions and source sizes were measured using a pinhole camera. For both gases the plasmas are elliptical with full widths at half maximum of 220×146 μm for argon and 325×143 μm for nitrogen, as shown in figures 3a and 3b. The numbers of photons in one pulse were $(7.23\pm0.58)\times10^7$ for argon source, and $(1.06\pm0.08)\times10^7$ for nitrogen.

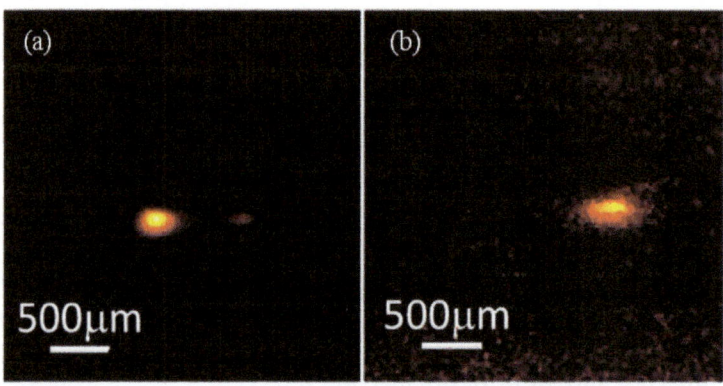

Figure 3 *Pinhole camera images of (a) the argon and (b) the nitrogen soft X-ray sources.*

4 CONCLUSIONS

Laser-plasma gas-puff target sources suitable for EUV and SXR microscopy have been presented. Microscopy with Fresnel zone plates, due to their high dispersion, requires monochromatic radiation, which to a good approximation is the quasi-monochromatic emission from an argon based EUV source at 13.84 nm. This source was used for microscopy experiments and provided a spatial resolution of about 50 nm.[19] Similarly a nitrogen based source can be used for SXR microscopy with the same type of optic. An argon based SXR source offers much higher photon flux through the broadband emission and, although not suitable for zone plate objectives, were employed in SXR microscopes with reflective, dispersion-free optics such as Wolter type I objectives. [32,33]

ACKNOWLEDGEMENTS

The research was partially performed in the frame of the EUREKA project Σ!3892 ModPolEUV supported by the Ministry of Science and Higher Education of Poland (decision Nr 120/EUR/2007/02), by the Foundation for Polish Science under the HOMING 2009 Programme (grant number HOM2009/14B), the EC's 7. Framework Program (LASERLAB-EUROPE - grant agreement n° 228334 and COST Action MP0601, also co-funded by Polish Ministry of Science and Education, decision number 816/N-COST/2010/0).

References

1 V. Bakshi, *Semiconductor International*, 2009, (Austin, Texas).

2 P.W. Wachulak, M.C. Marconi, R.A. Bartels, C.S. Menoni and J.J. Rocca, *J. Opt. Soc. Am.*, 2008, **B25**, 1811.

3 P.W. Wachulak, L.Urbanski, M.G. Capeluto, D. Hill, W.S. Rockward, C. Iemmi, E.H. Anderson, C.S. Menoni, J.J. Rocca and M.C. Marconi, *J. Micro/Nanolith. MEMS MOEMS*, 2009, **8**, 021206.

4 R.A. Bartels, A. Paul, H. Green, H.C. Kapteyn, M.M. Murnane, S. Backus, I.P. Christov, Y. Liu, D. Attwood and C. Jacobsen, *Science*, 2002, **297**, 376.

5 I.J. Kim, G.H. Lee, S.B. Park, Y.S. Lee, T.K. Kim and C.H. Namb, *Appl. Phys. Lett.*, 2008, **92**, 021125.

6 R.L. Sandberg, A. Paul, D.A. Raymondson, S. Hadrich, D.M. Gaudiosi, J. Holtsnider, R.I. Tobey, O. Cohen, M.M. Murnane, H.C. Kapteyn, C. Song, J. Miao, Y. Liu and F. Salmassi, *Phys. Rev. Lett.*, 2007, **99**, 098103.

7 R.I. Tobey, M.E. Siemens, O. Cohen, M.M. Murnane, H.C. Kapteyn and K.A. Nelson, *Opt. Lett.*, 32, 3 (2007)

8 G. Vaschenko, C. Brewer, F. Brizuela, Y. Wang, M. A. Larotonda, B. M. Luther, M. C. Marconi, J. J. Rocca, and C. S. Menoni, *Opt. Lett.*, 2006, **31**, 9.

9 T. Teresawa, T. Yamane, T. Tanaka, T. Iwasaki, O. Suga and T. Tomie, *Proc. SPIE*, 2009, **7271**, 727122.

10 H. Fiedorowicz, A. Bartnik, R.Jarocki, M. Szczurek and T.Wilhein, *Appl. Phys.*, 1998, **B67**, 391.

11 P.W. Wachulak, A. Bartnik, H. Fiedorowicz, P. Rudawski, R. Jarocki, J. Kastecki and M. Szczurek, *Nucl. Instrum. Methods Phys. Res.*, 2010, **B268**, 1692.

12 H. Fiedorowicz, A. Bartnik, R. Jarocki, R. Rakowski and M. Szczurek, *Appl. Phys.*, 2000, **B70**, 305.

13 A. Bartnik, H. Fiedorowicz, R. Jarocki, L. Juha, J. Kostecki, R. Rakowski and M. Szczurek, *Appl. Phys.*, 2006, **B82**, 529.

14 A. Bartnik, H. Fiedorowicz, R. Jarocki, J. Kostecki, R. Rakowski and M. Szczurek, *Appl. Phys.*, 2008, **B93**, 737.

15 A. Bartnik, H. Fiedorowicz, R. Jarocki, J. Kostecki, A. Szczurek and M. Szczurek, *Appl. Phys.*, 2009, **B96**, 727.

16 H. Fiedorowicz, A. Bartnik, R. Jarocki, J. Kostecki, J. Krzywinski, J. Mikołajczyk, R. Rakowski, A. Szczurek and M. Szczurek, *Journal of Alloys and Compounds*, 2005, **401**, 99.

17 P.W. Wachulak, A. Bartnik, H. Fiedorowicz, T. Feigl, R. Jarocki, J. Kostecki, R. Rakowski, P. Rudawski, M. Sawicka, M. Szczurek, A. Szczurek and Z. Zawadzki, *Appl. Phys.*, 2010, **B100**, 461.

18 P.W. Wachulak, A. Bartnik and H. Fiedorowicz, *Opt. Lett.*, 2010, **35**, 14, 2337.

19 P.W. Wachulak, A. Bartnik, H. Fiedorowicz, and Jerzy Kostecki, *Opt. Exp.*, 2011, **19**, 9541.

20 L.B. Da Silva, J.E. Trebes, R. Balhorn, S. Mrowka, E. Anderson, D.T. Attwood, T.W. Barbee Jr., J. Brase, M. Corzett, J. Gray, J.A. Koch, C. Lee, D. Kern, R.A. London, B.J. MacGowan and D.L. Mathews, *Science*, 1992, **258**, 269.

21 B. Niemann, D. Rudolph and G. Schmahl, *Opt. Commun.*, 1974, **12**, 160.

22 B. Niemann, D. Rudolph and G. Schmahl, *Appl. Opt.*, 1976, **15**, 1883.

23 G. Schneider, B. Niemann, P. Guttmann, D. Rudolph and G. Schmahl, *Synchr. Rad. News*, 1995, **8**, 19.

24 C. Jacobsen, S. Williams, E. Anderson, M.T. Browne, C.J. Buckley, D. Kern, J. Kirz, M. Rivers and X. Zhang, *Opt. Commun.*, 1991, **86**, 351.

25 C. Jacobsen, J. Kirz and S. Williams, *Ultramicroscopy*, 1992, **47**, 55.

26 J. Kirz, C. Jacobsen, S. Lindaas, S. Williams, X. Zhang, E. Anderson and M. Howells, *Synchrotron Radiation in the Biosciences*, 1994, (Oxford, Oxford University Press) p563.

27 G. Poletti, F. Orsini and D. Batani, *Solid State Phenomena*, 2005, **107**, 7.

28 A.C. Cefalas, P. Argitis, Z. Kollia, E. Sarantopoulou, T.W. Ford, A.D. Stead, A. Marranca, C.N. Danson, J. Knott and D. Neely, *Technical Report RAL-TR-98-007*, 1998, (Oxford, Rutherford Appleton Laboratory).

29 M. Kado, H. Daido, Y. Yamamoto, K. Shinohara and M.C. Richardson, *Proc. 8th Int. Conf. X-ray Microscopy*, 2005, *IPAP Conf. Series*, **7**, 41.

30 H. Fiedorowicz, A. Bartnik, R. Jarocki, M. Szczurek and T. Wilhein, *Appl. Phys.*. 1998, **B67**, 391.

31 R. L. Kelly, *J. Phys. Chem. Ref. Data*, 1987, **16**, suppl. 1.

32 P. W. Wachulak, A. Bartnik, L. Wegrzynski, J. Kostecki, R. Jarocki, T. Fok , M. Szczurek and H. Fiedorowicz, *NIM B,* 2013, **311**, 42–46

33 P. W. Wachulak, A. Bartnik, M. Skorupka, J. Kostecki, R. Jarocki, M. Szczurek, L. Wegrzynski, T. Fok and H. Fiedorowicz, *Applied Physics B*, 2013, **111**, 2, 239-247

NANOMETER SCALE IMAGING WITH TABLE-TOP EXTREME ULTRAVIOLET LASER

P.W. Wachulak,[1] A. Isoyan,[2] R.A. Bartels,[3] C.S. Menoni,[3] J.J. Rocca[3] and M.C. Marconi[3]

[1] Institute of Optoelectronics, Military University of Technology, 00-908 Warsaw, Poland
[2] Synopsys, Inc., 2025 NW Cornelius Pass Road, Hillsboro, OR 97124, USA
[3] NSF ERC for Extreme Ultraviolet Science & Technology and Department of Electrical and Computer Engineering, Colorado State University, Fort Collins, CO 80523 USA

1 INTRODUCTION

Decreasing the illumination wavelength provides the potential to improve the spatial resolution in all imaging techniques. Hence there is currently high interest in imaging using laboratory short wavelength sources. This chapter describes approaches to EUV imaging utilizing a compact table-top capillary discharge EUV laser. Various coherent imaging techniques, such as two and three dimensional holography, computer generated holograms (CGHs), EUV reconstruction and generalized Talbot self-imaging will be presented.

2 NANOMETRE SCALE IMAGING TECHNIQUES

In all coherent imaging experiments, described herein, a table-top discharge pumped capillary Ne-like argon laser was used.[1-3] Such lasers can be configured to produce pulses with energies of 0.1–0.8 mJ and durations of about 1.2 ns FWHM. They can be operated at repetition rates of several hertz, producing EUV pulses with average powers in excess of 1 mW with high degrees of spatial and temporal coherence. The lasers operate via the 46.9 nm $3p^1S0–3s^1P^1$ transition of Ne-like argon. An alumina capillary, 3.2 mm in diameter and 27 cm long, filled with argon is excited with a current pulse of about 22 kA, a 10–90% rise time of about 55 ns and a first half-cycle duration of about 135 ns. The capillary discharge laser (CDL) is based on high-gain amplification of spontaneous emission and generates a beam with high spatial and temporal coherence. The longitudinal coherence length is determined by the Doppler-broadened linewidth of the laser transition, $\Delta\lambda/\lambda \approx 10^{-4}$, corresponding to a coherence length of about 470 μm. The degree of spatial coherence increases with the length of the plasma column (capillary) and approaches full coherence for 36 cm long capillaries.[4]

Holography is a well developed imaging technique originally proposed in 1948 by Gabor as a new method of microscopy.[5] It has been used in many applications, especially since the development of lasers which provide the coherent illumination necessary for holography to become a practical imaging technique.

The acquisition of holographic images is a process consisting of two steps, recording and reconstruction. During the first step the interference pattern between the mutually coherent reference and object beams is stored in a medium which can be any material or device that records the incident intensity in a controllable way.

The potential for holography in the EUV spectral region was soon recognized, but it was not until 1970s, when the first images of very simple objects were obtained.[6,7] The first demonstration of holographic recording using an EUV/soft X-ray laser was performed at the Lawrence Livermore National Laboratory, using the NOVA laser facility.[8] Some early experiments also utilized synchrotron sources to image biological samples, nano structures and magnetic domains.[9,10] The first demonstration of holographic imaging using a table-top (high harmonic generation, HHG) source achieved 7 μm spatial resolution.[11] With a very similar set up the spatial resolution was later improved to 0.8 μm.[12] Time resolved holographic imaging was also implemented with HHG sources to study the ultrafast dynamics of surface deformation with a longitudinal resolution of less than 100 nm and a lateral resolution of less than 80 μm.[13]

In the experiments, described herein, a high spatial resolution photoresist was used to record large numerical aperture (NA) holograms[14,15] in the Gabor configuration, using an EUV capillary discharge laser source. Development of the photoresist gave a surface relief holographic pattern which was digitized using an atomic force microscope (AFM). The digitized hologram was used as input for a code that numerically reconstructed the image using a Fresnel propagator.[16,17,18] Holograms of AFM tips and carbon nanotubes and the corresponding reconstructions are shown in figure 1. The best resolution achieved to date was 45.8±1.9 nm, corresponding to 0.98 λ.

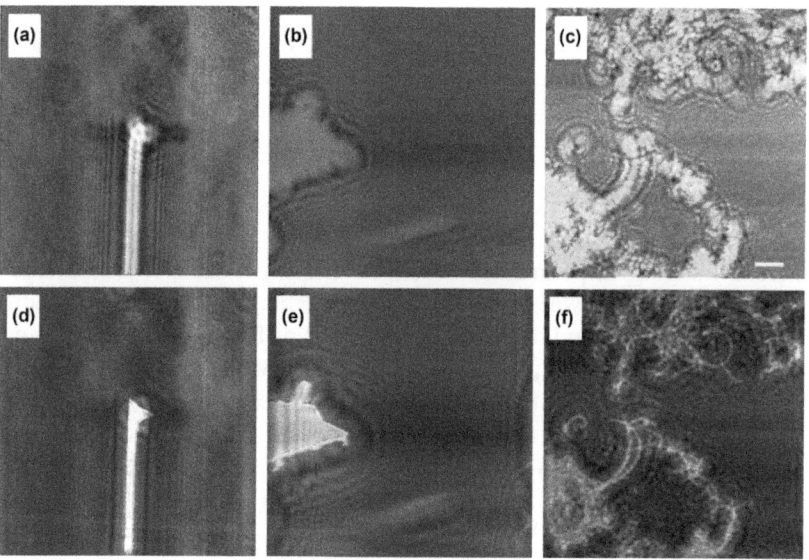

Figure 1 *EUV holograms of AFM tips (a,b) and carbon nanotubes (c) and the corresponding numerical reconstructions (d-f) with spatial resolutions of ≈380 nm, ≈164 nm and 46±2nm, respectively.*

Three dimensional imaging using numerical sections obtained from a single high NA hologram was also undertaken.[19,20] A tilted aluminium membrane with randomly deposited spherical markers was illuminated by a coherent beam in the Gabor in-line configuration. Three dimensional images were obtained from holograms recorded in photoresist. Digitized holograms were numerically reconstructed over the range of image planes by numerically varying the reconstruction distance, resulting in an optical sectioning of the image depths. A correlation algorithm was later applied to retrieve the depth information from a set of reconstructed images revealing the slope of the membrane with resolutions of about 160 nm laterally and 2.1 µm in depth.[20]

Another interesting approach to holographic imaging is the fabrication of "synthetic" or computer generated holograms (CGHs). Specially designed holographic masks replace classically recorded holograms, and the object does not have to physically exist. The hologram is constructed on the surface of a photoresist, this technique, also called holographic projection lithography, was proposed as a lithographic alternative technique to generate arbitrary nanostructures.[21,22] Initial results of the EUV reconstruction of computer generated nano-holograms, leading to nano-structures patterned in photoresist, are shown in figure 2.[23] Due to the restrictions of the electron beam fabrication method, binary holograms were designed and generated using a half-tone optimization algorithm to improve the image quality. The CGHs were made on a silicon nitride membrane spin-coated with a layer of hydrogen silsesquioxane (HSQ) photoresist that acted as an absorber for the electrons. The reconstructions, figures 2c and 2d, clearly resemble the initial patterns.The resolution was estimated to be about 230nm.

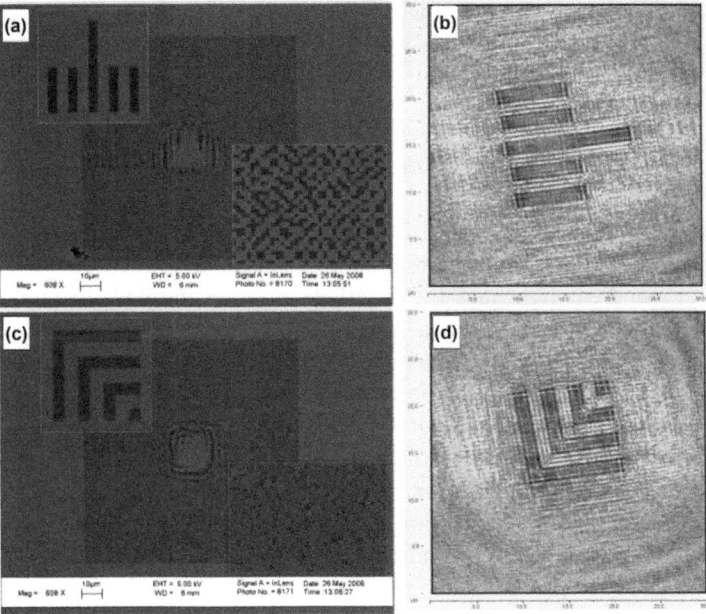

Figure 2 *(a,b) Two computer generated holograms of the objects shown at the top left of the images. The insets, at the bottom right of the images, show the binary nature of the holograms. (c,d) EUV reconstructions of the CGHs recorded in photoresist. The scale bar is 10 µm.*

Coherent EUV illumination also enabled an alternative approach based on coherent Talbot self-imaging. When a periodic structure is illuminated with a coherent source, self images of the structure are replicated at distances determined by its period and the wavelength of the illumination. This coherent imaging effect, also called coherent diffraction lithography, has been used to demonstrate the fabrication of photonic crystals.[24] In experiments, described herein, to demonstrate Talbot self-imaging at EUV wavelength of 46.9 nm,[25] three two-dimensional periodic masks were made using the electron-beam technique in thin HSQ photoresist layers and then illuminated by a fully spatially coherent

EUV beam from the CDL. To increase the NA of the mask, hence improving the resolution, the pattern was repeated multiple times in both orthogonal directions. At a certain distance, after illumination by the spatially coherent beam, the reconstructed image was formed by superposition of the diffracted orders coming from all the contributing patterns. This is an important advantage of this technique since any defects present in some cells are averaged out and contribute very little energy to the image power spectrum. The results of self-imaging at EUV wavelengths are presented in figure 3; the smallest features seen in figure 3c are 140 nm wide and were imaged with a theoretical spatial resolution of 84 nm.

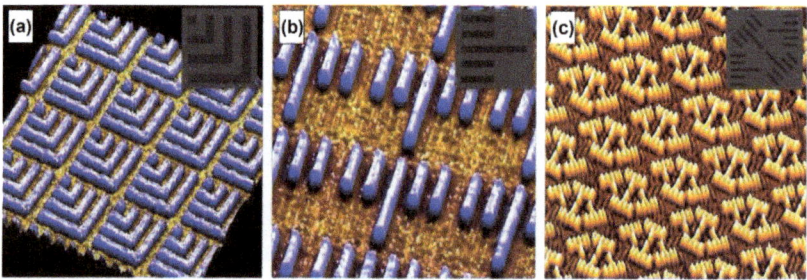

Figure 3 *False colour self images of the Talbot mask in the first Talbot plane. Insets show SEM images of single cells of the different Talbot masks.*

Another very promising technique is lens-less or diffraction microscopy (DM), where the coherent diffraction pattern of a non-crystalline specimen is recorded.[26] The specimen is illuminated by coherent light and the detailed and oversampled diffraction pattern is collected without any additional optics. Since phase information is lost a computerized algorithm iteratively solves for the phase and allows the reconstruction of the object shape. Since the first experimental demonstration,[27] short wavelength DM has successfully been applied to imaging of nanoscale materials and structures,[28] magnetic materials[29] and biological cells,[30] achieving spatial resolution as small as 50 nm using a wavelength of 1.5 nm.[29] Furthermore, in combination with free electron laser (FEL) sources this technique has the potential for single-shot imaging of large macromolecules, without the need for crystallisation.[31,32] The technique also eliminates aberrations from optical elements and the need for precise sample positioning, thus allowing for near diffraction-limited resolution in the EUV and soft X-ray wavelength range. However, it requires very bright and fully coherent illumination, normally meaning the necessity of using large, limited-access accelerator facilities such as third generation synchrotron sources and X-ray FELs.

In the experiments reported here,[33] a monochromatic and spatially coherent EUV beam from CDL illuminated the samples and highly over-sampled diffraction patterns were collected using beam blocks of different diameters to shadow the most intense central regions of the diffraction patterns and increase the dynamic range. Phase retrieval and the reconstruction were carried out by the guided hybrid-input-output (GHIO) algorithm.[33,34] To achieve the best spatial resolution the detector was placed as close to the sample as possible, increasing the NA of the system, and a novel field curvature correction was applied to the diffraction pattern. Phase retrieval allowed for the reconstruction of the object with a spatial resolution of 71 nm, approximately 1.5λ.

3 CONCLUSIONS

A range of coherent imaging techniques have been presented that allow imaging of nanometre scale objects at EUV wavelengths, with spatial resolutions approaching the wavelength used. Three different experiments in 2D holographic imaging were presented, allowing the capture, storage and reconstruction of objects from surface holograms with spatial resolutions of around 46 nm. In addition 3D information was extracted from high NA holograms at a transverse resolution of about 160 nm and a depth resolution of 2.1 μm. Two other imaging approaches were also discussed, namely reconstruction of computer generated holograms and generalised Talbot imaging at EUV wavelengths, providing spatial resolutions better than 90 nm. Finally, a very interesting and important imaging technique – diffractive lens-less imaging – was shown to be capable of achieving resolutions approaching 70 nm.

ACKNOWLEDGEMENTS

This work was supported in part by the NSF Engineering Research Center for Extreme Ultraviolet Science and Technology under NSF Award No. EEC-0310717 and by the NSF UW Nanoscale Science and Engineering Center, No. DMR-0425880. This work used the facilities and staff at the UW Synchrotron Radiation Center, NSF Grant No. DMR-0084402, the UW Center for Nanotechnology, CNTech and the Wisconsin Center for Applied Microelectronics WCAM. Electron Beam Lithography was performed at the Center for Nanomaterials, CNM, at Argonne National Laboratory.

References

1 J.J. Rocca, V. Shlyaptsev, F.G. Tomasel, O.D. Cortázar, D. Hartshorn and J.L.A. Chilla, *Phys. Rev. Lett.*, 1994, **73**, 2192.

2 B.R. Benware, C.D. Macchietto, C.H. Moreno and J.J. Rocca, *Phys. Rev. Lett.*, 1998, **81**, 5804.

3 C.D. Macchietto, B.R. Benware and J.J. Rocca, *Opt. Lett.*, 1999, **24**, 1115.

4 Y. Liu, M. Seminario, F.G. Tomasel, C. Chang, J.J. Rocca and D. Attwood, *Phys. Rev.*, 2001, **A6303**, 033802.

5 D. Gabor, *Nature*, 1948, **161**, 777.

6 J.W. Giles, *J. Opt. Soc. Am.*, 1969, **59**, 1179.

7 S. Aoki and S. Kikuta, *Jpn. J. Appl. Phys.*, 1974, **13**, 1385.

8 J.E. Trebes, S.B. Brown, E.M. Campbell, D.L. Matthews, D.G. Nilson, G.F. Stone and D.A. Whelan, *Science*, 1987, **238**, 517.

9 C. Jacobsen, M. Howells, J. Kirz and S. Rothman, *J. Opt. Soc. Am.*, 1990, **A7**, 1847.

10 S. Lindaas, H. Howells, C. Jacobsen and A. Kalinovsky, *J. Opt. Soc. Am.*, **A13**, 1788.

11 R.A. Bartels, A. Paul, H. Green, H.C. Kapteyn, M.M. Murnane, S. Backus, I.P. Christov, Y.W. Liu, D. Attwood and C. Jacobsen, *Science*, 2002, **297**, 376.

12 A.S. Morlens, J. Gautier, G. Rey, P. Zeitoun, J.P. Caumes, M. Kos-Rosset, H. Merdji, S. Kazamias, K. Casson and M. Fajardo, *Opt. Lett.*, 2006, **31**, 3095.

13 R.I. Tobey, M.E. Siemens, O. Cohen, M.M. Murnane, H.C. Kapteyn and K.A. Nelson, *Opt. Lett.*, 2007, **32**, 286.

14 P. Wachulak, R. Bartels, M.C. Marconi, C.S. Menoni, J.J. Rocca, Y. Lu and B. Parkinson, *Opt. Exp.*, 2006, **14**, 9636.

15 P.W. Wachulak, M.C. Marconi, R.A. Bartels, C.S. Menoni and J.J. Rocca, *J. Opt. Soc. Am.*, 2008, **B25**, 1811.

16 J.W. Goodman, *Introduction to Fourier Optics*, (New York: McGraw Hill) p66.

17 U. Schnars and W.P.O. Juptner, *Appl. Opt.*, 1994, **33**, 4373.

18 U. Schnars and W.P.O. Juptner, Meas. Sci. Technol., 2002, **13**, R85.

19 P.W. Wachulak, M.C. Marconi, R.A. Bartels, C.S. Menoni and J.J. Rocca, *Opto-Electronics Review*, 2010, **18**, 28.

20 P. Wachulak, M.C. Marconi, R. Bartels, C.S. Menoni and J.J. Rocca, *Opt. Exp.*, 2007, **15**, 10622.

21 M.R. Howells and C. Jacobsen, *Appl. Opt.*, 1991, **30**, 1580.

22 Y.-C. Cheng, A. Isoyan, J. Wallace, M. Khan and F. Cerrina, *Appl. Phys. Lett.*, 2007, **90**, 023116.

23 A. Isoyan, F. Jian, Y. Cheng, P. Wachulak, L. Urbanski, J. Rocca, C. Menoni, M.C. Marconi and F. Cerrina, in *Conference on Lasers and Electro-Optics/International Quantum Electronics Conference, OSA Technical Digest*, 2009, paper JFA7.

24 C. Zanke, M. Qi and H.I. Smith, *J. Vac. Sci. Technol.*, 2004, **B22**, 3352.

25 A. Isoyan, F. Jiang, Y.C. Cheng, F. Cerrina, P. Wachulak, L. Urbanski, J. Rocca, C. Menoni and M. Marconi, *J. Vac. Sci. Technol.*, 2009, **B27**, 2931.

26 J.W. Miao, H.N. Chapman, J. Kirz, D. Sayre and K.O. Hodgson, *Ann. Rev. Biophysics and Biomolecular Structure*, 2004, **33**, 157.

27 J.W. Miao, P. Charalambous, J. Kirz and D. Sayre, *Nature*, 1999, **400**, 342.

28 J. Miao, C. Chen, C. Song, Y. Nishino, Y. Kohmura, T. Ishikawa, D. Ramunno-Johnson, T. Lee and S.H. Risbud, *Phys. Rev. Lett.*, 2006, **97**, 215503.

29 S. Eisebitt, J. Luning, W.F. Schlotter, M. Lorgen, O. Hellwig, W. Eberhardt and J.Stohr, *Nature*, 2004, **432**, 885.

30 D. Shapiro, P. Thibault, T. Beetz, V. Elser, M. Howells, C. Jacobsen, J. Kirz, E. Lima, H. Miao, A.M. Neiman and D. Sayre, **PNAS**, 2005, **102**, 15343.

31 J.W. Miao, K.O. Hodgson and D. Sayre, **PNAS**, 2001, **98**, 6641.

32 H.N. Chapman, A. Barty, M.J. Bogan, S. Boutet, M. Frank, S.P. Hau-Riege, S. Marchesini, B.W. Woods, S. Bajt, W.. Benner, R.A. London, E. Plonjes, M. Kuhlmann, R. Treusch, S. Dusterer, T. Tschentscher, J.R. Schneider, E. Spiller, T. Moller, C. Bostedt, M. Hoener, D.A. Shapiro, K.O. Hodgson, D. van der Spoel, F. Burmeister, M. Bergh, C. Caleman, G. Huldt, M.M. Seibert, F.R.N.C. Maia, R.W. Lee, A. Szoke, N. Timneanu and J. Hajdu, *Nature Physics*, 2006, **2**, 839.

33 R.L. Sandberg, C. Song, P.W. Wachulak, D.A. Raymondson, A. Paul, B. Amirbekian, E. Lee, A. Sakdinawat, C.L. Vorakiat, M.C. Marconi, C.S. Menoni, M.M. Murnane, J.J. Rocca, H.C. Kapteyn and J. Miao, **PNAS**, 2008, **105**, 24.

34 J. Miao, T. Ishikawa, E.H. Anderson and K.O. Hodgson, *Phys. Rev.*, 2003, **B67**, 174104.

DEVELOPMENT AND OPTIMIZATION OF LASER-PLASMA EXTREME ULTRAVIOLET AND SOFT X-RAY SOURCES FOR MICROSCOPY APPLICATIONS

P. W. Wachulak, A. Bartnik, J. Kostecki, R. Jarocki, M. Szczurek and H. Fiedorowicz

Institute of Optoelectronics, Military University of Technology, 00-908 Warsaw, Poland

1 INTRODUCTION

Imaging in a routine way in the extreme ultraviolet (EUV) and soft X-ray (SXR) spectral regions requires efficient, compact and reliable sources. Hence the current interest in developing sources with sufficient flux and suitable coherence properties for practical imaging techniques. In this chapter compact, desk-top EUV and SXR sources, based on gas-puff targets, are presented. The EUV source was developed in the Institute of Optoelectronics and is capable of emitting quasi-monochromatic radiation at a wavelength of 13.8 nm with $\lambda/\Delta\lambda \approx 140$ and pulse energies of up to about 1.3 μJ at a repetition rate of 10 Hz. The source is debris-free and operates close to optimum wavelength for next generation lithography. These features make the source attractive for lithographic experiments, and quasi-monochromatic emission was used successfully in recent imaging experiments using a Fresnel zone plate objective.

The desk-top soft-X-ray source, based on a gas-puff target, was also recently studied in detail to determine its applicability for SXR microscopy. This source emits in the water-window spectral range at $\lambda = 2.88$ nm from a nitrogen gas target or in the 2–4 nm range of wavelengths from an argon gas target.

These debris-free sources are desk-top alternatives for free electron lasers and synchrotron installations. They can be successfully employed in high resolution microscopy, where various imaging methods and techniques are currently under development, capable of reaching spatial resolutions of tens of nanometres. One of these techniques is EUV microscopy, based on Fresnel zone plates, under development worldwide. In this chapter the first demonstration of a desk-top EUV transmission microscopy at 13.8 nm, with a spatial (half-pitch) resolution as good as 50 nm in a very compact setup, is presented.

2 SHORT WAVELENGTH SOURCE DEVELOPMENT AND OPTIMIZATION

The recent development of compact short-wavelength laboratory sources has promoted the study of material properties in the EUV region. The strong demand for such quasi-monochromatic sources results from the semiconductor industry striving for even smaller, faster and less power intensive computer chips but also motivated the development of compact, high-resolution imaging tools. Compact EUV sources provide the opportunity to perform experiments without the necessity to employ large photon facilities with limited user access. Thus these sources have huge potential impact on the speed of nanotechnological development since experiments previously restricted to large facilities can be performed in the laboratory environment. The importance of the development and commercial availability of EUV sources, in particular the lack of commercial tools for actinic inspection for the semiconductor industry, has already been noted by Intel.[1]

The usability of existing EUV sources has been proven by many examples of experiments involving compact table-top sources, including capillary discharge lasers for interferometric lithography (IL) and wavelength resolved holography,[2,3] high-harmonic generation sources[4,5] for lens-less imaging[6] and surface deformation studies,[7] optically pumped EUV lasers for microscopy using zone-plates,[8] and a xenon discharge produced plasma (DPP) source from Energetiq used for development of an actinic full-field EUV mask blank inspection tool at MIRAI-Selete.[9]

2.1 Quasi-monochromatic 13.8 nm Source Dedicated for Microscopy

One such compact source uses laser plasmas generated from gas-puff targets (LPGPT). Using X-rays in the wavelength range 6–20 nm, produced by irradiating a double-stream gas puff target with an Nd:YAG laser beam, direct and very efficient photo-etching of PTFE polymer was reported.[10] The gas-puff source was also used for studies of EUV fluorescence radiation from aluminium and silicon,[11] and for micro-structuring of PMMA polymer.[12] It was also shown that by employing a double stream gas-puff target the energy conversion efficiency from the infra-red pumping laser to EUV radiation emitted close to 13.5 nm can be improved significantly. The calculated EUV production can be as large as 8.7×10^{13} photons per steradian, in a 3 ns time duration pulses, corresponding to 8 mJ/2π of emitted energy and a conversion efficiency of 1.6%.[13]

In imaging using zone plate (ZP) objectives, however, monochromatic radiation is essential.[14] For this purpose a laser-plasma argon gas-puff target source was optimized for efficient EUV production.[15] This source, which was characterized and optimized for quasi-monochromatic emission, is based on a double nozzle gas-puff target and was later employed in applications requiring narrow bandwidth radiation, such as EUV microscopy with ZP objectives.[16,17] Using argon/helium gas-puff target and a collinear double stream electromagnetic valve, efficient EUV production was achieved in the 6–20 nm range. A Mo/Si ellipsoidal multilayer mirror (MLM) acted as a spectral filter and radiation collector to provide a reduced bandwidth of 13–14 nm, resulting in quasi-monochromatic radiation at 13.8 nm. The spectrum is shown in figure 1.

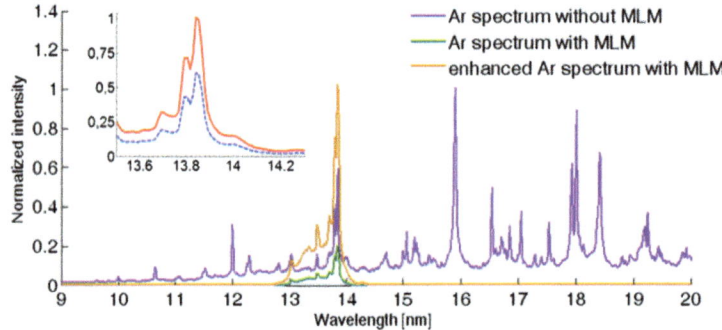

Figure 1 *Spectra of the LPGPT EUV source / MLM combination showing quasi-monochromaticity around λ = 13.8 nm. The result of further filtering with a 100 nm thick zirconium filter is shown as a dashed line.*

Argon plasmas emit radiation in the 13–14 nm range mostly in two strong Ar VIII lines, the transitions $2p^65d–2p^63p$ at 13.793 nm and $2p^65p–2p^63s$ at 13.844 nm, with $\lambda/\Delta\lambda \approx 140$ and a single pulse photon count of $(8.8\pm0.5)\times10^{10}$, measured in focus of the MLM. Using a pinhole camera it was found that the source was elliptical with major and minor axes of 1090 μm and 390 μm,[16] as shown in figure 2, imaged with the MLM optic.

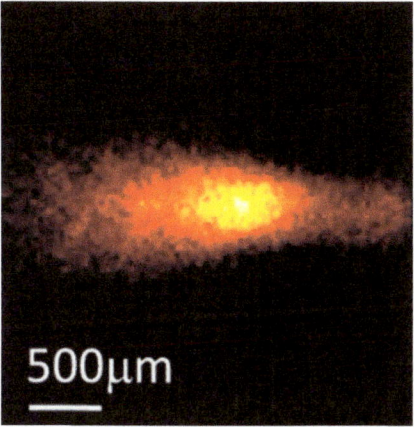

Figure 2 *Spatial distribution of the intensity of the LPGPT argon-based EUV source at 13.8 nm.*

2.2 Laser-plasma Gas-puff Target SXR Sources in the Water-window Range

X-ray sources emitting in the "water-window" region, $\lambda = 2.3–4.4$ nm,[18] are important for imaging of biological samples in near-natural environments. High contrast in this spectral range is obtained due to different absorption of different constituents; water has a relatively small absorption coefficient while carbon has much higher absorption, resulting very good image contrast.

Soft X-ray (SXR) microscopy has been successfully employed, mainly in transmission mode using diffractive optics, such as zone-plates,[19,20,21] and raster scanning the sample across the focused SXR beam.[22,23,24] Another technique is contact microscopy, where the sample is placed on a recording medium, such as a photoresist, and illuminated by the SXR beam to make an image in the surface of the recording medium.[25,26,27] However, in most of these important achievements synchrotron sources were used for sample illumination, due to their obvious advantages. The main motivation for development of compact, table-top SXR sources is to open the possibility to perform experiments without the necessity to employ large "photon factories" such as synchrotrons or free electron lasers. These sources are the state-of-the-art devices with tuneable, wide wavelength output ranges, high photon fluxes and brightnesses. Although desk-top sources cannot match these capabilities they are still important alternatives that can perform some of the experiments in a university laboratory, for example. While synchrotrons are always attractive SXR sources for cutting-edge experiments, compact EUV and SXR sources are demonstrably growing more important.

The suitability of gas puff targets for efficient soft X-ray generation in the water-window range has already been demonstrated, resulting in strong line emission from the 1–2 transitions in H- and He-like nitrogen ions at $\lambda = 2.478$ nm and 2.879 nm, respectively, using an Nd:glass laser with an energy of 5 J per pulse, 1ns in duration, for excitation[28] and, more recently, a very compact desk-top setup.[29]

This used a commercial Nd:YAG laser NL 303HT (Ekspla, Lithuania) with a smaller laser pulse energy of 737 mJ at $\lambda = 1064$ nm and a longer, 4 ns, pulse. The argon and nitrogen sources were demonstrated and are complementary, in that the broadband emission from an argon plasma with $\lambda = 2$–4 nm can be used for contact microscopy and the quasi-monochromatic emission from a nitrogen plasma, at $\lambda = 2.88$ nm, is suitable for imaging using diffractive optics. To spectrally narrow the broad emission from gaseous targets two filters were used, one 750 nm thick aluminium and the other 1 μm thick Mylar ($C_{10}H_8O_4$). The spectra for argon and nitrogen plasmas are shown in figures 3a and 3b.

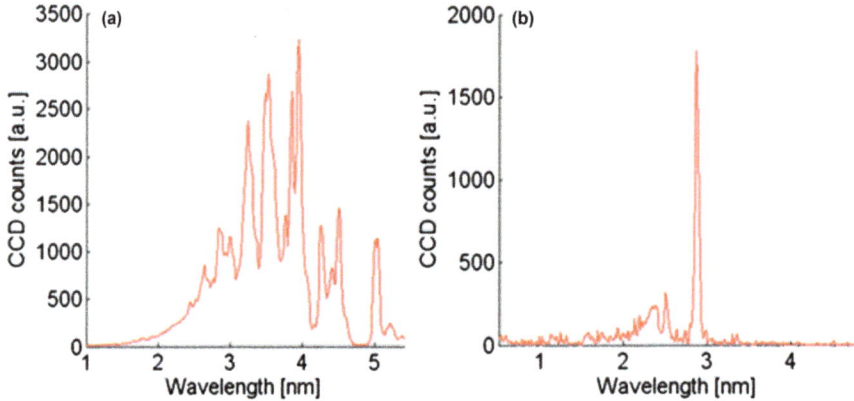

Figure 3 *(a) The quasi-continuous spectrum of argon in the 1-5.5 nm spectral range, including the water-window, and (b) the quasi-monochromatic nitrogen spectrum in the 0.5-5 nm spectral range.*

Argon plasmas emit in a broad spectral range in the water-window region. The dominant groups of spectral lines are Ar^{11+} $2s^22p^23d$–$2s^22p^3$ with $\lambda = 3.065$–3.274 nm, Ar^{10+} $2s^22p^33d$–$2s^22p^4$ with $\lambda = 3.353$–3.631 nm and Ar^{9+} $2s^22p^43d$–$2s^22p^5$ with $\lambda = 3.656$–3.888 nm.[30] Nitrogen plasmas, on the other hand, have quasi-monochromatic emission with one dominant spectral line, N^{5+} $1s2p$–$1s^2$ ($1P^0$–g1S) at $\lambda = 2.8787$ nm. Other nitrogen lines are N^{6+} $1s$–$2p$ (J–J 1/2–3/2 and 1/2–1/2), at $\lambda = 2.4779$ nm and $\lambda = 2.4785$ nm, N^{6+} $3p$–1s, $4p$–1s and $5p$–1s at wavelengths of 2.0910 nm, 1.9826 nm and 1.9361 nm respectively. The plasma spatial distributions and source sizes were measured using a pinhole camera. For both gases the plasmas are elliptical with full widths of half maximum of 220×146 μm for argon and 325×143 μm for nitrogen, as shown in figures 4a and 4b. The numbers of photons in one pulse were $(7.23\pm0.58)\times10^7$ for argon source, and $(1.06\pm0.08)\times10^7$ for nitrogen.

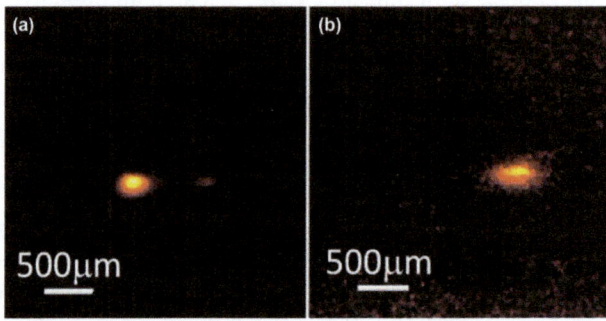

Figure 4 *Pinhole camera images of (a) the argon and (b) the nitrogen soft X-ray sources.*

3 EUV IMAGING BASED ON A LASER-PLASMA GAS-PUFF TARGET SOURCE

Future developments in nanoscience demand tools capable of capturing images at a nanometre scale spatial resolution. A direct path to improve the spatial resolution in photon based microscopy is to use short wavelength radiation. Various imaging methods and techniques are currently under active pursuit worldwide, one of them being EUV and SXR microscopy, based on Fresnel zone plates.[31]

The resolution of an optical system may be characterised by the Rayleigh criterion, which results from images of two point sources of quasi-monochromatic radiation of equal intensity in a noise free background producing two Airy intensity patterns at the image plane. If the point sources are sufficiently far apart, then the image will be formed by two distinct Airy patterns, one for each source. The point sources are said in this case to be resolved. If the sources are closer, then their Airy patterns will start to overlap. If the two sources are mutually incoherent, the intensities of the two Airy patterns will add. According to the Rayleigh criterion, the separation at which the two point sources are said to be just resolved corresponds to the coincidence of the first null of the Airy pattern from one source with the maximum of the pattern from the second source. This distance, which defines the resolution of the optical system under incoherent illumination in a straightforward way, is

$$\Delta_R = \frac{0.61\lambda}{NA}, \tag{1}$$

where λ is the illumination wavelength and NA is the numerical aperture of the system. Equation (1) is valid for incoherent illumination; in general

$$\Delta_R = \frac{k\lambda}{NA}, \tag{2}$$

where k may be in the range $\approx 0.34–1$ depending on the degree of coherence of the illumination, the illumination spectrum and the method of measuring the resolution.[32]

Equations (1) and (2) show that one way to improve the spatial resolution of an imaging system is to reduce the illumination wavelength; this was the motivation a laser-plasma source with wavelength $\approx 40–100$ times shorter than visible light. Using soft X-rays the best spatial resolution obtained so far is 12 nm using synchrotron radiation.[33] Demonstrations of actinic aerial EUV microscopes for mask inspection have been conducted using bending magnets at synchrotron facilities to provide the required short wavelength illumination.[34,35] These microscopes are capable of imaging small objects with spatial resolutions better than 100 nm,[36] but their size and operation costs are very high. The development of bright, compact, short wavelength sources provides a significant step forward in the commercialization of high resolution imaging tools in the near future. These devices can obtain images with exposures of a few seconds and spatial resolutions approaching those of synchrotron based microscopes.[8,37]

The work done so far in EUV and SXR imaging, using both coherent and incoherent sources, is very extensive and so it is only possible to present herein only a few examples. Images with 700 nm half-pitch resolution using an EUV recombination laser at $\lambda = 18.2$ nm were reported in the early 1990s,[38] and at the same time 75 nm resolution was reported employing a SXR laser at $\lambda = 4.48$ nm pumped by the fusion-class NOVA laser.[39] In the latter case image acquisition was limited to a few images per day due to the laser repetition rate. More recently, different approaches to sub-micrometre resolution imaging emerged due to the development of smaller scale short-wavelength sources such as high-order harmonics,[40] improved SXR lasers[41] and incoherent laser-plasma based sources.[42] Using radiation from a capillary discharge laser at $\lambda = 46.9$ nm, EUV images were obtained with a spatial resolution of 120–150 nm.[43] Radiation at 13.2 nm from a Ni-like cadmium EUV laser using 1J, 8 ps pulses for pumping provided 55 nm resolutoin in reflection mode[44] and sub-38 nm in transmission mode.[8] Using a capillary discharge laser, EUV microscopy in transmission mode was demonstrated with single laser pulse, leading to 54 nm half-pitch spatial resolution and temporal resolution of ~1 ns.[45] Quasi-monochromatic emission from an incoherent source based on liquid nitrogen, at $\lambda = 2.88$ nm in the water window, allowed demonstration of SXR microscopy with sub-50 nm ($\approx 17 \lambda$) spatial resolution.[46] As a final example, using a xenon gas discharge source, a Schwarzschild objective and a Fresnel zone plate for a second magnification step, EUV imaging was demonstrated with a spatial resolution of ≈ 100 nm.[47]

3.1 Experimental Setup

The microscope, described in detail in,[17] used an ellipsoidal Mo/Si multilayer mirror to focus EUV radiation onto an object (figure 5). A Fresnel zone plate objective was used to form the magnified images in the transmission mode on an EUV sensitive CCD camera. The use of a gas-puff target eliminates the debris production problem associated with solid targets. Quasi-monochromatic EUV radiation, required for the Fresnel optics, was produced by spectral selection of a single line emitting at 13.8 nm from argon plasmas or quasi-continuum emission in the 13–14 nm wavelength range from xenon plasmas. Two objects were imaged, a copper mesh of thickness ≈ 4 μm and a holey carbon foil coated with 70 nm of gold. EUV images, presented in section 3.2, of the samples were obtained with half-pitch spatial resolutions approaching 50 nm (3.7λ) in a very compact set up.

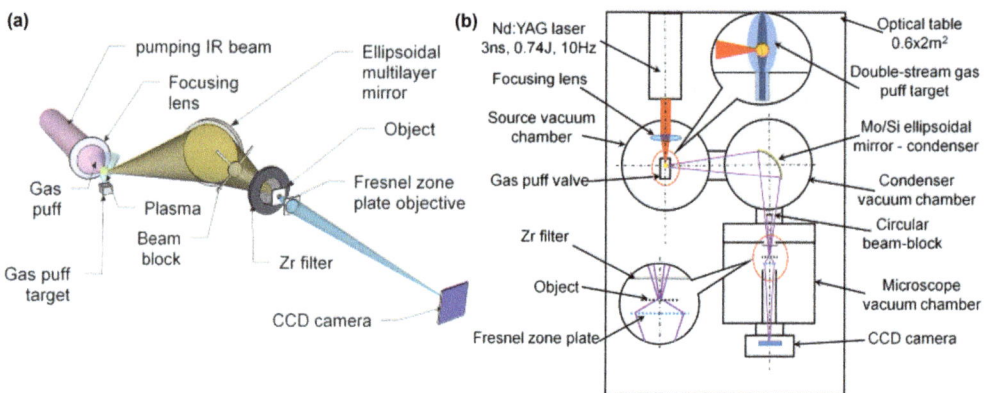

Figure 5 *(a) Schematic and (b) experimental arrangement of the EUV microscope (not to scale) using a laser-plasma EUV source based on a gas-puff target.*

The laser-plasma source used in these experiments was developed for EUV metrology applications in the MEDEA+ project[13] and later modified for quasi-monochromatic emission in the 13–14 nm wavelength range.[15] This source has the advantage over other compact sources in that is possible to change the working gas, thus allowing both the peak emission wavelength and the bandwidth to be varied.

Initially, to study the influence of bandwidth on the spatial resolution of the EUV microscope, argon and xenon plasmas were produced using Nd:YAG (Eksma) laser pulses with duration 4 ns and energy 0.74 J. The plasmas radiated over a very broad range of wavelengths, dominantly in the range 5–50 nm; by using additional spectral filtering the spectral emission could be shaped. The source operated at up to 10 Hz repetition rate, and a constant target chamber pressure of 2×10^{-3} mbar was maintained. The microscope was located inside the 24 cm diameter, 35 cm long vacuum chamber, and the entire system was installed on a single 2×0.6 m optical table.

EUV radiation from the plasma was collected, focused and spectrally filtered by an ellipsoidal off-axis 80 mm diameter Mo/Si multilayer mirror, which was corrected for the spherical aberrations. The multilayers were optimized for the 13.5 ± 0.5 nm (FWHM) wavelength range and an incidence angle of 45°. Its theoretical reflectivity at 13.5 nm is 37.7% for unpolarised radiation. The mirror was designed to image the plasma with a lateral magnification of unity, with both the object and image distances equal to 254 mm. It was developed in cooperation with REFLEX s.r.o. (mirror substrate) and the Fraunhofer Institut für Angewandte Optik und Feinmechanik (coating).

The laser plasma source was optimized for efficient EUV radiation generation from argon[15] and xenon[48] plasmas. The in-band ($\lambda = 13$–14 nm) photon flux from argon was previously measured to be $(8.8 \pm 0.5) \times 10^{10}$ photons per pulse in a horizontally elongated spot with FWHM width 1.09×0.39 mm, corresponding to 1.29 µJ per pulse. [15] For the xenon plasma, due to a much more efficient EUV photon production and to alleviate the risk of ablating the zone plate, the source was not particularly optimized but an approximately five times larger photon flux was obtained than for argon, with a similar plasma size. For both plasmas, to eliminate longer wavelengths ($\lambda > 18$ nm) a 100 nm thick, 10 mm diameter free-standing zirconium filter (Lebow) was used, positioned 4–5 mm upstream of the object. The EUV emission spectra were measured using a flat-field reflection grating spectrometer equipped with a 1200 line/mm grating with variable groove spacing (Hitachi), a 25.5 µm entrance slit and a 1300×400 pixels back illuminated CCD camera (Princeton Instruments). From the geometry of the spectrometer and the grating, the resolution of the spectrometer was estimated to be $\lambda/\Delta\lambda \approx 580$.

Figure 6 shows the emission spectrum from xenon plasmas in the wavelength range 9–20 nm. The quasi-continuous spectrum is complicated, covering the entire reflection band of the condenser. The main spectral transitions are $4p^5 4d^9$–$4p^6 4d^8$ and $4p^6 4d^7 4f$–$4p^6 4d^8$ in Xe XI at 11–13 nm, $4p^6 4d^7 4f$–$4p^6 4d^8$ and $4p^6 4d^7 5p$–$4p^6 4d^8$ in Xe XI at 13–14 nm and, predominantly, $4d^8 5p$–$4d^9$ and $4d^8 4f$–$4d^9$ transitions in Xe X, $4d^9 5p$–$4d^{10}$ in Xe IX and $3d^{10} 4p$–$3d^{10} 4s$ in Xe XXVI at wavelengths above 14 nm.[49,50] For the filtered xenon spectrum $\lambda/\Delta\lambda = 14$, with peak emission at $\lambda = 13.78$ nm. The $\lambda/\Delta\lambda$ value may, however, be an underestimate due to the relatively wide spectrometer spectral response of 0.024 nm, but even so the differences between the argon and xenon plasmas in terms of the character of their spectral emissions and $\lambda/\Delta\lambda$ are evident.

Another experiment concerned the study of the influence of object thickness on spatial resolution. Two different objects, with different thicknesses relative to the depth of focus (DOF) of the microscope, were used. The first was a 4 µm thick ($\approx 11 \times$ DOF) copper fine square mesh (G2000HS, SPI Supplies) with a period of 12.5 µm and a bar width of 5 µm. The second object was a Quantifoil holey 300M carbon foil supported on a steel mesh (SPI Supplies), 10 nm thick according to the manufacturer's specifications, with 1.5 µm diameter holes spaced on a square grid of period 2.5 µm. To improve the contrast the carbon foil was coated with a gold layer about 60 nm thick, having a transmission of about 4% at 13.84 nm; the total thickness was thus about 70 nm ($0.2 \times$ DOF). SEM images of the mesh and the perforated foil are shown in figure 7.

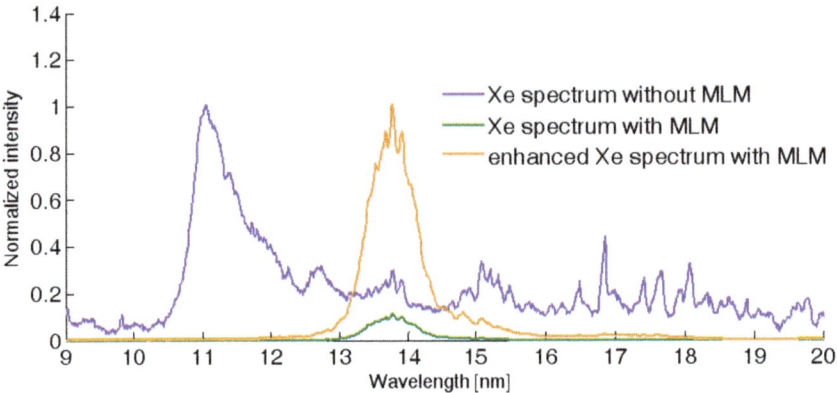

Figure 6 *Spectra of the xenon plasma EUV source.*

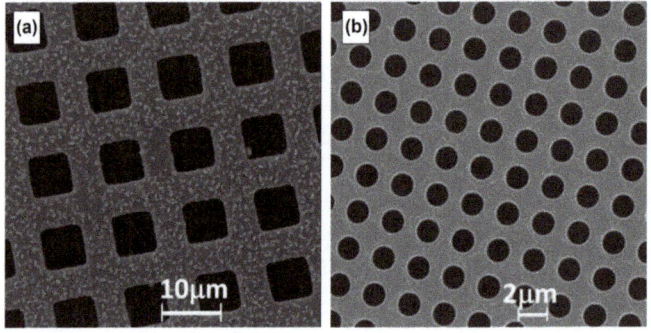

Figure 7 *SEM images of the two objects used for studies of the influence of object thickness on the EUV microscope spatial resolution: (a) a 4 µm thick copper mesh and (b) a holey carbon foil coated with a thin layer of gold, giving a total thickness of 70 nm.*

The objects, placed 254 mm from the mirror, were imaged using a ZP objective onto EUV sensitive CCD camera (iKon-M, Andor) with 1024×1024 13×13 µm pixels. The ZP was fabricated by Zone Plates Ltd. using electron beam lithography in a 220 nm thick polymethyl methacrylate (PMMA) layer spin coated on top of a 50 nm thick silicon nitride membrane. The ZP diameter was $D = 200$ µm, with 1000 zones and an outer zone width $d = 50$ nm. It was optimized for $\lambda = 13.5$ nm, and so the focal length

$$f = \frac{Dd}{\lambda} \tag{3}$$

was 740.7 µm with a numerical aperture

$$NA = \frac{\lambda}{2d} \tag{4}$$

of 0.135. For illumination at $\lambda = 13.84$ nm from argon plasmas the focal length was 722.5µm with $NA = 0.138$. For the spectrally broader xenon emission the ZP parameters are not so clear, because of the high dispersion ($f \propto 1/\lambda$, $NA \propto \lambda$); the DOF was ±0.4 µm, much smaller than the range of focal lengths. For both plasmas and both objects magnifications of 520–840× were used, adjusted by changing the camera to ZP distance and refocusing; details of the magnification, pixel size and field of view (FOV) are given in table 1. The geometrical numerical apertures of the collecting ellipsoidal mirror in the horizontal and vertical directions, $NA_H = 0.11$ and $NA_V = 0.15$, are similar to that of the ZP, thus providing incoherent illumination,[19] satisfying the condition (6)

$$\sigma_{H,V} = NA_{H,V} / NA_{ZP} \sim 1, \tag{6}$$

with $\sigma_H = 0.8$ and $\sigma_V = 1.1$. The ZP was mounted on a three axis translation stage driven by vacuum compatible stepper motor actuators (Standa). A piezoelectric, single axis flexure stage with a travel of 25 µm (model NF15A, Thorlabs) was used for precise adjustment of the object-ZP distance at a rate of 3 V/µm with a theoretical resolution of 10 nm. In order to provide conical illumination of the object and avoid stray light through the ZP a 12 mm diameter circular beam block was placed about 15 cm from the ZP. For the binary transmission mesh object using quasi-monochromatic radiation from argon plasmas 50 pulses were needed to obtain an image, at 2 Hz repetition rate. For quasi-continuous radiation from xenon the increased flux meant that an image could be obtained in just 10 pulses at the same repetition rate. The source can operate at up to 10 Hz repetition rate, but pressure buildup in the microscope chamber could cause reabsorption of the radiation in neutral gas. The CCD camera was cooled down to −20°C to decrease intrinsic noise during image acquisition.

Table 1 *Imaging details and resolution results for the two objects and two illumination bandwidths.*

Object	Mesh		Foil	
Thickness, t	11×DOF		0.2×DOF	
Illumination	Ar	Xe	Ar	Xe
Exposure time [s] for 2 Hz repetition rate	25	10	50	50
$\lambda/\Delta\lambda$	140	14	140	14
Magnification	840		523	
Image pixel size [nm^2]	15.9×15.9		25.9×25.9	
Field of view [µm^2]	16.3×16.3		26.4×26.4	
Knife-edge resolution, r_{KE}				
$\bar{r}_{KE}$ [nm]	73	152	51	140
σ_{KE} [nm]	5	12	11	15

3.2 Experimental Results

Both objects (mesh and foil) were imaged using both plasmas (argon and xenon). For each of the four cases the microscope alignment was optimized to provide uniform illumination in the entire FOV. In addition, during image acquisition the object to ZP distance was changed using the piezoelectric stage by 1 V per step, corresponding to about 330 nm (smaller than the DOF) over a range of ±20 µm from the ZP focal point, in order to obtain the sharpest possible image. From the entire set of images, for each object/bandwidth combination, the "sharpest" EUV image was chosen by observing the edge blurring for subsequent resolution measurements. The resolution of the microscope was assessed by the well-established knife edge (KE) test method. For incoherent illumination the 10-90% intensity transition across a sharp edge corresponds to the well-known Rayleigh resolution and to twice the half-pitch grating resolution of the optical system[31]

Figure 8 shows images of the mesh object obtained using (a) argon and (b) xenon plasma radiation. The magnified regions (c & d), where KE resolution measurements were carried out, show the resolution decrease resulting from the wavelenngth spread of the xenon illumination. Figure 9 shows the corresponding images of the holey carbon foil, where the blurring due to the 10× wider illumination bandwidth from the xenon plasma can also be observed.

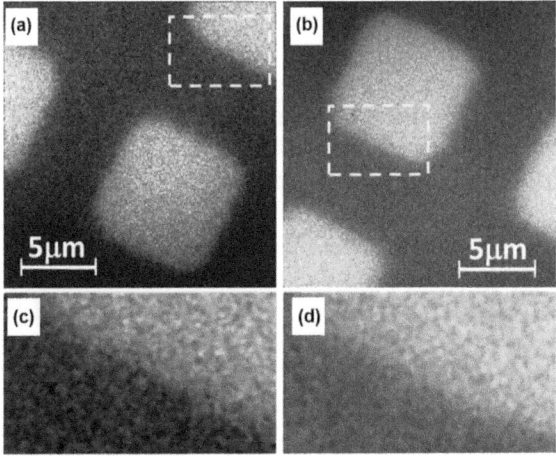

Figure 8 *EUV images of the copper mesh object using argon (a) and xenon (b) plasmas. Magnified subsections, indicated by the boxed regions (c,d) show the edge in more detail.*

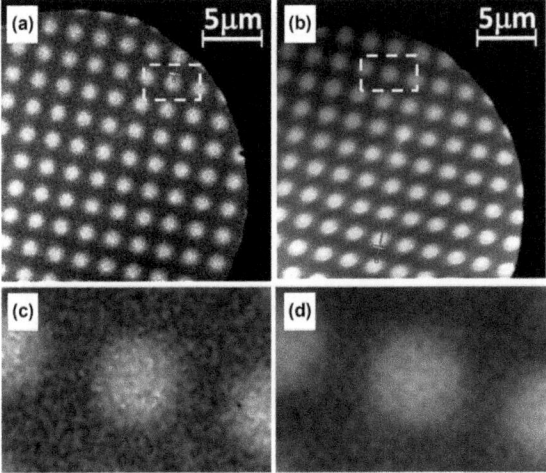

Figure 9 *EUV images of the carbon/gold foil object using argon (a) and xenon (b) plasmas. Magnified subsections, indicated by the boxed regions (c,d) show a single hole in more detail.*

Table 1 also shows the results of the half-pitch KE measurements $\bar{r}_{KE}$, based on six independent KE measurements for each object/bandwidth combination; σ_{KE} is the standard deviation. Lineouts of the EUV images were assessed by averaging five adjacent lines to improve the signal to noise ratio. The best KE half-pitch resolution was measured to be 51±11 nm for the perforated foil object and argon plasma illumination. The influence of the object thickness on the spatial resolution can also be determined, since for the thicker mesh object the resolution was measured to be 73±5 nm; this is in agreement with a previous measurement of 69±4 nm .[16] A more significant resolution change results from using for the broadband illumination from xenon plasmas; for the thin (carbon/gold) object the measured spatial half-pitch resolution was 140±15 nm and for the thick mesh object it was 152±12 nm . Summaries of the measurement results and statistics are presented in figure 10, which also shows, as a solid black line, the ZP theoretical resolution limit in terms the of half-pitch resolution 0.61d using incoherent illumination.[31]

Typical KE line-outs, obtained from the boxed regions of the images in figures 8 and 9 for each object/bandwidth combination, and the theoretical KE resolution limited by the ZP outer zone wide, can be seen in figure 10, which also shows also the full (10-90%) and half-pitch Rayleigh resolution measurements.

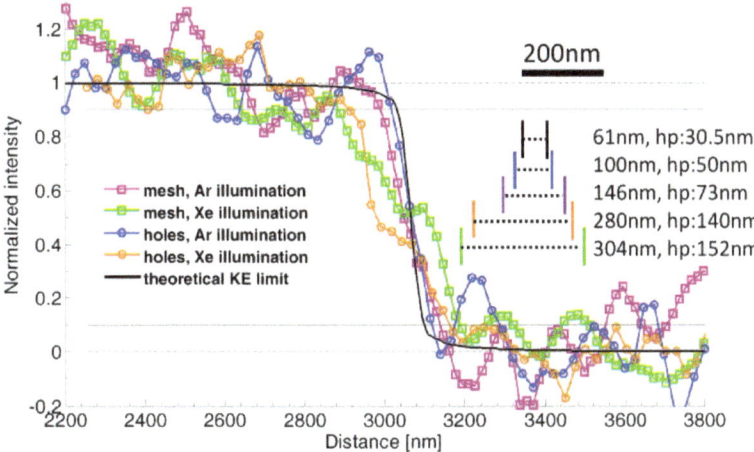

Figure 10 *Typical KE lineouts indicating 10-90% intensity transitions in the EUV images related to the Rayleigh resolution criterion for both objects and different illumination bandwidths. The theoretical KE limit is also shown.*

3.3 Discussion of the Results

The theoretical half-pitch resolution of the microscope can be expressed using

$$r_{KE} = \frac{k\lambda}{2NA_{ZP}} = kd = 0.5\Delta_R \tag{7}$$

where k, the resolution test specific constant, is illumination dependent[32] and NA_{ZP} is the numerical aperture of the ZP. For incoherent illumination ($k = 0.61$) this resolution is equal to 30.5 nm, better than the measured half-pitch resolution of 51 nm in the best case. This is because zone plates are highly dispersive diffractive elements. The theoretical resolution can only be achieved for monochromatic radiation, with

$$\frac{\lambda}{\Delta\lambda} > N_{ZP} \tag{8}$$

where N_{ZP} is the number of zones (in our case 1000). In the current case, even for quasi-monochromatic radiation from argon plasmas, this criterion is not satisfied, introducing achromatic blurring to the image. The loss in resolution is also related to the object thickness; if $t > DOF$ the resolution will be degraded. Hence the perforated foil object, with thickness much smaller than the depth of focus, allows study of the dependence of the illumination bandwidth on resolution.

To estimate the influence of both the EUV bandwidth and the object thickness on spatial resolution a Gaussian type point spread function (PSF) was assumed. Any image convolved with a PSF of non-zero width will suffer from resolution loss. The wider the PSF the more high resolution details, or high spatial frequency components, in the image will be removed. Moreover, many PSFs attributed to different resolution decreasing factors can be convolved leading to a total PSF of an optical system. Gaussian PSFs are neither the best, nor the most accurate estimates, but allow for quick de-convolution of the results to obtain estimates of the contribution of each factor on the spatial resolution. The results of deconvolution were calculated using

$$\sigma_{0.5G_{PSF}} \approx 0.845\sqrt{\sigma_{II}^2 - \sigma_I^2}, \sigma_{II} > \sigma_I \qquad (9)$$

where σ_{II} is the degraded HWHM (half-pitch) resolution and σ_I is the undegraded resolution. The numerical factor 0.845 is due to the difference between the Rayleigh resolution based on an Airy function and HWHM for the Gaussian function.

The deconvolution results are shown in table 2. For monochromatic illumination the theoretical half-pitch resolution achievable is 30.5 nm. For quasi-monochromatic argon plasma emission with $\lambda/\Delta\lambda < N_{ZP}$, the resolution is worse, equal to 51 nm. This can be attributed to a Gaussian PSF$_{Ar}$ with a FWHM equal of about 35 nm. For the xenon plasma the spatial resolution of 140 nm corresponds to a PSF$_{Xe}$ FWHM of about 113 nm.

Table 2 *Estimates of Gaussian PSF approximation by de-convolution based on the KE resolution measurements.*

	Experimental half-pitch resolution [nm]	Half-pitch resolution [nm] of degredation effect
Theoretical ZP resolution for monochromatic illumination, $\lambda/\Delta\lambda > 1000$	30.5	0
Quasi-monochromatic argon illumination, $\lambda/\Delta\lambda = 140$, thin object ($t = 0.2 \times DOF$)	51	PSF$_{Ar}$ = 35
Quasi-monochromatic argon illumination, $\lambda/\Delta\lambda = 140$, thick object ($t = 11 \times DOF$)	73	PSF$_{thick}$ = 45
Quasi-continuum xenon illumination, $\lambda/\Delta\lambda = 14$, thin object	140	PSF$_{Xe}$ = 113
Quasi-continuum xenon illumination, $\lambda/\Delta\lambda = 14$, thick object	152	PSF$_{thick}$ = 51

To quantify the influence of object thickness on spatial resolution the values obtained for the thick and thin objects can be compared. For an object much thicker than the DOF ($t = 11 \times DOF$ for the mesh object) the measured resolution for argon illumination of 73 nm corresponds to a PSF$_{thick}$ width of about 45 nm. For xenon illumination a comparable value of about 51 nm (ideally should be the same) was obtained from the measured resolution of 152 nm.

3.4 Single-Shot Operation

Another experiment was related to the possibility of single-shot operation to provide temporal resolution equal to the duration of the EUV pulse of about 3 ns. Using the copper mesh object a sequence of images was obtained with decreasing numbers of pulses, from 50 down to one, as shown in figure 11. The signal to noise ratio S/N clearly decreases with the number of pulses; although the main features of the mesh are visible even with a single pulse, the image is very noisy, indicating the necessity to increase the number of photons illuminating the CCD detector. The S/N can be improved by integrating the signal over small areas in the CCD, binning the image by 4×4 pixels for example, as can be seen in the rightmost image of figure 11. However, the equivalent pixel size after binning is larger and so the resolution is degraded..

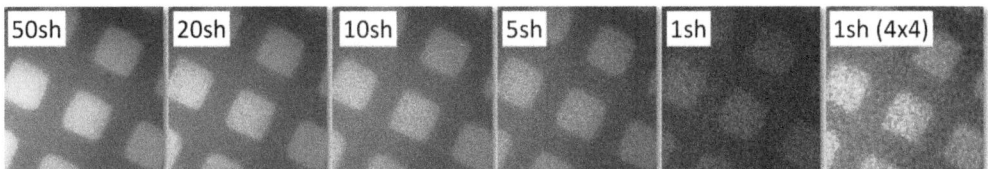

Figure 11 *Copper mesh images as the number of EUV pulses was decreased. The rightmost image in the sequence was obtained after integration of the signal from 16 CCD pixels (4×4 binning) to improve the S/N ratio.*

4 CONCLUSIONS

Several laser-plasma gas-puff target sources, suitable for EUV and SXR microscopy, have been presented. Due to the high dispersion of Fresnel zone plates, monochromatic radiation is needed for microscopy, and a good approximation is the quasi-monochromatic emission from argon based EUV sources at 13.84 nm. The use of such a source in microscopy experiments yielded a spatial resolution of about 50 nm, with exposure times of around 50 s,[17] comparable to larger table-top systems. Nitrogen based sources, which provide strong line emission, can be used for SXR microscopy with the same type of optic. Argon based SXR sources offer much higher photon fluxes and, although not suitable for zone plate objectives, could be successfully employed in microscopes with reflective, dispersion-free optics such as Wolter-type I objectives. [51,52]

A detailed analysis of the influence of the illumination bandwidth and object thickness on the spatial resolution was presented. Using either quasi-monochromatic radiation from argon plasmas or quasi-continuous radiation from xenon plasmas it was possible to quantify a bandwidth-related resolution degrading factor. Also, comparing results from the thick copper mesh and thin perforated foil objects the resolution degredation was related quantitatively to the object thickness and was expressed in terms of Gaussian PSF functions. The single-shot operation capability was mentioned addition.

The concept of practical, compact and high resolution EUV full-field microscopes that may be developed and experimentally implemented has been demonstrated. Such isntruments will allow studies of ultrafast processes at nanometre-scale resolutions with the potential to become enabling tools for a wide range of nanoscience and nanotechnology applications.

ACKNOWLEDGEMENTS

The research was partially performed in the frame of the EUREKA project Σ!3892 ModPolEUV supported by the Ministry of Science and Higher Education of Poland (decision   Nr 120/EUR/2007/02), by the Foundation for Polish Science under the HOMING 2009 Programme (grant number HOM2009/14B), the EC's 7. Framework Program (LASERLAB-EUROPE - grant agreement n° 228334 and COST Action MP0601, also co-funded by Polish Ministry of Science and Education, decision number 816/N-COST/2010/0).

References

1 V. Bakshi, *Semiconductor International*, 2009, (Austin, Texas).
2 P.W. Wachulak, M.C. Marconi, R.A. Bartels, C.S. Menoni and J.J. Rocca, *J. Opt. Soc. Am.*, 2008, **B25**, 1811.
3 P.W. Wachulak, L.Urbanski, M.G. Capeluto, D. Hill, W.S. Rockward, C. Iemmi, E.H. Anderson, C.S. Menoni, J.J. Rocca and M.C. Marconi, *J. Micro/Nanolith. MEMS MOEMS*, 2009, **8**, 021206.
4 R.A. Bartels, A. Paul, H. Green, H.C. Kapteyn, M.M. Murnane, S. Backus, I.P. Christov, Y. Liu, D. Attwood and C. Jacobsen, *Science*, 2002, **297**, 376.
5 I.J. Kim, G.H. Lee, S.B. Park, Y.S. Lee, T.K. Kim and C.H. Namb, *Appl. Phys. Lett.*, 2008, **92**, 021125.
6 R.L. Sandberg, A. Paul, D.A. Raymondson, S. Hadrich, D.M. Gaudiosi, J. Holtsnider, R.I. Tobey, O. Cohen, M.M. Murnane, H.C. Kapteyn, C. Song, J. Miao, Y. Liu and F. Salmassi, *Phys. Rev. Lett.*, 2007, **99**, 098103.
7 R.I. Tobey, M.E. Siemens, O. Cohen, M.M. Murnane, H.C. Kapteyn and K.A. Nelson, *Opt. Lett.*, 32, 3 (2007)
8 G. Vaschenko, C. Brewer, F. Brizuela, Y. Wang, M. A. Larotonda, B. M. Luther, M. C. Marconi, J. J. Rocca, and C. S. Menoni, *Opt. Lett.*, 2006, **31**, 9.
9 T. Teresawa, T. Yamane, T. Tanaka, T. Iwasaki, O. Suga and T. Tomie, *Proc. SPIE*, 2009, **7271**, 727122.
10 A. Bartnik, H. Fiedorowicz, R. Jarocki, L. Juha, J. Kostecki, R. Rakowski and M. Szczurek, *Appl. Phys.*, 2006, **B82**, 529.
11 A. Bartnik, H. Fiedorowicz, R. Jarocki, J. Kostecki, R. Rakowski and M. Szczurek, *Appl. Phys.*, 2008, **B93**, 737.
12 A. Bartnik, H. Fiedorowicz, R. Jarocki, J. Kostecki, A. Szczurek and M. Szczurek, *Appl. Phys.*, 2009, **B96**, 727.
13 H. Fiedorowicz, A. Bartnik, R. Jarocki, J. Kostecki, J. Krzywinski, J. Mikołajczyk, R. Rakowski, A. Szczurek and M. Szczurek, *Journal of Alloys and Compounds*, 2005, **401**, 99.
14 A.G. Michette, G.R. Morrison and C.J. Buckley (eds), *X-Ray Microscopy III, Springer Series in Optical Sciences*, 1992, **67** (Berlin, Springer)
15 P. W. Wachulak, A. Bartnik, H. Fiedorowicz, T. Feigl, R. Jarocki, J. Kostecki, R. Rakowski, P. Rudawski, M. Sawicka, M. Szczurek, A. Szczurek, Z. Zawadzki, *Applied Physics B*, 2010, **100**, 3, 461-469
16 P. W. Wachulak, A. Bartnik and H. Fiedorowicz, *Optics Letters,* 2010, **35**, 14, 2337-2339
17 P. W. Wachulak, A. Bartnik, H. Fiedorowicz, and J. Kostecki, *Optics Express*, 2011, **19**, 10, 9541–9550

18 L.B. Da Silva, J.E. Trebes, R. Balhorn, S. Mrowka, E. Anderson, D.T. Attwood, T.W. Barbee Jr., J. Brase, M. Corzett, J. Gray, J.A. Koch, C. Lee, D. Kern, R.A. London, B.J. MacGowan and D.L. Mathews, *Science*, 1992, **258**, 269.

19 B. Niemann, D. Rudolph and G. Schmahl, *Opt. Commun.*, 1974, **12**, 160.

20 B. Niemann, D. Rudolph and G. Schmahl, *Appl. Opt.*, 1976, **15**, 1883.

21 G. Schneider, B. Niemann, P. Guttmann, D. Rudolph and G. Schmahl, *Synchr. Rad. News*, 1995, **8**, 19.

22 C. Jacobsen, S. Williams, E. Anderson, M.T. Browne, C.J. Buckley, D. Kern, J. Kirz, M. Rivers and X. Zhang, *Opt. Commun.*, 1991, **86**, 351.

23 C. Jacobsen, J. Kirz and S. Williams, *Ultramicroscopy*, 1992, **47**, 55.

24 J. Kirz, C. Jacobsen, S. Lindaas, S. Williams, X. Zhang, E. Anderson and M. Howells, *Synchrotron Radiation in the Biosciences*, 1994, (Oxford, Oxford University Press) p563.

25 G. Poletti, F. Orsini and D. Batani, *Solid State Phenomena*, 2005, **107**, 7.

26 A.C. Cefalas, P. Argitis, Z. Kollia, E. Sarantopoulou, T.W. Ford, A.D. Stead, A. Marranca, C.N. Danson, J. Knott and D. Neely, *Technical Report RAL-TR-98-007*, 1998, (Oxford, Rutherford Appleton Laboratory).

27 M. Kado, H. Daido, Y. Yamamoto, K. Shinohara and M.C. Richardson, *Proc. 8th Int. Conf. X-ray Microscopy,* 2005, *IPAP Conf. Series*, **7**, 41.

28 H. Fiedorowicz, A. Bartnik, R.Jarocki, M. Szczurek and T.Wilhein, *Appl. Phys.*, 1998, **B67**, 391.

29 P.W. Wachulak, A. Bartnik, H. Fiedorowicz, P. Rudawski, R. Jarocki, J. Kastecki and M. Szczurek, *Nucl. Instrum. Methods Phys. Res.*, 2010, **B268**, 1692.

30 R. L. Kelly, *J. Phys. Chem. Ref. Data*, 1987, **16**, suppl. 1.

31 D. Attwood, *Soft X-Rays and Extreme Ultraviolet Radiation*, 1999, Cambridge, Cambridge University Press.

32 J. M. Heck, D. T. Attwood, W. Meyer-Ilse, E. H. Anderson, *Journal of X-Ray Science and Technology*, 1998, **8**, 95

33 W. Chao, J. Kim, S. Rekawa, P. Fischer and E. H. Anderson, *Opt. Express*, 2009, **17**, 20, 17699

34 K. A. Goldberg, Proc. SPIE 6730, 67305E-1-12 (2007).

35 H. Kinoshita, K. Hamamoto, N. Sakaya, M. Hosoya, and T. Watanabe, *Jpn. J. Appl. Phys. Part 1*, 2007, **46**, 6113

36 K. A. Goldberg, P. Naulleau, I. Mochi, E. H. Anderson, S. B. Rekawa, C. D. Kemp, R. F. Gunion, H.-S. Han, and S. Huh, *J. Vac. Sci. Technol. B*, 2008, **26**, 2220

37 P.A.C. Takman, H. Stollberg, G.A. Johansson, A. Holmberg, M. Lindblom, and H.M. Hertz, *J.Microsc.*, 2007, **226**,175–181

38 D. S. DiCicco, D. Kim, R. Rosser, and S. Suckewer, *Opt. Lett.,* 1992, **17**, 2, 157

39 L. B. Da Silva, J. E. Trebes, S. Mrowka, T. W. Barbee, Jr., J. Brase, J. A. Koch, R. A. London, B. J. MacGowan, D. L. Matthews, D. Minyard, G. Stone, T. Yorkey, E. Anderson, D. T. Attwood, and D. Kern, *Opt. Lett.,* 1992, **17**, 754

40 M. Wieland, C. Spielmann, U. Kleineberg, T. Westerwalbesloh, U. Heinzmann, and T. Wilhein, *Ultramicroscopy,* 2005, **102**, 93

41 M.Kishimoto, M.Tanaka, R.Tai, K.Sukegawa, M. Kado, N.Hasegawa, H.Tang, T.Kawachi, P.Lu, K. Nagashima, H.Daido, Y.Kato, K.Nagai, and H. Takenaka, *J. Phys. IV,* 2003, **104**, 141

42 I.A. Artioukov, A.V. Vinogradov, V.E. Asadchikov, Y.S. Kasyanov, R.V. Serov, A.I. Fedorenko, V.V. Kondratenko and S.A. Yulin, *Opt. Lett.,* 2005, **20**, 2451

43 G. Vaschenko, F. Brizuela, C. Brewer, M. Grisham, H. Mancini, C.S. Menoni, M.C. Marconi, J.J. Rocca, W. Chao, J.A. Liddle, E.H. Anderson, D.T. Attwood, A.V. Vinogradov, I.A. Artioukov, Y.P. Pershyn, V.V. Kondratenko, *Opt. Lett.,* 2005, **30**, 16, 2095

44 F. Brizuela, Y. Wang, C. A. Brewer, F. Pedaci, W. Chao, E. H. Anderson, Y. Liu, K. A. Goldberg, P. Naulleau, P. Wachulak, M. C. Marconi, D. T. Attwood, J. J. Rocca, and C. S. Menoni, *Opt. Lett.,* 2009, **34**, 3, 27

45 C. A. Brewer, F. Brizuela, P. Wachulak, D. H. Martz, W. Chao, E.H. Anderson, D.T. Attwood, A. V. Vinogradov, I. A. Artyukov, A.G. Ponomareko, V.V. Kondratenko, M.C. Marconi, J.J. Rocca, C.S. Menoni, *Optics Letters,* 2008, **33**, 518, this article was also published in *Virtual Journal for Biomedical Optics,* 2008, **3**, 4, and in *Virtual Journal of Nanoscale Science & Technology,* 2008, **17**, 15

46 K. W. Kim, Y. Kwon, K.Y. Nam, J.H. Lim, K.G. Kim, K.S. Chon, B.H. Kim, D.E. Kim, J.G. Kim, B.N. Ahn, H.J. Shin, S. Rah, K.H. Kim, J.S. Chae, D.G. Gweon, D.W. Kang, S.H. Kang, J.Y. Min, K.S. Choi, S.E. Yoon, E.A. Kim, Y. Namba and K.H. Yoon, *Phys. Med. Biol.,* 2006, **51**, N99-N107

47 L. Juschkin, R. Freiberger, and K. Bergman, *J. Phys.: Conf. Ser.,* 2009, **186**, 012030

48 R. Rakowski, A. Bartnik, H. Fiedorowicz, F. de Gaufridy de Dortan, R. Jarocki, J. Kostecki, J. Mikołajczyk, L. Ryc, M. Szczurek, P. Wachulak, *Applied Physics B*, 2010, **101**, 773

49 S.S. Churilov, Y.N. Joshi, J. Reader, *Opt. Lett.* 2003, **28**, 16, 1478

50 G. O'Sullivan, *J. Phys. B: At. Mol. Phys.,* 1982, **15**, L765-L771

51 P. W. Wachulak, A. Bartnik, L. Wegrzynski, J. Kostecki, R. Jarocki, T. Fok , M. Szczurek and H. Fiedorowicz, *NIM B,* 2013, **311**, 42–46

52 P. W. Wachulak, A. Bartnik, M. Skorupka, J. Kostecki, R. Jarocki, M. Szczurek, L. Wegrzynski, T. Fok and H. Fiedorowicz, *Applied Physics B*, 2013, **111**, 2, 239-247

SUBJECT INDEX